THE USE OF SURFACTANTS IN THE PETROLEUM INDUSTRY

PRIMENENIE POVERKHNOSTNO-AKTIVNYKH VESHCHESTV
V NEFTYANOI PROMYSHLENNOSTI

ПРИМЕНЕНИЕ ПОВЕРХНОСТНО-АКТИВНЫХ ВЕЩЕСТВ
В НЕФТЯНОЙ ПРОМЫШЛЕННОСТИ

THE USE OF

SURFACTANTS

IN THE

PETROLEUM INDUSTRY

Edited by

P. A. Rebinder

Authorized translation from the Russian

c/b

CONSULTANTS BUREAU
NEW YORK
1965

c.1

The original Russian text — the Proceedings of the Second All-Union Conference on the use of Surface-Active Substances in the Petroleum Industry — was published by Gostoptekhizdat in Moscow in 1963 for the Ufa Petroleum Scientific-Research Institute and the Council on the Problems of Physicochemical Mechanics, Surface Phenomena, and Surface-Active Substances of the Academy of Sciences of the USSR. The Soviet editors have revised and corrected the original text and have chosen seven papers from the First All-Union Conference on the Use of Surface-Active Substances in the Petroleum Industry and Their Production for inclusion in this edition.

Library of Congress Catalog Card Number 64-23247

Printed in the United States of America

PREFACE TO THE ENGLISH EDITION

In the past ten years, extensive developments have taken place in the Soviet Union in work on the use of surface-active substances in the technological processes of petroleum recovery: in drilling, the development of petroleum deposits, the operation of oil wells, the removal of water and salts from petroleum, the prevention of corrosion of equipment in the petroleum industry, etc.

In order to summarize the results of the work which has been carried out, to examine the prospects in different fields, and to develop further the theory and practice of the use of surfactants, the Ufa Petroleum Scientific Research Institute and the Council on Problems of Physicochemical Mechanics, Surface Phenomena, and Surface-Active Substances of the USSR Academy of Sciences convened three All-Union Conferences.

The reports presented at these conferences have been published in two volumes with the title "The Use of Surface-Active Substances in the Petroleum Industry." The first volume was published in 1961 and the second in 1963.

The third volume is at present being prepared for publication.

The English translation covers chiefly the second volume. Since some reports presented at the first conference may be of interest to workers in the petroleum industry in the USA, the papers from the second volume have been supplemented by some articles from the first volume.

The reports presented at the third conference are not included in this book. We hope that the translation of this collection of papers will make it possible for a wide circle of workers in the petroleum industry of the USA to become familiar with the work carried out in the USSR on the use of surfactants.

Academician P. A. Rebinder
Doctor of Technical Sciences, Professor G. A. Babalyan
Candidate of Technical Sciences I. I. Kravchenko

CONTRIBUTORS

F. S. Abdullin	Candidate of Technical Sciences, Laboratory Director, Stavropol' Section, GrozNII
M. A. Akhmetshin	Laboratory Director, TF VNII
G. A. Babalyan	Doctor of Technical Sciences, Professor, Director, UFNII
A. I. Babayan	Senior Scientist, NV VNIIGG
N. F. Bergshtein	VNII NP
R. G. Blyum	Laboratory Director, TsNIL Perm'
V. A. Brovenko	Engineer, UFNII
R. I. Cherkasov	Laboratory Director, VNII Neftekhim
V. I. Dronov	Candidate of Chemical Sciences, Laboratory Director, BF AN SSSR
R. S. Enikeev	Senior Scientist, BF AN SSSR
I. B. Genkin	Section Head, UFNII
V. A. Glebov	No affiliation
I. E. Gol'dshtein	No affiliation
V. P. Goncharov	Laboratory Director, NV VNIIGG
A. A. Gonik	Candidate of Technical Sciences, Section Head, UFNII
E. G. Ivanova	Junior Scientist, UFNII
A. L. Kalitaeva	VNII NP
E. G. Khalilov	Junior Scientist, IRNGM AN AzSSR; UFNII
Ch. T. Khaliullin	Senior Engineer, TsNIPR Arlanneft'; UFNII
F. G. Khatmullin	Post-Graduate Research Student, BGU
V. I. Khokhlov	VNII NP
A. D. Khudyakova	VNII NP
V. L. Komarov	Candidate of Technical Sciences, Laboratory Director, UFNII
A. F. Koretskii	Physical Chemistry Institute, AN USSR
E. K. Kovalenko	Engineer, UFNII
I. I. Kravchenko	Candidate of Technical Sciences, Laboratory Director, UFNII
V. V. Kukin	Senior Scientist, Kuibyshev NII NP
V. G. Kunov	TsNIL Perm'

U. Sh. Kurbanaev	Senior Scientist, UFNII
S. A. Lebedev	Senior Scientist, UFNII
R. S. Lerner	Senior Scientist, VNII BT
D. N. Levchenko	Candidate of Technical Sciences, Laboratory Director, VNII NP
L. N. Malysheva	Senior Scientist, UFNII
I. L. Markhasin	Candidate of Technical Sciences, Section Head, UFNII
M. Z. Mavlyutova	Laboratory Director, UFNII
V. A. Merkulov	Section Head, Kuibyshev NII NP
I. D. Mineeva	Senior Scientist, NV VNIIGG
V. T. Morgunov	Senior Scientist, Kuibyshev NII NP
R. B. Morozov	Senior Scientist, UFNII
R. A. Musaev	Junior Scientist, IRNGM AN AzSSR
Ya. I. Nel'kenbaum	Laboratory Director, UNPZ im. "XXII s'ezda KPSS"
N. M. Nikolaeva	Candidate of Chemical Sciences, Senior Scientist, VNII NP
K. F. Oshchepkov	Laboratory Director, VNII NP
T. S. Osmolovskaya	No affiliation
G. T. Ovnatanov	Section Head, VNIGNI
A. A. Petrov	Candidate of Technical Sciencies, Laboratory Director, Giprovostokneft'
V. I. Portnov	Laboratory Director, UFNII
P. A. Rebinder	Academician of the USSR Academy of Sciences, Professor, Doctor of Physicomathematical Sciences, Doctor of Chemical Sciences
G. V. Rudakov	UFNII
N. A. Rybinskaya	Junior Scientist, UFNII
N. F. Semenko	No affiliation
R. B. Shaevich	Senior Scientist, GrozNII
K. V. Sharina	Engineer, UFNII
I. E. Shevaldin	No affiliation
E. A. Shklyaruk	VNII NP
N. I. Skomorovskaya	Senior Scientist, UFNII
Yu. S. Smirnov	Senior Scientist, Giprovostokneft'
Ya. G. Sorkin	Candidate of Chemical Sciences, Director, UNPZ im. "XXII s'ezda KPSS"
T. M. Stolbova	Senior Scientist, UFNII
N. D. Tairov	Candidate of Technical Sciences, Laboratory Director, IRNGM AN AzSSR

A. B. Taubman	Physical Chemistry Institute, AN USSR
V. I. Tokunov	Senior Engineer, UFNII
E. M. Tosunov	Laboratory Director, GrozNII
O. I. Tselikovskii	Junior Scientist, NII transneft', UFNII
E. N. Umrikhina	Senior Scientist, UFNII
S. G. Yakhin	Junior Scientist, UFNII
S. N. Yatrov	Candidate of Technical Sciences, Director, ITI EI Neftegaz
S. Z. Zaripov	Candidate of Technical Sciences, Laboratory Director, UFNII
G. V. Zhovner	Senior Scientist, GrozNII
K. F. Zhigach	MINKh i GP

CONTENTS

PARTIAL LIST OF SURFACTANTS

This list was compiled by members of the editorial staff of Consultants Bureau.
(The composition of substances marked *** is not given in this volume and not known to the translators)

Alkamon OS-2	A cationic quaternary salt of the dimethylaminomethyldiglycolic ethers of high fatty alcohols (saturated alcohols): $$[R-OCH_2CH_2OCH_2CH_2OCH_2\underset{\mid \atop CH_3}{N}(C_2H_5)_2]^+ C_2H_5SO_3^-$$ where $R = C_{14-18}H_{29-37}$.
Alkamon OS-2N	Similar to Alkamon OS-2 except that the alcohols are unsaturated. It is cationic.
Amines C, C_I, and C_{II}	High-molecular aliphatic amines with normal alkyl chains, containing 8-18 carbon atoms. Coconut oil is used to prepare Amine C_{II} and stearic acid to prepare C_I.
Arquad 2C	A cationic dialkyldimethylammonium salt: 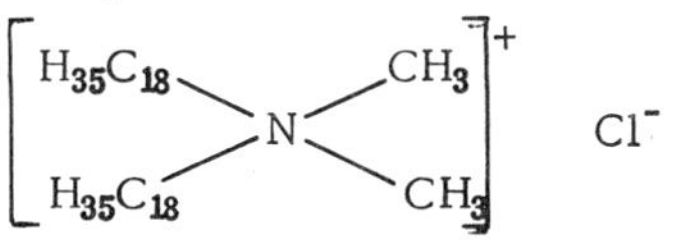
Arquad S-50 and T-50	Cationic derivatives of the monoalkyltrimethylammonium chloride: 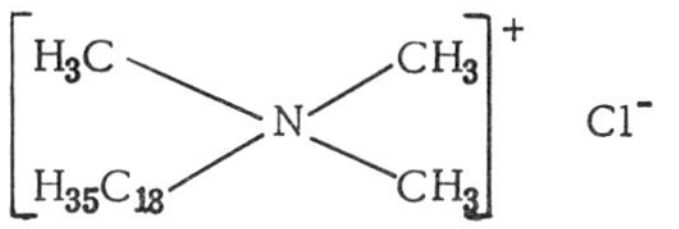
Arquads S and T	Also cationic derivatives of mono- and dimethylalkylammonium chlorides (alkyl 14-18C).
Azolyat	Purified petroleum sulfonic acids.
Bunigal	***
Ca-NSD	The Ca Salts of sulfonated dextrin with a low degree of nitration.
ChNPZ-59	Ethoxylated product from di(mono)glyceride of o-phthalic acid.
DF	Ethoxylated dodecylphenol.
Diamine dioleate	$\lvert R-NH_2-C_3H_6-NH_3 \rvert^{2+} \lvert C_{17}H_{35}COO \rvert^{2-}$
Disolvan	Probably the brand name for the series designated as 4400, 4411, 4422, and 4433, said to be made by Anorgana Munchen.
DK	Ethoxylated dodecylcresol.
DS	"Soviet Detergent"—probably refined alkylaryl sulfonate.

DS–Na	***
DS–RAS	***
ESTs	Sulfoethylcellulose.
Ethomide HT/60	An oxyethylated amide.
Gamlen	***
Gipan	Copolymer of sodium acrylate, acrylamide, and acrylonitrile – saponification product of polyacrylonitrile and caustic.
GIPKh-37	***
GKZh-94	A nonionic silicoorganic compound.
GMFN	Sodium hexametaphosphate.
K_{150}	***
Karbozolin O	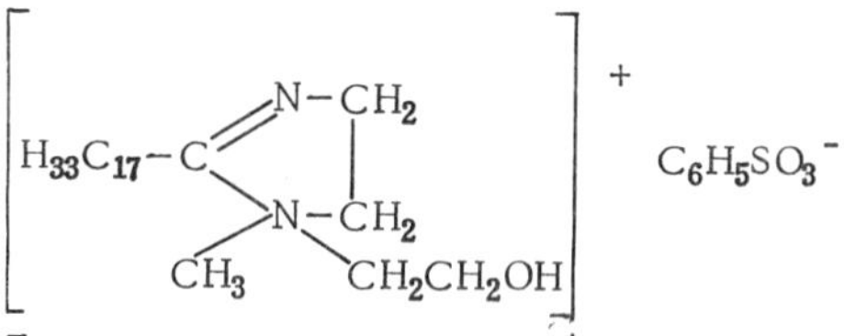
Karbozolin S	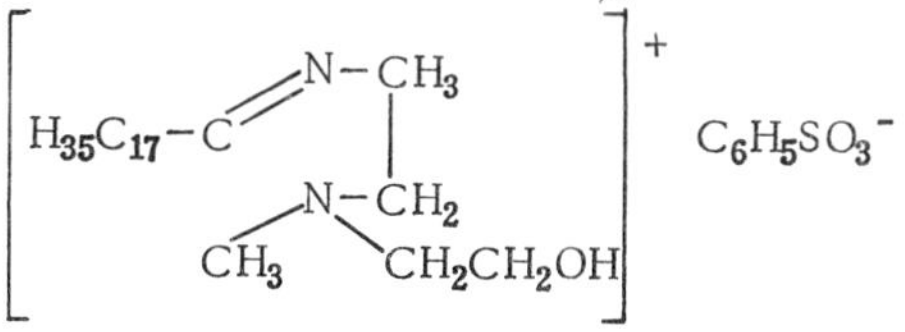
Katapin A	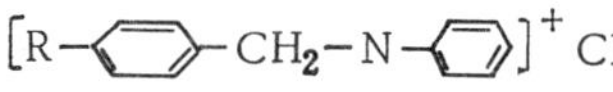 ; R has 12 to 18 carbon atoms.
Katamin A	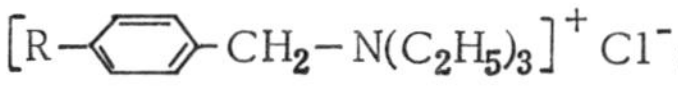 ; R has 12 to 18 carbon atoms.
$KAUFE_{14}$	Product of coal tar phenol fraction alkylated with fraction of cracked naphtha and ethoxylated (14 ethoxy groups).
$K_{10-16}E_6$	An oxyethylated aliphatic acid.
$KOSE_{28.6}$ and $KOSE_{15.5}$	A nonionic condensation product of alcohols obtained by the Oxosynthesis and containing 12 to 16 carbon atoms, with 28.6 or 15.5 moles respectively of ethylene oxide.
KMTs	Carboxymethylcellulose.
KShchR	Carbonate-alkali reagent.
KSSB	Condensed sulfite-alcohol waste.
Leveling Agents A and A 20	The quaternary ammonium salts of diethylaminomethyl derivatives of the polyethyleneglycol ethers of isooctylphenols.
$LKFE_{6.1}$	***
NChK	"Neutral black contact"–a crude sulfonate from acid treating of petroleum distillates. Usually sodium salt, but appears to be ammonium salt in present book.
Novost'	Anionic alkyl sulfates.
NSSK	Sodium sulfonates.
OF	Ethoxylated isooctylphenol.

OGPK	***
OK	Ethoxylated isooctylcresol; OKO may be identical to OK but is at least similar.
OKPN	***
OMZhK	Ethoxylated monoglycerides of fatty acids.
OP-4–OP-45	The reaction product from n moles of ethylene oxide and a mixture of mono- and dialkylphenols, where the alkyl groups are C_8 to C_{10}:
	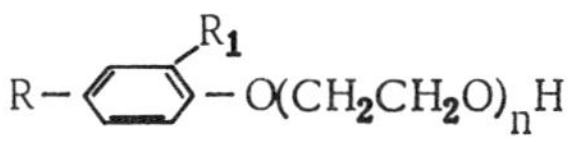
OPS	***
OS	***
Oxidat	***
Oxamid ST-15	An oxyethylated amide.
OZhK	An oxyethylated aliphatic acid.
Petrov Contact	Petroleum aromatic and cyclic sulfonic acids and their salts; similar to NChK.
Progress	An alkyl sulfate.
Proksanol 146	***
Separol 221	***
Sapal	***
$S_{10}E_{8.3}$	An oxyethylated alcohol.
Shebekino alcohols	***
SMS	A copper sulfonate obtained by reacting sodium sulfonates (SNS) with $CuSO_4 \cdot 5H_2O$.
SNS	Sodium salts of sulfonated shale resins.
Sodium Detergent DS	An alkyl aryl sulfonate.
SPV-5	***
SSB	Sulfite-alcohol waste.
Starch-reagent	***
Stearox-6	An oxyethylated aliphatic acid.
STs	Cellulose sulfate.
Substance P	A water-soluble alkyl sulfonate
Sulfamide OE-10	An ethylene oxide sulfamide:
	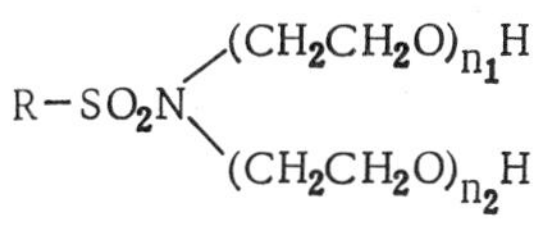

	R is an acid residue obtained by the oxidation of soft parafins; $n_1 + n_2 = 10$; for Sulfamide OE-5 $n_1 + n_2 = 5$.
Sulfoalmonite salts	***
Sulfonate	Alkyl sulfonates.
Sulfonol NP-1	Dodecylbenzene sulfonate made by sulfonating 250-320°C fraction of alkylated product from benzene and propylene tetramer.
Sulfonol NP-5	***
UFE_8	Ethoxylated coal tar phenol fraction, with 8 ethoxy groups.
UShchR	Carbonate-alkali reagent.
VNII NP-58	An ethoxylated alkylphenol.
Wetting Agent DB	***
Wetting Agent NB	The sodium salt of monobutylnaphthalenesulfonic acid.
Wetting Agent ND	***
Xylital S-15, Xylital KS-59	Oxyethylated derivatives of xylitol.
Nekals	Metal salts of esters of intermediate-length (6-8 carbons) sulfocarboxylic acids OR metal salts of sulfonated alkylnaphthalenes and polymerized alkylnaphthalenes.
Mylonaft	A mixture of sodium salts of naphthenic acids (43), water, and small amounts of mineral oil and mineral salts.

SURFACTANTS: THEIR IMPORTANCE AND APPLICATIONS IN THE PETROLEUM INDUSTRY

P. A. Rebinder

I

Surface-active substances (surfactants) are acquiring considerable importance in all branches of the national economy. This is due to the fact that these substances concentrate at different surfaces of separation, forming extremely thin adsorption layers which produce marked changes in the molecular nature and properties of the surfaces. This leads, in the first place, to a change in the kinetics of the processes involved in the transfer of substances across surfaces of separation between phases, and in the second place to equally important changes in the conditions of molecular interaction between the phases in contact with one another. These include cohesion at the phase boundary, coagulation of particles and coalescence of drops of the dispersed phase in a given disperse system, adhesion and friction conditions, and all the conditions of interaction involving molecular forces.

Thus, the addition of very small quantities of surfactants may change the course of physicochemical processes and the conditions of interaction between phases. This makes it possible to control technological processes.

The chemistry and technology of the production of synthetic surfactants has developed rapidly in the past 30-40 years, particularly in the USA, Britain, and Germany. Industry in the Soviet Union, however, has been occupied by more immediate problems and in this field has fallen behind the technically developed capitalist countries. At the present time in the USA there are already more than a thousand patented names for synthetic surfactants produced by different competing firms.

Analysis of this varied assortment shows that its variety is in no way related to any real technical advantage but is to a considerable extent the result of capitalist competition and reflects the advantages to individual owners of businesses and their organizations.

The task of industry and science in the Soviet Union and other socialist countries is to select from several dozen reagents a range of absolutely necessary surfactants of different types on the basis of a varied range of applications, which will satisfy the requirements of all the main branches of the national economy, particularly the petroleum industry. This problem is fundamental for the physical chemistry of surface phenomena and disperse systems—a new branch of science which is developing the problems of the application of surfactants.

The studies of L. G. Gurvich have played an important part in this connection. A number of his papers published in 1912-1916 put forward theories regarding the chemical nature of adsorption forces as unsaturated secondary valences in the surface layers of liquids and solids. The American physicist and physical chemist Irving Langmuir, who is acknowledged as the founder of the modern theory of surfactants, based his famous publication of 1917 on the theories of L. G. Gurvich.

L. G. Gurvich, who carried out extensive scientific studies in the field of petroleum technology, subsequently showed the important part played by surface phenomena and particularly by the adsorption of surfactants in petroleum industry practice and in the processes of petroleum treatment. His book, "The Scientific Principles of Petroleum Processing," was reprinted many times in a number of countries.

Work by Soviet scientists subsequently acquired ever-increasing prominence in the development of this field of the science of surface phenomena, together with studies of the structure of surface-active molecules by Harkins (USA) and Hardy (Britain), the structure and properties of monomolecular layers of surface-active substances by Adam and Rideal (Britain), and two-dimensional migration by Volmer (Germany). Of particular importance is the work of A. N. Frumkin (from 1919) and his school on the thermodynamic properties and equation of state of adsorption layers and on the relationship between the surface activity and structure of molecules and ions and its importance in electrode processes.

In order to determine which surfactants should be produced by our industry in fairly large quantities, it is necessary first of all to solve the problem of the mechanism and efficiency of the action of these substances for each particular example of their application. Quantitative estimation of the efficiency of the action of these additives should be carried out by means of measurement of physical quantities, and account should be taken of the optimum physicochemical conditions.

The action of surfactants (the same substance may produce directly opposite effects, depending on its concentration and other conditions of application) should be classified not according to the branches of the national economy, since their use in different branches is often based on the same physicochemical principle, but rather should be related to the molecular mechanism of the action of the surfactants. This criterion characterizes the essential features of the action of various surfactants and determines their optimum composition and conditions of application.

II

The magnitude of the adsorption—the excess mass of surfactant on 1 cm^2 of surface (Γ in moles/cm^2)—for the formation of an adsorption layer is directly related to the decrease which it produces in the surface tension (σ in ergs/cm^2)—the specific free surface energy as the excess free energy concentrated at the boundary of separation between the phases on 1 cm^2 of surface. This relationship is given by the familiar thermodynamic equation of Gibbs:

$$-\frac{d\sigma}{d\mu} = \Gamma, \tag{1}$$

where μ is the chemical potential of the adsorbed substance, determined from the formula

$$\mu = RT \ln a + \mu_0,$$

where μ_0 is a constant which depends only on T ($\mu = \mu_0$ when a = 1), and a is the thermodynamic activity.

The latter is found from the formula

$$a = fc,$$

where f is the activity coefficient and c the concentration of the substance in one of the phases forming the surface in equilibrium with the adsorption layer.

From (1) in the range of small values of c ($f \approx 1$, $a \approx c$) characteristic of strongly surface-active substances, for which all the effect up to saturation of the monomolecular adsorption layer is concentrated in the range of low concentrations,

$$-\frac{d\sigma}{dc} = G = RT\frac{\Gamma}{c}. \tag{2}$$

Thus, the surface activity ($G = -d\sigma/dc$), i.e., the ability to decrease the surface tension by adsorption at a given temperature, is proportional to Γ/c and a coefficient RT, the same for all surfactants.

In vacuo, any substance in the form of vapor is to some extent surface-active towards the surfaces of practically nonvolatile liquids or solids, i.e., it is adsorbed on these surfaces. In the case of a mixture of gases (or vapors) or liquid solutions, for adsorption (in the simplest case) from a two-component system, however, one of the components will be surface-active with respect to the given boundary of separation between the phases (this component will form the adsorption layer) and it is convenient to regard this component as the dissolved substance. Thus, for example, in aqueous solutions at the boundary with air (or with its own vapor), surface-active properties are shown by organic substances whose molecules (or ions) are asymmetrically constructed from a polar group and a hydrocarbon radical (alcohols, amines, phenols, carboxylic acids, sulfonic acids, etc.) and their alkali metal salts. These can produce a marked decrease in the surface tension of water, and in the case of sufficiently high homologs they can give surface activity values as high as required when the sign of the activity is positive.

In actual fact, in this case,

$$\Gamma = (c_s - c)\,h > 0, \tag{3}$$

$$-\frac{d\sigma}{dc} = RTh\,\frac{c_s - c}{c}, \tag{4}$$

where c_s is the concentration of surface-active substance in the adsorption layer, c the corresponding concentration in the volume (moles/cm^3), and h is the thickness of the adsorption layer.

For a sufficiently high surface activity

$$c_s \gg c, \quad -\frac{d\sigma}{dc} = RTh\,\frac{c_s}{c}.$$

Surface tension measurements show that highly surface-active substances of the ordinary soap type (Na oleate) decrease the value of σ for water at 20° from 72.8 to 30 ergs/cm^2 at $c = 10^{-6}$ moles/cm^3 ($1 \cdot 10^{-3}$ M), which gives $G = 4 \cdot 10^7$ ergs$\cdot$cm^{-2}/mole$\cdot$cm^{-3}. Here $RTh \approx 1.2 \cdot 10^3$ ergs$\cdot$cm/mole, so that c_s, the concentration of surface-active substance in the surface (adsorption) layer, is greater than its concentration in the volume by a factor of $3 \cdot 10^4$, i.e., by a factor of thirty thousand!

A different situation is observed in aqueous solutions of ordinary salts—inorganic electrolytes; at the water—air boundary these salts are surface-inactive with respect to water, i.e., their presence increases slightly the surface tension of pure water. This effect is nonsymmetric, however, relative to the action of surface-active additives: this means that surface-inactive substances can produce only a small increase in the surface tension of water for unit concentration; this is a direct consequence of Gibbs' thermodynamic equation.

Moreover, we can readily calculate from this equation the upper limit of this increase, its highest possible value for $d\sigma/dc > 0$, $\Gamma < 0$, and $c_s < c$ (the adsorption is negative—the concentration in the surface layer is always lower than that in the volume of the solution). From $d\sigma/dc = RTh\,(c - c_s)/c$ when $c_s \to 0$ (or $c_s \ll c$) it follows that

$$\mathrm{Lim}\left(\frac{d\sigma}{dc}\right)_0 = \frac{\sigma_c - \sigma_0}{c} = RTh,$$

i.e., the increase in surface tension at $c = 10^{-3}$ mole/cm^3 (1 M) for any inactive salt does not exceed the highest value $cRTh \approx 10^{-3} \cdot 8.3 \cdot 10^7 \cdot 300 \cdot 3 \cdot 10^{-8} \approx 0.8$ erg/cm^2 ($h \approx 3 \cdot 10^{-8}$ represents the dimensions of the water molecules forming the adsorption layer).

Thus, in aqueous solutions of inactive substances—ordinary salts—water must be regarded as a surfactant. At the boundary between two liquid phases or between a liquid and a solid, ordinary inorganic electrolytes may be adsorbed positively, i.e., they may behave as surface-active substances.

Here we examine only typical organic surfactants with an asymmetric (diphilic) molecular structure consisting of polar and nonpolar groups. These substances are always surface-active to some extent at a water—air boundary and to a greater extent at other boundaries of separation. In the estimation of the surface-active properties of these substances, the balance between the polar group and the hydrocarbon (nonpolar) radical is of decisive importance. At a water—air boundary this radical must be regarded as the source of the surface activity. In this case, the surface activity will be greater the longer the hydrocarbon chain and the less hydrophilic the polar group.

The greatest surface activity is shown by substances of the type $C_nH_{2n+1}A$ or parasubstituted alkylaryl derivatives, e.g., $C_nH_{2n+1}C_6H_4A$, where A represents a polar group: $-OH$, $-COO(H, Me)$, $-SO_2O(H, Me)$, $-O-SO_2O(H, Me)$, $-NH_2$, etc.

The actual adsorption in this case, unlike adsorption on solid surfaces, is "passive" in character and involves the displacement of the hydrocarbon chains (and rings) from the aqueous medium as a result of the strong attraction of the water molecules for one another; this is expressed in the higher value of σ for water ($\sigma = 72.8$ at $t = 20°$), compared with all organic liquids, and the lowest value of σ in the saturated hydrocarbons ($\sigma \approx 20$).

The dissolution in water of a strongly hydrated polar group and the explusion of the hydrocarbon chain from the water, followed by the attraction of its surface, correspond to the minimum free energy of the system: this corresponds to so-called horizontal orientation of the molecule of surface-active substance in the adsorption layer with its polar group "submerged" in the water and its chain floating on the water surface. With increase in the degree of filling of the adsorption layer the hydrocarbon chains become orientated, approaching the vertical position more closely and undergoing increasing interaction with one another.

The work which must be done to transfer one molecule of surface-active substance $C_nH_{2n+1}-A$ (where n is the number of links) from the adsorption layer to the bulk of the water (or solution) is

$$W = \varphi_1 N + W_0, \tag{5}$$

where $\varphi_1 N = 625$ kcal/mole (N is Avogadro's number) is the isothermal work required to move the water molecules apart by means of a hydrocarbon (straight) chain to a distance corresponding to one CH_2 link (i.e., the free energy of reaction for the corresponding number of water molecules), and W_0 is a constant which depends on the nature of the polar group.

By using Boltzmann's equation and formula (3), we can calculate the adsorption and surface activity:

$$W = kT \ln \frac{c_s}{c}, \quad c_s = c\, e^{W/kT}, \quad k = R/N,$$

$$\Gamma = ch\,(e^{W/kT} - 1), \tag{6}$$

$$G = \frac{\sigma_0 - \sigma_c}{c} = RTh\,(e^{W/kT} - 1). \tag{7}$$

In order to make practical use of their surface-active properties, however, the surfactants should be sufficiently soluble in water, since it is only in this case that it is possible to achieve, in addition to a high surface activity $G = -(d\sigma/dc) \cong (\sigma_0 - \sigma)/c$, a low surface tension for the solution $\sigma \leq 30$ ergs/cm^2. For this purpose it is necessary to alter the molecular balance towards an increase in the hydrophilic character of the polar group, e.g., by changing from carboxylic acids to their alkali metal salts or to sulfonic acids, or by decreasing the length of the hydrocarbon chain to an appropriate extent. These two methods are not equivalent, however.

The change from the higher members of a homologous series to the lower members, i.e., decrease in the length of the hydrocarbon chain with the same polar group, produces a sharp increase in the solubility in water, which is inversely proportional, in approximately geometrical progression, to the surface activity (the lower C_1-C_3 homologs have an unlimited solubility in water at room temperature, while the higher homologs from C_{11} upwards are practically insoluble). The surface activity increases with increase in chain length by a factor of 3-3.5 on going to each subsequent homolog (changing by one CH_2 link) in accordance with the Duclos-Traube rule derived theoretically by Langmuir. From Eq. (7), for homologs which are not too low and are present at low concentrations ($c_s \gg c$ and $\Gamma = che^{W/kT}$), we have in fact

$$G = RTh\, e^{W/kT}. \tag{8}$$

From (4) for two subsequent homologs (n and n+1) with identical values of c

$$\frac{\Gamma_{n+1}}{\Gamma_n} = \frac{G_{n+1}}{G_n} = e^{\varphi_1 W/RT} \approx 3.5, \tag{9}$$

since in the range of adsorption layers far from saturation, i.e., with horizontal orientation, the layer thickness h is independent of n.

An increase in the hydrophilic character of the polar group with retention of a fairly long hydrocarbon chain, e.g., by changing from carboxylic acids to their alkali metal salts—i.e., to soaps—leads, however, not only to an increase in the true solubility in water but also to the production of a marked colloidal solubility, as shown by the work of McBain and other authors. At a solution concentration greater than a certain critical value (the critical concentration of micelle-formation, which decreases with decrease in temperature), the excess soap-like surfactant forms micelles in the surrounding aqueous solution; these micelles form a new (colloidal) phase. These micelles are produced as a result of the cohesion of hydrocarbon chains by means of van der Waals' forces with the formation of a nonpolar nucleus and a hydrophilic sheath consisting of polar groups. The production of these micelles and the formation of the adsorption layer leads to a minimum free energy of the system: the sheath of polar groups is hydrated in the aqueous medium, while the hydrocarbon chains leave this medium and enter the new colloidal phase. In practice, however, this takes place only at n = 10-11 (beginning with the laurate C_{11}). This is the most important example of the spontaneous formation of lyophilic colloidal (two-phase) systems, which are sometimes referred to as semicolloidal systems and are characterized by thermodynamic stability, i.e., by equilibrium curves giving the size distribution of the micelles in the mobile equilibrium: micelles of different size $\rightleftharpoons$ molecules (ions). This phenomenon is independent of the ionic nature of the soap-like surfactant, i.e., whether it is anion- or cation-active, a weak electrolyte (soaps of carboxylic acids) or a strong electrolyte with surface-active ions or micelles formed by these ions (alkali metal salts of alkyl and aryl sulfates or sulfonates), or nonionogenic, i.e., forming a purely molecular solution containing uncharged micelles (ethers formed by polyethylene glycol with cetyl alcohol or an alkylphenol).

The formation of the hydrocarbon nucleus of the micelles corresponds to removal of the nonpolar parts of the molecule or ion from the aqueous medium as a result of their insolubility. The chains of saturated hydrocarbons themselves in water, however, would form microdroplets which would be aggregatively unstable due to the high value of the free surface energy σ_{12} at the water—hydrocarbon boundary; the value in this case amounts to approximately 50 $ergs/cm^2$ (e.g., for the water—heptane system), irrespective of chain length. These droplets would coalesce with one another as in a nonstabilized emulsion, and would flow together to form a layer of liquid hydrocarbon on the water surface.

The polar groups which are joined to the hydrocarbon chains by main valences and form the sheath of the micelles or "ultramicrodroplets," decrease considerably the surface tension σ_{12} at the boundary with the aqueous medium. If σ_{12} decreased to zero, this would correspond to the spontaneous breakdown (dissolution) of these micelle-droplets with the disappearance of the second (colloidal) phase and the complete formation of a true solution of the surfactant, which would be present in this solution in the form of molecules (ions) or their associates, forming no separate phase.

Thus, the two-phase thermodynamically stable colloidal solutions of soaps, dyes, and other surface-active semicolloids are completely analogous to the "critical" spontaneously produced emulsions whose theory was put forward by Volmer. In these colloidal-disperse thermodynamically stable emulsions, the surface tension at the droplet—medium boundary retains a positive value which is, of course, at the same time sufficiently small to permit the spontaneous dispersion of the bulk (macro) phase into ultramicrodroplets with dimensions of 10^{-6} cm (100 A).

This spontaneous dispersion takes place as a result of the fact that the increase in the surface energy is compensated by the increase in entropy, resulting from the more regular distribution of the material of the given dispersed phase throughout the whole volume of the medium. The surface tension at the boundary between the two phases σ_{12} should then decrease below the critical value $\sigma_m = \gamma RT/N\delta^2$, where δ is the arbitrary average size of the equilibrium ultramicrodroplets, and γ is a dimensionless factor ($\gamma \approx 15\text{-}30$). For ordinary temperatures and $\delta \approx 10^{-6}$ cm, $\sigma_m \approx 0.1$ erg/cm^2.

Disperse systems which satisfy this condition are lyophilic colloidal systems with a maximum stable degree of dispersion, forming a contrast to lyophobic disperse systems which are thermodynamically fundamentally unstable, i.e., which show spontaneous aggregation and coagulation or, in the case of emulsion droplets, coalescence. At the same time, these lyophobic emulsions, even those which in practice are stabilized to the maximum extent by structured adsorption layers, always have a low degree of dispersion and the droplets have dimensions of not less than 1-0.1 μ (10^{-4}-10^{-5} cm).

The satisfactory bulk properties of aqueous solutions of soap-like surfactants, i.e., substances which have sufficiently long hydrocarbon chains and sufficiently hydrophilic polar groups and form lyophilic colloidal systems, are confirmed by the regular features of the characteristic phenomenon of solubilization, sometimes called induced (colloidal) solubility.

Solubilization is the greatly increased dissolution of nonpolar or slightly polar substances in micellar solutions of soap-like surfactants. A typical example is the solubilization of saturated hydrocarbons (octane, dodecane) by aqueous soap solutions. Hydrocarbons are practically insoluble in water. In true solutions of surfactants they may give a slight increase in solubility (hydrotropy effect) due to the increase in the polarity of the solvent. Solubilization takes place, however, only in micellar solutions of surfactants, i.e., in two-phase colloidal solutions of surfactants at a concentration greater than the critical concentration of micelle formation.

Solubilization is due to the equilibrium distribution of the hydrocarbon between the aqueous medium, in which its concentration is negligibly small, and the colloidal phase—the hydrocarbon nuclei of the soap micelles. This is confirmed by the equilibrium character of the solubilization and the fact that its value, calculated as moles of solubilized hydrocarbon per mole of soap, is constant over a fairly wide range of concentrations, at least in the range of identical micelle configuration. In soap solutions which are not too highly concentrated, these micelles are spheroidal as was suggested, independently of Hartley, in our papers.

In concentrated soap solutions the spheroidal micelles are converted to the thermodynamically more stable layered micelles; these micelles were considered by McBain and are the prototype of the elementary lamellar soap crystallites. In the solution, these lamellar micelles may be joined to one another by their perimeters, i.e., the hydrocarbon fringe, with the formation of a disordered three-dimensional framework analogous to the coagulation structures formed by anisometric particles, e.g., in colloidal suspensions of bentonite clays.

As shown in our laboratory by Z. N. Markina, the formation of these frameworks is readily detected from the sharp increase in the highest (limiting) viscosity of the system corresponding to the practically unbroken structure. A soap solution of this type, e.g., 0.8 M sodium oleate solution, forms a gel and its limiting viscosity amounts to 10^6 cP, which is approximately one million times (!) greater than the viscosity of water at a temperature of 20°. In the solubilization of 0.22 mole of dodecane per mole of soap in this solution, the lamellar micelles become spheroidal, the new micelles cannot form a three-dimensional structure, and the gel liquefies—the limiting viscosity of the system decreases by a factor of 200 thousand (to 5 cP), the new value being only five times greater than the viscosity of water, evidently due entirely to the Einstein effect involving the filling of part of the volume by soap micelles which move in the volume independently of one another.

Solubilization is characteristic not only of micellar-colloidal solutions of soap-like substances but also of true solutions of surface-active polymers, as shown by D. L. Talmud and other workers, particularly for the case of aqueous solutions of proteinaceous substances capable of exhibiting equilibrium absorption of considerable quantities of hydrocarbons from the vapor state. This takes place as a result of the formation of globules by these polymer macromolecules with polar (hydrophilic) and nonpolar (hydrophobic) groups in a long flexible chain. This globule formation has been studied in detail by P. I. Zubov in V. A. Kargin's laboratory. Solubilization under these conditions has recently been studied by V. A. Pchelin and co-workers.

When they undergo adsorption on a water—air or water—liquid hydrocarbon boundary, these macromolecules lie in the plane of the boundary with their polar groups in the water and their nonpolar groups in the second phase; in the bulk of the aqueous solution, however, they form globules by coiling up in such a way that the hydrophilic groups are directed outwards towards the water, while the hydrophobic groups are directed inwards. A single macromolecule of a surface-active polymer of this type forms a "globule" which is analogous to the micelle in the colloidal phase of a soap solution and is capable of dissolving a hydrocarbon to reach an equilibrium state, showing the phenomenon of solubilization. It should be borne in mind that in solutions of ordinary low-molecular surfactants of the soap type, solubilization with its characteristic regular features is possible only when micelles are formed as ultramicrodroplets of a second (colloidal) phase.

Solubilization is extremely characteristic of the action of surfactants as wetting agents; the stable washing away of oily liquid contaminants is associated to a considerable extent with their solubilization in the soap micelles. This also corresponds to the fact that the wetting action is observed only at concentrations higher than the critical concentration of micelle formation..

When a liquid hydrocarbon is introduced into an aqueous soap solution in excess (in quantities greater than the quantity which can undergo equilibrium solubilization), the entire excess of nonpolar liquid immediately and in continuously increasing quantities forms a lyophobic emulsion which has a comparatively low degree of dispersion and is stabilized by an adsorption layer of soap, which acts as emulsifier. An interesting change can be brought about in the opposite direction: an ordinary direct emulsion of mineral oil or an individual liquid hydrocarbon in water, stabilized by a sodium soap (sodium oleate), is treated with excess soap in a form which enters the oil phase, e.g., in the form of an acidic soap (with excess aliphatic acid), forming a colloidal solution in the nonpolar phase.

As shown in our studies and later by Lawrence, the thermodynamically stable micelles is this solution correspond to the opposite orientation of the surface-active molecules with their polar groups directed inwards and the hydrocarbon chains directed into the surrounding medium. These micelles give reverse solubilization of water in oil by absorbing water by means of their hydrophilic nuclei to reach an equilibrium state, corresponding to an increase in the total solubility of water in the hydrocarbon (nonpolar) phase as a result of colloidal solubility.

If we isolate a nonpolar liquid phase containing a dissolved surface-active soap-like substance and gradually add water to this phase, we find that when a certain critical volume of water has been added, a characteristic inversion of phases takes place in the colloidal system, readily detected from the sudden increase in electrical conductivity: the water, instead of being the inner dispersed phase, becomes the outer dispersion medium in the form of thin films (due to its relatively small volume) separating the soap micelles with their sheath of polar groups and their hydrophobic nucleus in which all the hydrocarbon liquid (oil) is solubilized. This type of gel-like colloidal system (hydrogel) is well known in practice and is referred to as a soluble oil or emulsol if oil in excess of the quantity which can be solubilized is present in the form of large droplets stabilized by soap.

Gel-like concentrates of oils and surfactants in water can be readily diluted without loss of stability by any given quantity of an equilibrium aqueous medium, i.e., an aqueous soap solution which in the case of a soluble oil forms stable colloidal solutions of the oil in water, and in the case of an emulsol forms the stable aqueous emulsions which are used extensively in the metal-treatment and mechanical engineering industries for the preparation of active aqueous lubricating and cooling liquids containing different quantities of oils and surfactants for processes involving the treatment of metals by pressure and cutting.

On a newly formed metal surface which has not yet been oxidized and is therefore always hydrophobic, these oil emulsions or colloidal solutions automatically form a thin oily film as a result of selective wetting and coalescence of the oil droplets, which flow over the metal surface. This film, particularly in the case of activated oils (e.g., sulfur-containing oils), ensures maximum lubricating action during boundary lubrication and adsorption softening of thin surface layers on the metal, and this greatly facilitates the processes of metal treatment. At the same time, the formation of this oily film protects the surface of the treated metal and the machine tool from corrosion (protective lubricating films).

The phenomena of direct and inverse solubilization (of hydrocarbons in water and of water in hydrocarbons in the presence of sufficient quantities of soap-like surfactants), and also the changes from one type of such system to another with reversal of the phases, indicate clearly the two-phase character of micellar soap solutions. These phenomena are of considerable practical importance, since they form the basis of processes of polymerization and copolymerization in emulsions with the production of synthetic latexes—polymer dispersions convenient for the manufacture of various articles.

Inverse solubilization (of water in oils) in the presence of suitable surfactants which undergo colloidal dissolution in oil, with displacement of the balance towards hydrophilic groups, is of considerable importance in the food industry, particularly in the production of margarine emulsions.

III

Surfactants can be divided into four main groups, depending on the mechanism of their action.

The first group includes substances which are surface-active at a liquid—gas boundary, primarily the water—air boundary, but which do not form colloidal phases (structures) either in the volume or in the adsorption layers and are low-molecular substances showing true solubility in water. These include the lower and intermediate homologs of surface-active homologous series (the alcohols form a typical example), which are weak wetting agents when added to water. They decrease the surface tension of water from 72.8 to 50-30 ergs/cm^2 at 20° and thus facilitate the flow of water in a thin film over hydrophobic surfaces which are difficult to wet. These surfactants are also always weak foam-producing agents which increase the stability of the free (two-sided) liquid films in the foam by means of the so-called Marangoni-Gibbs effect.

When liquid flows from the film local differences in surface tension as a result of extension of the adsorption layer restrict this flow of liquid and increase the time required for rupture of the film by several tenths of a second. Substances of the first group, of the terpineol type, are widely used as foam-producing agents in flotation. In flotation processes the froth should be unstable and readily broken down, since otherwise the selectivity of flotation is decreased and control of the process becomes impossible (the stability of the froth, even when a weak foam-producing agent is used, is always increased as a result of mineralization by particles undergoing flotation, which adhere to the bubbles and are carried with them into the froth).

Surfactants of the first group are most widely used as antifoaming agents to displace a less surface-active foam stabilizer as a result of the development of a three-dimensional structure in the adsorption layers and films of the foam. These antifoaming agents decrease considerably the stability of the froth. At the same time, the antifoam action of practically water-insoluble, strongly surface-active substances is based on their ability to flow over the water surface, covering it with a saturated monomolecular layer. These layers, as shown at an early date by Hardy, decrease the stability of films, bubbles, and froth practically to zero by decreasing the surface tension to the minimum constant value.

Antifoaming agents are acquiring ever-increasing importance in all industries where the production of stable (e.g., proteinaceous) froth breaks down the course of a technological process, in high-pressure steam boilers, and in the flushing liquids used for drilling deep wells.

In cases of soap-like surfactants which undergo structure-formation in the bulk of the aqueous solution and in the adsorption layers, the foam-forming power and stabilization of the froth increase sharply (i.e., we change to a third group—stabilizers) with an increase in the wetting action as a result of the production of hydrophilic properties on the solid surfaces in the usual way. In the range of low surfactant concentrations, however, the wetting action breaks down as a result of the formation, on the solid surfaces, of hydrophobizing adsorption layers which are chemically fixed by means of a polar group to the corresponding regions of the solid surface with their hydrocarbon chain directed outwards.

The second group includes substances which are surface-active at the boundary between two immiscible liquids or on solid surfaces of separation but which do not form structures either in the bulk of the solution or in the surface layers. These substances are best described as dispersing agents; by undergoing adsorption and thus decreasing the free surface energy of the liquid or solid, they facilitate the process of formation of new surfaces—dispersion in the given medium. These processes include the spraying of liquids, emulsification, and dispersal of solids.

An extremely important and characteristic feature is the decrease in the strength of adsorption of solids—the increase in the ease of dispersion under the influence of external forces by the action of adsorbed substances. In this case, new surfaces develop at surface defects (imperfections in the structure), and their development is facilitated by adsorption. The limiting case is spontaneous adsorption dispersion resulting from a decrease in the surface energy to very low values (less than σ_m) under the influence of a surface-active medium. Examples include spontaneous emulsification under the influence of large quantities of added surfactants and the "dispersion" (colloidal dissolution) of coagulation aggregates which are held together by weak van der Waals' forces and thus break down readily into individual primary particles under the influence of adsorption.

Any substances which are adsorbed on the surface of particles from the surrounding dispersion medium may act as dispersing and peptizing agents. The dispersing agents usually employed, however, act to some extent as stabilizers of the highly dispersed suspension produced. In aqueous media, therefore, the dispersing agents which find extensive application in fine grinding and crushing and are used generally in the preparation of pigments and dyes are always colloidal structure-forming surfactants (or surface-active polymers) with some hydrophilic properties.

The surfactants of the first group which are adsorbed on a water—air boundary are almost always surface-active on liquid and solid surfaces, i.e., they act to some extent as dispersing agents (often without showing a stabilizing action). The reverse, however, is not frequently observed. Thus, for example, mineral electrolytes may be strongly adsorbed by solid surfaces and act as dispersing agents (peptizing agents) but in accordance with both theories are surface-inactive at a water—air boundary. The same may be said of carbohydrates (glucose and saccharose) and other organic substances with a large number of polar groups in the molecule.

Strongly surface-active substances which are not stabilizers may act as demulsifiers of stable emulsions, i.e., they may facilitate their separation into layers as a result of coalescence of the droplets: these demulsifiers are adsorbed more strongly than the stabilizer and displace the latter from the surface of the droplets, but the demulsifiers themselves are unable to preserve the aggregative stability of the emulsions, i.e., they are unable to prevent the coalescence and merging of the droplets.

When adsorbed on solid surfaces, e.g., on the surfaces of the particles of pigments or fillers, surfactants of the second group may produce a marked change in the molecular nature of the solid surface, i.e., in the conditions of selective wetting at the water—oil boundary between two antipolar liquids. This orientated adsorption of surfactants leads to the production of hydrophobic properties on initially hydrophilic solid surfaces and to the production of hydrophilic properties on initially hydrophobic surfaces. In this case the effect leading to the production of hydrophobic properties is particularly marked; it is intensified by chemical bonding, leading to fixation of the polar groups of the surfactants on corresponding regions of the solid surfaces.

Sufficiently long hydrocarbon chains, directed outwards, either prevent aqueous wetting of the surface or lead to selective displacement of water from the surface by a nonpolar liquid (oil). These reagents which produce hydrophobic properties consist primarily of those used as flotation collectors. They serve primarily to decrease the aqueous wettability of solid particles of various types of minerals by either selective chemical adsorption or an appropriate surface chemical reaction. It is these particles which adhere to the bubbles of air in the suspension (pulp) in the flotation machine with the formation of a contact angle whose highest hysteresis value determines the intensity of adhesion (force of separation).

The reagents which produce hydrophobic properties on nonoxidized metals and sulfides are surfactants with specific chemically adsorbed metallophilic polar groups containing divalent sulfur or phosphorus. These include alkyl and aryl xanthates and thiophosphates. The reagents which produce hydrophobic properties and act as collectors for the so-called oxidized minerals (the carbonates, sulfates, and oxides of heavy metals and the insoluble salts of the alkaline earth metals calcium and barium) are anion-active substances, chiefly carboxylic acids or their water-soluble soaps, which form insoluble coatings of the corresponding calcium or barium soaps on the surface of the mineral. The reagents which produce hydrophobic properties on silicon dioxide and a number of silicates, aluminosilicates, minerals with acidic groups on their surface, or glasses, are cation-active compounds of the organic base type and their salts (tetraalkylammonium or alkylpyridinium bases and their halides).

Reagents producing hydrophobic properties also play an important part in the production of pigments by the paste method, which eliminates drying and grinding, i.e., mechanical dispersion of the pigment, by making it possible to obtain the pigment by a condensation method involving precipitation in aqueous medium, e.g., as the result of a double decomposition reaction in the colloidal dispersed state. The reagent used to produce hydrophobic properties in this process ensures transfer of the pigment particles from aqueous medium to the oily binding state with complete removal of water as a result of selective wetting. The production of hydrophobic properties by adsorption in glass fiber and a number of synthetic fibers plays an important part in the preparation of these fibers and also during subsequent treatment and spinning, by producing a stable lubricating film on the surface of the fibers and preventing rupture or damage to the surface. The production of hydrophobic properties is also extremely important in the finishing treatments applied to fabrics and a number of other materials.

The orientation which results from the adsorption of any organic surfactant from a liquid hydrocarbon medium on the surface of the particles of a suspension and leads to the production of hydrophobic properties, leads to stabilization of these particles by preventing their aggregation; this is particularly marked if the particles are initially hydrophilic, i.e., oleophobic.

In aqueous suspensions this production of hydrophobic properties as a result of adsorption gives the opposite effect—loose cohesion of particles by means of the hydrocarbon chains directed outward; this process is usually called flocculation.

Flocculation forms the basis of a number of ore-dressing processes, accompanies flotation, and is of considerable importance for increasing the rate of settling of suspensions (concentrates) and their filtration from aqueous medium. Flocculation in this way by means of surfactants is much more effective for the above purposes than coagulation by ordinary electrolytes. Surface-active polymers whose macromolecules may be adsorbed on different particles of the suspension, binding them together in a loose aggregate, are particularly effective as special flocculating reagents.

It would seem that adsorption of surfactants, which may lead to the production of hydrophilic properties in solid particles in water, particularly when these particles are initially hydrophobic, should lead to the stabilization of suspensions. These effects, however, are often not clearly defined, since they are masked by electrical phenomena at the surface of the particles, involving the formation of diffuse double layers of ions or their compression with increase in the concentration of the electrolyte (in the ionic strength of the solution).

Surfactants which produce a gel-like structure in the adsorption layer and in solution have a strong stabilizing action on suspensions, i.e., they prevent the coagulation or aggregation of the particles, irrespective of the concentration of the solid disperse phase, by ensuring maximum regularity in its distribution in the surrounding medium. These surfactants now come into the third group and may be described by the general term stabilizers.

In addition to the production of a structural-mechanical barrier to the mutual approach of the particles in the form of a gel-like protective sheath, an important condition for stabilization is that the outer surface of this sheath should be hydrophilic and that no aggregation by means of the outer surfaces of these sheaths (secondary coagulation) should take place. This is in fact the mechanism of action of these reagents which have a strong stabilizing action on suspensions, emulsions, and froths, and ensure maximum possible stabilization and complete aggregative stability of these lyophobic systems. A stabilizing action may also be shown by comparatively weak surface-active substances; even where adsorption is weak these reagents may form strongly structured protective sheaths. Examples are provided by glucosides (saponin), polysaccharides, and high-molecular compounds of the protein type. These compounds, which are present in solution in the form of large micelles or macromolecules and act as effective stabilizing agents for suspensions, emulsions, and froths, cannot, however, act as dispersing agents; because of their size, penetration of the particles of these reagents into the surface defects (narrow microcracks) is difficult and the rate of their adsorption, like the rates of ordinary diffusion and migration over the surfaces, is reduced considerably. At the same time, the surface activity of these substances is comparatively low, due to the more symmetrical distribution of the polar and nonpolar groups in the large particles.

Stabilizers not only prevent the ordinary aggregation of particles, i.e., ordinary coagulation or coalescence, but also prevent the development of coagulation structures, since their adsorption blocks the points of cohesion of the particles and thus prevents their close approach. Suspension stabilizers are therefore also adsorption plasticizers. When added in very small quantities they decrease the stability of the structure (the structural viscosity).

Thus, the addition of a plasticizer (stabilizer), by breaking down the three-dimensional framework, decreases the quantity of liquid medium which is not bound by molecular forces but is mechanically held in the cells of the structure. At the same time, there is a decrease in the water uptake and oil capacity of the solid dispersed phase, i.e., in the minimum volume of liquid necessary for the production of a uniform sufficiently mobile paste with maximum concentration per unit volume of solid dispersed phase. It is for this reason that the addition of surfactants, as well as the surface activity of the binding agent itself, ensures minimum oil capacity of pigments in paint systems, thus improving the quality of the protective covering while retaining ease of application.

The addition of surface-active plasticizers to cement mixes and concretes leads to a sharp decrease in the water uptake, i.e., they ensure conversion to stiff but at the same time uniform mixtures by facilitating regular mixing. The density and durability (frost resistance) of the concrete after setting is increased, leading to an increase in the strength and to a considerable saving in cement (by 10-20%) without loss of strength. At the same time, the addition of plasticizers may decrease considerably the rate of the initial stage of crystallization and formation of clusters of crystallites of the new hydrate phase during the setting of cement, gypsum, or lime in building materials, leading to an increase in the induction period of crystallization structure-formation. This decreases the intensity of heat liberation resulting from the exothermic effect accompanying hydration, which often leads to the production of dangerous thermal stresses.

In hydrotechnical construction the use of small quantities of added surface-active plasticizers makes it possible to lay massive concrete in blocks of much larger dimensions (the dimensions of these blocks are limited by the intensity of heat liberation, i.e., by the temperature differences produced between the core of the block and its surface).

In the Soviet Union the material used as surface-active plasticizer is a standard product from waste material from the cellulose-paper industry—sulfite-alcohol waste (SSB), which is the residue remaining after the distillation of alcohol from the rejected sulfite-cellulose extract. The active constituents of this plasticizer are calcium lignosulfonates and substances of the carbohydrate type. Although this plasticizer is extremely effective, particular in its optimum combination with electrolytes, the subsequent development of research work has shown that it should be replaced by synthetic surfactants of the alkyl sulfate and alkylaryl sulfonate types or by nonionogenic substances (polyethylene glycol ethers, alkyl phenols, or cetyl alcohol) or by copolymers in their optimum combinations with stabilizers of the carboxymethylcellulose type and modifying electrolytes. These formulations should also ensure uniform air uptake and even foaming in concrete mixes, i.e., the formation of a closed porosity in the initial stage of the mobile mixture as a result of the equilibrium distribution of small air bubbles which are not connected to one another. For this purpose technical soaps are usually employed (the quantities added amount to several hundredths of 1% of the weight of cement), although these are not sufficiently effective plasticizers. The development of this closed porosity increases the durability and more particularly the frost resistance of concrete by decreasing its water permeability (by closing the open pores).

When the air uptake is even greater, we obtain "warm" concrete with its low specific gravity and high thermal insulation properties for house building.

The addition of surface-active plasticizers also plays an important part in the modern ceramic industry and in the production of cement (clinker): by liquefying the highly dispersed suspensions or slurries, they make it possible to concentrate them, i.e., to reduce the water content to the maximum possible extent, and this intensifies and decreases the cost of the technological processes.

In building and road-construction materials based on bitumens, particularly in the various types of asphalt-concrete additives, plasticizers are acquiring considerable importance, since they lead to a sharp increase in the cohesion between the bitumen and the dispersed mineral filler and between the bitumen and the stone material. This increase in adhesion is due to the production of hydrophobic properties on the mineral surfaces as a result of chemically fixed adsorption with the formation of a normally directed adsorption layer of surfactant. Siliceous surfaces, such as quartz sand, are made hydrophobic in the process, as already mentioned, by cation-active substances, but with preliminary activation by alkaline earth cations, e.g., by treatment with lime water.

The production of hydrophobic properties can also be achieved by means of anion-active substances—carboxylic acids and their soaps, which are moreover more universal activators, producing hydrophobic properties primarily in carbonate rocks (limestones and dolomites). The addition of surfactants decreases sharply the bitumen uptake, ensures regular mixing of the dispersed components of the bitumen mixture, and activates the surface of the mineral granules, making them active bitumen fillers. This ensures a sufficiently high mobility in the asphalt mixture, necessary for its compact packing, and the production of a strong, high-quality, water-resistant asphalt concrete, and also makes it possible to carry out asphalting work at all seasons.

The addition of surfactants for the production of hydrophilic properties and as plasticizers producing hydrophilic properties is being used on an ever-increasing scale for the strengthening and compacting of soil structures in geological engineering and building construction and also for the production, in soils, of structures which increase their fertility and productivity and prevent soil erosion.

The fourth group of surfactants consists of the wetting agents, which occupy a leading place as far as volume of practical application is concerned. Their function is to wash a variety of contaminants from surfaces by converting these contaminants to the state of a stabilized suspension (or emulsion) and thus preventing their reattachment to the surfaces from which they were washed.

Wetting agents should exhibit a complex set of properties characteristic of surfactants of the previous three groups: they should produce a marked decrease in the surface tension at the water—air boundary, i.e., they should be highly surface-active, showing a wetting action and at the same time producing hydrophilic properties. By forming three-dimensional micellar structures in the volume of the solution and more particularly in the surface layers, wetting agents should not only be dispersing agents but should have a strong stabilizing action as well on suspensions and emulsions (emulsifiers). They should also ensure the general solubilization of hydrocarbon impurities and oily contaminants in the nuclei of the micelles.

At low concentrations in extremely dilute aqueous solutions, all the ordinary soap-like wetting agents (with the exception of the nonionogenic materials) can produce hydrophobic properties and can be used as flotation collectors under suitable conditions. To produce a wetting action, therefore, these surfactants are used only in sufficiently high concentrations above the critical concentration of micelle formation, although these concentrations are in general fairly low.

Moreover, surface-active wetting agents are always used in practice not individually but in the form of complex mixtures including, in addition to the main surface-active agent itself, additional surface-active structure-forming stabilizers (protective colloids or thickeners of the carboxymethylcellulose type) and electrolytes including alkaline electrolytes, which create a favorable medium for the wetting action and for the optimum efficiency of the surfactant.

The use of small quantities of added surfactants in the petroleum industry is more varied than in any other branch of the national economy. At the same time, the synthesis of the main types of organic surfactants is closely related to the development of petroleum chemistry and petroleum treatment.

All surfactants with their varied applications can be classified on the basis of two most important features: on the basis of chemical features we can distinguish anion-active, cation-active, and nonionogenic substances, and on the basis of colloidal-structural features we can distinguish substances which are present in true (molecular or ionic) solution in a given medium (solvent) and therefore do not exhibit a wetting action, and wetting agents which form micellar or, at high concentrations, gel-like structures in the volume of the solution and in the surface adsorption layer, i.e., which undergo colloidal dissolution in the form of molecular or ionic micelles.

Wetting agents (detergents) may be substances from any of the three classes: anion-active, cation-active, and nonionogenic, the necessary condition being that they have a polar group with a sufficiently high polarity (sufficiently hydrophilic) and also a sufficiently long hydrocarbon radical. It is for this reason that the higher homologs of the alcohol and carboxylic acid series are not wetting agents and do not exhibit colloidal solubility in water as do the paraffins. In the transition from carboxylic-acid to alkali-metal-salt form, however, the hydrophilic and ionogenic character of the carboxylic polar group increases sharply and soap-like properties are produced.

The even more hydrophilic sulfonic acid group increases the colloidal solubility in water, not only in the alkali metal alkyl sulfates or alkylaryl sulfonates but also in the corresponding acids themselves or their alkaline earth metal salts, which to a considerable extent form the basis of the modern synthetic wetting agents which replace soaps made from fats. Ionogenic character leads to an increase in the hydrophilic properties of the polar group but is not a necessary condition, since wetting (soap-like) substances may also be nonionogenic. Examples include the polyethylene glycol ethers of alkyl phenols and various fat-sugars. This class also includes surface-active nonionogenic polymers, e.g., the block copolymers of polypropylene and polyethylene glycols, in which the former forms the hydrophobic part of the compound (insoluble in water), while the latter forms the hydrophilic part; these new wetting agents are produced in the USA with the general name of "pluronics."

Let us consider in more detail the main fields of application of surfactants in the processes of petroleum recovery and treatment.

1. The structured flushing liquids which have to be used in the drilling of deep exploratory or operational wells should exhibit thixotropic properties, i.e., they should be capable of undergoing ready reversible breakdown and restoration of the three-dimensional structure in a stream. This ensures that these flushing liquids are capable of producing suspensions even with large drilled rock particles and also the particles of dispersed agents added to increase the specific gravity during drilling under complicated conditions. At the same time, thixotropic structure formation ensures that these flushing liquids have a low viscosity at sufficiently high rates of flow in the well.

It is also necessary to add surfactants to aqueous and petroleum-based flushing liquids to act as stabilizers which liquefy the thixotropic structures, i.e., decrease the viscosity; this is particularly important where additives have been used to increase the density. The stabilizers used in aqueous flushing liquids are weakly surface-active substances of the protective colloid or hydrophilic polymer type which at the same time produce hydrophilic properties and have a strong structure-forming action, particularly in the adsorption layers. These include the alkali humates (alkaline extracts of peat or brown coal), lignosulfonates from sulfite-cellulose extracts (sulfite-alcohol waste) and the cellulose derivatives—carboxymethylcellulose, ethylcellulose, cellulose sulfates, or sulfonated cellulose ethers. These are all water soluble and produce structured aqueous solutions and protective (stabilizing) adsorption sheaths on the surface of the particles of the dispersed phase (clays, drilled rocks, and materials added to increase the density).

The surface-active stabilizers used in petroleum flushing liquids may be surfactants of the hydrophobic type—naphthenic and asphaltenic acids, the aluminum soaps of carboxylic acids, etc., in accordance with the general regular features of the stabilization of suspensions in hydrocarbon media established by ourselves with E. K. Venstrem and other co-workers. The aqueous and petroleum-based flushing liquids of varied composition are based on the extensive physicochemical studies of K. F. Zhigach and others. At the same time, flushing liquids of the aqueous type should also include electrolytes (phosphates, sodium silicate, water glass) which together with the surfactant ensure optimum conditions for the decrease in adsorption strength of the rocks being drilled.

The introduction into the flushing liquids of materials which are able to undergo sufficiently rapid adsorption on the newly formed rock surface in the zone of preexisting imperfections under the end face during the development of these surfaces, leads to an increase in the mechanical rates of drilling in wells in hard rocks, to a marked decrease in wear in the cutting tool, and to the formation of a fine fraction of drilled rock particles capable of forming a structured flushing liquid in water alone without the addition of clay to the well (work by K. F. Zhigach and L. A. Shreiner, and subsequently by S. N. Yatrov). Studies by K. F. Zhigach and co-workers have shown that the adsorbed substances in flushing liquids which decrease hardness during drilling in hard rocks also bring about self-sharpening of the cutting edge of the drill bit, thus reducing considerably the number of raising and lowering operations.

Scientifically based (optimum) formulations for flushing liquids must include substances added to decrease hardness (dispersing agents), particularly when hard rocks have to be drilled. Even in cases where clays are being drilled, however, formulations of this type facilitate the production of a high-quality liquid in the well itself.

2. When surfactants of the stabilizer type are used in aqueous flushing liquids it is necessary to prevent stable foam formation, which prevents normal operation of the liquid. For this purpose, small quantities of substances which are strongly surface-active at the aqueous solution–air boundary are added in small quantities as antifoaming agents. These include the intermediate homologs of the alcohol series (C_5-C_9) in the form of suitable technical materials and waste products. These substances operate by displacing the less surface-active stabilizer (foam-forming agent) from the adsorption layer in the foam films without producing a structure in the adsorption layers or forming stable films.

3. When a well is being made operational, an important influence on the rate of establishment of the well is shown by the processes in which the pores of the zone next to the end face, particularly when this zone has a highly porous structure, are freed from the aqueous flushing (clay) liquid; in this case it is necessary to ensure ready displacement of the water by the inflowing petroleum. For this purpose the end face in carbonate rocks is treated with hydrochloric acid containing corrosion inhibitors (to protect the equipment), which are also surfactants exhibiting a specific action on the surface of the metals. Dispersing and wetting agents are also used.

A process which is acquiring considerable importance involves flushing of the zone next to the end face with comparatively concentrated aqueous solutions of surfactants which may produce a considerable decrease (almost to zero) in the surface tension at the water–petroleum boundary of separation in the pores of the reservoir. Let us consider the dimensionless "capillary criterion"

$$K = p_\sigma / p_H = \frac{2\sigma_{12}\cos\theta}{r(\varrho_1 - \varrho_2)\,gH},$$

where θ is the contact angle measured in aqueous medium; ρ_1 and ρ_2 are the densities of petroleum and water respectively; g is the force due to gravity; r the effective pore radius; H is the effective height characterizing the stratal pressure; $p_H = (\rho_1 - \rho_2)gH$; and p_σ is the capillary pressure.

Sharp decrease in σ_{12} leads to a favorable result–to a decrease in the capillary pressure p_σ, when it prevents the displacement of water by petroleum in the zone next to the end face under the influence of the stratal pressure.

Industrial tests in the United States, together with data obtained by a number of Soviet workers, have shown that treatment with surfactants can lead to a considerable increase in the rate of establishment of wells and can also increase petroleum output by a factor of 2-3 in wells which are already operational, and decrease considerably the quantity of water and salts in the petroleum as a result of demulsification in the well.

A closely related problem is the opposite but no less important task of displacing petroleum from the finely porous structure of the reservoir by means of forced water inside or outside the contour. In the control of the processes of the mutual displacement of water by petroleum or of petroleum by water in the pores of the reservoir, it is particularly important that the liquid being displaced should move in the form of drops (in the emulsified state) without direct reaction with the rock surface forming the walls of the pores in the reservoir. In this case the dispersion of the liquid being displaced, decrease in the dimensions of the drops, and increase in the ease of their deformation during passage through narrow pores, together with stabilization of the drops to prevent their merging (coalescence), are extremely important. The dispersion and stabilization and hence the passage of the liquid being displaced through the porous medium are facilitated by a decrease in surface tension at the petroleum–water boundary between the two liquid phases. In this case it is desirable that the small quantities of surfactant used should not be used up in the formation of bonds with the highly developed surface of the rocks forming the reservoir (particularly in the case of clay or carbonate minerals). A decrease in the viscosity of the petroleum also facilitates its displacement by water. All these problems are being examined in work by G. A. Babalyan and co-workers and are described in his articles.

In the processes involved in the mutual displacement of petroleum and water in the pores of a reservoir, the adsorption of surfactants on the solid surface (in the pores) close to the meniscus is of considerable importance. This adsorption may lead to the inversion of the selective wetting: with increase in petroleum recovery the reservoir should be made hydrophobic close to the meniscus as a result of orientated adsorption, particularly with chemical fixation of the polar groups—anion-active on alkaline earth metal carbonate rocks or cation-active on quartz sandstones or silicate rocks.

In the case of forced flooding—the displacement of petroleum (particularly petroleum occurring in the form of films) by water—the walls of the pores of the reservoir close to the menisci should be made hydrophilic. This is achieved by the production, close to the menisci, of a sufficient concentration of soap or any surface-active wetting agent, e.g., by simple alkalization of the water used for forced flooding: by reacting with the naphthenic acids present in the petroleum, the alkali leads to the formation of sodium soaps close to the water—petroleum surface of separation.

In the application of surfactants, however, considerable difficulties are introduced by the extensive consumption of the materials as a result of their adsorption by the extensive rock surface at a distance from the menisci where this adsorption produces no effect. Attempts are therefore being made to achieve formation of surfactant at the actual petroleum—water boundary during the displacement process (as in alkalization of the water, for example) by using weak nonionogenic surfactants which are adsorbed by solid surfaces, and by limiting their action to the production of a decrease in surface tension at the petroleum—water liquid boundary of separation when the capillary pressure prevents migration in the required direction.

Extensive experimental studies of the physicochemical processes involved in the mutual displacement of water and petroleum in porous rock reservoirs have been carried out with a variety of models in M. A. Kapelyushnikov's laboratory and subsequently by M. M. Kusakov and K. E. Zinchenko with the assistance of the present author, M. M. Kusakov, et al., M. A. Geiman and V. B. Shneerson, G. A. Babalyan, I. I. Kravchenko, I. L. Markhasin, F. I. Kotyakhov, and also by A. T. Kochmarev and G. T. Ovnatanov under industrial conditions. In these studies the influence of added surfactants was also investigated and it was shown that in the displacement of petroleum by water under conditions of forced flooding, the process usually does not involve displacement of the petroleum by the aqueous solution from the rock surface, which would take place by the familiar mechanism involving inversion of selective wetting.

In this process, in actual fact, the main bulk of the petroleum is washed out, leaving a film of petroleum on the surface of the hydrophobized reservoir surface. At first the petroleum is forced from the pores in the water-free state and its place in the volume of the pores is occupied by water, but afterwards the rest of the petroleum is washed gradually from the surface of the rock particles in the form of a direct emulsion of petroleum in water, and here the added surfactant functions as an emulsifying wetting agent. Extremely successful tests with cation-active substances (dimethyldilaurylammonium chloride) have recently been carried out in Romania. The use of surface-active additives for treatment of the zone next to the end face in wells has proved to be advantageous in spite of the high cost of the materials.

Even when no petroleum—water menisci are formed, however, e.g., under conditions of petroleum filtration or petroleum transport involving flow through pipes, the addition of surfactants together with increase in temperature is acquiring considerable importance (the effect of heating the zone next to the end face in wells has been studied by F. A. Trebin and A. B. Sheinman). The action of peptizing and stabilizing surface-active additives, which are adsorbed on the paraffin crystallites in the initial nucleation stage of their formation and growth to produce a crystallization structure, involves blockage of the nuclei, decrease in the rate of their growth, and prevention of growth (i.e., retention of the paraffins and ceresins in solution) and prevention of the formation of deposits and blockages even at comparatively low temperatures.

In the absence of visible deposits, in the initial stage of their production, the introduction of surface-active additives produces a sharp decrease in the structural viscosity of petroleums and petroleum products, which is particularly high at low temperatures. These surfactants are depressors which lower the solidification point of lubricating oils and other petroleum products. The adsorption mechanism of their action has been studied by the Soviet workers K. S. Ramaiya and, later, T. P. Zhuze, P. I. Sanin, and others.

An analogous effect is produced by a different type of additive which improves the viscosity temperature curve for lubricating oils by preventing the development of three-dimensional structures at low temperatures and also by surface-active additives which prevent the formation of carbon deposit on the cylinder walls in internal combustion engines.

Demulsifying agents are acquiring particular importance in the petroleum industry: these are surfactants added to increase the rate of separation into layers in the emulsions formed by water and crude petroleum, as far as possible in the actual well during recovery, in order to achieve the production of water-free and salt-free petroleum. The problem of finding demulsifiers for emulsions of both types, like the problem of finding surface-active additives to increase the quality and degree of utilization of cement in cementation processes and additions for increasing the lubricating power of lubricants and liquids combining lubricating and cooling action, used for the treatment of metals by pressure and cutting, is a general problem which extends far beyond the range of interests of the petroleum industry alone. It should be pointed out, however, that at the present time it is generally accepted that the use of any method of demulsification—thermal, mechanical, or electrical—is advantageous only when the optimum quantity of demulsifier is added to the petroleum being treated.

SURFACTANTS IN PETROLEUM DRILLING AND RECOVERY PROCESSES

P. A. Rebinder

In all branches of the new technology, the use of small quantities of so-called surface-active substances (surfactants) for the intensification of technological processes and for increasing the quality of production is acquiring considerable importance [1, 2, 3]. Surfactants are substances which are adsorbed on definite surfaces of separation between phases in contact with one another and which produce, on these surfaces, layers with a thickness corresponding to one molecule—adsorbed monomolecular layers. Thus, even very small quantities of added surfactants may produce a very marked change in the conditions of molecular interaction at the surfaces of separation and in the velocity of phase changes and the transfer of a substance from one phase to another.

The influence of surfactants is acquiring particular importance in the many fields of technology based on the principles, regular features, and conclusions reached in the new interdisciplinary science – physicochemical mechanics [4]. This field of physicochemical science has developed in the last ten years at the boundaries of physical and, more particularly, colloid chemistry—the physicochemistry of surface phenomena and disperse systems, the physics of solids, mechanics, and a number of branches of the technology of the production of building, constructional, and other materials. The two basic problems of physicochemical mechanics are: 1) control of the deformation and breakdown of solids during processing and, in particular, during the drilling of wells in different rocks; and 2) control of the processes by which three-dimensional structures with required mechanical properties are formed in disperse systems, e.g., in the structured flushing liquids used for drilling deep wells. The study of this problem is providing a scientific basis for selecting the optimum technological conditions for the production of various high-quality materials with required properties and disperse structure, usually a fine granular structure. These materials include building materials (cement concretes and asphalt concretes), materials based on soils and bitumens, materials based on polymers (rubber and plastics), fibrous and finely porous materials, ceramics and metal ceramics, refractories, and heat-resistant and inert alloys for the new technology.

These problems may be said to be related: the first is concerned with overcoming the strength of solids, while the second is concerned with developing and increasing their strength. In both types of process the addition of surfactants is acquiring considerable importance and is making it possible to control processes of breakdown and structure formation as well as mechanical, thermal, and other effects.

Let us consider the chief applications of surfactants in the petroleum industry—in the drilling of test wells and operational wells, in the actual processes of petroleum recovery, and in the removal of water and salts from the petroleum during its preparation for processing and transportation [1, 2, 3]. Incidentally, it should be noted that all the chief types of the most effective surfactants are obtained or can be obtained as products of the chemical treatment of petroleum. This applies particularly to the anionogenic soap-like alkyl sulfates and alkylaryl sulfonates (the Na, Ca, Mg, and Al salts and the corresponding acids) and the nonionogenic alkyl derivatives of polymerized ethylene oxide, or copolymers of ethylene oxide and propylene oxide.

Surfactants in Well-Drilling Processes. Surfactants, including electrolytes which are adsorbed by the surfaces of the rocks being drilled, always decrease the hardness of the rocks to a greater or lesser extent: the addition of surfactants to the flushing liquid—water or a nonaqueous medium—leads to an increase in the mechanical rate of drilling and an increase in the durability of the cutting tool [5, 6, 7, 8].

Since these reagents (alkaline electrolytes, sulfonic acid derivatives, carboxymethylcellulose, sulfonated cellulose ethers, etc.) bring about an all-round improvement in the quality of clay liquids and natural flushing liquids [9], they are widely employed at present in these materials.

Improving the Quality of Flushing Liquids. The highly dispersed thixotropic suspensions used as flushing liquids for drilling wells under complicated conditions cannot be used without the addition of surfactant stabilizers to peptize the particles of the solid disperse phase. These additives are of particular importance for heavy solutions in the presence of coagulating electrolytes in high concentrations in highly mineralized stratal waters. In these cases the flushing suspensions undergo considerable thickening, and in order to decrease their viscosity stabilization is necessary, i.e., coagulation and cohesion of the particles, particularly those of the finest fractions, must be prevented.

The addition of surfactants and structure-forming reagents (coagulating electrolytes) in the optimum proportions makes it possible to produce the required thixotropic properties in flushing suspensions formed by particles of drilled rock.

Improving the Properties of Cements Used for Cementation. The addition of surfactants and electrolytes ensures an initial (induction) period of appropriate duration before the start of hardening, i.e., before the development of a crystallization structure in the cement or cement-sand mix (or cement suspension for injection to strengthen oil-well walls). At the same time, the hardening process, i.e., the increase in strength under the influence of the additives, takes place at a greater rate when packing (or penetration of the suspension) is complete.

Control of the Mutual Displacement of Water and Petroleum in the Pores of the Oil-Bearing Stratum. For water-flooding and displacement of the residual petroleum from a stratum, or for its elution, as shown by G. A. Babalyan and co-workers [15, 16], it is necessary to use surfactants. These additives should increase the selective wetting of the rock by water by producing permanent hydrophilic properties in the rock and simultaneously decreasing the interfacial tension between the water and the petroleum. Suitable surfactants are the nonionogenic, soap-like polyethyleneglycol ethers or alkylphenols, which also emulsify the petroleum in water and displace it in the emulsified state as droplets of emulsion.

An effect acquiring particular importance is the opposite of that described above—the production of hydrophobic properties at the end face of the well by flushing with an aqueous or petroleum solution of surfactant, usually an anionogenic compound, to intensify the recovery of petroleum from the well and to reduce the time required to make the well operational.

The addition of surfactants to the hydrochloric acid used to loosen ("freshen") the end face of the well in carbonate strata ensures that the action of the hydrochloric acid is effective and penetrating, since it retards the chemical dissolution of the rock. At the same time, the inhibitors added to the hydrochloric acid to prevent corrosion of the drilling equipment also include surfactants which undergo specific adsorption (under certain electrochemical conditions) on the surface of the metal (steel). Anionogenic surfactants and, to an even greater extent, cationogenic surfactants not only produce hydrophobic properties in carbonate, aluminosilicate, and quartz-containing rocks but also prevent clays from swelling in water, so that the corresponding strata retain a fairly high penetrability.

Of considerable industrial importance are the surfactants used as demulsifiers for removing water and salts from petroleum, i.e., for bringing the moisture content to an appropriate level. Crude petroleum usually contains up to 30-60% of water in the emulsified state in the form of a stable (stabilized), highly dispersed emulsion of the inverse type (water, oil). This water is often highly mineralized and forms a saturated salt solution. Work by N. N. Serb-Serbina and A. M. Smirnova in our laboratory, by B. P. Tonkoshurov [10], and by V. G. Ben'kovskii and co-workers has shown that the water droplets in the petroleum emulsion are stabilized by a natural highly dispersed solid emulsifier consisting of carbenes and carbides made hydrophobic by the resins and asphaltenes of the clays.

The particles of this emulsifier in the petroleum adhere to the surface of the water droplets to produce structured sheaths—structural-mechanical barriers with high strength, preventing the approach and coalescence of the droplets.

The action of the demulsifiers involves the strong adsorption of the surfactants on the surface of the water droplets and on the surface of the particles of solid emulsifier, displacement of the latter from the water–petroleum surface, and prevention of the adhesion of the particles of natural emulsifier to the surface of the water droplets. The structural-mechanical barrier is thus removed from the surface of the water droplets and the adsorbed layer of demulsifier, in spite of the considerable decrease in the interfacial tension (which,however, does not reach very low values), does not produce a stable emulsion, i.e., it does not prevent the coalescence of the water droplets in the petroleum.

A. B. Taubman and co-workersA. F. Koretskii [11], S. A. Nikitina, and later M. A. Filatova [12] have shown that the addition of small quantities of a demulsifier to a stable emulsion causes the emulsion to separate rapidly into layers and also leads to a sharp decrease in the strength of the structure of the stabilizing layer on the surface–its limiting shearing stress.

A. A. Petrov (Giprovostokneft') [13] and D. N. Levchenko (VNII NP) [14] have suggested the use of nonionogenic surfactants as highly effective demulsifiers. These surfactants decrease considerably the surface tension at the petroleum–water boundary and move preferentially from the petroleum to the aqueous phase as a result of the hydrophilic properties of an OP with a sufficiently long oxyethylene chain in the molecule. The demulsifiers are strongly adsorbed on the petroleum–water boundary and displace the emulsifier from the surface of the droplets, but do not themselves stabilize the emulsion of water in petroleum to a sufficient extent. It should be noted, however, that in the absence of a natural strong emulsifier all demulsifiers form emulsions of low stability, chiefly of the direct petroleum–water type.

Similar direct emulsions, but with a very high stability, are formed in the presence of wetting agents based on soap-like surfactants of the hydrophilic type. These surfactants are widely used to wash traces of petroleum products from various containers (barges, tanks, etc.) [11].

Literature Cited

1. P. A. Rebinder, Surface-Active Substances (Izd. Znanie, Moscow, 1960).
2. P. A. Rebinder, Khimicheskaya nauka i promyshlennost'. IV, 5, 554 (1955).
3. P. A. Rebinder, Collection, The Use of Surface-Active Substances in the Petroleum Industry (Gostoptekhizdat, Moscow, 1961).
4. P. A. Rebinder, Physicochemical Mechanics–A New Field of Knowledge (Izd. Znanie, Moscow, 1958).
5. P. A. Rebinder, L. A. Shreiner, and K. F. Zhigach, Surface Reducers in Drilling (Izd. AN SSSR, Moscow-Leningrad, 1944).
6. K. F. Zhigach, Collection, The Use of Surface-Active Substances in the Petroleum Industry (Gostoptekhizdat, Moscow, 1961), p. 69.
7. M. Sh. Akmullin, K. F. Zhigach, and L. A. Shreiner, Izv. vyssh. uchebn. zav. Neft' i gaz, No. 9, 29 (1960).
8. M. Sh. Akmullin, Izv. vyssh. uchebn. zav. Neft' i gaz, No. 2, 35 (1962).
9. M. Sh. Akmullin and K. F. Zhigach, Izv. vyssh. uchebn. zav. Naft' i gaz, No. 8, 31 (1961); No. 6, 33 (1961).
10. B. P. Tonkoshurov, N. N. Serb-Serbina, and A. M. Smirnova, Principles of the Chemical Demulsification of Petroleum (Moscow, 1946).
11. A. B. Taubman and A. F. Koretskii, Collection, The Use of Surface-Active Substances in the Petroleum Industry (Gostoptekhizdat, Moscow, 1961), 105.
12. M. A. Filatova, S. A. Nikitina, A. B. Taubman, and P. A. Rebinder, Dokl. AN SSSR, 140, 4, 874 (1961).
13. A. A. Petrov, Collection, The Use of Surface-Active Substances in the Petroleum Industry (Gostoptekhizdat, Moscow, 1961), 112.
14. D. N. Levchenko, Collection, The Use of Surface-Active Substances in the Petroleum Industry (Gostoptekhizdat, Moscow, 1961), 162.
15. G. A. Babalyan, Problems of the Mechanism of Petroleum Recovery (Aznefteizdat, 1956).
16. G. A. Babalyan, I. I. Kravchenko, I. L. Markhasin, and G. V. Rudakov, The Physicochemical Principles of the Use of Surface-Active Substances in the Exploitation of Oil-Bearing Strata (Gostoptekhizdat, 1962).

THE STATE AND DEVELOPMENT OF WORK ON THE USE OF SURFACTANTS IN THE PETROLEUM INDUSTRY

G. A. Babalyan

The scale of application of surface-active substances (surfactants) in the Soviet Union increases every year. The use of surfactants gives considerable technological advantages and also results in a considerable saving of materials in spite of the present relatively high cost of these reagents. The development of this field of petroleum science and practice was greatly assisted by the work of L. G. Gurvich, begun as early as 1911.

Subsequently, in the Soviet period, the development of accurate physicochemical theories of petroleum recovery processes and methods for their control is associated with the work of Academicians A. N. Frumkin, P. A. Rebinder, and their colleagues and pupils.. P. A. Rebinder is the physical chemist who first undertook the direct development of the scientific principles of the applications of surfactants in the petroleum industry. Extensive use was not made of this work in petroleum practice, however, chiefly because of the inadequate development of petrochemical synthesis and the lack of the required range of surfactants.

With the development of petroleum chemistry and the synthesis and production of surfactants, various fields of applications of surfactants began to attract ever-increasing attention from workers in the petroleum industry. In a series of studies, the essential physicochemical features of the technological processes were revealed and methods were developed for controlling and improving these processes.

In February 1960, the First All-Union Conference on the Synthesis, Production, and Application of Surface-Active Substances was held in the UFNII. This conference laid the foundations for the systematization and coordination of scientific research work in this field.

In July 1962, the Second All-Union Conference on the Problems of Using Surface-Active Substances in Petroleum Recovery was held in the UFNII. The conference summarized the activities of the Institutes and outlined ways for the further development of research and experimental work. In April 1962, a Conference for the Exchange of Experience in the Application of Surface-Active Substances in the Petroleum and Chemical Industries was held in Saratov. Studies on the application of surfactants, particularly in the petroleum industry, are being carried out in UFNII, VNII, VNII NP, VNIIBT, VNIGNI, MINKh and GP, IGiRGI, VZPI, AzNII DN, IRGNI AN Azerb. SSR, the Institute of Chemistry of the AN Azerb. SSR, Giprovostokneft', Kuibyshev NII NP, TatNII, VNIING, NVNIIGG, VNIIneftekhim, GrozNII, NII Transneft', KFVNII, and many TsNIL's and TsNIPR's. The Physical Chemistry Institute of the USSR Academy of Sciences is carrying out scientific studies of a general character in this field, and these are providing a basis for a number of new applications of surfactants.

Work is being carried out in the following directions.

Increasing the Recovery of Petroleum From Strata and Increasing the Rate of Their Development

It is now established that a vast quantity of petroleum is not being extracted. Low recovery of petroleum leads to a decrease in the rate of development and a decrease in its efficiency. Considerable importance is thus attached to the problem of increasing the petroleum recovery coefficient, both in the USSR and in other countries. Problems associated with the addition of surfactants to water forced into the stratum in order to increase petroleum recovery are being studied in the UFNII, Kuibyshev NII NP, AzNIIDN, IRGNI AN Azerb. SSR, IGiRGI, and TatNII. Up to the present time, however, this work has not proceeded beyond the stage of laboratory experiment. Plans have been made to carry out industrial tests on surfactants for this purpose in the Arlanskii deposit in 1963.

The Use of Surfactants to Improve the Properties of Flushing Liquids

In the complex variety of problems encountered in the technology of oil-well drilling, one of the decisive factors ensuring the success of the process is the use of flushing liquids satisfying certain conditions. As is well known, flushing liquids should not only ensure removal of the drilled rock but should also prevent the possibility of ejection of petroleum, gas, and water, cave-in, collapse, engulfing, etc., during the drilling process. Even higher requirements are made of flushing liquids for drilling deep wells in which pressures of 500-1000 atm and temperatures of 150-300°C may be reached.

Laboratory studies and industrial tests have shown that the properties of flushing liquids can be successfully controlled by the addition of suitable surfactants. Work in this field is being carried out by the UFNII, VNIIBT, TatNII, and NVNIIGG.

It has been shown possible to use synthetic mixtures of hydrocarbons, products of the condensation of phenols and formaldehydes, gipan, and oxidized hydrolyzate lignin to control the rheological properties of clay suspensions. Work is also being carried out to find other heat-resistant reagents based on copolymers of acrylic and methacrylic acids, oxyethylated alkylphenols, etc.

Industrial tests are being carried out with sulfonol and the sodium salts of sulfonated shale resins (SNS), which improve a number of the properties of flushing liquids, while studies are also being made of the use of synthetic surfactants based on cellulose esters and sulfonated cellulose ethers (cellulose sulfate, sodium sulfoethylcellulose, and the calcium salts of sulfonated dextrin with low degrees of nitration), aliphatic acids, the vat residues of these acids, and oxidized petrolatum to prevent complications during drilling and to increase the durability of drilling-tool bearings.

Joint stabilization by a combination of different reagents: starch, SSB, KMTs, etc., has now been used successfully for drilling under complicated conditions or with highly mineralized drilling solutions.

Surfactants are now being widely employed to stabilize and control the structural, mechanical, and rheological properties of natural flushing liquids formed from the drilled rock. As is well known, such liquids can be produced only when surfactants are used.

The aqueous suspensions of nonclay rocks formed naturally in oil wells during drilling can be converted to stable drilling liquids with a sufficiently high concentration of solid phase and controllable parameters only when surfactants and electrolytes are used as stabilizers and structure-forming reagents respectively. The optimum proportions of surfactants and active additives required to stabilize aqueous suspensions of different nonclay rocks (carbonates, sulfate-halide argillites, and aleurolites) have been established. Experimental studies on the use of new synthetic surfactants for the stabilization of drilling solutions have been extended. Synthetic surfactants are generally more effective stabilizers than the reagents widely employed in drilling practice.

Mention must be made of work on the use of surfactants based on cellulose esters and sulfonated cellulose ethers for the stabilization of aqueous suspensions of nonclay rocks. It has been shown that cellulose sulfate (STs) and sulfoethylcellulose (ESTs) are highly effective. The addition of small quantities of these surfactants (1.5-2%) ensures practically complete stabilization of mineral suspensions of nonclay rocks. It should be emphasized that STs and ESTs stabilize the suspensions effectively in the presence of the soluble salts in highly mineralized water, and also have a high heat resistance. Good results have also been obtained in tests on the calcium salts of sulfonated dextrin with a low degree of nitration (Ca–NSD). The production of these surfactants has not yet been organized, however. The following conclusions may be reached on the basis of this work.

It is necessary to extend considerably research work on the use of synthetic surfactants for the stabilization of flushing liquids and to continue work on the synthetis of high-molecular surfactants based on cellulose esters and sulfonated cellulose ethers for the stabilization of drilling solutions. It is also necessary to develop the theory and practice of the use of surfactants and electrolytes for controlling the properties of flushing liquids formed from nonclay rocks. This is one of the most promising methods for improving drilling liquids and decreasing their cost.

The Cementation of Oil Wells

In the cementation of oil wells, surfactants and electrolytes are being used on an ever-increasing scale to control the time taken for the setting and hardening of the cement mix. A retarding action is shown by sulfite-alcohol waste, the sodium salts of carboxymethylcelluloses, tartaric acid, etc. An accelerating action is shown by the active additives (inorganic electrolytes): calcium chloride and sodium silicate; tests have also been carried out with aluminum sulfate, aluminum chloride, mixtures of synthetic hydrocarbons, and other synthetic surfactants.

A promising technique is the use of surfactants to clean the walls of oil wells before lowering and cementing the encasing columns. In the United States, for example, surfactants of the wetting-agent type are used for this purpose (a mixture of an alkylaryl sulfonate with nonionic wetting agents and sodium pyrophosphate, cationogenic quaternary ammonium compounds—salts of alkylamines and tetraalkyls). We have also obtained satisfactory results in experimental work with sulfonol in the Bashkir Republic and Kuibyshev Oblast'. Work in this field must be extended, since preliminary studies have indicated its considerable importance, particularly for the cementation of gas wells.

Decreasing the Hardness of Rocks

The addition of surfactants to flushing liquids facilitates directly the breakdown of rocks by acting as adsorption hardness-reducing agents (the Rebinder effect). By being adsorbed on the surface of the rock, the surfactants decrease considerably the free surface energy and thus decrease the work required to break down the rock.

Certain data indicate a considerable increase in the rate of drilling when nonionogenic surfactants are added to drilling solutions or to water.

Thus, according to data obtained by the UFNII, the use of aqueous solutions of the nonionogenic surfactants OP-10 and UFE_8 while drilling the productive horizon in the Arlanskii, Shkapovskii, and Tuimazinskii deposits increased the mechanical velocity by a factor of almost 2 compared with the use of a clay suspension treated with calcined sodium carbonate, and by 50% compared with the use of water alone as flushing liquid.

The Prevention of Absorption of Flushing Liquid, Cave-In, and Collapse, and Methods of Combating These Effects

The use of surfactants and electrolytes to prevent the absorption of flushing liquid, cave-in, and collapse has not yet been sufficiently developed. Only a few isolated attempts have been made to use surfactants to control the properties of drilling liquids in absorption zones and to strengthen the walls of oil wells. This work has not been summarized, however. At the same time, there is no doubt that the use of surfactants to solve these complex problems appears extremely promising.

The Use of Surfactants to Open Up Productive Strata and Establish New Wells

It has been established that the opening of productive strata by means of water or drilling liquids with high water loss leads to a decrease in the oil permeability of the zone next to the end face of the well. The establishment of such wells is difficult and in a number of cases becomes impossible. The reason for this is that the water or filtrate from the drilling liquid penetrates into the zone next to the end face, displaces the oil from the end face of the well, and leads to swelling of the clays.

In the reverse displacement of water by oil, a considerable proportion sticks in the zone next to the end face and prevents the entry of oil into the well. The addition of suitable surfactants to the water or clay suspension leads to a sharp decrease in the surface tension of the water at its boundary with the oil, a decrease in the forces of molecular interaction between the solid surface and the water, and dispersal of the water in the pore space, leading to more ready displacement of the oil from the zone next to the end face. Work in this field is being carried out by the UFNII, VNIIneft', TatNII, Kuibyshev NII NP, AzNIIDN, GrozNII, VNIGNI, and TFVNIIneft'.

In spite of the considerable volume of work which has been carried out and in many cases the high efficiency of surfactants for this purpose, the results have not yet been summarized and no clear estimate has been made of the efficiency of different types of surfactant for different geological conditions. Different methods are used to apply and test surfactants in different regions and this makes it difficult to generalize the results of this work.

Substances used as additives to drilling suspensions and water include nonionogenic surfactants such as OP-10, OP-7, and UFE_8, and anionogenic surfactants such as Sulfonol, Azolyat, SNS, etc. Industrial tests on these surfactants have been carried out in more than 100 wells. The time required to establish a well has been decreased considerably (to 2-8 h) and the output of the wells has been increased by an average of 30%.

The use of nonionogenic surfactants has assisted in the detector of a gas deposit in the Yuzhno-Borisovskii area in Siberia and has helped oil workers to obtain an additional supply of petroleum in the Parfenovskii area, and the first Cambrian oil in the Azov area.

The Use of Surfactants to Intensify the Recovery of Petroleum

It is known that the output from wells can be increased by treating the zone next to the end face with aqueous or hydrocarbon solutions of surfactants. In order to increase the productivity of oil and gas wells, water- and oil-soluble surfactants are added to the zone next to the end face in the form of aqueous or oil solutions, or in the form of a mixture of oil and an aqueous solution of the surfactant. The surfactants are adsorbed on the surface of the granules and at the oil—water interface, and decrease the surface tension. By selecting suitable surfactants it is possible to facilitate the displacement of connate water by the oil and hence to increase the water permeability of the zone next to the end face and accelerate the establishment of the well.

In flooded wells, the introduction of suitable surfactants makes it possible to increase the forces of adhesion between the oil and the rock and thus to decrease the phase permeability for water and to increase it for the oil. This leads to a decrease in the inflow of water to the well and an increase in the inflow of petroleum. Work in this field is being carried out in the UFNII, TatNII, GrozNII, TFVNIIneft', VNIIneft', and the AzNIIDN.

These institutes have carried out a considerable amount of work on the use of surfactants to treat the zone next to the end face of wells in a number of oil fields in the Bashkir Republic, the Tatar Republic, the Turkmen Republic, Azerbaidzhan, etc. For this purpose, use was made of water-soluble surfactants, chiefly of the ionogenic type, mixed with hot oil—sulfonol, SNS, etc.—or in individual cases, the nonionogenic surfactants OP-10 and UFE_8.

Surfactants are used effectively for the acid treatment of wells. They decrease the rate of the reaction between hydrochloric acid and carbonate rocks and thus make it possible for the acid to penetrate to a greater distance from the end face of the well. In this case the spent hydrochloric acid solution has a reduced surface tension at its boundary with the petroleum and is thus readily displaced by the petroleum. At the same time, the surfactants actively inhibit the corrosion of the equipment.

Work on the use of surfactants in acid treatment is being carried out in the UFNII, TatNII, GrozNII, and NVNIIGG. It has been found that the most effective surfactants for acid treatment are the cationic surfactants Katapin A and Karbozolin O, the nonionogenic surfactant UFE_8, and the anionogenic surfactant Sulfonol, etc.

Industrial tests showed that the output of petroleum after acid treatment with the nonionogenic surfactant UFE_8 was considerably greater than that when formalin was used as corrosion inhibitor.

In recent years, in addition to hydrochloric acid treatment and hydraulic rupture of the stratum, treatment of the zone next to the end face of the wells with surfactants has been used in petroleum industry practice to establish driving wells. The use of surfactants to increase the pickup of driving wells is possible both in the case of wells outside the contour, drilled into the water-saturated stratum, and in the case of wells inside the contour, drilled into a stratum with the same or different degree of oil saturation.

For wells outside the contour, the most effective surfactants are again those of the nonionogenic type; in suitable concentrations, they decrease the swelling of the clays and their use under industrial conditions gives an increase in the pickup of driving wells by an average of several dozen percent. This work is carried out on an extensive scale in the United States. The pickup of partings of the section with low permeability is increased to a particularly marked extent. Partings which previously did not take up water begin to take up water after treatment with surfactants, as a result of which the pick-up of the wells increases considerably and the driving pressure is reduced.

In the case of wells inside the contour in the absence of bottom water, the increase in permeability when suitable surfactants are added is due chiefly to a decrease in the quantity of residual oil and to swelling of the clay particles. Experiments showed that a decrease in the quantity of residual oil and hence an increase in the water permeability are produced by the addition of nonionogenic surfactants which increase the wettability of the rock by water and decrease the surface tension at the boundary with the oil to a much greater extent than when ionogenic surfactants are added.

When bottom water is present, irrespective of whether or not the water-saturated section contains residual oil, it is again advantageous to use nonionogenic surfactants. The increase in the pick-up of driving wells inside the contour may be considerably greater than the increase in the pick-up of wells outside the contour particularly where the oil has a high activity or a high concentration of asphaltic and tarry substances and where the strata themselves have a low permeability, are nonuniform, and have a high carbonate content.

Treatment of the zone next to the end face with surfactants is also used as a method of combating the flow of water into wells. An aqueous or petroleum solution of ionogenic surfactants is used for this purpose. In both cases the surfactants chosen are those which produce hydrophobic properties on the surface of the solid particles, preventing the capillary seepage of water into the oil-saturated zone. Laboratory studies indicate that this method should give good results in the case of the flooding of wells because of cone formation and in the case of breakthrough of water into a well along partings of high permeability. Hydrophobic properties in the solid surfaces are best produced by using petroleum solutions of ionogenic surfactants.

These problems are being studied in the IRGNI of the AN Azerb. SSR, UFNII, KFVNII, NVNIIGG, and Kuibyshev NII NP.

The work has now passed the stage of laboratory studies. Industrial tests have begun. In particular, the UFNII has started work on the production of hydrophobic properties as a result of the intensive injection of water by the introduction of hydrophobizing surfactants into driving wells.

Industrial tests on the use of surfactants to produce hydrophobic properties in the zone next to the end face are being carried out in Kuibyshev Oblast', the Bashkir Republic, Saratov Oblast', and Krasnodar Krai. In Krasnodar Krai aerated liquids stabilized with surfactants are used for this purpose.

The Use of Surfactants for the Removal of Water and Salts From Petroleum

The petroleum recovered in the USSR forms stable emulsions with a considerable water content. These emulsions are broken down by heating, the action of electricity, filtration, centrifuging, etc. The use of highly active surfactants as demulsifiers, however, makes it possible to remove water and salts from petroleum to the required levels. Until recently the petroleum industry in the USSR carried out the removal of water and salts from petroleum by means of the comparatively inefficient reagent NChK (neutralized black contact), in which the concentration of surfactants did not exceed 10-35%. The search for new and highly effective demulsifiers was begun in the USSR only in 1956. These studies, which were carried out in the UFNNI, VNII NP, Giprovostokneft', GrozNII, TatNII, and other Institutes, showed that oxyethylated alkylphenols OP, higher aliphatic acids, alcohols, and other alkyl derivatives with sufficiently long hydrocarbon chains have a high demulsifying power, which greatly exceeds that of NChK. The best results under laboratory conditions were shown by OP-32, 34, Shebekino alcohols condensed with 20-24 molecules of ethylene oxide, OZhK, OKO, Disolvan, Reagents 4400 and 4411, KS-59, OPS, $KAUFE_{14}$, oxyethylated alcohols, and alkylphenols, Gamlen, Sulfonol, UFE_8, VNII NP-58, Wetting Agent ND, Alkomon, OS-20, OP-29, and Leveling Agent A.

Positive results under industrial conditions were given by Disolvan 4411 and 4422 on Devonian petroleum, OZhK on Volgograd petroleum, OKO-100 on Romashkinskii petroleum, Sulfonol, $KAUFE_{14}$, OZhK, and a demulsifier based on pyrolysis tar on Ozeksuatskii and Zhirnovskii petroleums, OP-10, OK-59, and OPS on petroleum from Gor'kii Oil Refinery, OP-10, and Sulfoalmonite salts on Mukhanovskii petroleum, and $KAUFE_{14}$, UFE_8, and OGPK on Romashkinskii petroleum. It is now time to summarize the results obtained for the extensive introduction of the most effective nonionogenic surfactants as demulsifiers.

The Use of Surfactants in the Transport and Storage of Petroleum and Petroleum Products

The process of decreasing the viscosity of petroleum and viscous petroleum products which are to be transported is of considerable importance for reducing the cost of pumping and decreasing the depth to which pipelines have to be buried.

The addition of surfactants of the depressor type and additives improving the viscosity index in petroleum, black oil, and other petroleum products decreases their structural viscosity and lowers their solidification point. A decrease in viscosity is achieved by introducing into the product to be transported surfactants which are polar dispersing agents and stabilizers such as organic acids, alcohols, phenols, and certain salts of organic acids with metals of groups II and III. It is known that the addition of 0.001-0.002% of aluminum sulfonate to Artemovskii petroleum decreases its viscosity by 33%. Analogous data were obtained for the addition of 0.2% of aluminum naphthenate to black oil.

Work in this field is still in the initial stages of development; it is being carried out chiefly in the NIItransneft'. Here, tests are being carried out on salts of aliphatic acids and Sulfonol in the form of aqueous solutions to decrease the hydraulic resistance encountered in the transport of highly viscous petroleum products (black oils) and petroleum through pipelines. Studies are also being made of various surfactants in order to decrease the rate of volatilization of the light fractions of petroleum and petroleum products during storage in reservoirs. Surfactant 4411 has now been used successfully under winter conditions to force thickened petroleum into the outlet lines when oil is being collected in industry.

The Use of Surfactants to Combat Corrosion of Equipment in the Petroleum Industry

Surfactants are widely employed to increase the corrosion resistance of equipment and to increase its useful life. In particular, surfactants are widely used as inhibitors to combat corrosion of the underground equipment of oil wells (tubes, drill rods, and deep pumps) and to eliminate the intensive internal corrosion of oil and gas pipelines, containers used for the settling and storage of petroleum, etc.

The retardation of the corrosion of a metal by a given surfactant is brought about in different ways and may involve a retardation of the anodic or cathodic process of electrochemical corrosion and a change in the selective wetting of the metal surface by hydrocarbons at the boundary with the electrolyte (water, salt solutions, etc.) with the formation of a stable protective hydrocarbon film. The use of surfactants to combat corrosion is of interest, since they make it possible to increase considerably the useful life of equipment without altering its design and without using corrosion-resistant alloys or coatings. Work in this field is being carried out by the UFNII, UzNIIDN, Gipromorneft', the Chemistry Institute of the AN Azerb. SSR, VNIIneftekhim, GIPKh, etc.

The water-soluble inhibitor UFE_8 is extremely effective under conditions of corrosions by hydrogen sulfide. Unlike formalin, this inhibitor has a high chemical stability, is not oxidized by air, and can be used in compressor wells, where the liquid is raised by means of air.

Under laboratory conditions, good results have been achieved in overcoming corrosion by hydrogen sulfide in gas pipelines by using oil-soluble surfactants such as GIPKh-37, Diamine-Dioleate, and Amine C.

Laboratory studies are also being carried out to develop protective pigments using surfactants. These pigments are designed for painting the surface of marine equipment used in the petroleum industry in the most dangerous zone of periodic wetting.

Study of the Physicochemical Properties of Surfactants

Work is being carried out in this field in the UFNII, Giprovostokneft', VNII NP, MINKhiGP, IRZhI AN Azerb. SSR, VNIIneft', NIItransneft', etc. Studies are being made of the adsorption of surfactants from aqueous and hydrocarbon solutions on solid adsorbents, the surface activity of surfactants, the influence of surfactants on the structural and mechanical properties and stability of disperse systems, the properties of cement suspensions, the stability of crumbling rock, the coalescence of water droplets in petroleum and petroleum droplets in water, etc. It can now be claimed that, on the basis of work carried out in the petroleum industry on the application of surfactants, a new progressive branch of petroleum science has been established—industrial physicochemistry.

The effective utilization of surfactants in different technological processes requires further development of the theoretical principles of the application of these materials. The volume of work in this field is inadequate. Development of work in the field of adsorption is particularly necessary. In particular, there is a lack of clarity in certain methodological problems in the determination of the magnitude of the adsorption of surfactants from aqueous and hydrocarbon solutions, the influence on adsorption of pressure, temperature, composition, and structure of surfactants, the nature of the adsorbents, etc. The kinetics of the adsorption of surfactants have been little studied. All these features are to a considerable extent holding back the effective utilization of surfactants. Work in the scientific research institutes in most cases is purely empirical in character.

There are no unified methods for testing surfactants and determining their efficiency, as a result of which differences are observed in the results of experimental work and it is difficult to generalize these results.

The Institutes of the Academy of Sciences of the USSR and Union Republics must be drawn into these studies. The solution of these physicochemical problems will also provide considerable assistance to organizations working in the field of synthesis and will show the way in which synthetic work must be directed in order to satisfy the requirements of the petroleum industry.

It should be noted that the volume of industrial experimental work on the testing of surfactants in the petroleum industry is still quite inadequate.

Some points must be made regarding the synthesis of surfactants. No work is being carried out on the synthesis of surfactants designed specially for the technological processes of petroleum recovery. The only exception is the synthesis of surfactants for the demulsification of petroleum and the stabilization of flushing liquids. In view of the increasing demand for surfactants in the petroleum industry and the development of petroleum chemistry, this situation cannot be regarded as normal.

One effective means of increasing petroleum recovery and the rate of development of deposits is by the addition of surfactants to the forced water. The economic efficiency of this application of surfactants, however, is determined not only by the technological effect but also by their adsorption, i.e., their consumption. Thus, certain workers in the United States have pointed out that the use of surfactants is economically advantageous only when the adsorption does not exceed 10^{-4} g/g. The solution of these problems—determination of the minimum rational consumption of surfactant and choice of the appropriate surfactants—will help the petroleum industry to become one of the biggest users of surfactants and this will lead to some extent to a change in the profile of petrochemical production.

The special preparation of surfactants for the petroleum industry is necessary, as is already done for the textile industry, for the concentration of ores by flotation and in other branches of the national economy. It would be appropriate to organize work on the synthesis of surfactants in one or two Institutes. In certain aspects of the technology of the application of surfactants in the recovery of petroleum, we are lagging considerably behind the United States. In order to eliminate this gap we must develop scientific foundations for the extensive utilization of a variety of surfactants in the petroleum industry, and in this we must receive assistance from the organizations concerned with synthesis.

We must recognize the inexhaustible possibilities of surfactants for improving existing technological processes and creating new processes. The rapid development of petroleum chemistry and the extension of the range of surfactants provide a basis for extending the range of their application, covering, for example, a number of problems of the gas industry. Thus, surfactants can be used to increase the productivity of gas wells and the rate of drilling of gas wells, to increase the capacity of underground gas-storage systems, to solve problems of separation, etc. All this points to the need for more attention to be paid to the synthesis and production of surfactants for the petroleum industry.

THE PHYSICOCHEMICAL PRINCIPLES OF THE USE OF SURFACTANTS IN THE DRILLING OF WELLS IN ORDER TO INTENSIFY THE RECOVERY OF PETROLEUM AND INCREASE THE PETROLEUM OUTPUT FROM STRATA

G. A. Babalyan

I. Introduction

In the Soviet Union today, surfactants are finding ever-increasing application in the petroleum industry. At the present time, scientific research institutes working in the field of petroleum recovery are carrying out studies on the use of surfactants for increasing petroleum output from reservoirs, opening up strata, preventing collapse during the drilling of wells, improving the conditions for the establishment of oil wells and forcing wells, increasing their productivity and pickup, and preventing the formation of emulsions in oil wells, and also for the demulsification of petroleum, the improvement of methods for the hydraulic rupture of oil-bearing strata, the acid treatment of the zone next to the end face in wells, cementation of wells, prevention of paraffin wax deposition and corrosion of equipment in the petroleum industry, geophysical measurements, etc. In most cases the characteristics of the above processes are related to the behavior of the rock–petroleum–water system.

In the present paper we shall restrict ourselves to an examination of the influence of surfactants on this system during the displacement of petroleum by water or of water by petroleum from a porous medium.

II. Principal Factors Influencing the Displacement of Petroleum by Water or Water by Petroleum from a Porous Medium

In the case under consideration, the pore space may contain free drops of petroleum and water, i.e., drops which are not attached by molecular forces to the solid surface; free volumes of petroleum and water, i.e., volumes which again are not attached to the solid surface by molecular forces and which form a dispersion medium in a certain volume of the pore space; and petroleum and water in the form of films on the surface of the solid particles, together with drops of liquid phases attached to these particles and forming, in the general case, finite contact angles of selective wetting, i.e., menisci. In order to obtain a complete picture of the influence of surfactants on the displacement of one of the above liquids by the other from a porous medium, let us consider the elements of this process.

The Characteristics of the Displacement of a Free Drop of Petroleum by Water or of a Free Drop of Water by Petroleum

For the sake of simplicity we shall first of all exclude from the discussion the possibility that attachment of the drops of petroleum or water to the solid surfaces and coalescence of these drops take place. We shall assume that two drops of the same petroleum (or different petroleums) with identical dimensions, undergoing displacement by different water specimens with practically identical viscosities but different values for the surface tension at the petroleum–water boundary, have to pass through a porous medium via a series of identical constrictions which are arranged in series and have dimensions smaller than the dimensions of the drop.

We shall use the following symbols:

$\sigma_{p.w_1}$ and $\sigma_{p.w_2}$—the surface tensions of water at the boundary with different petroleums;

T_1 and T_2—the respective times required for displacement of the drops through the above constrictions;

k_{p_1} and k_{p_2}—the relative permeabilities of the porous medium for the petroleums during displacement of their drops by water;

k_{w_1} and k_{w_2}—the same for water;

Q_1 and Q_2—the water factors, i.e., the ratio of the volume of forced water to the volume of petroleum recovered;

η_1 and η_2—the output of petroleum from the porous medium due to the displacement of drops 1 and 2; and

μ_1 and μ_2—the viscosities of the petroleums.

Let us assume that $\sigma_{p.w_1} < \sigma_{p.w_2}$ and $\mu_1 = \mu_2$ and that the pressure difference acting on the drops increases gradually. Here the following cases are possible.

1. The pressure difference acting on the drops is less than that required to deform the drops so that they pass through the constriction, i.e., the drops are immobile. In this case $k_{p_1} = k_{p_2} = 0$, $k_{w_1} = k_{w_2}$, $T_1 = T_2 = \infty$, $Q_1 = Q_2 = \infty$, and $\eta_1 = \eta_2 = 0$.

An analogous phenomenon may also be observed when the viscosities of the petroleums are different. Thus, in the case under consideration, surface tension and the viscosities of the petroleum have no influence on the fundamental characteristics of the process involved in the displacement of drops of petroleum by water.

2. The pressure difference acting on the drops gradually increases and overcomes the resistance of the first drop (since $\sigma_{p.w_1} < \sigma_{p.w_2}$), which begins to move and passes through the constriction; at this stage the second drop is still immobile.

In this case $k_{p_1} > k_{p_2} = 0$, $k_{w_1} > k_{w_2}$ (since the moving drop offers less resistance to the stream of water than the immobile drop), $Q_1 < Q_2 = \infty$, $T_1 < T_2 = \infty$, and $\eta_1 > \eta_2 = 0$.

3. The pressure difference acting on the drops is so great that both drops begin to move and pass through the constriction.

In this case $k_{p_1} > k_{p_2} > 0$, $k_{w_1} > k_{w_2}$, $Q_1 < Q_2 < \infty$, $T_1 < T_2 < \infty$, and $\eta_1 = \eta_2$.

The smaller value of T_1 is due to the fact that the first drop is deformed more rapidly under the influence of the applied pressure difference. This also explains the higher value of k_{p_1}. We have $k_{w_1} > k_{w_2}$, since the rapidly moving first drop offers less resistance to the water stream than the slowly moving drop. We have $Q_1 < Q_2$, since the quantity of water which can slip past the first drop is smaller. This reveals the influence of surface tension on the characteristics of the displacement process in the second and third cases.

4. Let us now assume that at a sufficiently high pressure difference (when both drops are moving) $\sigma_{p.w_1} = \sigma_{p.w_2}$ and $\mu_1 < \mu_2$. We shall then have $T_1 < T_2$, $k_{w_1} > k_{w_2}$, $Q_1 < Q_2$, and $\eta_1 = \eta_2$.

We have $T_1 < T_2$, since the time required for deformation of the first drop is less than that required for deformation of the second drop. Since the first drop is displaced more rapidly, $k_{w_1} > k_{w_2}$, while $Q_1 < Q_2$ and $\eta_1 = \eta_2$, since the force required for deformation of the drop is in this case determined by the capillary pressure $2\sigma/r$ and not by the viscosity of the petroleum.

Another case is possible where the dimensions of the drops are much less than the dimensions of the constrictions under consideration. The drops can then pass freely through the constrictions and, irrespective of whether $\sigma_{p.w_1}$ and $\sigma_{p.w_2}$ or μ_1 and μ_2 are equal or unequal,

$$T_1 = T_2, \quad k_{p1} = k_{p2}, \quad k_{w1} = k_{w2}, \quad Q_1 = Q_2, \quad \eta_1 = \eta_2.$$

In this case neither the surface tension nor the viscosity of the petroleum have any influence on the characteristics of the displacement process. This displacement of the drops is apparently analogous to the movement, in the pore space of solid particles which are not in contact with the main skeleton of the soil, i.e., particles which are freely suspended in the stream. The dimensions of the drops which can move freely in the pore space can be determined for this case (by analogy with the process of undermining in soil by noncontacting solid particles) as a first approximation, from the familiar relationship [5]

$$\frac{D}{d} = 8 - 12,$$

where D is the average diameter of the granules making up the soil skeleton, and d is the diameter of the drop.

In this case the rate of displacement is determined only by the dimensions of the drop. The smaller its diameter, the greater the rate at which it moves in the pore space and hence the smaller the rate at which water slips past the drop and the smaller the quantity of water required to displace a given quantity of dispersed petroleum.

It follows from the foregoing that for drops whose dimensions are greater than the dimensions of the pores, other conditions being equal, decrease in surface tension (e.g., as a result of the influence of added surfactants) will produce the following results:

a) a decrease in the work required to move the free drops of petroleum from pore to pore,

b) a decrease in the critical force required to drive these drops through the pore constrictions,

c) an increase in the rate of displacement of the drops and a decrease in the quantity of water required for their displacement,

d) an increase in the possibility of displacing a large quantity of dispersed free petroleum from the pores, and

e) an increase in the relative permeability of the porous medium for free, dispersed petroleum and water.

All that has been said above regarding the displacement of free drops of petroleum by water is applicable to an equal extent to the displacement of free drops of water by petroleum or of drops of petroleum and water by gas and the reverse process of displacement of gas bubbles.

The conclusions reached show that in order to achieve the maximum possible displacement of the free drops of water by petroleum or of the free drops of petroleum by water and to increase the rate of the displacement process, it is necessary to reduce to a minimum the surface tension at the boundary between the liquids. A decrease in $\sigma_{p.w}$ can be achieved by the use of water-soluble or petroleum-soluble surfactants. In this case, however, allowance must be made for the fact that the adsorption layers of surfactant on the outer surface of the drops may be structured layers, i.e., they may exhibit mechanical strength and increased (structural) viscosity in the unbroken structure. This feature may to some extent hinder deformation and hence the displacement of large drops, in spite of the decreased surface tension. Low surface tension, however, will lead to the formation of drops with small dimensions, and the strength (high viscosity) of the structure of the adsorption layers ensures complete stabilization of the small droplets of petroleum in water, i.e., stability towards coalescence and prevention of phase reversal, and this facilitates the displacement of petroleum.

As pointed out above, the materials undergoing displacement in the pore space include not only the free drops of petroleum but also the drops adhering to the solid particles and the petroleum covering these particles in the form of a film. In order to determine the cases in which it is advisable to use water-soluble or petroleum-soluble surfactants, let us also examine the displacement of adherent drops and films of petroleum from a porous medium.

The Displacement of Adherent Drops and Films of Petroleum by Water

Petroleum in the form of a film exhibits a highly increased resistance to flow. This is due primarily to the low thickness of the film. Moreover, petroleum consists of a structured (particularly at sufficiently low temperatures) colloidal dispersed system containing, in true solution, macromolecular compounds (products of the oxidative polymerization of hydrocarbons) which may also form three-dimensional networks. All this leads to a greatly increased viscosity, particularly at low velocity gradients, in the range of unbroken structures

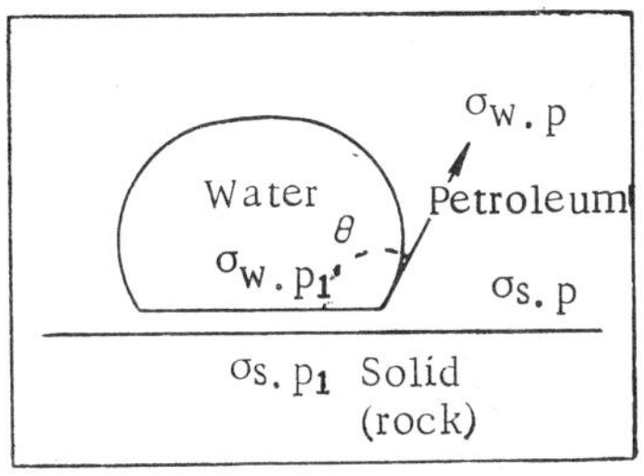

Fig. 1. Drop of water on the film.

where the bending elasticity (strength) is also manifested. The formation of a film of petroleum is undoubtedly related to the adsorption of surfactants, dissolved in the petroleum, on the solid surfaces. The thickness of the actual adsorbed layer, however, is many times smaller than the thickness of the film of petroleum [6, 8]. The surfactants dissolved in petroleums may be present either in true solution or in colloidal solution [9]. The displacement of the film of petroleum from the solid surface, if rupture of the film by water does not take place, is difficult. In this case displacement is brought about only as a result of a slight decrease in the thickness of the film under the influence of tangential forces as the stream of water moves over the surface of the film and as a result of the removal of petroleum particles from this surface. Experiment shows that in most cases the film of petroleum is broken by water, i.e., it is displaced from the actual solid surface (in the case where this is hydrophilic) by a selective wetting mechanism [10]. For example, we have observed the rupture by water of films of a number of petroleums from the Apsheron peninsula, on glass and quartz plates. Most of these petroleums contained 1.5-2.5% naphthenic acids but almost no asphaltenes. Glass and quartz plates, wet with these petroleums, were immersed in water. The petroleum films were broken by the water, and drops of different size formed on the surface of the plates; some of these drops were torn from the surface, while others remained attached to the surface. It was established that the greater the concentration of active components in the petroleum, the slower the rupture of the film of petroleum and the smaller the quantity of petroleum torn from the surface. In analogous experiments with petroleum from a Carboniferous formation containing a large quantity of asphaltenic tar materials, the film on the glass plate was not broken either by fresh mains water (hard water) or by highly mineralized hard stratal water, even when the plates were left in these waters for a period of many days.* Under the conditions of the pore space, rupture of a petroleum film is even more difficult. This is indicated by experiments on capillary impregnation. In the case of sand completely saturated with highly active petroleums from the Apsheron productive stratum, whose films were broken on a glass plate, no capillary impregnation by water took place, i.e., no rupture of the films on the sand granules took place [4]. An analogous phenomenon was observed with Tuimazinskii petroleum from the Devonian stratum and with Arlanskii petroleum from the Carboniferous formation [3]. The principal components of petroleums preventing the capillary impregnation of water are apparently macromolecular naphthenic acids in the case of Bakinskii petroleums, and asphaltenes in the case of Devonian and Carboniferous petroleums.

On the basis of these experimental studies, together with published data [11-13], it may be concluded that rupture of the film by water is possible only when its thickness exceeds a certain value, which depends on the properties and concentrations of the active components in the petroleum. Moreover, the higher the thickness and viscosity of the film, i.e., the greater (other conditions being equal) the concentration of active components in the petroleum, the greater will be the time required for its rupture, the greater the dimensions of the drops of petroleum formed on the solid surface, and the lower the probability of rupture of the film in general. The possibility of removal of the drops from the solid surface is determined by the adhesion perimeter of the drop, the magnitude of the wetting angle, the dimensions of the drop formed, the radius of curvature of its upper surface, and the surface tension [7]. The greater these quantities, the greater the strength of adhesion of the drop.

The rupture of a petroleum film and the strength of adhesion of the drops of petroleum formed on a solid surface after rupture depend to a considerable extent on the quality of the water and the presence of added surfactants. When a drop of water lies on a residual adsorption film of petroleum (Fig. 1) equilibrium may be produced in the petroleum–water–solid surface system when†

* The experiments were carried out by the author with I. I. Kravchenko.

† In accordance with the theories developed by P. A. Rebinder [10], it is assumed here and subsequently that there is no hindrance–hysteresis of wetting of the friction type along the perimeter of the three-phase boundary, i.e., it is assumed that the contact angle θ corresponds to true thermodynamic equilibrium.

$$\sigma_{s.p} - \sigma'_{s.w} = \sigma_{s.p} \cos\theta, \quad (1)$$

where $\sigma'_{s.w}$ is the surface tension at the solid (covered with a thin film of petroleum)—water boundary.

$$\sigma'_{s.w} = \sigma_{s.p_1} + \sigma_{w.p_1},$$

where $\sigma_{s.p_1}$ and $\sigma_{w.p_1}$ are the surface tensions at the solid phase—residual petroleum adsorption film and water—residual petroleum adsorption film boundaries respectively, and $\sigma_{s.p}$ and $\sigma_{w.p}$ are the surface tensions at the solid phase—petroleum and water—petroleum boundaries respectively.

It is obvious that $\sigma_{s.p_1} \geq \sigma_{s.p}$ and $\sigma_{w.p_1} \geq \sigma_{w.p}$; with increase in the thickness of the residual petroleum film, $\sigma_{s.p_1}$ and $\sigma_{w.p_1}$ approach the limiting (minimum) values $\sigma_{s.p}$ and $\sigma_{w.p}$ and then it follows from (1) that when the petroleum film is sufficiently thick, $\sigma_{s.p_1} = \sigma_{s.p}$, $\sigma_{w.p_1} = \sigma_{w.p}$, $\cos\theta = -1$, and $\theta = 180°$, i.e., no contact angle is formed and the drop of water does not adhere to the solid surface.

The condition for the formation of a contact angle $\theta < 180°$ is:

$$\sigma_{s.p} + \sigma_{w.p} \geq \sigma_{s.p_1} + \sigma_{w.p_1}.$$

When a surfactant is added to the water it is adsorbed at the free petroleum—water and water—petroleum film surfaces of separation. As a result there is a marked decrease in the values of $\sigma_{w.p}$ and $\sigma_{w.p_1}$. The sign of the selective wetting changes and becomes positive, i.e., the contact angle becomes less than 90° and the drop of water spreads over the surface. With decrease in θ the thickness of the residual adsorption film of petroleum under the water drop decreases. After rupture of the petroleum film, the water drop with finite contact angle adheres to the rock surface.

The smaller the contact angle with a residual adsorption film of petroleum under the water drop, the smaller will be the thickness of this residual film, the greater the probability of formation of rupture centers and the greater the number of such centers and hence the greater the number of petroleum drops formed on the solid surface. The decrease in the thickness of the residual adsorption film of petroleum and its rupture when surfactants are added to the water are confirmed by experiments which we have carried out. When nonionogenic surfactants (OP-10 and $KAUFE_{14}$) are added to the water, the film of Carboniferous petroleum on the glass plate is broken.* Rupture is preceded by an irregular decrease in the thickness of the film on the glass surface. The irregular decrease in thickness leads to the production of film rupture centers on the glass, and the formation of petroleum drops of different dimensions on the glass surface (Fig. 2a and b).

Fig. 2. a) Film of Carboniferous petroleum on glass in mains water after seven days. b) State of Carboniferous petroleum film 2 h after the addition of 0.05% OP-10 to the water.

* The experiments were carried out by the author with I. I. Kravchenko.

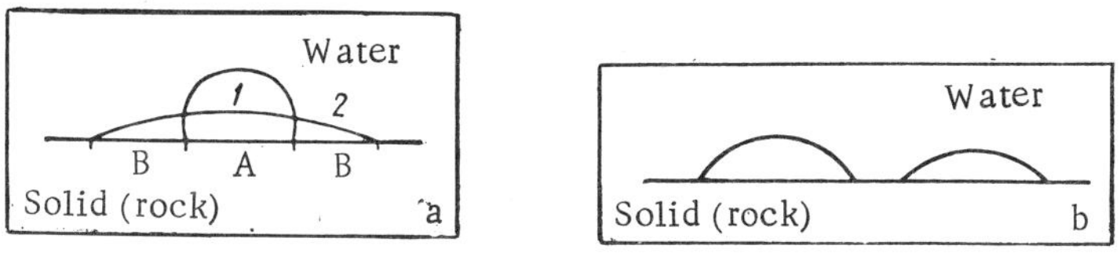

Fig. 3. a—1) Initial state of the petroleum drop; 2) state of the drop in the initial moment after hydrophobization of the solid surface. A) The more hydrophilic section of the solid surface; B) hydrophobized section of the solid surface. b—Dispersed droplets of petroleum in water and the formation of new fine droplets (the number of drops may also be greater).

Figure 2a shows a film of Carboniferous petroleum on glass in mains water after seven days, and Fig. 2b shows the same film 2 h after the addition of 0.05% OP-10 to the water. When this nonionogenic surfactant is added to the water, the petroleum film is broken by the water not only on the planar surface of glass but also on the surface of a porous sand medium. When this surfactant is added, capillary impregnation of the petroluem-saturated sand by water begins. As pointed out above, this does not take place when the water contains no surfactant [3].

Rupture of the petroleum film may be accompanied not only by physical adsorption of the surfactant on the solid surface but also by chemical adsorption (chemosorption) and the formation of a new, more hydrophobic surface. In this case the contact wetting angle may prove to be greater, in spite of the decrease in $\sigma_{W.p}$.

We have observed this in the case of the rupture of films of a number of Bakinskii petroleums on a glass surface by water containing various surfactants. If the rupture of the petroleum film by water is accompanied by the chemosorption of surfactant on the solid surface and hence the formation of a new, more hydrophobic surface, the resulting petroleum droplets in contact with the new surface over this surface; the petroleum, as it were, flows from the hydrophilic surface to the hydrophobic surface. When the petroleum drops spread over the surface, a certain quantity of its active components are transferred from the volume to the surface layer on to the newly formed surface of contact formed by the petroleum with the solid phase and water. As a result, the concentration of active components in the volume of the petroleum droplet decreases. The petroleum flows from the center of the drop to the periphery in the form of branched streams over the most hydrophobic surface and is broken down in the process to form even finer droplets. When the petroleum flows over a solid surface previously wet by petroleum droplets, a new thin petroleum film is formed which has a smaller thickness and undergoes rupture more readily. In this way the film is broken again at the most hydrophilic point of contact between the solid surface and the petroleum (Fig. 3a and b), i.e., at the surface initially wet by the drop.

The process involving the flow and subdivision of drops on a solid surface with hydrophobization of the surface as a result of chemosorption can also be attributed to the tendency of the surface energy of the solid phase—petroleum drop—water system to decrease. On the surface initially wet by petroleum, the free energy of the petroleum—solid surface interface is greater than the free energy of the petroleum—new solid surface interface, so that the flow of petroleum from the drop can take place only in one direction—from the center to the periphery.

In the case under consideration, in spite of the high values of the contact wetting angles, drops with small dimensions apparently have a lower strength of adhesion to the solid surface, due to the low values of $\sigma_{W.p}$, and are readily removed from the surface by a stream of water. This, together with the small volume of the adherent drops and the possibility of foam rupture, is responsible for the high petroleum output from rocks when the water is treated with surfactants which produce a marked decrease in $\sigma_{W.p}$ but at the same time increase the contact wetting angle [8].

Thus, it is possible to select water-soluble surfactants which decrease the dimensions of the free drops and adherent drops, the strength of attachment of the latter to the solid surface, and the capillary pressure in the drops, and also facilitate rupture of the petroleum film by water and capillary impregnation of an oil-saturated porous medium by water. All this improves the characteristics of the displacement of petroleum by water; there is a decrease in the quantity of water and time required for displacement, and an increase in petroleum output.

Water in the form of a film also has an increased viscous resistance to flow. The possibility and rate of rupture of a residual adsorption film of water by petroleum depend on the properties of the petroleum, the water, and the rock. The lower the degree of hydration of the solid surface, the greater will be the probability of rupture of the film by petroleum, the smaller the dimensions of the water drops formed on the solid surface, and the lower the strength of adhesion of the drops to the surface. Thus, water-soluble surfactants, if they are to facilitate the displacement of water in the form of a film by petroleum, should decrease the hydration of the solid surface and the surface tension of water at its boundary with petroleum, since this leads to an increase in the rate of rupture of the water film, and the resulting drops have small dimensions.

Under these conditions the quantity of water displaced from a porous medium by petroleum increases. This reveals another characteristic feature of water-soluble surfactants. Their presence in water makes it possible to displace a greater quantity of water in the form of a film from the pore space, i.e., to decrease residual water saturation in the displacement of water by petroleum.

Equilibrium in an oil drop – water – solid phase system, when a residual adsorption film of water is present under the petroleum drop (Fig. 4), is reached when

$$\sigma'_{s.p} - \sigma_{s.w} = \sigma_{w.p}\cos\theta, \qquad (2)$$

where $\sigma'_{s.p}$ is the surface tension at the solid (covered with a thin film of water)–petroleum boundary.

$$\sigma'_{s.p} = \sigma_{s.w_1} + \sigma_{p.w_1},$$

where $\sigma_{s.w_1}$ and $\sigma_{p.w_1}$ are the surface tensions at the solid phase–water film and petroleum–water film boundaries respectively.

It is obvious that $\sigma_{s.w_1} \geq \sigma_{s.w}$ and $\sigma_{p.w_1} \geq \sigma_{w.p}$; with increase in the thickness of the residual film of water, $\sigma_{s.w_1}$ and $\sigma_{p.w_1}$ approach the limiting (maximum) values $\sigma_{s.w}$ and $\sigma_{w.p}$ respectively; it then follows from (2) that when the film of water is sufficiently thick, $\sigma_{s.w_1} = \sigma_{s.w}$, $\sigma_{p.w_1} = \sigma_{w.p}$, $\cos\theta = +1$, and $\theta = 0°$, i.e., no contact angle is formed and the petroleum drop does not adhere to the solid surface.

The condition for the formation of a contact wetting angle $\theta > 0°$ is:

$$\sigma_{s.w_1} + \sigma_{p.w_1} \leq \sigma_{w.p} + \sigma_{s.w}.$$

When a surfactant is added to the hydrocarbon liquid, the values of $\sigma_{p.w_1}$ and $\sigma_{p.w}$ decrease. The angle θ increases when $\sigma_{p.w_1}$ decreases, but decreases when $\sigma_{w.p}$ decreases. The decrease in the thickness of the film of water under the drop will therefore depend on the intensity of the changes in the above terms in Eq. (2).

If the addition of surfactant leads to a marked decrease in $\sigma_{p.w_1}$ and a slight decrease in $\sigma_{p.w}$, the value of θ will increase and the thickness of the film of water under the petroleum drop will decrease. If, however, there is a marked decrease in $\sigma_{w.p}$ and only a slight decrease in $\sigma_{p.w_1}$, the value of θ will decrease and the thickness of the film of water will increase. The quantity of water displaced by the petroleum will be greater in the first case than in the second case.* Decrease in the thickness of the film of water will be accompanied by an increase in the probability of its rupture by petroleum, i.e., in the probability that the film of water on the solid surface will be converted to free drops and drops adhering to this surface. Rupture of the film of water is accompanied by adsorption of the active components of the petroleum on

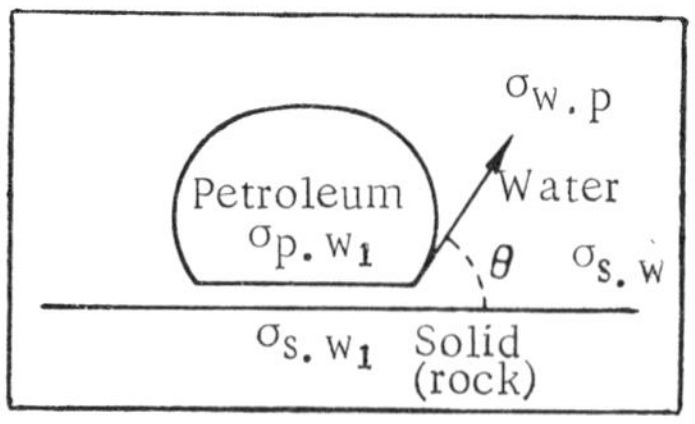

Fig. 4

*Here we do not consider the possibility that part of the surfactant dissolves in water and is adsorbed on the interface with $\sigma_{s.w_1}$ and $\sigma_{s.w}$.

the solid surface with a corresponding decrease in the surface tension at its boundary with the petroleum $\sigma_{s.p}$. In this case, equilibrium conditions are established when

$$\sigma_{s.p.} - \sigma_{s.w} = \sigma_{w.p} \cos\theta. \tag{3}$$

It follows from this equation that the increase in θ or, in other words, the displacement of water from the solid surface, is greater when the ratio $(\sigma_{s.p}-\sigma_{s.w})/\sigma_{w.p}$ is decreased.

Thus, the best conditions for the displacement of water in the form of a film may be achieved by using petroleum-soluble surfactants which are readily adsorbed on a solid surface and weakly adsorbed at the petroleum–water boundary.

These individual elements of the process involved in the displacement of petroleum by water and of water by petroleum from a porous medium make it possible to establish how the petroleum output or water output from a porous medium will change when surfactants are added to the water or to the petroleum, but it is impossible to reach any definite conclusions regarding the possible changes in the rate of dispacement and the quantity of water required to displace petroleum or the quantity of petroleum required to displace water, since these characteristics of the displacement process are alsoinfluenced to a considerable extent by the rate of coalescence of the drops of petroleum in water and of the drops of water in petroleum,and by the rate of adhesion of these drops to the surface of the solid particles forming the porous medium.

On the basis of earlier studies of this problem [1, 2], we have reached the following conclusions. Other conditions being equal, the greater the rate of coalescence and adhesion of drops of petroleum to a solid surface in water, the greater the time required for the displacement of the petroleum by water. With increase in the rate of adhesion of the drops of petroleum to a solid surface and their coalescence, there is an increase in the rate of displacement of water by petroleum. The time required for displacement of the water by petroleum decreases.

The rate and probability of adhesion of drops of petroleum increase with increase in the concentration of substances in the petroleum with a strong hydrophobizing action on the solid surface, i.e., with increase in the contact angle of wetting and also with increase in the dimensions of the drops; the rate of coalescence, however, increases with decrease in the strength of the adsorption layer of surfactant on the surface of the petroleum drops. In order to decrease the rate of coalescence and adhesion of drops of petroleum to a solid surface, it is necessary that the water-soluble surfactants produce the maximum possible decrease in the value of θ and dispersion of the petroleum, and also increase the strength of the adsorption layer of surfactant on the surface of the drops. For a given decrease in $\sigma_{w.p}$ accompanying the addition of water-soluble surfactants, the time required for displacement of the petroleum by water is greater, the greater the value of θ.

Summary

1. Both petroleum-soluble and water-soluble surfactants can increase the quantity of water displaced by petroleum and the rate of displacement. These surfactants may therefore be used successfully to decrease the quantity of residual water in a porous medium; this is required in order to improve various technological processes of petroleum recovery, e.g., in order to increase the flow of petroleum into a well and ensure rapid establishment of the well.

The greatest decrease in the quantity of residual water will be given by those water-soluble surfactants which not only produce a marked decrease in $\sigma_{w.p}$ but also are chemically adsorbed on the solid surface to form a new, more hydrophobic surface.

2. The addition of water-soluble surfactants to water makes it possible to displace a greater quantity of petroleum from a porous medium by the water, to increase the rate of the displacement process, and to decrease the quantity of water required for displacement. These surfactants can therefore be used successfully to increase the output of petroleum from reservoirs, to increase the rate of development of reservoirs, and to decrease the cost of the petroleum. The best results will be given by those surfactants which produce a marked decrease in $\sigma_{w.p}$ and θ.

Water-soluble surfactants exhibiting these properties can also be used in all technological processes of petroleum recovery where rapid displacement of petroleum by water from a porous medium is necessary. A characteristic feature of these surfactants is that they give an increase in the relative permeability of a porous medium for petroleum and water and in the total permeability of the porous medium for these liquids. These surfactants can therefore be used extremely successfully to increase the rate of establishment of wells for the intensive extraction of acid-treated petroleum from the zone next to the end face in oil wells during acid treatment.

3. Water-soluble surfactants facilitate the capillary impregnation of petroleum-saturated rock by water and can therefore be used to achieve more regular displacement of petroleum from nonuniform reservoirs, i.e., from reservoirs made up of strata with different permeabilities, and this is of considerable practical importance. The use of these surfactants is particularly important in those cases where the stratal petroleum has produced marked hydrophobic properties on the surface of the solid particles so that no capillary impregnation by water takes place when the water contains no surfactant. In most cases this occurs when the stratum contains small quantities of connate water, highly active petroleums, and petroleums containing a large quantity of asphalt materials.

4. Petroleum-soluble surfactants, which produce a marked decrease in the relative permeability of a porous medium for water and increase its water output, can be used in the technological processes of drilling and petroleum recovery where a decrease in the residual water saturation of the rock is required, e.g., in order to decrease the quantity of residual water in the zone next to the end face in wells and to decrease the filtration of water from emulsion solutions into strata during drilling.

5. Petroleum-soluble surfactants increase the possibility of capillary impregnation of water-saturated rock by petroleum. This process is accompanied by a decrease in surface energy which, according to P. A. Rebinder [12], leads to a decrease in the hardness of the drilled rocks. Petroleum-soluble surfactants can therefore also be used as additives to emulsion liquids and petroleum-based liquids in order to increase the rates of drilling.

Literature Cited

1. Sh. N. Aliev and G. A. Babalyan, The Mobility of Oil Pool Outlines (Aznefteizdat, 1956).
2. G. A. Babalyan, Problems of the Mechanism of Petroleum Recovery (Aznefteizdat, 1956).
3. G. A. Babalyan, L. A. Nikolaeva, and I. L. Markhasin, "The capillary absorption of water into oil-bearing Carboniferous and Devonian rocks," Collection, The Use of Surface-Active Substances in the Petroleum Industry (Gostoptekhizdat, Moscow, 1961).
4. G. A. Babalyan and Z. F. Rzabekov, Tr. AzNII, Vol. 7 (1957).
5. N. M. Bochkov, The Mechanical Undermining of Soil (Izd. NKTP SSSR, Moscow-Leningrad, 1936).
6. B. V. Deryagin and M. M. Kusakov, Izv. AN SSSR, ser. khimicheskaya, 5 (1937).
7. B. N. Kabanov and A. N. Frumkin, ZhFKh, 4, 5 (1933).
8. M. M. Kusakov, K. S. Zinchenko, and P. A. Rebinder, Dokl. AN SSSR, 28, 5 (1940).
9. S. S. Nametkin, The Chemistry of Petroleum (Izd. AN SSSR, Moscow, 1955).
10. P. A. Rebinder, et al., The Physical Chemistry of Flotation Processes [in Russian] (Metallurgizdat, Moscow-Leningrad, 1933).
11. P. A. Rebinder and K. A. Pospelova, Introductory article to V. Clayton's book: Emulsions (Izd. inostr. lit., Moscow, 1950).
12. P. A. Rebinder, L. A. Shreiner, and K. F. Zhigach, Reagents for Decreasing Hardness During Drilling (Izd. AN SSSR, Moscow-Leningrad, 1944).
13. A. N. Frumkin, ZhFKh, 12, 4, 5, 6 (1938).

THE USE OF SURFACTANTS IN INCREASING THE OUTPUT OF PETROLEUM FROM RESERVOIRS

G. A. Babalyan, I. L. Markhasin, I. I. Kravchenko, and G. V. Rudakov

In the past ten years the attention of research workers has been concentrated on the way in which the displacement of petroleum by water from a porous medium depends on a number of physicochemical factors. This problem is related to purely applied questions and is of considerable practical and theoretical interest. Solution of this problem will make it possible to find surfactants which will ensure the most rapid and complete extraction of petroleum from strata with different characteristics and to improve the technological processes of drilling and petroleum recovery.

One of the main factors responsible for low petroleum output from reservoirs is the presence of polar components in the petroleums. The elimination of their undesirable influence is one of the most important problems in the modern theory and practice of the displacement of petroleum by water.

In view of the discovery of extensive Carboniferous petroleum provinces in the USSR and particularly in the Bashkir SSR, together with the specific characteristics of their reservoirs and of the petroleum itself, it is necessary to solve a wide range of new problems on the intensification of petroleum recovery. The petroleums from most of these deposits are characterized by a high viscosity and a high concentration of asphaltic tar materials. For example, the petroleum from the Arlanskii deposit, which is the most characteristic petroleum of the Bashkir Carboniferous formation, has the following characteristics (Table 1).

The viscosity of the petroleum under stratal conditions varies from 16 to 40 cP [1]. The clay content of the Carboniferous reservoirs in the northwestern regions of the Bashkir SSR varies from 1 to 15%.

TABLE 1

Properties	Arlanskii Carboniferous formation
Density	0.9014
Sulfur content, %	3.13
Tars (silica gel), %	16.2
Asphaltenes, %	-3.8
Oils, %	82.3
Paraffin wax, %	2.9
Viscosity, cP	58.3
Surface tension of petroleum at petroleum–water boundary, ergs/cm^2.	30.1
Contact angle of wetting (petroleum drop–water–quartz)	41.5

TABLE 2

Porous medium	Perme-ability with respect to air, darcys	Porosity, %	Residual water, % of pore space volume	True rate of filtration, m/month	Quantity of water filtered, in pore space volumes	Petroleum output, % of initial petroleum saturation		Duration of water-free output, h	Duration of total output,	% water in out-flowing stream at the moment when the experiment was stopped
						Water-free	Total			
Disaggregated sandstone from the Arlanskii deposit	4.0	36.9	18.5	10.0	1.4	45.5	60.1	32.6	163.3	95-98
Quartz sand	0.7	28.1	18.0	9.0	0.7	38.7	50.2	35	74	95-98
The same	2.3	37.5	23.4	10	2.6	56.5	68.9	31	233	95-98
" "	3.3	31.6	0	15	3.2	12.4	41.9	7	218	95-98

Note. The experiments were carried out with L. N. Malysheva.

TABLE 3. The Displacement of Kerosene Mixed with Different Petroleums by Hard Water Subjected to Vacuum Treatment

Experiment No.	Porosity, %	Permeability, darcy	Concentration of residual water, %	True rate of filtration, m/month	Liquid displaced	Thickness of layer, % of thickness of porous medium		Output from initial satuation, %			Duration of water-free period, days	Duration of total period, days	Total duration of water filtration, days	Total quantity of water used, cm^3	Water content of outflowing stream when experiment was stopped, %
						Oil-saturated layer	Water-saturated layer	Water-free period	Water-containing period	Total period					
5	33.4	0.5	–	20	Kerosene	75	25	53.0	3.4	56.4	1.6	1.7	4	1755	–
7	30.2	0.5	–	12.6	Kerosene + 5% petroleum	75	25	0	6.8	6.8	0	4	16	989	99.5
9	33.3	0.5	20	11.5	The same	75	25	62.5	7.0	69.5	2.65	5.7	10	1851	99.2
11	35.7	0.5	20	8.3	" "	75	25	33.4	15.6	49.0	0.96	13.5	26	4000	88.5

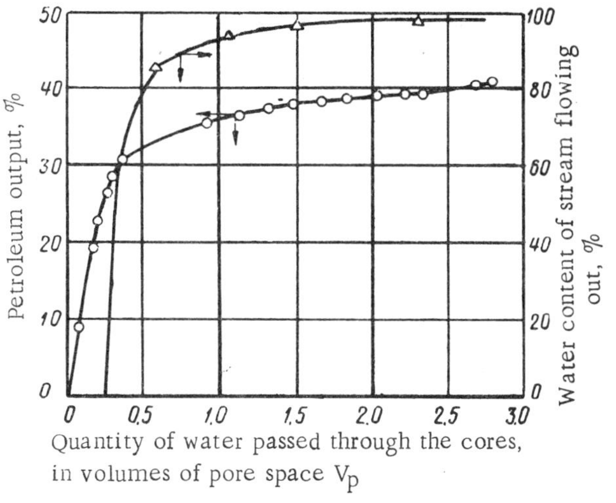

Fig. 1

We have as yet no extensive experience in the working of Carboniferous deposits. Determination of the chief characteristics of the process involved in the displacement of a deep specimen of Arlanksii petroleum by stratal water from noncemented quartz sand (permeability 3.3 darcy, porosity 31.6%) under conditions close to stratal conditions for a true rate of filtration equal to 15 m/month shows that petroleum output in the absence of connate water remains very low, while the water content of the stream of liquid flowing out reached 98%(Fig. 1).

Analogous experiments with quartz sands of different permeability and on disaggregated sandstone from the Arlanskii deposit were carried out at a pressure higher than the saturation pressure with the same petroleum, taken from the trap, with and without residual water in the sands. The results of the experiments are given in Table 2. The water-free and total petroleum recovery increases sharply with increase in the quantity of residual water.

Practical experience shows that the volume of water extracted in the final stage of the treatment of strata in many cases does not exceed three volumes of the pore space in its oil-bearing section. On this basis it may be concluded that if appropriate measures are not taken, the petroleum output from strata of the Carboniferous formation will be very low, particularly in the section where the quantity of residual water is small. This situation is also encouraged to a considerable extent by the high viscosity of Arlanskii petroleum and the nonuniformity of the strata. With increase in these factors the coefficient of coverage of the stratum by the forced water decreases. A low coefficient of coverage is also produced by a low rate of capillary impregnation. Carboniferous and Devonian petroleums produce such marked hydrophobic properties on the surface of the solid particles of the reservoirs that capillary impregnation of the rock by water does not take place in the absence of residual water. It begins with the appearance of this water in the rock.

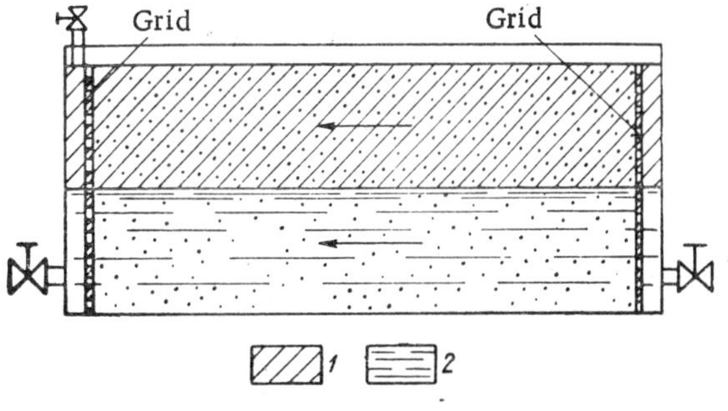

Fig. 2. 1) Kerosene; 2) water.

TABLE 4

Sand	Concentration of fractions, %				Carbonates, %	Carbonates in fraction IV, %	Specific surface, cm^2/g	Porosity, %
	I	II	III	IV				
	0.25 mm	0.25-0.1 mm	0.1-0.01-mm	0.01 mm				
Quartz	2.7	77.7	17.3	2.3	0	0	579	32.6
Natural	5.1	59.7	21.9	13.3	6.0	53.4	715	34.8

When there is a large number of partings in a section with different permeabilities, decrease in the rate of capillary impregnation facilitates the breakthrough of water through the individual partings with the greatest permeability, leading to premature flooding of the wells with water, the consumption of large quantities of water in the displacement of petroleum, and a decrease in petroleum output. When capillary impregnation by water takes place, the coefficient of coverage of the deposit by water in this case is found to be very much greater than that in the absence of capillary impregnation, so that the petroleum output is also greater.

This is confirmed by experiments on models of a porous medium consisting of sand (Fig. 2) in which the conditions of a water-containing deposit were created. The upper section of the model contained kerosene and the lower section contained water. The experiments were carried out with and without residual water (Table 3).

The experiments showed that in the displacement, in this case, of nonpolar kerosene the process takes place without the appearance of water at the outlet from the model. Almost all the output (53%) is obtained in the water-free period. When 5% of highly active petroleum from the Podkirmakinskii formation of the Apsheron peninsula deposits is added to the kerosene, water appears immediately at the outlet from the model, and the petroleum output decreases sharply to very low values (~7%).

A quite different picture is observed under the same conditions when the oil-saturated section of the model contains residual water. In this case almost all the output is obtained in the water-free period (62.5%). The conditions for displacement are much better in the presence of residual water than in its absence, even when the kerosene contains 25% petroleum. This is due to the production of intensive capillary impregnation by water. This is indicated by special experiments on capillary impregnation for the case under consideration [2].

A low coefficient of petroleum output from the vast petroleum reserves in the Carboniferous deposits of the northwestern Bashkir SSR cannot be tolerated, since each percentage increase in petroleum output amounts to several ten million tons of additional petroleum recovered. The problems of increasing petroleum output from Carboniferous reservoirs, intensifying petroleum recovery from these reservoirs, and reducing the cost of development are thus of primary importance. The same is true to some extent of Devonian deposits.

The Use of Surface-Active Substances

The surfactants used to improve the petroleum-flushing properties of the water pumped into a stratum for forced flooding must satisfy the following requirements: they must increase the water-free and total output of the stratum, decrease the quantity of water and time required for displacement, dissolve readily in the forced water and in stratal waters, and be readily desorbed, i.e., readily removed by means of water, from the surface of the solid particles and the residual petroleum; they should also be inexpensive. The quantities required should be small. It is desirable that the surfactants react with the active components of the petroleum and form chemical compounds facilitating the intensive displacement of petroleum.

Studies of petroleum output give grounds for assuming that the action of the surfactant is to convert the maximum quantity of petroleum in the pore space during displacement to the free state, i.e., to a state in which the petroleum is not attached to the solid surface by molecular forces [3]. This state is reached when the petroleum film on the solid surface is broken rapidly and when the contact angles of the petroleum drops formed on the surface rapidly acquire equilibrium values, while the angles themselves, the dimensions of the drops, and the area of their adhesion to the solid surface are all small. Under these conditions, removal of the pe-

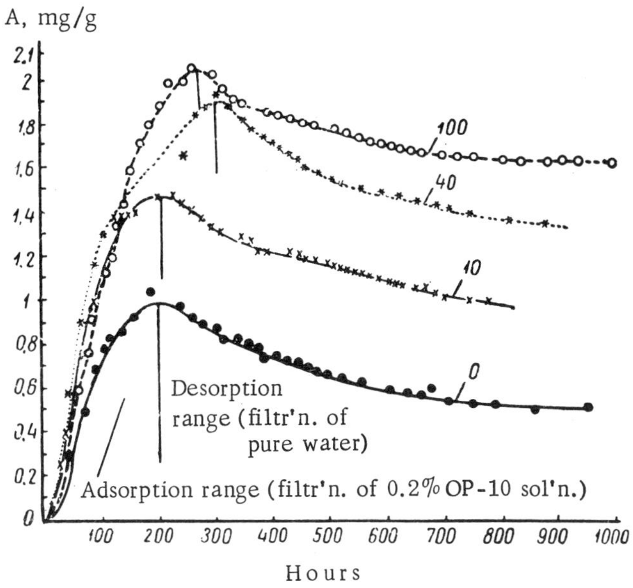

Fig. 3. The figures on the lines give the percentage of water in the pore space.

troleum drops from the solid surface is facilitated and the possibility of their reattachment to the surface is decreased. These conditions are satisfied when the added surfactants decrease sharply the surface tension at the water—petroleum boundary, the contact angle of wetting, and the time required for coalescence of the water drops in the petroleum and their adhesion to the solid surface, and increase the time required for coalescence and adhesion to the solid surface for the drops of petroleum in water. In most cases the surfactants used only partly satisfy the above requirements, so that their influence on the displacement process is different under different conditions.

We have studied the influence of the addition of both nonionogenic and ionogenic (anion-active and cation-active surfactants. Most of the anion-active substances tested (NChK, Sulfonol NP-1, Azolyat A, Azolyat B, "Novost'" powder, "Progress" liquid, etc.) cannot be used to increase petroleum output since they react with the salts dissolved in hard stratal waters to form insoluble precipitates. It is known [4] that ionogenic surfactants are adsorbed on solid surfaces to a greater extent than nonionogenic surfactants. The quantity of surfactant adsorbed on purified quartz sand is less than the quantity adsorbed on natural sand with the same permeability.* The characteristics of the sands used in our experiments are given in Table 4.

The high adsorption by natural sand is apparently due to the presence of carbonates and clays. In the case under consideration, the magnitude of the adsorption is characterized by the volume of filtered aqueous surfactant solution of a given concentration. This volume for quartz sand is smaller than that for natural sand by a factor of approximately five. The magnitude of this volume is inversely proportional to the concentration of surfactant in the solution.

During the forced flooding of a petroleum-bearing stratum, the aqueous surfactant solution may pass through sections of the stratum containing no petroleum, flooded sections containing only residual petroleum, and partially flooded and oil-saturated sections containing residual water in large quantities or in such small quantities that it may be neglected. For example, in forced flooding outside the contour, the water passes through all the above types of section; in forced flooding on the edge of the contour the water passes through the last three types of section; and in forced flooding inside the contour the water passes through the last two types of section. On this basis, experiments were carried out in order to establish the conditions under which the adsorption is at a minimum. As a result of experiments carried out with petroleums from productive sections of the Apsheron peninsula it was established† that the quantity of surfactant used is at minimum when the sand

*The studies were carried out by G. A. Babalyan and V. M. Baryshev.

†The experiments were carried out by G. A. Babalyan and V. M. Baryshev.

TABLE 5. Characteristics of the Displacement of Petroleum by Water Containing Different Surfactants

Displacing agent	Porosity, %	Air permeability, darcys	Duration of water-free output, h, min	Duration of total output, h, min	Water-free petroleum output, %	Total petroleum output, %	Increase in output, %	Quantity of filtered water, cm^3	Quantity of filtered water, in pore space volumes
Mains water	35	3.45	3-20	22-00	20.5	54.0	–	463	4.87
Mains water + 0.05% UFE_8 .	35	3.45	8-30	28-00	48.0	53.0	–	238	2.5
Mains water + 0.05% $KAUFE_{14}$	35	3.45	7-35	33-00	45.0	60.0	1.10	265	2.8
Mains water + 0.05% OP-10	35	3.45	4-52	18-20	30.0	63.0	1.16	402	4.23

Note. No residual water was present.

contains no residual water. The quantity of surfactant used increases by approximately 25% when residual water is present, by 50% when residual petroleum is present, and by 100% in the case of total saturation of the sand by water. These ratios are the same for both quartz and natural sands. Analogous results were also obtained in the displacement of Carboniferous petroleums from quartz sand (Fig. 3). The maximum on the curves corresponds to the limiting value of the adsorption and the start of desorption of the surfactant by water containing no surfactant. Thus, the experiments show that the adsorption of surfactant and hence the quantity of surfactant used increases when the surfactant is pumped into the region outside the contour containing no petroleum.

The quantity of surfactant used is slightly lower when the surfactant is pumped into the region on the edge of the contour and is even smaller when the surfactant is pumped into a series of wells sectioning the petroleum-bearing deposits.

The experiments show that the nonionogenic surfactants synthesized by VNIINeftekhim and NIOPIK are washed from the solid surface and from the surface of the residual petroleum. The quantity of water which must be passed through the porous medium to remove a surfactant which is weakly held by a solid surface in experiments with Bakinskii petroleums, however, was greater by a factor of 1.5 than the quantity of water required to establish adsorption equilibrium during the displacement of the petroleum, and greater by a factor of 2.4 than the quantity of water required in experiments with Carboniferous petroleums.

In the United States, tests have been carried out under laboratory conditions with nonionogenic, anionic, and cationic surfactants. Thus, 56 different surfactants have been tested in the laboratory [5]: of these, 30 were nonionogenic, 21 anionic (chiefly sodium and potassium salts), and 5 cationic (chlorides). These surfactants were tested in experiments on the displacement of petroleum by the centrifuge method at a concentration of 0.1%. The quantity of petroleum displaced when nonionogenic surfactants were added to the solution was greater than that displaced when cationic and anionic surfactants were used. The petroleum recovery when cationic surfactants were used was greater than that when anionic surfactants were used. The total recovery of Oklahoma petroleum from sand amounted to 78.1% for nonionogenic surfactants, 71.2% for cationic surfactants, and 68.7% for anionic surfactants.

Guereca and Butler [6] also carried out petroleum recovery tests with all three types of surfactant, using the centrifuge method. They obtained best results with nonionogenic surfactants. Cationic and anionic surfactants were generally less effective. Hawkins [4] mentions the advantages of nonionogenic surfactants associated with their less extensive reaction with rocks. In the USSR, data have been obtained in

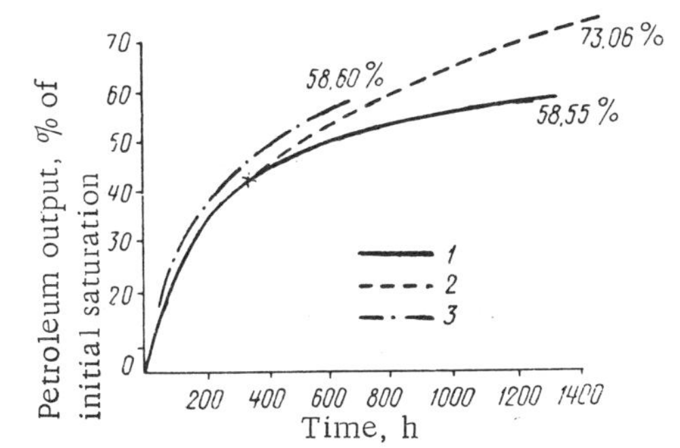

Fig. 4. 1) Displacement by sea water; 2) displacement by sea water containing 0.1% $KAUFE_{14}$; 3) displacement by sea water containing first 0.3% and then 0.05% $KAUFE_{14}$. Natural sand, air permeability 2.3 darcy, no residual water, highly active petroleum, adsorption equilibrium established at point x, rate of filtration 10 m/month.

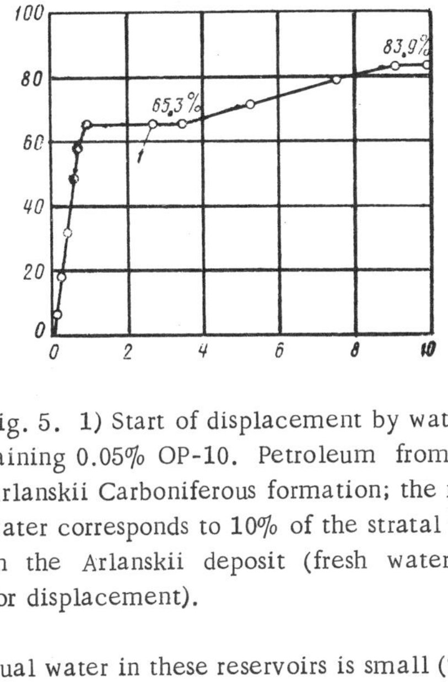

Fig. 5. 1) Start of displacement by water containing 0.05% OP-10. Petroleum from the Arlanskii Carboniferous formation; the residual water corresponds to 10% of the stratal water in the Arlanskii deposit (fresh water used for displacement).

laboratory studies on the use of aqueous surfactant solutions for the displacement of petroleums from the Bakinskii deposits, chiefly petroleums from the Podkirmakinskii formation, which are characterized by a high surface activity and low petroleum recovery [7, 8, 9], and also for the displacement of Carboniferous and Devonian petroleums. The results of experiments on the displacement of Carboniferous petroleums by water containing added surfactants at constant pressure from sand containing no residual water are given in Table 5.

It can be seen that good results are obtained with nonionogenic surfactants of the oxyethylated alkylphenol type—OP-10 and $KAUFE_{14}$. A considerable increase in petroleum output when surfactants are used is also observed in the case of Bakinskii petroleums (Fig. 4). A high initial concentration of surfactant gives best results. This has also been observed in other experiments [10].

When residual water was present in the case of the highly active Bakinskii petroleum, the use of surfactants had no effect, whereas in the case of Carboniferous petroleum an effect was observed but this was slightly less than that observed in the absence of residual water. This is not due to a decrease in the intensity of the action of the surfactant, however, but is to some extent the result of an improvement in the conditions for the displacement of petroleum when residual water appears in the porous medium.

Thus, for Bakinskii conditions the most acceptable surfactant was a nonionogenic surfactant of the $KAUFE_{14}$ type synthesized by the VNIINeftekhim, while for Carboniferous petroleum the most acceptable surfactant was OP-10 synthesized by NIOPIK, although $KAUFE_{14}$ also gives fairly good results in the latter case.

Studies on the use of OP-10 for different conditions of Carboniferous petroleum displacement are being continued in the UFNII. Already, on the basis of laboratory data obtained, the question of industrial tests on OP-10 for increasing petroleum output for Carboniferous reservoirs is being raised, since it is known that the quantity of residual water in these reservoirs is small (2-8%). In a discussion of industrial tests on surfactants mention must be made of the conclusions reached in other countries regarding the use of surfactants for this purpose.

Thus, the nonionogenic surfactant Triton X-100, which is a product of the reaction between m-octylphenol and ethylene oxide, has been tested in forced flooding experiments under industrial conditions in concentrations of 0.2% [10]. In the experimental section where forced water containing added surfactant was used, 1350 m^3/ha of petroleum was obtained from 19 wells in two years, whereas in the section where forced water containing no surfactant was used, only 1220 m^3/ha of petroleum was obtained in five years.

Breston and Johnson [11] have described industrial tests in the Keiber-Coulter deposits with a nonionogenic surfactant of the Igepal-CA types at a concentration of 0.043%. The tests gave an additional output of petroleum whose value exceeded the cost of the surfactant used by a factor of 2.78. In the Tibbet field, however, the value of the additional petroleum output amounted to only 16% of the cost of the surfactant used.

Torrey [12] gives some details of the conditions of application of surfactants in the Humboldt and Chanute, Kansas (Bartsville sandstone) and Mirando City, Texas (Mirando sandstone) fields. Measurements made six months after the injection of surfactants showed that the penetration of water into the stratum was much more regular and that the pickup of the forcing wells increased.

The literature describes several attempts to summarize industrial experience in the use of surfactants. These summaries cannot claim to be complete, however, since the quantity of available data is small and the authors do not have sufficiently clear ideas regarding the mechanism of the action of surfactants under different conditions. Thus, some authors consider that surfactants should have no significant influence on petroleum output, since the forced water containing these substances mixes with the residual water so that the latter forms a fringe in front of the forced water front. It is considered [13] that the displacement is brought about chiefly by this fringe of water with low flushing power. This is essentially the main argument put forward by those who oppose the use of surfactants to increase petroleum output. There is insufficient justification for this argument, however. In the first place, in the displacement of particularly viscous petroleums from reservoirs characterized by a more or less marked nonuniformity, no piston-like conditions are observed. At the moment when water appears in the wells, the quantity of petroleum remaining in the region of the stratum to which the water has penetrated is very high. At the same time, all the residual water ahead of the front of forced water containing surfactant is displaced. In the second place, the displacement process is accompanied by intensive mixing of the forced water with the residual water moving ahead of the forced water front. Intensive mixing begins even at rates of filtration equal to 2-3 m/month. This is indicated by experimental data on the displacement of one specimen of water by another from a porous medium [3]. In the third place, those who oppose the use of surfactants make no allowance for the fact that the surfactants facilitate the capillary impregnation of the petroleum-saturated rock by water, particularly when the rock contains a small quantity of residual water, and thus increase the coefficient of coverage of the stratum by the forced water and hence the petroleum output. From this viewpoint the use of surfactants becomes particularly advantageous when the petroleum produces marked hydrophobic properties in sandstone rock and thus prevents capillary impregnation of the rock by water. This characteristic is shown in particular by Devonian and Carboniferous petroleums, due to the presence of asphaltenes. It may thus be concluded that the more varied the stratum and the greater the ability of the petroleum to produce hydrophobic properties on the solid surface, the more effective will be the use of surfactants.

Experiments show that the production of an additional large quantity of petroleum is also possible when surfactants are used to displace the residual petroleum formed after forced flooding by water with low flushing power (Fig. 5). Analogous data have been obtained in a number of experiments. Industrial studies of the pick-up profile of forcing wells before and after the use of surfactants show indirectly that surfactants lead to a considerable increase in the coefficient of coverage of the stratum by forced water. When surfactants are used, the water begins to penetrate into partitions which previously were not penetrated. For example, Torrey [12] states on the basis of industrial experiments that surfactants increase the rate of filtration of water in partitions with low permeability and facilitate the more regular displacement of the water front in nonuniform strata. In addition, surfactants decrease the degree of swelling of clay minerals.

The theoretical basis of the influence of surfactants on the displacement of petroleum by water has been developed to a considerable extent by Soviet scientists. This has been assisted by the physicochemical study of surface phenomena by Academician P. A. Rebinder and his school [14-17] and other Soviet scientists, and by special studies of the nature of the phenomena accompanying the displacement of petroleum by water, gas, and their mixtures. The conditions under which the use of surfactants is most advantageous have already been established.

In conclusion, mention must be made of certain characteristic features of methods for carrying out industrial tests and for economic evaluation of the use of surfactants. Industrial studies should reveal the efficiency of surfactants as quickly as possible and experimental work should be carried out by the least tedious and most inexpensive methods. For this purpose it is advisable to have strata at the minimum possible depth and to drill specially a small number of wells at short distances from one another. Unfortunately there are no such strata in the Bashkir SSR and it is necessary to use strata of the Carboniferous formation at a depth of 1200-

1300 m. A whole range of problems must be considered in this case in order to decrease the cost of experimental work and to obtain rapid results from industrial tests.

In the investigation of economic efficiency it is necessary to take account of the possibility of producing fringes of water containing surfactants so that forced flooding can subsequently be carried out with ordinary water, since 25-50% of the adsorbed surfactant undergoes desorption, i.e., it is washed out when pure water is forced into the stratum. It is thus quite possible to create a fringe of this type.

At the present time the cost of one ton of surfactant of the OP-10 or $KAUFE_{14}$ type and other nonionogenic surfactants is still too high. It is to be assumed that their cost will decrease considerably with the growth and improvement of the petrochemical industry. The present-day cost cannot therefore be made the basis for economic calculations. These calculations must be based on the costs for mass production of surfactants. If the use of surfactants makes it possible to increase petroleum output even by several percent in Carboniferous and Devonian deposits, the costs associated with their use will be rapidly recovered.

To summarize, it may be stated that laboratory studies indicate that it is now time to undertake industrial tests on surfactants in order to increase petroleum output from collectors.

Literature Cited

1. E. I. Sukhankin, UFNII Fund [in Russian] (1959).
2. G. A. Babalyan and Z. F. Rzabekov, Tr. AzNII, Vol. 7 (1957).
3. G. A. Babalyan, Problems of the Mechanism of Petroleum Recovery (Aznefteizdat, 1956).
4. R. R. Hawkins, Oil and Gas J., 54, 25 (1955).
5. H. N. Dunning, R. T. Johansen, and Lung-Hsiao, Petroleum Engineer, January (1954).
6. R. A. Guereca and H. S. Butler, Producers Monthly, 19, 3 (1955).
7. G. A. Babalyan and V. M. Baryshev, AzNII DN Fund (1957).
8. V. M. Baryshev, E. I. Ibragimov, P. P. Pavlov, and S. G. Movsesyan, ANKh, No. 2 (1955).
9. G. A. Babalyan, V. M. Baryshev, and E. I. Ibragimov, ANKh, No. 8 (1951); ANKh, Nos. 1-2 (1952).
10. H. I. Schneider, H. L. Greenwald, D. L. Felley, and R. F. Lollar, Oil and Gas J., 7 (1958).
11. J. N. Breston and W. E. Johnson, "Experiments with wetting agents in the Bradford industrial deposits," Producers Monthly, 16, 1 (1951).
12. P. D. Torrey, Mines Magazine, 17, 10 (1957).
13. R. L. Slobod, "Review of methods for increasing petroleum output," Producers Monthly, 22, 4 (1958).
14. P. A. Rebinder, The Physical Chemistry of Flotation Processes and Wetting Phenomena (Metallurgizdat, Moscow-Leningrad, 1933).
15. P. A. Rebinder, Izv. OMEN AN SSSR, ser. khimicheskaya, No. 5, p. 705 (1936).
16. P. A. Rebinder, Kolloidnyi zhurnal, 20, 5, 527 (1958).
17. P. A. Rebinder, "Surface-active substances, their importance and applications in the petroleum industry," Collection, The Use of Surface-Active Substances in the Petroleum Industry (Gostoptekhizdat, Moscow, 1961).

THE POSSIBILITY OF USING SURFACTANTS TO INCREASE THE PETROLEUM YIELD FROM FISSURED TRAPS

G. A. Babalyan and E. K. Kovalenko

Petroleum in fissured traps may be contained in both the porous part of the rock and the fissures, or in the fissures only [20, 18]. Figure 1 shows sections through petroleum-saturated cores, illustrating the connections between pores, cavities, and fissures [10]. The liquid seeps out of the pores and cavities into the fissures, and from them into the well.

Figure 2 [8, 7, 18] shows the forms of fissures in various rocks. The fissures vary in extension, roughness, tapering, and openness. They may be classified as macro- or microfissures, depending on how open they are. The length of a fissure depends on its openness and the lithology of the rock. The fissures are more pronounced in places. They may either widen locally into cavities, or over a greater distance into channels. Figure 3 [8, 9, 7, 18] shows photographs of microfissures with local widening in rocks of various origins and structures. Micro- and macrofissures divide the rocks into blocks of various sizes and shapes. The microfissures are more tortuous and are less oriented with respect to the strata. Fissures may be vertical, inclined at some angle to the horizontal, or nearly horizontal.

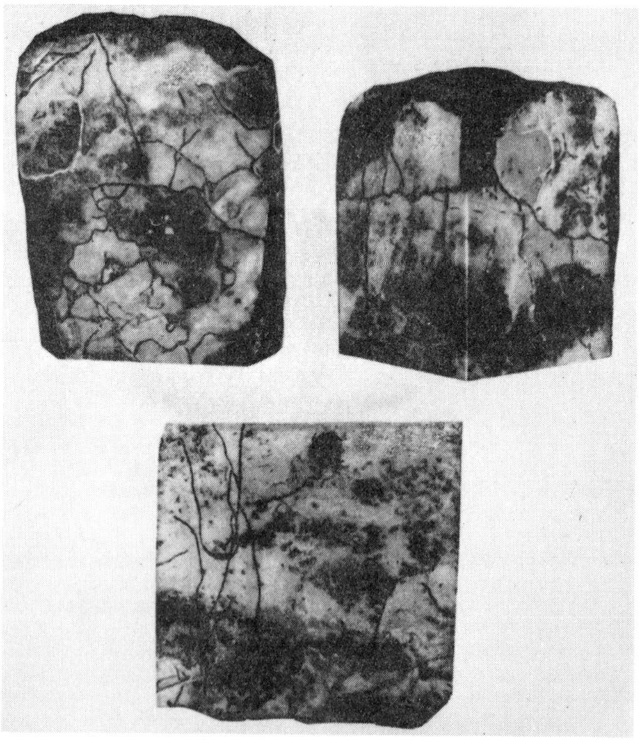

Fig. 1. Sections of petroleum-saturated cores showing connections between pores and cavities with fissures.

Fig. 2. Some forms of fissuring in various rocks. a) Dolomite, stylolite oriented perpendicular to the stratum (Irkutsk region, Lower Cambrian); b) limestone; a fissure with bitumen intersects the mineral fissure (Yuzhno-Minusinsk region, Devonian); c) aleurolite, coarse grain, with open microfissure passing through a closed inclusion filled with calcite (Yuzhno-Minusinsk depression, Middle Devonian, Abakanskian formation, Bystryanka, well 9-r); d) marl with infrequent organic residues, with closed fissure filled with calcite, through which passes an open microfissure (Yuzhno-Minusinsk depression, Middle Devonian, Bystryanka); e) argillite, open microfissure (Yuzhno-Minusinsk region, Devonian); f) medium-grain sandstone with infrequent open microfissures, enveloping grains of plastic material and cutting through cement grains (Yuzhno-Minusinsk region, Ilemorovskian series, Middle Devonian, Bidzha).

The main reserves of petroleum, now being extracted from fissured traps, are based on carbonate rocks. Although carbonate and other rocks are microheterogeneous, and contain inclusions of such other minerals as gypsum, calcite, anhydrite, clay, pyrites, bitumen, and various metallic oxides (Fig. 4) [8, 7, 18], we may consider the petroleum displacement mechanism as being the same in principle from all forms of fissure. With simultaneous seepage of petroleum and water, or of petroleum, water, and gas, the physical properties of these phases have a pronounced effect upon the overall extraction process, as do the fissured rocks themselves and the extent to which the fissures are open.

Let us consider the various cases for simultaneous seepage of petroleum and water.

Let us suppose that the walls of a fissure are completely covered by a layer of petroleum which is not ruptured by water. Inside the fissure the petroleum is the disperse phase and the water the dispersion medium (Fig. 5). There is no leakage in or out through the walls of the fissure. Both the coalescence of free oil drops with the continuous oil film and the reverse process of breakup of oil drops take place during the simultaneous seepage of oil and water.

a

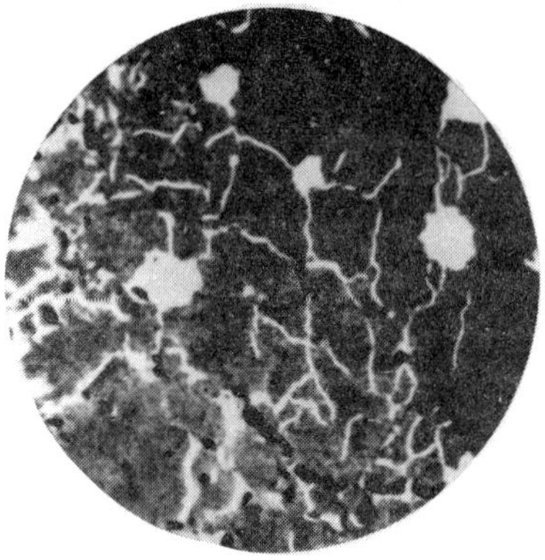

b

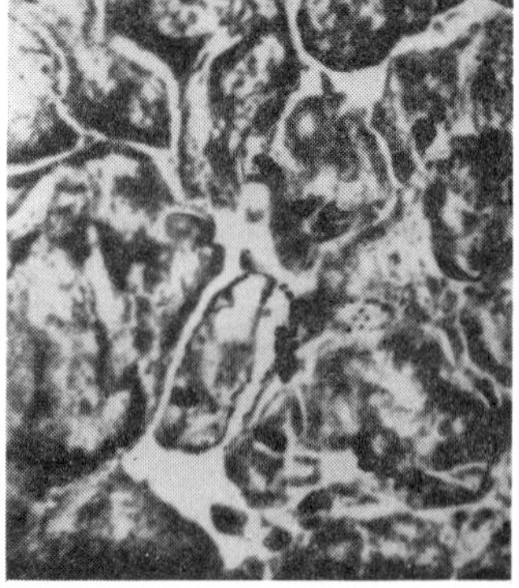

c

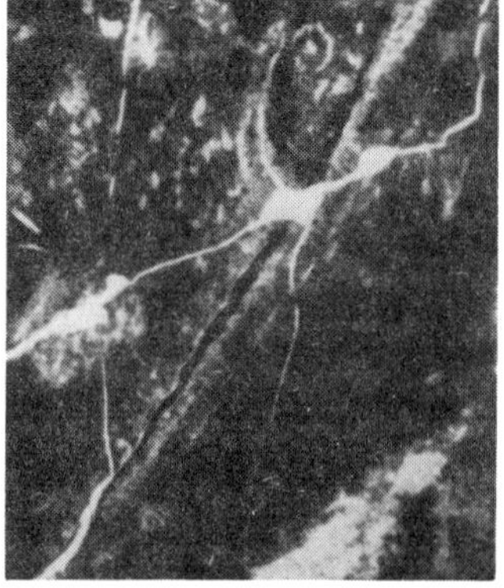

d

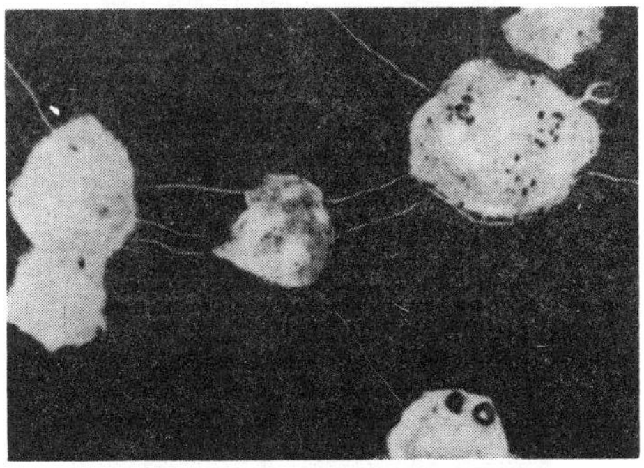

e

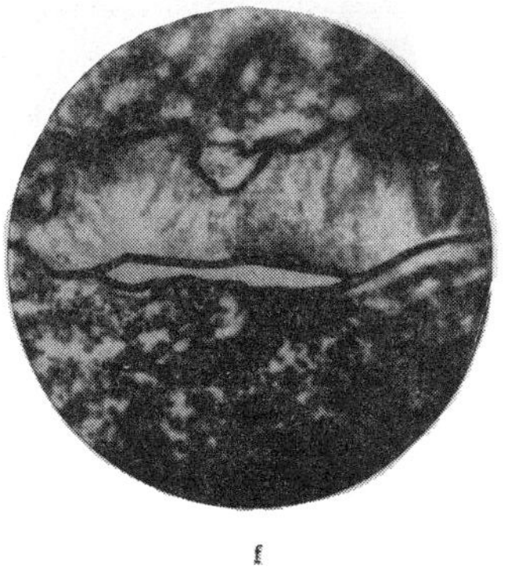

f

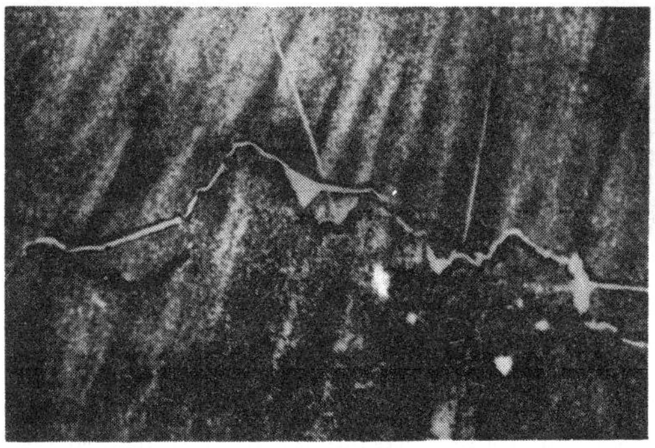

g

Fig. 3. Microfissures with local cavities and broadenings in rocks of various origins and structures. a) Cavities within skeletal residues of organisms, with small prisms of brown bitumen (black in color) in organogenic detrital limestone (Sakmaro-Artinskian deposits, Grachevka, well 585); b) dolomite, hollow pores communicating with microfissures (Irkutsk region, Lower Cambrian); c) hollow cavities of irregular shape with connecting channels or microfissures between algal residues (algal limestone, Usol'skian formation, Atovka, well R-4); d) cavities in the form of broadenings of microfissures in organogenic silty limestone (Sakmarno-Artinskian deposits, Grachevka, well 583); e) cavities connected by microfissures with small prisms of saline clay in fine-grain saline dolomite (Usol'skian formation, Bokhan, well R-1); f) subsequently incrusted cavities in dolomite of irregular grain size, with small prisms of brown bitumen (black color) (Usol'skian formation, Atovka, well R-4); g) cavities in the form of broadenings of open microfissures in fine-grain dolomite (Usol'skian formation, Bokhan, well R-1).

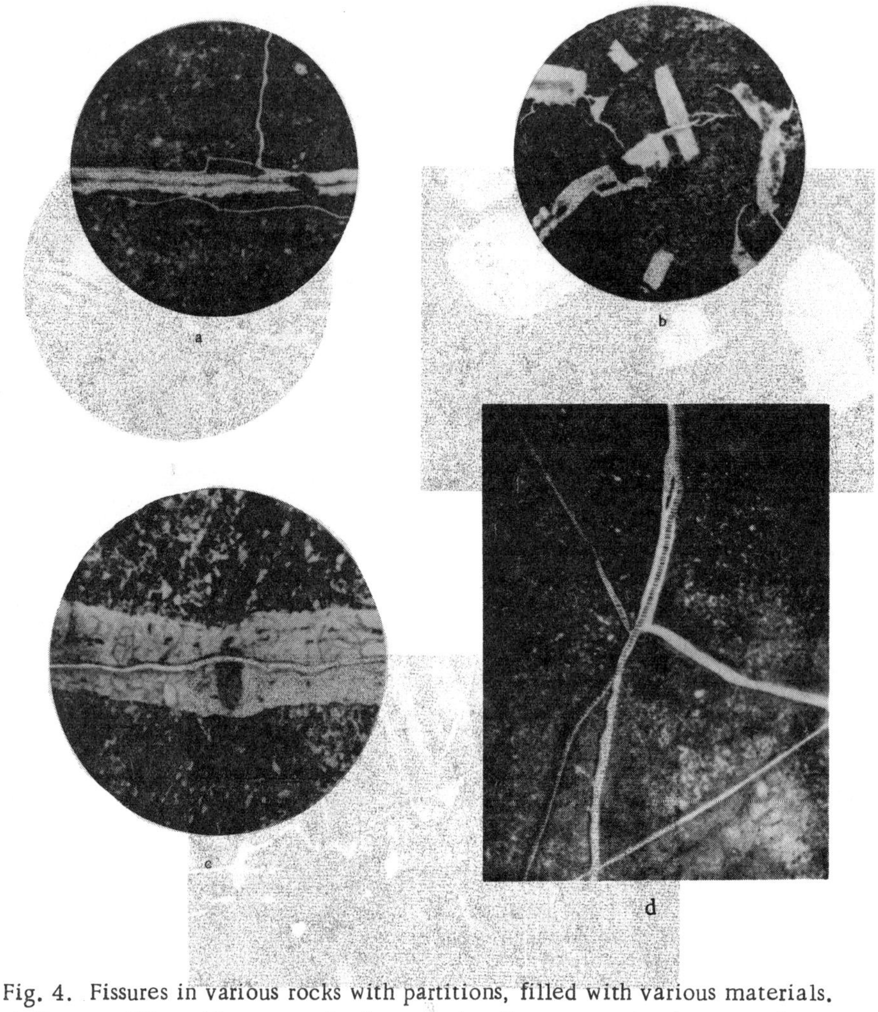

Fig. 4. Fissures in various rocks with partitions, filled with various materials. a) Silty argillite with a network of open microfissures, passing between the walls of a closed fissure with calcite and a parallel closed fissure some distance away (Yuzhno-Minusinsk depression, Middle Devonian, Abakanskian formation, Bystryanka, well 9-R); b) dolomite of varying grain size with small prismatic grains of anhydrite, with open microfissures intersecting the anhydrite grains and passing through the enclosing rock in the form of slightly branched fissures (Irkutsk amphitheater, Lower Cambrian, Motskaian formation, Osa, well 1-R); c) sandstone, open fissure within mineral fissure (Yuzhno-Minusinsk region, Devonian); d) limestone, fissure with oil within mineral fissure or intersecting it (Karabulak, Upper Cretaceous).

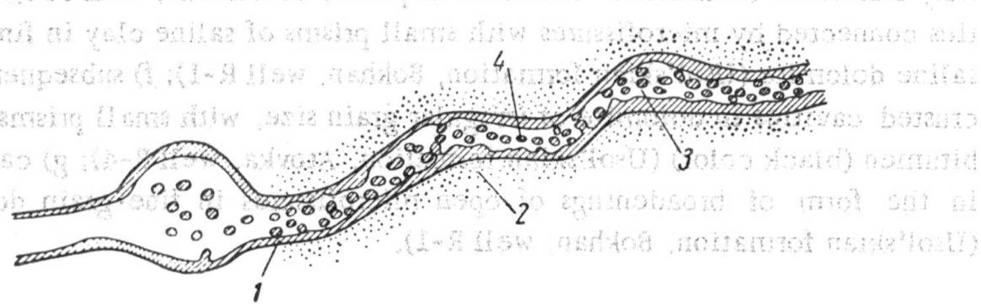

Fig. 5. Movement of a mixture of petroleum and water in a fissure when there is a continuous oil film on the walls of the fissure. 1) Continuous oil film; 2) porous part of stratum; 3) water; 4) dispersed freely in oil.

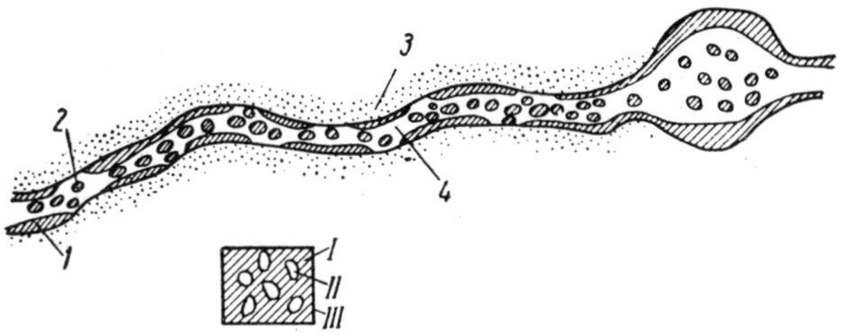

Fig. 6. Movement of a mixture of petroleum and water in a fissure when the oil film has been displaced by water or individual parts of the fissure walls. 1) Continuous oil film; 2) dispersed freely in oil; 3) porous part of stratum; 4) water. I) Oil film; II) part from which oil has been displaced by water; III) surface of fissure wall.

In addition to the movement of the free oil drops in the water, there is also a flow of the continuous oil film along the walls in the direction of seepage. It is clear that the dimensions of the broken-up drops must decrease with increasing rate of water movement, and with a decrease in the viscosity of the petroleum and of the surface tension at the petroleum–water boundary ($\sigma_{w.p}$). The surface from which the drops break up and the number of these drops increase with increasing rate of water movement and with decreasing $\sigma_{w.p}$. The coefficient of coalescence, which is the ratio of the number of coagulating drops to the total number of drops impinging against the surface of the continuous oil film, decreases with an increase in the coalescence time and with the rate of movement of the drops, and also with a decrease in the drop size and the volume of the free oil. The dimensions of the free oil drops decrease, but the number of drops increases with an increase in the rate of water movement and in the coalescence time of the drops.

The rate of movement of the free oil drops in the water is greater than that of the continuous oil film. Thus, the thickness of the continuous film should be as small as possible in order to speed up the displacement process. A small thickness in the film should be produced by reducing the coefficient of coalescence of the drops with the continuous film, by increasing the time of coalescence of the free drops with each other, by reducing the drop size, and by a relatively high flow rate of the oil in the continuous layer itself. A high dispersion of the oil and a high flow rate in the continuous film can be achieved by reducing the surface tension at the water–petroleum boundary and the force of adhesion of the oil to the solid surface.

Let us now consider the case when the surface of the walls of the fissures is wetted by both water and petroleum. The oil at the wall surface may now be the dispersive medium (Fig. 6) or the disperse phase (Fig. 7). As in the previous case, a high displacement rate and a higher petroleum yield will be achieved when the smallest possible amount of free oil adheres to the wall surface. The greater the dispersion, and the less the contact angle for wetting (always relative to the aqueous phase), the less will be the volume of oil adhering to the solid surface.

Thus, in order to increase the petroleum yield from fissured traps and to speed up their rates of development, and regardless of the openness of the fissures and the state of the petroleum on the walls (whether the petroleum is the disperse phase or the dispersive medium), it is necessary to create conditions such that the oil does not adhere to the fissure walls and has as high a dispersion as possible, i.e., conditions which stabilize emulsions of the oil-in-water type. This can be achieved by using nonionic surface-active substances (surfactants), such as OP-10, which is known to make substances hydrophilic, i.e., it promotes selective wetting by water, strongly opposes adhesion of petroleum to a solid surface, promotes conditions for breaking it away from the surface, and assists dispersion of petroleum in water [2, 3, 13, 15]. In this case the consumption of surfactant will be inconsiderable since adsorption will only occur on the fissure walls and in the surrounding zones.

Let us now consider the case when parts of the walls are preferentially wetted by water and other parts by petroleum, and there is an inflow of petroleum from the porous part of the stratum. It is supposed that there is no free water in the pores of the stratum. Under these conditions of petroleum inflow any interstitial water remains immobile.

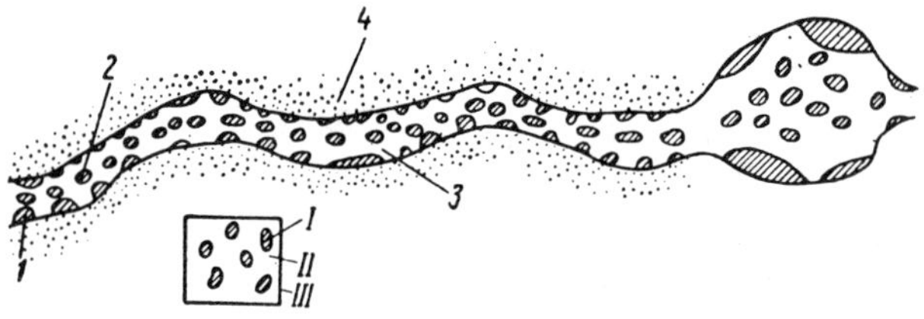

Fig. 7. Movement of a mixture of petroleum and water in a fissure when the continuous oil film has been displaced from the fissure walls, but there are individual drops of oil on the walls. 1) Oil drops adhering to walls; 2) freely dispersed in oil; 3) water; 4) porous part of stratum. I) Adhering oil drops; II) water; III) surface of fissure wall.

The flow of oil into the fissures may result from elastic stresses in the stratum and from the pressure of gas evolved from the petroleum in the porous part of the stratum. Let us accept that the water in the actual fissures remains the dispersive phase. It is clear that the influx of oil into a fissure will be determined (under otherwise similar conditions) by the phase permeability of the walls for oil. It will be greater the larger the area of wall surface wetted by oil, i.e., the greater the oil saturation of the pore walls. The walls may be wetted by adhesion of free oil and of oil flowing into the fissures. Part of the influx oil will mix directly with oil adhering to the walls, but another part will by-pass this when entering the fissure.

We will consider the factors which favor an increase in the petroleum yield and an increase in the rate of petroleum displacement by water. To increase the yield it is necessary to intensify the flow of oil into the fissures, and to achieve this the wall surface must be preferentially wetted by oil. However, to increase the rate of displacement of petroleum from the fissures, it is necessary to establish conditions such that the oil readily breaks away from the walls without exposing the wall surface to water. This can be achieved by the use of a water-soluble surfactant of a type to make the surface of the fissure wall hydrophobic, and also to reduce greatly the surface tension at the water–oil interface. Such a reduction in surface tension favors the breakaway of drops from the continuous oil film on the walls [1]. Rendering the fissure walls hydrophobic should make possible the formation in the fissures of a hydrophobic (inverse) emulsion of the water-in-oil type [16, 19]. The surfactant added to make the walls hydrophobic, together with that in the volume of the fissure, should make it possible to break the emulsion, i.e., the water droplets emulsified in the oil should be able to coalesce. The use of such a surfactant can produce a marked effect when the fissures are wide open and the main problem is to release the petroleum from the porous part of the stratum. This is because the more open the fissures the less is their hydraulic resistance to the flow of hydrophobic emulsion, and the lower the possibility of formation of such emulsion. Thus, it is important to increase the intensity of oil flow from the porous part of the stratum into the fissures.

In order to make the surface of the fissure walls hydrophobic it is necessary to inject a certain quantity of water, petroleum, or other hydrocarbon solvent, containing an appropriate surfactant. This will particularly increase the influx of petroleum into the fissures when the petroleum contains dissolved gas. In order to render a high proportion of the fissure wall surface hydrophobic it is important that the aqueous or hydrocarbon solution of surfactant should be injected at a high rate. It may be convenient to inject only liquid hydrocarbon with the appropriate surfactant added, or petroleum alone if it contains the necessary surfactant.

The use of a water-soluble surfactant to render the fissure walls hydrophobic may be convenient, since this will make the walls hydrophobic and also strongly impede inversion of the hydrophilic emulsion in the fissure to a hydrophobic emulsion. In this case a cationic or anionic surfactant should be used.

According to Rebinder [16], the ability to form emulsions of the inverse water-in-oil type increases with increasing γ, and the ability to form emulsions of the oil-in-water type increases with increasing $1/\gamma$, i.e., with decreasing γ:

$$\frac{\tau_{w.p} V_p}{\tau_{p.w} V_w} = \gamma,$$

where $\tau_{w.p}$ is the life of a water drop in petroleum; $\tau_{p.w}$ is the life of a petroleum drop in water; V_w is the concentration of water throughout the volume of the mixture; V_p is the concentration of petroleum throughout the volume of the mixture.

The two extreme cases are interesting, when $\beta = \tau_{w.p}/\tau_{p.w} \to 0$ ($\tau_{p.w} \gg \tau_{w.p}$), and when $1/\beta \to 0$ ($\tau_{w.p} \gg \tau_{p.w}$).

In experiments carried out under such conditions, which are generally rare, only one type of emulsion is formed whatever the volume ratio of the phases. Only an oil-in-water emulsion forms, or, strictly speaking, persists, when $\beta \approx \infty$, and only a water-in-oil emulsion when $\beta \approx 0$. Drovetskaya [11] showed that the liquid which selectively wets the wall tends to become the dispersive medium, i.e., the external phase. Thus an emulsion of the direct oil-in-water type tends to form at a surface which is normally hydrophilic, i.e., is preferentially wetted by water.

The presence of a hydrophobic phase can lead to phase inversion, i.e., to the formation of a water-in-oil type emulsion on the wall [11]. Normally the ratio β is very high for water–oil systems, so that formation of a hydrophilic emulsion in the fissure space is only possible at very low concentrations of oil in water. On this basis, it may be concluded that a hydrophilic emulsion will only be formed in the fissure space when the influx of petroleum from the porous part of the stratum is small, and the influx of water is high. This can occur when the stratum water is under pressure.

Influx of oil and water from the pores into the fissures may take place in the initial stages of development of a stratum, provided that the latter is highly saturated with water, and that gas is absent or only evolved to a very slight extent.

If the petroleum in the porous part of the stratum and in the fissures is mainly mixed with water, with water as the dispersive medium and oil as the disperse phase, then it is clear that in order to intensify the influx into the fissures it is necessary to increase the phase permeability of the walls to both oil and water. The surfactant to be used should therefore prevent the rendering of the fissure walls hydrophobic and prevent the formation of a hydrophobic emulsion in the fissures. OP-10 is a suitable surfactant in this case. Experiments have shown that it increases the phase permeability of porous materials for both oil and water, increases the petroleum yield, and reduces the displacement time [2]. Displacement of oil into the fissures by means of water from the porous part of the stratum is possible, as, for example, by flooding the trap. In order to displace as much as possible of the oil from the porous part of the stratum under these conditions, it is essential to produce the maximum possible increase in the permeability of the porous part of the stratum to both oil and water. This can be achieved by using nonionic surfactants of the type OP-10. In stagnant regions of the stratum, where after flooding there may be equality of pressure in the fissures and pores in separate parts of the stratum, the influx of petroleum into the fissures will be mainly the result of capillary attraction of water into the pores, subject to the action of the force of gravity [5, 6, 14, 4]. In this case a surfactant should be used that markedly improves capillary penetration.

These conclusions, based on analysis of the physical nature of the phenomena occuring in fissured traps with simultaneous movement of petroleum and water, make it possible to lay down the conditions under which the process of flooding a trap can be carried out most effectively.

When the trap contains petroleum only in fissures, the stratum slopes gently, and the fissures are relatively open, the best variant to apply is flooding with low rates of water movement, since a high velocity leads to an intense breakthrough of water and this takes place mainly through the most permeable fissures. The intensity of the breakthrough increases with an increase in the viscosity of the hydrocarbon liquid evolved and with its rate of displacement [17], and will also depend on the activity of the petroleum [6]. Figure 8 shows the development of contact in the displacement of oil by water in a single fissure with an opening of 0.9 mm. The breakthrough of water may lead to bypassing of some parts of the fissures and their isolation.

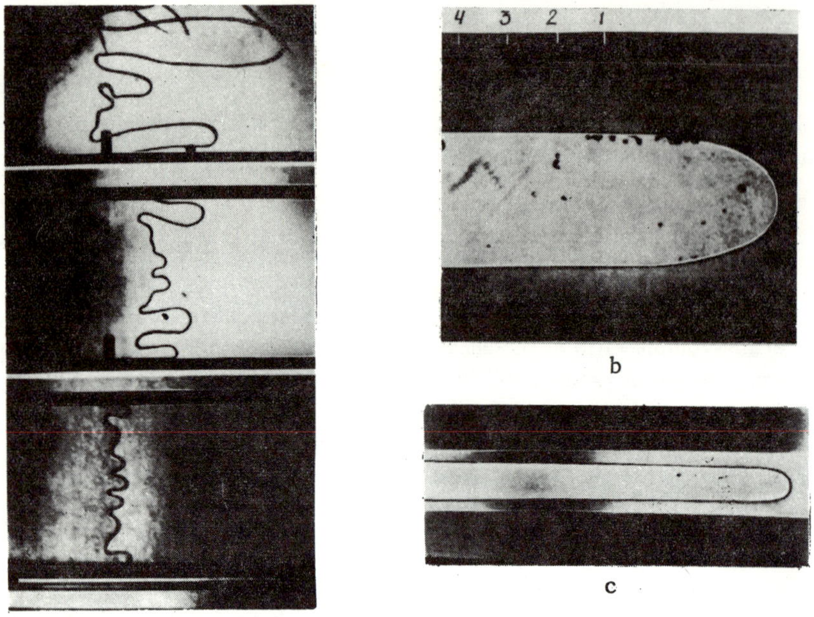

a b c

Fig. 8. The development of contact during displacement in model fissures. a) The development of tongues of the displacing liquid; b) the phenomenon of dispersion at the boundary between phases; c) the character of the advance of contact displacement in long fissures.

Flooding may be carried out at a higher tempo when the trap stratum is steeply inclined, since the uniform advance of the water and its more complete coverage of the fissure system make it possible for the liquids to separate under the influence of gravity. In general, the more open the fissures and the steeper the slope of the trap stratum, the higher is the permissible flooding rate. An increase in development tempo and in the coverage coefficient of the fissures can be achieved by using a water-soluble nonionic surfactant; this helps wet the walls of the fissures with water and prevents formation of a hydrophobic water in oil emulsion. The separation of water and petroleum under gravity will then proceed more intensively than in the absence of surfactant. It should be taken into account that the separation is retarded in less open fissures. Thus, introducing a surfactant not only makes it possible to increase the petroleum yield, to obtain a more complete coverage of the fissure system by water, and to improve the petroleum-eluting properties of the water, but also makes it possible to use higher flooding rates.

When the trap contains petroleum in fissures and in pores, the worst conditions for elution are observed with gently sloping strata. It has been explained above that penetration of water into the fissures leads to a marked reduction in the permeability to petroleum of the walls and reduces the influx of petroleum from the pores into the fissures. Use of a high water-injection rate is undesirable in this case. The conditions for flooding improve with increasing angle of slope of the trap formation, because of the more rapid gravitational separation of petroleum and water and the more uniform penetration of water into the porous part of the stratum. The rate of water movement into the fissures decreases with increasing rate of gravitational separation. The smaller the fissures and the greater the permeability of the porous part of the stratum and coverage coefficient, the slower is the rate of water movement.

The use of a surfactant, to improve the wetting of the fissure walls by water, and thus to increase their permeability to water and petroleum, and also to facilitate the gravitational separation of the liquids, prevents the formation of a hydrophobic water in oil emulsion and therefore becomes very important in connection with the flooding of such strata. When, even at very low flooding rates, the water bursts rapidly through the fissures into the wells, it may be desirable to use a cationic or anionic surfactant to render the fissure walls hydrophobic.

This has been observed particularly in the flooding of reef massifs. Reefs consist of interconnected oil-saturated inclusions of high porosity in the main mass of the relatively impermeable porous part of a trap, and evidently contain only small stocks of petroleum. If the main petroleum reserve is in the porous inclusions of high permeability, then it follows that a surfactant solution should be used to improve wetting of the rock by water, when the reef is flooded by injecting water into the petroleum-bearing part. If the main petroleum reserves are in the relatively impermeable porous part of the reef, then, with the above method of flooding, it is possibly an advantage to use a surfactant to render the rock hydrophobic and thus increase the coverage coefficient during flooding, and also to increase the influx of petroleum into the pores when flooding is stopped and the pressure is subsequently reduced.

Displacement of Petroleum and Water from Fissured Traps by Gas

Petroleum can be displaced by gas, either evolved from the petroleum or injected from outside. The first process occurs when dissolved gas is present, the second when there is repressurization by gas or by water and gas. When dissolved gas is present and the buried water is immobile, petroleum and gas move into the fissures. The disperse phase in the fissures may be either gas or petroleum.

Let us suppose that petroleum is contained in the fissures only, and that there is no influx or outflow through the walls. The gas in the fissure space is the dispersive medium and the oil is the disperse phase. The fissure walls are completely covered by oil. In this case, in order that the gas evolved from solution produce the most complete and rapid displacement of petroleum, it is essential that the petroleum should break away from the walls to the maximum extent and that the thickness of the residual layer on the walls should be as small as possible. Other things being equal, the thickness of the petroleum layer is greater the more active the oil and the more readily the wall surface is wetted by oil.

If we now take into account influx of petroleum from the pores of the stratum as the result of gas evolution, then, if the gas is the dispersive medium in both the pores and the fissures, i.e., mist conditions prevail with petroleum as the disperse phase, it is clear that, other things being equal, it is desirable to have the highest possible permeability of the porous part of the stratum to both oil and gas. It is known [12] that, the less active the petroleum, the greater is the total permeability of the porous medium for oil and gas, and the greater are both the petroleum yield and the rate of displacement of oil by gas. The total permeability, and also the phase permeabilities of the porous part of the stratum for both oil and gas, are less the larger the amount of buried water.

Therefore, in order to increase the influx of petroleum and gas into the fissures, it is necessary to reduce the amount of buried water in the parts of the stratum surrounding the fissures walls. To achieve this, it might appear to be desirable to inject into the fissures a small amount of hydrocarbon-soluble anionic or cationic surfactant mixed with gas. However, it is undesirable to render the fissure walls hydrophobic, since this will increase the adhesion of petroleum to the walls and thus reduce the phase permeability of the walls to gas and reduce the rate of displacement of petroleum by gas in the fissures.

When gas is the dispersive medium in the fissures, the rate of movement of the gas, and of the free droplets of petroleum carried with it, is much greater than the rate of movement of the petroleum adhering to the walls. Thus, to increase the tempo of development of a trap, it is essential that the petroleum should be largely in the free state and should be carried, in the form of droplets, by gas into the fissures.

Adhesion to the fissure walls increases with increasing activity of the petroleum, so that transition of petroleum to the free state is retarded. A petroleum-soluble surfactant is, therefore, inappropriate. The correct procedure is to inject with gas a highly concentrated aqueous solution of surfactant, which will reduce the wetting of the fissure walls by petroleum and will permit a reduction in the amount of buried water in the part of the formation surrounding the fissure walls. The introduction of this solution facilitates the washing away and breakaway of petroleum from the fissure walls and the conversion of petroleum to the free state. It may be desirable to inject carbon dioxide as solution or as gas, since this will reduce wetting of the rock by petroleum and will decrease the amount of buried water in the pore space.

The gas in the fissures and in the porous part of the stratum can be the disperse phase, with the petroleum as the dispersive medium. This corresponds to the initial conditions for development, when the amount of gas evolved from the petroleum is only slight. This is the most favorable case from the point of view of influx of petroleum into the fissures, and it is difficult to suggest any economical means for speeding up the influx of petroleum into the fissures; the rate of such influx is determined by the phase permeability of the fissure walls to petroleum, and this permeability is already high if the amount of buried water is small. If the viscosity and activity of the petroleum are high and gas is the dispersive medium, then the tempo of development can be raised by injecting with the gas a small quantity of a highly concentrated solution of surfactant in water, which greatly reduces the adhesion of petroleum to the fissure walls. It may be desirable to inject some carbon dioxide with the gas. In order that the petroleum should be highly dispersed, it is important that the coalescence rate of the petroleum drops in the gas should be low.

The most complicated process to consider is the displacement of petroleum and water by gas pressure. Let us consider the case when a mixture of water and petroleum is displaced from the fissures by gas, and when there is no influx or loss through the fissure walls. We will suppose that gas is the dispersive medium in the fissure space, that petroleum and water are both disperse phases, and that parts of the fissure walls are wetted well by petroleum and other parts by water. At the fissure walls, in one case petroleum will be the continuous phase, the dispersive medium, with water present as droplets; in the other case water will be the continuous phase, with petroleum present as droplets. There can also be cases when part of the water and petroleum is present as separate droplets on the walls, and another part is present as separate continuous films along the whole length of a fissure. It is clear that phase inversion at the walls will be determined by the intensity of attachment to the walls of water and petroleum.

The more hydrophilic the walls, the greater will be the volume of water attached to them. Thus, in the displacement of petroleum and water by gas, an increase in the hydrophilic character of the walls will favor inversion at the walls to water as the dispersive medium, and there will be a corresponding increase in the volume of petroleum converted into the free state in the fissures.

A water-soluble surfactant reduces the surface tensions of water, $\sigma_{w.p}$ at the boundary with petroleum and $\sigma_{w.g}$ at the boundary with gas, as well as the angle of contact, and thus reduces the adhesion of petroleum to the walls and increases its rate of movement along the fissures, i.e., the phase permeability of the fissures for petroleum is increased. The addition of such a surfactant may also render the fissure walls hydrophobic. This, together with the reduction in $\sigma_{w.p}$ and $\sigma_{w.g}$, provides conditions favoring the break away of water from the walls, i.e., the conversion of water to the free state. The reductions in the surface tensions of water at the boundaries with petroleum and gas increase the rate of movement of the water layer along the walls, under the driving influence of the gas layer and petroleum droplets in contact with it. The result is to increase the rate of water movement along the fissures, i.e., the phase permeability for water is increased. Thus, by using a suitable surfactant, it is possible to increase the total permeability of the fissures to water, petroleum, and gas.

Addition of an oil-soluble surfactant dissolved in a nonpolar hydrocarbon liquid increases the surface tension at the gas interface. It also increases the strength of attachment of petroleum to the fissure walls and increases the angle of contact. The effect of all this is to increase both the volume of petroleum attached to the walls and the wall surface covered by petroleum. A greater volume of water is detached from the wall surface, converted into the free state, and carried along by the gas. Thus, addition of an oil-soluble surfactant, dissolved in a hydrocarbon liquid, reduces the phase permeability of the fissures for petroleum and, in this case, increases it for water. There is also the possible case when both petroleum and water flow into the fissures from the porous part of the stratum.

If water is the dispersive medium and petroleum the disperse phase at the fissure walls, then part of the petroleum entering the fissures will combine with petroleum droplets at the walls, and part of the entering water will coalesce with the continuous layer on the walls. The volumes of petroleum and water entering the fissures will depend on the phase permeabilities of the walls for these liquids. If the petroleum in the porous part of the stratum is transported by water, then it is desirable to increase the permeability of the walls for both liquids, for example by injection of a water-soluble nonionic surfactant with gas, which reduces the wetting of the walls by petroleum. If, however, both petroleum and water are transported by gas in the porous part of the stratum, then it is desirable to inject with gas a water-soluble surfactant which facilitates the wetting of the fissure walls by petroleum.

The above analysis of the phenomena involved in the simultaneous movement of water and petroleum, or of petroleum, water, and gas, in fissured rocks, shows that there are extensive possibilities for the use of both ionic and nonionic surfactants, to increase the rate of development and the petroleum yield from fissured traps of various types. On this basis, extensive tests should be carried out, both in the laboratory and under practical conditions, on the effectiveness of various types of surfactant when used in the flooding or gas repressurization of fissured traps.

Literature Cited

1. G. A. Babalyan, The Physicochemical Basis of the Simultaneous Filtration of Mutually Nonmiscible Liquids, TrudyUFNII, No. 9 (Gostoptekhizdat, 1962).
2. G. A. Babalyan, I. I. Kravchenko, I. L. Markhasin, and G. V. Rudakov, The Physicochemical Basis of the Use of Surfactants in the Development of Petroleum-Bearing Strata (Gostoptekhizdat, 1962).
3. G. A. Babalyan, "The physicochemical principles of the use of surface-active substances in the drilling of wells in order to intensify the recovery of petroleum and increase the petroleum output from strata," Collection, The Use of Surface-Active Substances in the Petroleum Industry (Gostoptekhizdat, Moscow, 1961).
4. G. A. Babalyan, S. G. Movsesyan, and I. L. Markhasin, "The displacement of petroleum from two-layer strata," Neft. khoz., No. 4 (1954).
5. G. A. Babalyan and Z. F. Rzabekov, The Capillary Penetration of Water into Oil-Saturated Strata. Trudy AzNII DN, No. VII (Azerneftneshr., Baku, 1958).
6. G. A. Babalyan, Problems in the Mechanism of Oil Recovery (Aznefteizdat, Baku, 1956).
7. L. P. Gmid, "Results of the lithological and petrographic investigation of fissured rocks of the Palaeozic Bashkir Pre-Ural, Yuzhno-Minusinsk Depression, and Irkutsk Amphitheater," Fissured Rocks and Their Trapping Properties, Trudy VNIGRI, No. 121 (Gostoptekhizdat, Leningrad, 1958).
8. L. P. Gmid and V. N. Kalacheva, "Lithological factors and their effects on the trapping properties of Lower Cambrian carbonate rocks of the Irkutsk Amphitheater," The Fracturing of Rocks and Fissured Traps, Trudy VNIGRI, No. 193 (Gostoptekhizdat, Leningrad, 1962).
9. L. P. Gmid, "The trapping properties of carbonate rocks of the Sakmaro-Artinsk beds of the Grachevsk formation of the Bashkir Pre-Ural, and the effects on them of the processes of dolomitization and sulfation," The Fracturing of Rocks and Fissured Traps, Trudy VNIGRI, No. 193 (Gostoptekhizdat, Leningrad, 1962).
10. V. K. Gromov, "Fissured petroleum traps of platform-like Bashkir formations," Proceedings of the Second All-Union Conference on the Problems of Fissured Traps (Groznyi, 1962).
11. R. M. Dvoretskaya, "The effects of selective wetting on the process of emulsion formation," Kolloidnyi zhurnal, 10, 5 (1948).
12. A. N. Kerimov and G. A. Babalyan, "The effects of petroleum activity on the main indices of the process of its displacement by air," Izv. vuz., Neft' i gaz, No. 8 (1958).
13. I. L. Markhasin, G. A. Babalyan, and N. M. Chistyakova, "The problem of the coalescence of petroleum droplets in water," Collection, The Use of Surface-Active Substances in the Petroleum Industry (Gostoptekhizdat, Moscow, 1961).
14. J. Silvain and Pirson, "Performance of fractured oil reservoirs," Bulletin of the American Association of Petroleum Geologists, 37, No. 2 (February, 1953).
15. P. A. Rebinder, "Surface-active substances, their importance and application in the petroleum industry," Collection, The Use of Surface-Active Substances in the Petroleum Industry (Gostoptekhizdat, Moscow, 1961).
16. P. A. Rebinder, Introductory article to the book by V. Clayton: Emulsions (Izd. inostr. lit., Moscow, 1950).
17. P. G. Safman and G. Taylor, "The penetration of liquid into a porous medium or slotted model containing a more viscous liquid," Mekhanika, Periodical Collections of Translations from Foreign Literature, No. 4 (Moscow, 1960).
18. E. M. Smekhov, Rules for the Development of the Fracturing of Rocks and Fissured Traps, Trudy VNIGRI, No. 172 (Gostoptekhizdat, Leningrad, 1961).

19. A. V. Taubman and A. F. Koretskii, "Some characteristic features of the influence of dispersed solid phase on the emulsification and demulsification of petroleum products," Collection, The Use of Surface-Active Substances in the Petroleum Industry (Gostoptekhizdat, Moscow, 1961).
20. A. A. Trofimuk, "Determination of the capacity of fissured traps from technological data," Proceedings of the All-Union Conference on Fissured Traps for Petroleum and Gas (Gostoptekhizdat, Leningrad, 1961).

GEOLOGICAL CONDITIONS FOR THE USE OF SURFACTANTS TO IMPROVE THE PETROLEUM YIELD FROM THE ARLANSK FIELD

T. M. Stolbova

In the course of the last two years UFNII has carried out investigations in the field of surfactants and their effects on the rate of displacement of petroleum by water. The work showed that the most effective results were obtained when water, enriched with the surfactant OP-10, was injected into a stratum with the minimum content of buried water. Injection should proceed into the oil-saturated part of the stratum. The best effect was obtained in experiments with petroleum from the Arlansk field [1]. The indication was that the use of surfactant solution to displace the petroleum should increase the coefficient of petroleum recovery by about 10% and reduce the cost considerably. Raising the coefficient of petroleum recovery by 10% for such large fields as that at Arlansk would be equivalent to discovering two new large fields. On this basis it was decided that the first field trials on the use of surfactants should be carried out jointly by Bashneft' and the NPU of Arlanneft' in one part of the Arlansk field. The geological characteristics of the Arlansk field are described below.

A Brief Account of Characteristics of the Geological Structure

The Arlansk petroleum field is in northwestern Bashkiria, within the limits of the Birsk anticline. The main petroleum reserves are associated with terrigenous deposits of the so-called "coal-bearing stratum" of the Lower Carboniferous (Fig. 1).

Industrial surveying of the Arlansk region was started in 1955. Further exploratory drilling showed a considerable increase in the oil-bearing area of the Arlansk region in the southeast and in the north.

The southeastern part of the field, known as the Novo-Khazinsk region, and the northern part, known as the Kasevo-Berezovsk region, were contoured during the period 1957 to 1961. The petroleum content contour to the north extends outside the administrative boundaries of Bashkiria into the territory of the Udmursk Autonomous Soviet Socialist Republic, where it is known as the Vyatsk petroleum field. Commercial accumulations of petroleum in the Arlansk field appear also in the carbonaceous deposits of the Kashirskian and Podol'skian levels of the Middle Carboniferous and the Tournaisian strata of the Lower Carboniferous (Fig. 1).

A geological section of the Arlansk field shows it to consist mainly of Palaeozoic deposits, laid down on the eroded surface of the Upper Bavlinskian beds, and covered by a thin coating of Quaternary rocks. The total thickness of the Palaeozoic rock complex is 2000 m.

In conformity with the unified scheme of 1960, the main commercial level (the coal-bearing stratum) belongs to the Vizeiskian stratum of the Lower Carboniferous.

The Arlansk field is confined to the Ivanaevsk arch, extending through the whole Birsk anticline in its center, from the northwest to the southeast.

The structure of the Devonian deposits in the region of the whole Birsk anticline shows itself in the form of a monoclinal merging of layers in the easterly direction. The accumulation of petroleum is associated with

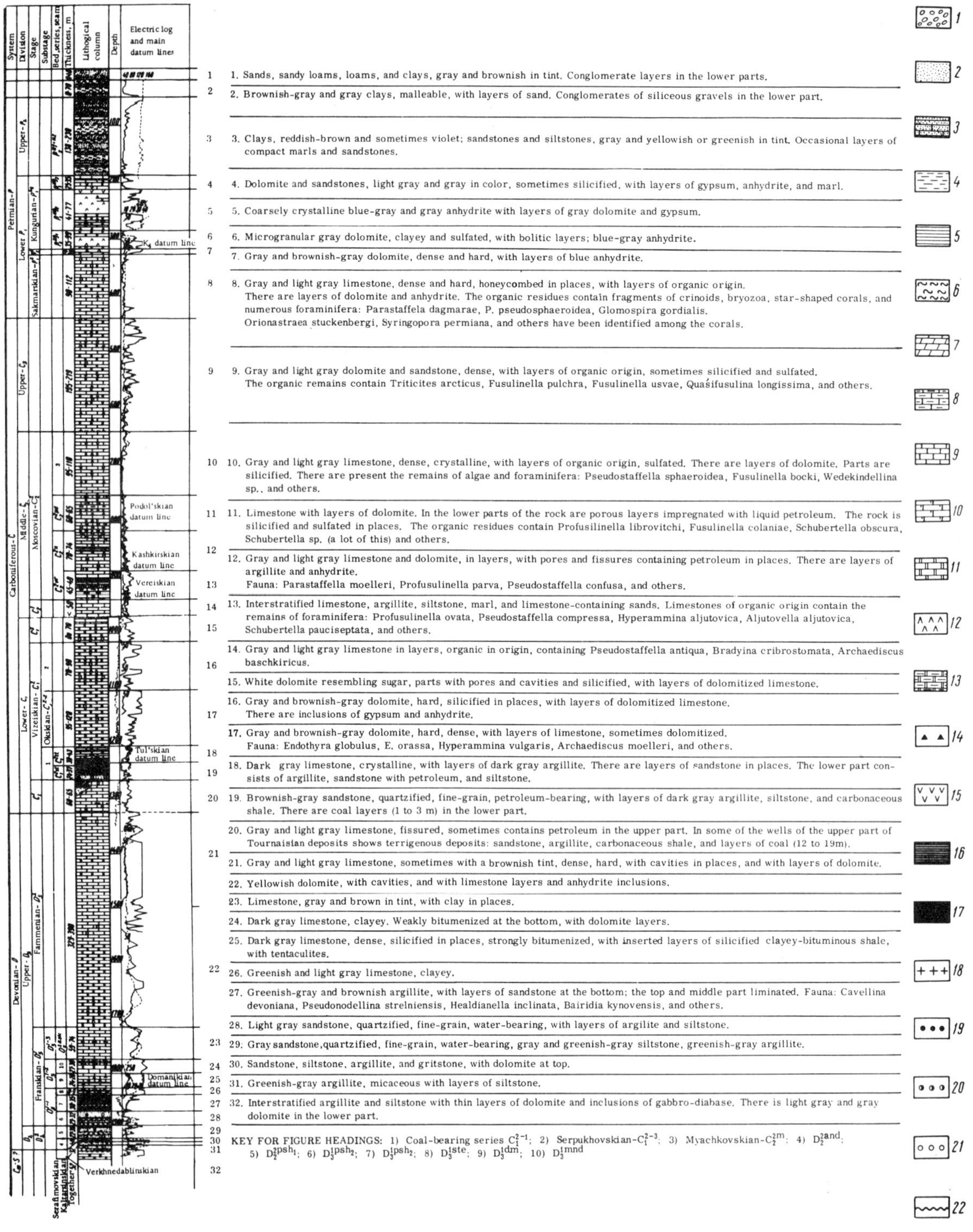

Fig. 1. Composite geological-geophysical sections. 1) Gritstone and gravel; 2) sand and sandstrone; 3) calcareous sandstone; 4) siltstone; 5) argillite; 6) clay; 7) marl; 8) limestone with clay; 9) limestone; 10) dolomitic limestone; 11) dolomite; 12) anhydrite; 13) dolomite with clay; 14) silicification; 15) gypsum; 16) carbonaceous shale with clay; 17) coal; 18) gabbro-diabase; 19) commercial petroleum content; 20 noncommercial petroleum content; 21) water content; 22) erosion.

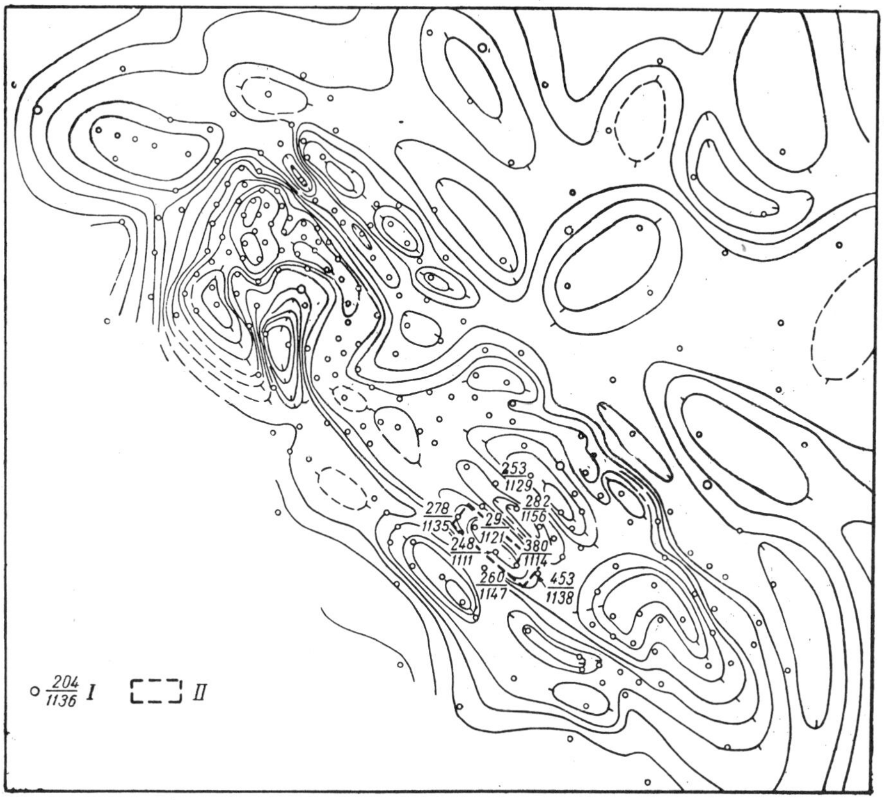

Fig. 2. Structural plan of the top of the Carboniferous series (Arlansk region). I) Bore holes (number as numerator, absolute depth of roof as denominator); II) experimental area.

a well defined gross upheaval of the platform type, outlined by a contour at −1170 m. The length of the tectonic zone, outlined by the −1170 isohypse, is about 90 km, between Kasevo-Berezovsk and the Dyurtylinsk area. The mean width of the zone is 15 to 20 km, and the extent of the upheaval increases from 25 m in the north and 45 m in the Arlansk region to 72 m in the Novo-Khazinsk region.

The structure is asymmetric. The mean slope of the rock towards the western edge is 40' for the Kasevo-Berezovsk and Arlansk regions, but increases to 2° in the Novo-Khazinsk region. The slope of the rock towards the northeastern edge does not exceed 15 to 20' over the whole zone. Up to the present, a large number of local domes and upheavals, connected by one water saturation ceiling have been established by drilling within the boundary of the zone under consideration. Their contours are shown in Fig. 2.

A relatively small, closed, submerged region, completely saturated with water (in the vicinity of bore holes 11, 282, and others), was found in the Arlansk area, within the limits of the inner petroleum-bearing contour, in the lower strata. From the top of the Vereiskian level to the datum line K_4 the structural plan in the vicinity of the Arlansk field is clearly defined and, in general, repeats the structural plan of the coal-bearing strata of the Lower Carboniferous.

The Structure of the Production Level

A section through the coal-bearing series of the Lower Carboniferous of the Arlansk field shows interstratification of sandy, aleurolitic, and argillitic rocks of varying thickness. The section shows layers of limestone to the south of the field, and of carbonaceous shale in the other parts.

In spite of the similarity of the geological section throughout the whole region, there are considerable differences in the quantity of the sandstone trap, the structure of the petroleum deposit, the position of the

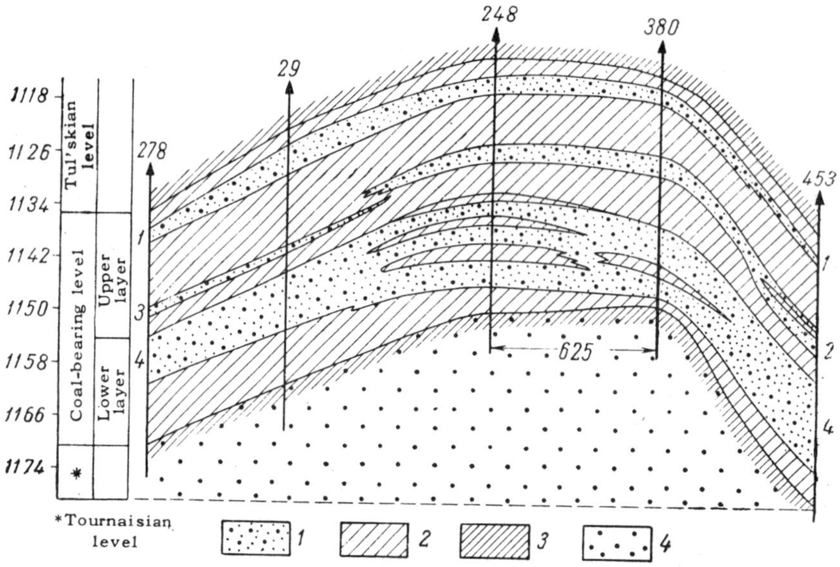

Fig. 3. Geological profile along the line of boreholes 278, 29, 248, 380, and 453. 1) Sandstone; 2) aleurolitic-argillitic rock; 3) limestone; 4) commercial oil content.

water saturation ceiling, and in the commercial productivity in the various parts of the field. In accordance with the features listed above, the whole region of the field can be divided into three parts: Novo-Khazinsk in the south, Arlansk in the center, and Kasevo-Berezovsk in the north. In the vicinity of Arlansk the section through the productive strata can be divided into two levels, in accordance with the lithological characteristics: an upper layer of sandy shale and a lower layer of sandstone (Fig. 3).

The upper layer consists of alternating sandstone, aleurolitic, and clay strata of different thicknesses. The sandstone strata in this part of the section are relatively thin, whereas the clay strata are lenticular and show relatively low trapping properties. There is a development of one to five sandstone seams. The total thickness of these seams varies from 0 to 11 m. The mean porosity is 20% according to core and geophysical data, and the permeability is 0.418 darcys according to core data.

The lower layer is sandstone; it consists of sandstone strata with some disjointed thin intercalations of dense rock. Its thickness varies from 2 to 22 m, and there are separate regions where the trap is completely displaced by aleurolitic and argillaceous rocks. According to core data, the mean porosity in the petroleum-saturated part is 23%, and the permeability is 1.2 darcys.

In the vicinity of Arlansk, the water saturation ceiling is mainly confined to the lower strata. Over nearly all the area the water ceiling surface is located at an absolute depth of −1180 m; however, it is at −1188.5 m under the Saklovsk dome, and at −1183 m east of the Ashitovsk dome. The mean petroleum reserve per hectare of surface is 10000 tons.

The Geological and Commercial Characteristics of the Experimental Region

The experimental region is confined to the Nagaevsk dome, which has the highest absolute ceiling level to the strata in the whole of the Arlansk region: −1111 m at bore hole 248, −1114 m at bore hole 380, and −1121 m at bore hole 29. The steepest inclination of the rock is 72 m in 1 km in the western limb (bore holes 248 and 260). The amplitude of the upheaval is 25 to 45 m. The dimensions of the fold enclosed by the −1130 contour are 2.2 · 0.750 km (Fig. 4).

In section the Carboniferous formation of the region shows four sandstone strata, which are conditionally designated as 1, 2, 3, and 4 from above downwards (Fig. 3). The three upper strata correspond to the upper production level, and the lowest stratum to the lower production level.

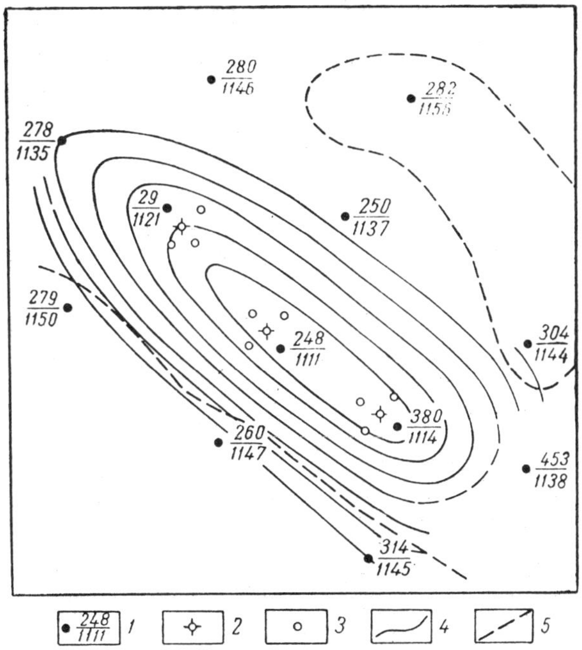

Fig. 4. Structural map of the ceiling of the Carboniferous strata (Nagaevsk dome). 1) Bore holes used for exploitation (number as numerator, absolute depth as denominator); 2) bore holes planned for exploitation; 3) pressure bore holes planned; 4) isohypses at 5 m intervals; 5) inner contour of petroleum content.

Stratum 1 is developed in all three wells of the anticline which were proposed for experimental work (Fig. 4). However, its character alters considerably from bore hole to bore hole. There is a reduction in thickness to the south, from 5.2 m for well 29 to 2.2 m for wells 248 and 380. There are also changes in the properties of the trap rock; the porosity is 18.5% for well 29, 20.7% for well 248, and 17.4% for well 380; the oil saturation is 89% for well 29, undetermined for well 248, and 76.5% for well 380. Similarly the reserves, related to the area of 2 hectares over which the investigation will be carried out, differ considerably: 14,570 tons for well 29, 6450 tons for well 248, and 4900 tons for well 380.

Stratum 2 is developed in wells 248 and 380, where its thickness is 2.2 and 2.8 respectively, but it is missing to the north in well 29, where it is replaced by siltstone and argillites. The trapping properties of the sandstone vary. At well 248 the porosity is 25.5%, but at well 380 it is 19%, i.e., 6.5% less, and the oil saturation is 82.5% at well 248 and 80% at well 380. This means that the best trapping properties are in the center of the structure, around well 248. The reserves for the area of the experimental region are 0, 8815, and 7240 tons for wells 29, 248, and 380 respectively.

Stratum 3 is only developed in well 29, and is completely replaced to the south by dense rocks, siltstones, and argillites. Its trapping properties are poor; the sand is probably argillaceous and consolidated. The porosity is 16.2%, the oil saturation 66%, and the reserves 2150 tons for 2 hectares.

Stratum 4 is developed in all the wells of the experimental region. It is the thickest stratum and shows the best trapping properties as compared with the other strata. Thus, its thickness is about the same in all the wells intended for the experimental work, namely 9, 9.4, and 9.8 m for wells 29, 248, and 380 respectively.

This stratum shows the best trapping properties in well 29 (as is also the case with stratum 1). Here it has a porosity of 29.5% and an oil saturation of 88.5%. It should be noted that in well 248, occupying the central position in the anticline, the stratum is split up into three sections by two intercalations of dense rock, 0.8

Well No.	Total petroleum reserve in tons per 2 hectares area	Total withdrawal, tons	Withdrawal rate on March 1, 1962, tons/day	Connecting pipe diameter, mm	Perforated interval
29	56,720	112,533	57	12	All of strata
248	47,265	43,391	98	13	The same
380	47,141	14,710	55	9	" "

and 3.6 m thick. The porosity varies in the sections from 24 to 25.7%, and the oil saturation from 72.5 to 85.5%, i.e., by 13%, within the limits of the same stratum. The dense intercalations taper away within a short distance and do not isolate the three sections of the stratum from each other, so that the reserves can be calculated for the stratum as a whole, using mean values for the porosity and oil saturation, weighted with respect to thickness.

In well 380 there is also an intercalation of carbonaceous clay rock, 1.2 m thick. The porosity of the stratum is different above and below the intercalation, 26.2 and 21% respectively, but the oil saturation is much the same, 86.2 and 86.5% respectively. The reserves of this region for wells 29, 248, and 380 are 40,000, 32,000, and 35,000 tons respectively. The total reserves for all the strata, in the 2-hectare region, for wells 29, 248, and 380, are 56,720, 47,265, and 47,141 tons respectively. Thus, the steady petroleum reserve per hectare of the experimental region exceeds the mean reserve per hectare of the Arlansk field as a whole.

It should be noted that, in determining the parameters of the strata, geophysical data only were available to calculate the porosity and the oil saturation, because no cores were taken from the experimental region. There is no information as to the permeability from cores in the experimental region, but mean values of the permeability of the oil-saturated sandstone in the Arlansk field are 0.420 darcys for the upper strata and 1.240 darcys for the lower stratum.

The experimental region is located within a petroleum-bearing contour. The water saturation ceiling, as over the whole of the Arlansk field, is confined to the lowest stratum 4, and the upper layers are completely petroliferous over the whole area. The water ceiling below the Nagaevsk dome is at an absolute depth of −1178.6 to −1180 m, i.e., at 27, 34, and 33 m below the base of stratum 4 in wells 29, 248, and 380 respectively, and the planktonic part is located at distances of 500, 375, and 500 m from wells 29, 248, and 380 respectively at the western limb, and 875, 1000, and 600 m at the eastern limb.

The characteristics of the production strata of the experimental region, according to the data from investigation of the exploitation wells, are as follows:

Primary formation pressure in the experimental region (from well 29 at a depth of 1180 m, measured in 1958)	143 atm
Formation temperature	24°
Saturation pressure	83.0 to 89.0 atm
Gas factor	15 to 18 m^3/ton
Viscosity of petroleum under conditions in strata	13 to 16 cP
Mean permeability of strata (from pressure reduction curve recorded in wells 29, 248, and 380)	1.3 darcys
Conductance of strata	90 darcys
Productivity coefficient	6.0 tons/day · atm
Flowing formation pressure (measured March, 1962):	
In well 29	110 atm
In well 380	115 atm
Face pressure (measured March, 1962):	
In well 29	85 atm
In well 380	95 atm

In March, 1962 the wells under investigation were allowed to discharge with a face pressure somewhat greater than the saturation pressure. A total fall in formation pressure of 27 to 30 atm did not lead to any changes in the productivity and permeability of the strata. Pressure reduction curves, measured in 1958 and in 1962, gave the same value for the permeability. The minimum gushing pressure at the Arlansk field was found to be 85 to 90 atm, equal to the petroleum gas saturation pressure.

Well 29 was brought into use on January 26, 1958, well 248 on January 28, 1961, and well 380 on June 23, 1961. The table shows petroleum yield data and the quantities of petroleum withdrawn from each well, since exploitation started down to March 1, 1962.

It is clear from the table that, up to March 1, 1962, the amount of petroleum taken from well 29 was considerably greater than the calculated balance reserves for an area of two hectares, outlining the exploitation well of the experimental region. Thus, geological analysis of the structure of the experimental region leads to the following conclusions.

1. The experimental region is inside the contour of petroleum content.

2. There are four sandstone trapping strata in the well sections of the experimental region, and their parameters differ from well to well.

3. Strata 1 and 4 are the best defined. It would be possible to commence the experiments with strata 1 or 4. It is proposed to begin with stratum 4.

The petroleum reserves per 2 hectares area are 32,000 and 35,000 tons for wells 248 and 380 respectively, and the geological characteristics of these wells are sufficiently similar.

4. Strata in the vicinity of these wells are being exploited at pressures greater than the saturation pressures. The conditions for petroleum displacement from the strata will, therefore, be similar to the conditions for displacement in laboratory tests.

5. On completion of an experiment, the wells can be switched over from one stratum to another to carry out further work with surfactants. Finally they can be switched over to higher-lying carbonaceous levels for exploitation or for experimental work.

6. To evaluate the effectiveness of using surfactant, it will be necessary to drill two sets of wells in the vicinity of wells 248 and 380, on a five-point system. Water with surfactant will be injected into one set, and water without surfactant into the other. The proposed two sets of wells would make it possible to carry out the first systematic investigations and to determine the technological effectiveness of the use of surfactant. Further sets could possibly be drilled later for a more complete evaluation.

7. To estimate the effectiveness of using surfactant in the upper layer, it is proposed to drill one set of wells in a five-point system round well 29. These wells will penetrate the whole section but will be used only for the upper layer. Their boring will constitute the third section of the work. It is expected that the tests on the injection of surfactant into the lower layer, by means of the two previously mentioned sets of wells, will have been completed by the time that the experimental work on the third set of wellshas been finished; the wells of the two previous sets will then be switched over to the upper layer. Thus, the tests with surfactant in the upper layer will be carried out with three sets of wells, provided that the tests in the lower layer are positive. It is clear that stratum 1 should be used as experimental stratum for the upper layer, since it is intersected by all the wells. Thus, in the upper stratum there would be two locations with surfactant and one without.

8. It will be possible to determine the effectiveness of using surfactant if the latter is injected at the four sides of a well, and petroleum is taken off from the center.

Finally, by investigation of the wells it should be possible to establish a first approximate contour of the distribution of water and of the range of aqueous strata in the experimental region (with injections of both surfactant solution and water).

After carrying out the above program, it may be necessary, in order to assess the effectiveness of the use of surfactant, to drill appraisal wells with continuous coring.

Literature Cited

1. G. A. Babalyan, I. L. Markhasin, and others, The Development of New Methods for Increasing the Petroleum Yield from Strata by the Use of Surface-Active Agents, Report on Subject No. 24-61, UFNII Productions (1962).

A PROCEDURE FOR ESTABLISHING THE EFFICIENCY OF THE USE OF SURFACTANTS FOR INCREASING PETROLEUM RECOVERY

Ch. T. Khaliullin

The method used to establish the efficiency of the use of surfactants for increasing petroleum recovery during flooding inside the contour in the region of wells 29, 248, and 380 in the Arlanskii oil field is based on a comparison of petroleum and water extracted on an experimental assembly. In order to compare the results obtained by displacing petroleum with water and with a surfactant solution, it is necessary to create a minimum of two flooding centers in the stratum being studied. Water is pumped into one center and the surfactant solution into the other. The estimate of the efficiency of the surfactant will be greater, the greater the number of centers compared. The distribution of the flooding centers in the stratum and also the distribution of the wells in each center should be such that the geological and physical parameters of the stratum—porosity, permeability, water-saturation, and thickness—are identical within the region covered by the centers.

A five-point distribution of the wells in the flooding centers is recommended, i.e., one operation well in the center is surrounded in regular fashion by four forcing wells. This system of arranging the wells in the flooding center gives more reliable results in the estimation of the efficiency of application of surfactants compared with the two-point (one operational well and one forcing well in the flooding center) or four-point (one operational well surrounded in regular fashion by three forcing wells) systems. The four forcing wells in a system with five-point distribution of the wells in the flooding center make it possible to create around the operational well a region in which the petroleum is displaced by the water only into the operational well, while beyond this region the petroleum is displaced by the water away from this operational well.

We are thus faced with the problem of "cutting off" the flooding center from the rest of the stratum. The quantity of petroleum which could be extracted by the operational well from the flooding center obviously amounts to

$$Q_i^t = hS_i m(1 - \rho_{i.w}), \qquad (1)$$

where h is the thickness of the stratum; m the porosity of the stratum; $\rho_{i.w}$ the initial water saturation; and S_i the area in the flooding center from which the petroleum will be displaced towards the operational well.

The area S_i is found by constructing a map for the isobars and current lines in the center at the moment when flooding begins. In the construction of the map for the isobars in the flooding center of the stratum, allowance must be made for the interference from all the wells working in the stratum and for the wells in the flooding center. The construction of the map for the isobars in the flooding center of the stratum should be carried out in two stages.

In the first stage the isobar map for the proposed flooding center is constructed with allowance for the interaction of the surrounding wells only. For this purpose it is necessary to have information on the current stratum pressure on the basis of a study of the wells in the center, and also to have information on the end-face pressures of the wells surrounding the center. In the second stage the isobar map for the flooding center is constructed on the basis of the reaction of the wells in the flooding center only in a stratum of real dimensions.

In the second stage, the problem of constructing the map for the isobars in the flooding center can be solved with sufficient accuracy (for the given stratum parameters) either analytically or by constructing an electrical model, since at the initial moment in the region of the stratum being studied planar filtration of petroleum alone is assumed. The isobar maps obtained are used to construct an isobar map for the flooding center, which takes account of the interaction of all the working wells in the stratum, including the wells of the flooding center, at the moment when flooding starts by the method of superposition.

The isobars of both maps are combined algebraically, while the rates of filtration of the petroleum are combined geometrically. In this way we find S_i, the area of the region in which petroleum is displaced into the operational well of the flooding center at the moment when displacement starts. The area in which petroleum is displaced into the operational well of the flooding center decreases and tends to zero as the displacement front moves towards the operational well. Necessary conditions for the correct estimation of the ratio of the petroleum extracted from the center Q_i^p to the petroleum which could be extracted from the center Q_i^t are: continuity of the process of displacement of the petroleum by water and preservation of the given pressure drops in the flooding area and of the operational conditions in the neighboring wells until the center is completely flooded. The quantity of petroleum extracted from the center when flooding is complete can be represented as:

$$Q_i^p = hS_i m (1 - \varrho_{i.w} - \varrho_{r.o}), \tag{2}$$

where $\rho_{r.o}$ is the residual oil saturation.

The petroleum recovery coefficient of the flooding center is then found from the relationship

$$\eta = \frac{Q_i^p}{Q_i^t} = \frac{1 - \varrho_{o.w} - \varrho_{v.o}}{1 - \varrho_{o.w}}. \tag{3}$$

From a knowledge of the petroleum recovery from flooding centers when water and the surfactant solution are used, it is possible to determine the efficiency of the use of the surfactant solution for the displacement of petroleum.

While the stratum is being flooded, parameters such as the time taken for the displacement front to move from the forcing wells to the operational well in the center, and the nature of the flooding of production in the operational well, can also give some information regarding the influence of surfactants on petroleum recovery. These parameters only indirectly characterize the influence of surfactants on petroleum recovery from the stratum. Interpretations may be made and conclusions drawn from these data only after extensive industrial testing has been undertaken.

THE TECHNOLOGICAL INVESTIGATION OF PETROLEUM AND INJECTION WELLS WITH THE OBJECT OF DETERMINING THE EFFECTIVENESS OF THE USE OF SURFACTANTS TO INCREASE THE PETROLEUM YIELD

F. S. Abdulin and V. I. Portnov

The technological investigation of injection and petroleum wells involves detailed study of the trapping properties of the strata, in order to control and regulate the process of development of flooded strata, and determination of the effectiveness of all the technological processes associated with injection of water treated with surfactant.

The Problem of Determining the Effectiveness of Opening Up a Productive Stratum with a Surfactant

It is well known that when a productive stratum is opened up by drilling the drilling mud and its filtrate penetrate the facial zone. This can lead to a fall in the permeability of the stratum, and many wells are developed with a low productivity. For this reason, trials have been carried out on the opening up of productive strata with high-quality clay suspension with a low water loss, or using nonfiltering liquids based on petroleum [2].

Aqueous surfactant solutions have recently been used for this purpose. The result is to create a depression at the stratum, so that the water which has previously infiltrated is readily drawn out again into the well, and this preserves unchanged the characteristic parameters of the petroleum-bearing stratum [2].

Nowadays an estimate of the effectiveness of the opening up of a productive stratum, when various drilling liquids are used, is normally obtained by comparing the yield from a bored well with the yield from accompanying petroleum wells whose facial zones have similar geological characteristics, or, in favorable cases, by comparing the specific productivities of petroleum wells in the course of their exploitation. However, in the overwhelming majority of cases it is difficult to determine the productivity coefficients of petroleum wells, for various technological reasons, so that it is impossible to obtain an objective estimate of the effectiveness of any particular drilling liquid.

To estimate the effectiveness of the use of surfactant, the state of the facial zone of the stratum is best determined by comparing the permeability coefficient of a remote zone with that of the facial zone. A value for the permeability of the remote zone is given by the rectilinear part of the curve for the restoration of facial pressure, while the permeability coefficient of the facial zone is determined from the initial part of the curve with allowance for the influx of liquid, or from the productivity coefficient of the well.

The wells should be tested after steady conditions have been achieved, from shut-down to practically complete restoration of the facial pressure; this makes it possible to estimate the presence of lithological limitations to the stratum remote from the shaft of the well, and to determine a stratum pressure close to the true value. It is recommended that the wells should be tested using highly sensitive depth manometers.

Where possible, IPG-UFNII stratum test equipment should be used to investigate the wells. The results should be compared with the permeability estimated from the restoration-of-facial-pressure curve and the values determined from core samples.

Technological Investigation of Injection Wells

Technological investigations of injection wells are carried out to estimate the state of the facial zone of the stratum and to determine the effectiveness of using surfactant. Special observations are also made on dynamic changes in the acceptance profile, on coefficients of productivity and permeability during the processes of injecting aqueous surfactant solution and water without surfactant, on the curve for restoration of facial pressure after injection of surfactant into the well, and periodically during the process of injecting the water. The restoration in facial pressure curves should be obtained using standard recording manometers with an appropriate scale. The results of these tests are compared with the results of the original tests, carried out after the opening up of the stratum, and also with each other.

It is well known that the penetration of water into a stratum proceeds very unevenly because of heterogeneities in the production stratum, with respect to both thickness and extent, possible fissuring in the facial zone of the stratum, and considerable differences between the viscosities of the injected water and the formation petroleum. It is for this reason that a large amount of undisplaced petroleum may remain in the stratum or in separate intercalations. Thus, investigations on the injection wells should be carried out using facial flowmeters. This makes it possible at the proper time to work with a steady acceptance profile for the injection wells and, consequently, to increase the final petroleum yield of the stratum.

In order to determine the fissuring of the rock of the facial zone of the stratum it is necessary to investigate injection wells under the different conditions of injection and withdrawal, at a fixed steady facial pressure, and with simultaneous recording of the acceptance profile according to layers. From these results it is possible to construct indicator curves for the injection wells and for the separate layers of the stratum. With this object in view, it is useful to investigate injection wells by means of their restoration in facial pressure curves under the different conditions of injection and withdrawal. Allowing for the results of the acceptance profile investigations, a correction can be introduced in treating the injection well data obtained from the restoration in facial pressure curve. For observations on the dynamic acceptance profile, tests on the injection wells are preferably carried out without shutting down the well.

The quality and quantity of the injected water and the concentration of surfactant are checked by daily measurements of the consumptions of the injection wells with meters connected to the pumping station, by daily measurements of the injection pressure with standard manometers, and by daily sampling the water for determination of mechanical impurities, iron content, and surface tension. From this data it is possible to construct graphs for the operation of wells, with the application of curves for the changes in stratum and face pressure, the yields and productivity coefficients of the wells, the contents of mechanical impurities and iron in the water, and the surface tension of the latter.

In order to solve some of the problems associated with the adsorption of surfactant in the facial zone of injection wells, drilled down to the water-bearing stratum, it is useful to carry out special tests on the injection and withdrawal of 200 to 300 m^3 of surfactant solution as follows:

1) before the tests are carried out, the restoration in facial pressure curves of the injection wells are investigated, using a flowmeter;

2) an injection of 200 to 300 m^3 of surfactant solution, of appropriate concentration, is made into the facial zone of the stratum, and driven in by a quantity of water equal in volume to that of the compressor pump tubes and sump;

3) time is allowed for the surfactant solution to react with the rock; the time may be one to three or four days;

4) the well is opened for discharge, andsamples of liquid are taken at intervals for measurement of surface tension;

5) restoration in facial pressure curves are measured again, using a flowmeter, and the results are compared.

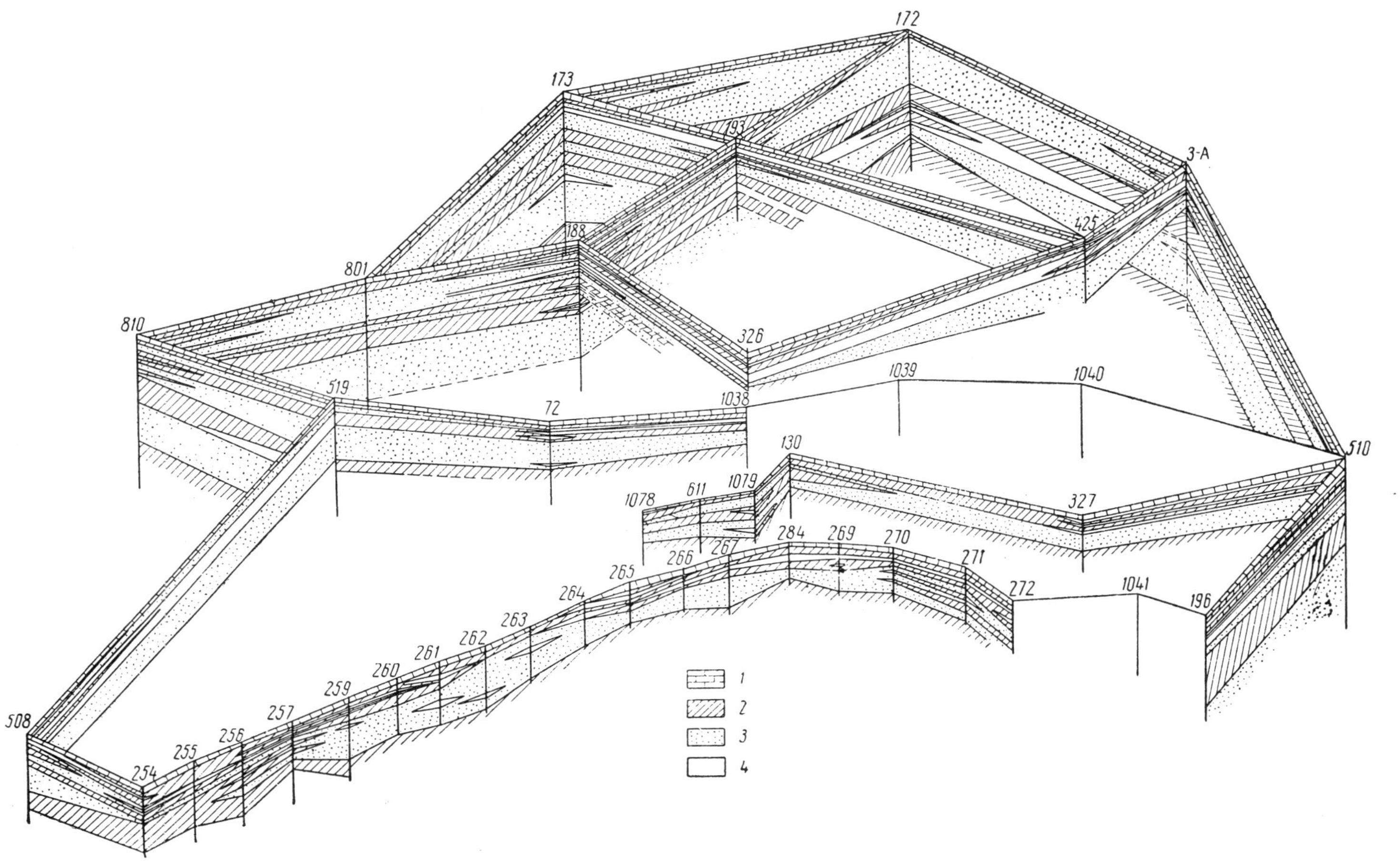

Fig. 1. 1) Limestone; 2) clay; 3) sandstone; 4) siltstone.

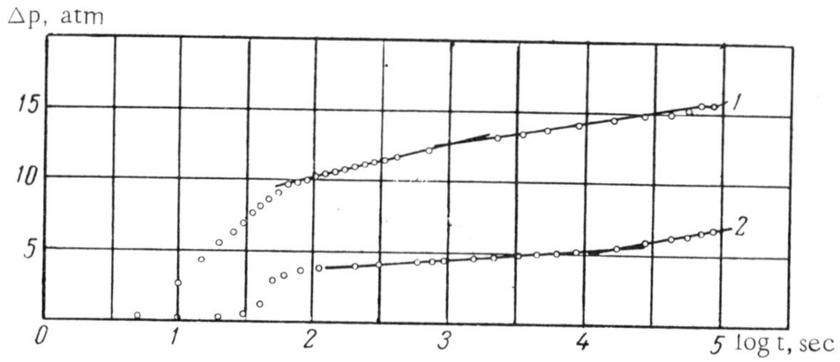

Fig. 2. 1) For well 193; 2) for well 172.

Technological Investigation of Petroleum Wells

As with injection wells, tests are carried out on the petroleum wells with the following objects in view: a) observations of changes in the permeability of the facial zone; b) determination of the behavior of all the seams and parts of the stratum; c) elucidation of the dynamics of changes in productivity coefficient and stratum pressure; d) observation of the appearance of water in the well and its content of surfactant. In order to check the effectiveness of the methods used for intensifying the petroleum yield, it is essential that regular determinations should be made of the parameters of the petroleum stratum, together with systematic measurements of the facial temperature in the wells. Periodic investigations should also be made of the restoration in facial pressure curves, using precision depth manometers of type DGM-4, remote reading manometers type DLMP2/M, and depth flow meters of UFNII design [4]. The results obtained can be shown in the form of graphs for changes in permeability, temperature, percentage of water in petroleum, and temperature with time.

Checking the Rates of Movement of Injected Water and the Development of the Stratum

The results of numerous investigations on Devonian injection wells have shown that most of the restoration in facial pressure curves have one rectilinear region. This shows that the trapping properties of the stratum are constant for a considerable distance around the injection well. However, there are also wells whose restoration in facial pressure curves are characterized by two or even three rectilinear regions. The permeability coefficients, calculated from the slopes of these regions, differ markedly from each other; the permeability coefficient decreases with increasing slope and increases with decreasing slope [4].

On the basis of analogous investigations [1], it has been suggested that the character of the restoration in facial pressure curve reflects the effect either of various lithological changes in the production stratum, or of differences between the viscosity of water and petroleum under the stratum conditions. More recent technological investigations have shown that if with repeat measurements of the curve the points of intersection of the rectilinear regions move to the right, they are due to a difference in the viscosity of the liquid moving through the stratum. If the point of intersection of the rectilinear regions of the curve always remains in the same position, there is some sort of lithological discontinuity in the stratum [1].

In order to decide which of these factors controls the character of the restoration in facial pressure curve, it is necessary to construct a geological profile, or block diagram, along the length of the wells investigated, or to carry out repeat investigations on these wells after a prolonged injection. Investigations on the heterogeneity of the stratum and on the movement of injected water were carried out in Bashkiria by NPU Tuimazaneft' in 1954, and by NPU Arlanneft' in 1959 [4]. Investigations in the same field were also carried out on the interaction between wells by the sonic method. Profiles of acceptance and productivity were also constructed on the basis of investigations with an RGD-3 well flowmeter.

TABLE 1

Well No.	Date of test	Thickness of productive stratum at well section	Q = yield or consumption of well before test, m^3/day	First slope	k_1 = permeability from slope, darcys	Second slope	k_2 = permeability from slope, darcys	Third slope	k_3 = permeability from slope, darcys	Fourth slope	k_4 = permeability from slope, darcys	Operating conditions before shutdown
172	Oct. 8-9 1954	16.8	410	0.785	0.66	1.84	0.28	–	–	–	–	Withdrawal
193	Aug. 7-8, 1954	5.2	146.2	2.50	0.25	1.47	0.41	–	–	–	–	"
406	Sept. 4-5, 1954	9.2	87	1.70	0.12	3.33	0.06	3.6	0.05	–	–	"
406	Oct. 15-16, 1954	9.2	62.2	1.26	0.12	2.26	0.06	3.0	0.05	–	–	"
406	Oct. 23-26, 1957	9.2	365	8.60	0.10	12.50	0.07	16.05	0.05	10.5	0.08	Injection
161	July 12, 1953	8.3	86	0.525	0.62	–	–	–	–	–	–	Swabbing
161	June 29, 1954	8.3	67	0.427	0.60	0.95	0.27	–	–	–	–	Withdrawal
161	Aug. 15-19, 1958	8.3	415	1.90	0.56	3.94	0.28	–	–	–	–	"
161	June 24-29, 1956	8.3	480	2.15	0.57	4.45	0.28	–	–	–	–	"

Technological Observations Carried Out at the Bashneft' Field Injection Wells of NPU Tuimazaneft'

Well 172 is located in the northeastern part of the Tuimazinsk beds (Fig. 1). The productive stratum DI is a sandstone, 16.8 m thick. Development of the well was begun in August 1949, using all the contemporary methods: prolonged swabbing, torpedoing the face with large missiles, repeated acid treatment, etc. At the start, a well usually showed a high acceptance of 1000 to 1600 m^3/day. However, after 15 to 20 days, this dropped to 150 to 250 m^3/day.

The well was tested in October 1964,by the method of restoration in pressure curves (Fig. 2) The curve shown in Fig. 2 has two clearly defined linear regions. Calculations, presented in Table 1, showed that the productive stratum in the vicinity of well 172 consisted of two zones with different permeabilities. The first zone was characterized by a permeability of 600 mD (millidarcys), while the second zone, farther from the well, had a permeability of only 280 mD. Another well, 193, at a distance of 1500 m from well 172, also gave a restoration-in-facial-pressure curve with two well defined rectilinear regions. However, here, in contradistinction to well 172, the first region had a steeper slope than the second.

Comparison of the calculated results for the two wells showed that the permeability coefficients were similar for the first (nearest) part of the stratum around well 193 and for the first part of the stratum around well 172. Judging by the times for the restoration of facial pressure on both curves, it is possible to make an estimate of the dimensions of both zones around these wells. The values for the better zone and for the worse zone of the stratum, deduced from the two curves, coincide.

From the block diagram given in Fig. 1 it is clear that the thickness of the sandstone decreases in all directions away from well 172 (see wells 173, 193, 326, 1040, and 3-A).

Comparison of the curves for injection and withdrawal at well 172 showed that they were similar in shape; both yield and consumption were high initially, but then dropped rapidly. The results indicated that well 172 was in the center of a productive stratum, which was lenticular in shape. The initially high consumptions of the well, observed in repeated tests, could clearly be explained by supposing that a closed reservoir was first filled with injected water by elastic storage in the stratum rock and in the liquid saturating it. Because of this characteristic of the stratum, application of fracturing to well 172 did not give a positive effect.

In view of these conclusions, the application of all possible measures to increase the acceptance was abandoned at well 172, except for drainage of the facial zone by the method of repeated short gushes. Up to the present, this well has been utilized with a mean acceptance of 200 to 250 m^3/day.

Well 406. The facial zone of the productive stratum at well 406 shows an alternation of sandstone with clay intercalations (Fig. 3). The well was brought into use after a double hydraulic fracturing of the stratum in 1954, with a mean acceptance not exceeding 300 m^3/day. It has continued to give this acceptance up to the present time.

Starting in 1954, the well was tested by the method of the restoration in facial pressure curve. Three tests were made: two withdrawal tests with an interval of 1.5 months in between them, and one injection test three years later (Fig. 4). Curves 1 and 2, obtained on the basis of observations on pressure restoration during single days, both have three well defined rectilinear regions. However, curve 3, obtained while the well was shut down for 17 days, shows a fourth rectilinear region.

Examination of the curves shows that the points of intersection of the rectilinear regions occur at the same times. The values of the permeability, calculated from the slopes of the rectilinear regions, also coincide (Table 1). Thus, in this case it cannot be that differences in the viscosity of oil and water are responsible for the appearance of regions with various slopes on the curves, and, in all probability, the character of the restoration in pressure curves is due to heterogeneity of the stratum. This is confirmed by the block diagram of Fig. 3, which shows pronounced tapering of the upper sandstone seams in all directions, and, on the other hand, improvement of trapping properties in the lower seam.

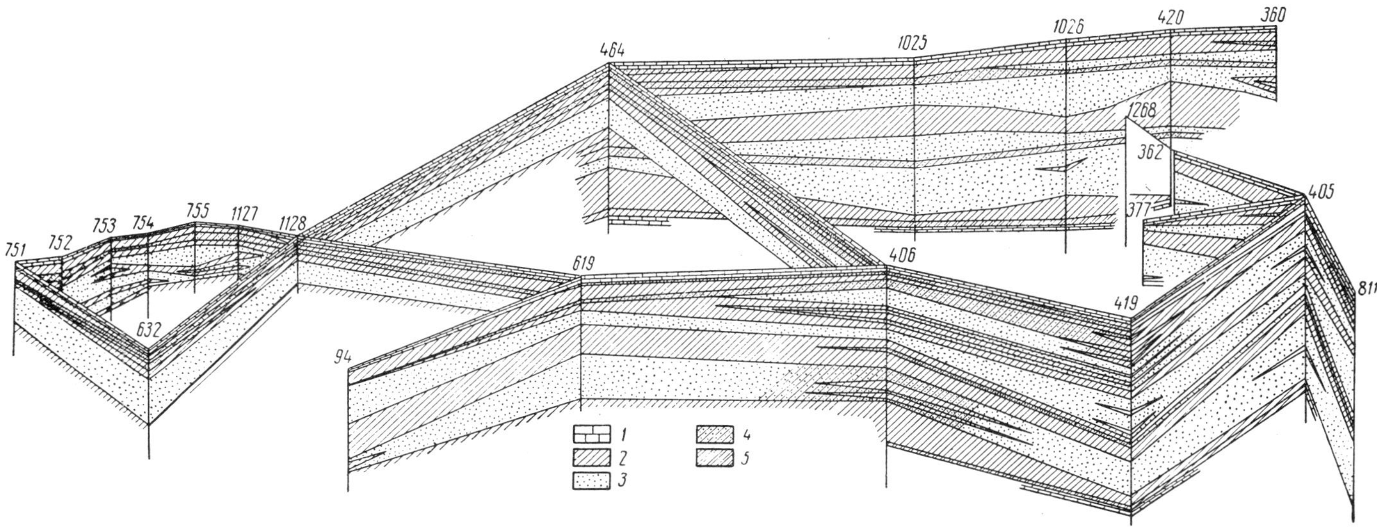

Fig. 3. 1) Limestone; 2) clay; 3) sandstone; 4) siltstone; 5) marl.

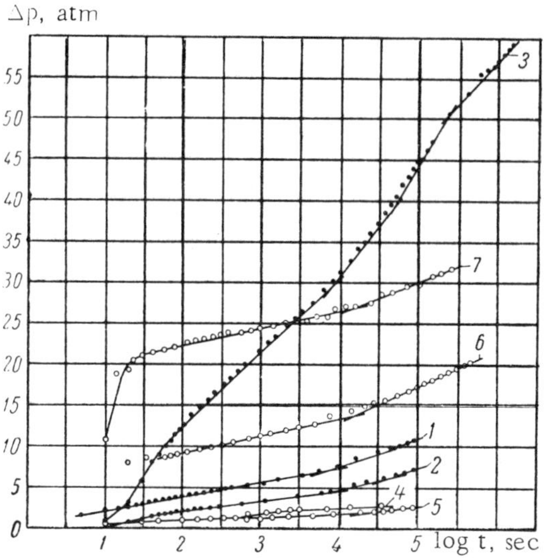

Fig. 4

Curve No.	Date of test	Operating conditions before shutdown
	For well 406	
1	Sept. 4-5, 1954	Withdrawal
2	Oct. 15-16, 1954	"
3	Oct. 23-26, 1957	Injection
	For well 161	
4	July 12, 1953	Swabbing
5	June 29, 1954	Withdrawal
6	June 24-28, 1956	"
7	Aug. 15-18, 1958	"

On this basis, we can explain the mechanism of formation of the various rectilinear regions in the curves for well 406. At the moment when the well is closed, a redistribution of the stratum pressure begins simultaneously in all the seams and is completed after various times, depending on the distances between the tapering of the separate seams and the well axis.

The reduction in slope of the fourth recilinear region, as compared with the second and third such regions, can be explained by the development of better trapping properties in the stratum in the parts furthest from the injection well. This is clear from the block diagram; there is an increase in the thickness of the lower sandstone seam in the direction of wells 464, 1025, 419, 619, 94, 1128, and 632.

Well 161. A similar picture was observed with well 161 (Fig. 5), where the curves for restoration of facial pressure, obtained as the results of repeated tests over a time interval of five years, showed two rectilinear regions. As in the case of well 406, the boundaries of these regions were at the same times (Fig. 4). The permeability coefficients, calculated from the corresponding slopes of the rectilinear regions, were practically the same (Table 1). The behavior of well 161 was precisely the same as that of well 172. Figure 5 shows that the well is located where there is a change in the trap parameters. The absence of a second rectilinear region from curve 4, obtained by swabbing, was the result of premature completion of the test.

Wells 81 and 769. On the basis of the restoration in pressure curves, there was complete tapering of the productive stratum round the injection wells 81 and 769. These wells were therefore listed as nonproducing for geological reasons.

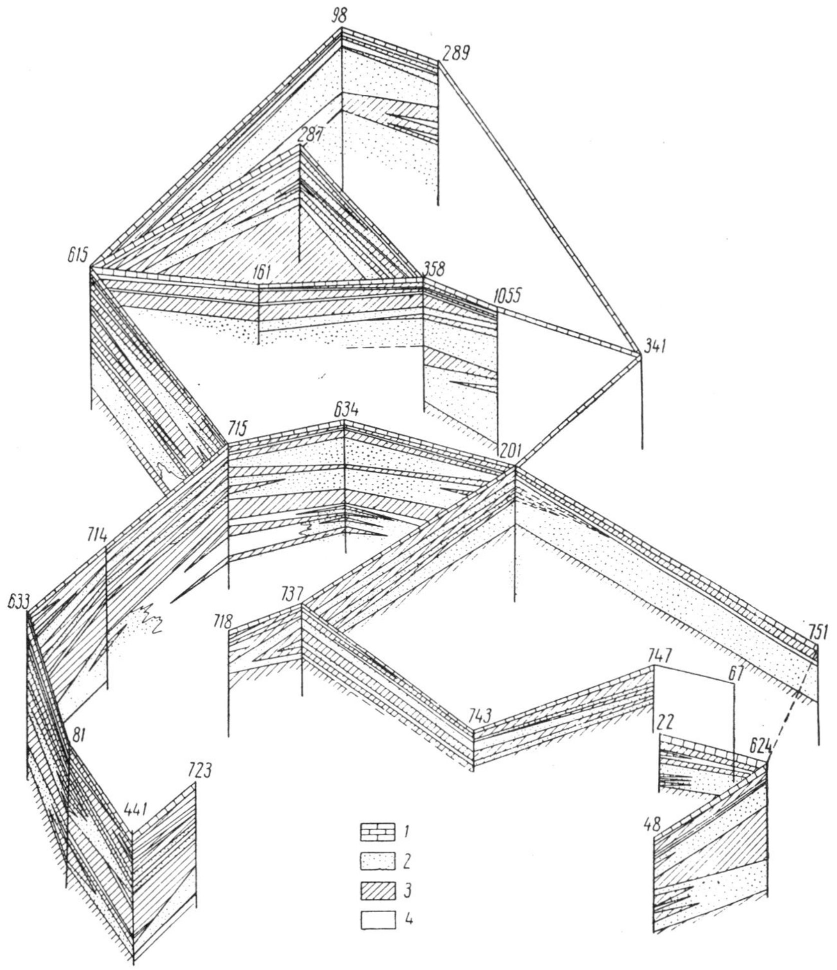

Fig. 5. 1) Limestone; 2) sandstone; 3) clay; 4) siltstone.

The above examples show that, by means of periodical tests by the curves for the facial pressure reduction method, it is possible to make an approximate determination of the presence of lithological restrictions remote from the well axis.

Injection Wells of NPU Arlanneft'

Well 14. The productive stratum or injection well 14 consists of two sandstone seams, of total thickness 4.8 m, which have been opened up by 440 PK-103 shot. The well was opened up in 1959 with a capacity of 1000 m^3/day at 90 atm, but after 1 to 1.5 months the acceptance had dropped to 100 to 150 m^3/day and the well was opened up again.

The well was tested five times between November 1959 and July 1960, under different pumping-out ditions. All the curves for restoration of facial pressure had two well defined rectilinear regions (Fig. 6). However, the times for the points of intersection moved to the right with time. It is clear that the character of the restoration-in-facial pressure curves was not controlled by a lithological restriction in the stratum, but by the difference in viscosity between water and the oil which saturated the stratum, which became more significant as the stratum zone occupied by water increased around the well. Table 2 shows the results of calculations of the permeability from the first rectilinear regions of the restoration-in-facial pressure curves, with allowance for the viscosity of the issuing water. It is clear from Table 2 that the various values almost coincide, and that the mean permeability is 200 to 220 mD. However, the permeabilities calculated from the second rectilinear regions of the curves, with the same correction for viscosity, are less by a factor of 5 to 10. The viscosity of the stratum petroleum from the Arlansk field is 15 to 20 cP.

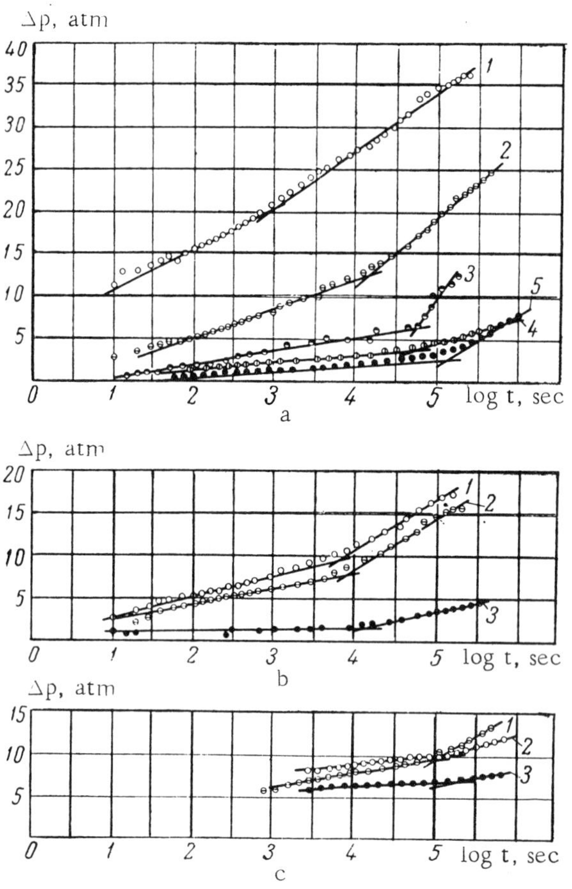

Fig. 6. a) For well 14. The dates of the tests were: 1) Nov. 5-9, 1959; 2) May 22-28, 1960; 3) June 29, 1960; 4) Sept. 22-30, 1960*; 5) July 6-18, 1960. b) For well 221. The dates of the tests were: 1) Feb. 6, 1960; 2) April 1, 1960; 3) Sept. 14, 1960. c) For well 11. The dates of the tests were: 1) Sept. 15, 1960; 2) July 1-10, 1961. For well 254: the date of the test was: Oct. 14, 1961.

TABLE 2

Date of test	Flow prevailing before shutdown m^3/day	k = Permeability coefficient of stratum, darcys		
		k_1 at $\mu = 1$ cP	k_2 at $\mu = 1$ cP	k_3 at $\mu = 16$ cP
Nov. 5-9, 1959	864	0.23	0.063	
May 22-28, 1960	584	0.24	0.090	
June 29 to July 1, 1960	45	0.23	0.022	0.35
Sept. 22-30, 1960	122	0.24	0.056	
July 6-18, 1961	71.3	0.20	0.022	0.35

* There seems to be an inconsistency between this date and information in the text. The date has been copied directly from the Russian. – Ed.

TABLE 3

Date of test	Flow rate established, m^3/day	Productivity coefficient, m^3/day · atm	Permeability, darcys	
			From first region	From second region
Feb. 6, 1960	447	25.8	0.247	0.096
April 1, 1960	257	16.2	0.242	0.072
Sept. 14, 1960	40	8.6	0.253	0.034

If the viscosity of the stratum petroleum (16 cP) is used in the calculation, then the permeability obtained for the second part of the withdrawal zone approximates that derived for the first part. This confirms the assumption that there is an increase in the size of the zone occupied by water as the result of an advance in the injection front. It is clear that the initial high acceptance of the well can be attributed to the filling of this zone with water, as the result of elasticity of the rock stratum and liquid and that the low acceptance prevailing later is due to the slow movement of the viscous petroleum away from the well.

In considering the curves for the restoration of facial pressure at well 14, it should be noted that in the first test of Nov. 5-9, 1959 the rectilinear region rapidly altered its direction. There was practically no transition zone around the point of intersection. The obvious explanation is that initially there was hardly any transition zone between the aqueous and petroliferous parts of the stratum, but that an intermediate zone was produced later by mixing, or by uneven movement of the water injection front caused by heterogeneity in the permeability of the stratum across its thickness. In our view, this also clearly explains the variable values for the permeability coefficient, calculated from the second rectilinear regions of the curves.

Well 221. The productive stratum consists of two monolithic seams of sandstone at 1283.9 to 1289.2 m and 1292 to 1303.2 m. The well was cased with 168-m pipe and perforated over the interval 1284 to 1289 m with 150 PK-103 shot, and over the interval 1293 to 1303 m with 330 PK-103 shot. The casing was sealed off at 1328 m; swabbing of the well was started in April, 1960. With a water density of 1.18 g/cm^3 the static level was at 90 m below the surface.

The level dropped only to 135 m below the surface during the swabbing process. A test on the well showed an initial high acceptance of 600 to 700 m^3/day at p = 100 atm, but this dropped to 100 to 150 m^3/day in the course of 10 to 15 days. As with injection well 10, the volume of water injected into the stratum was equal to the volume withdrawn from the well. The well is now classified as nonproductive. It has been tested three times by the method of constructing a restoration-in-facial-pressure curve during withdrawal (Fig. 6).

It is clear from Fig. 6 that all the curves have two well defined rectilinear regions. The intersection points of these curves also shift to the right. The permeability coefficients calculated from the first rectilinear regions are all the same, but those calculated from the second rectilinear regions are less by a factor of 2.5 to 7 (Table 3).

It is clear that, in this case, the character of the restoration-in-facial-pressure curves was affected by movement of the injected water front in the stratum. There is therefore a sound basis for returning to this well and attempting to start it up by injection. The need for further opening-up of the well is demonstrated by the coring diagram data, the investigations on the acceptance and productivity profiles, and by observations on the productivity coefficient of the well.

Conclusions

For an objective appraisal of the effectiveness of applying surfactant to a part of a stratum it is necessary to carry out a complex study of the wells. The first step should be to make a geological and geophysical investigation of the state of the facial zones of injection and production wells while they are being opened up and brought into use.

It is recommended that orientative experiments on the adsorption of surfactant should be carried out on individual wells of new oil fields. This involves injection of a definite volume of surfactant solution into the facial zone over a period of one to three days. A strict check on the change in surfactant concentration of the solution must then be carried out during the subsequent withdrawal period.

After a careful study of the state of the facial zone of the stratum, further investigations should be made by the method of restoration-in-pressure curves. The results give a qualitative estimate of the movement rate of the water–petroleum boundary, and they also show the presence of any well defined lithological changes in the more remote zone of the stratum. Use of the sonic method, with highly sensitive piezographs, may give a qualitative picture of the movement of the water–petroleum boundary in various directions.

After flooding of the petroleum wells, it is necessary to drill appraisal bore holes, between the injection and production wells, to get continuous core samples from the productive part of the stratum. These are required to establish how completely the petroleum has been displaced by water, with and without surfactant.

Literature Cited

1. F. S. Abdulin, Handbook on the Investigation of Injection Wells by Means of the Restoration in Facial Pressure Curve (Ufa, 1958).
2. G. A. Babalyan, I. I. Kravchenko, I. L. Markhasin, and G. V. Rudakov, The Physicochemical Basis of the Use of Surface-Active Agents in the Development of Petroleum-Containing Strata (Gostoptekhizdat, 1962).
3. V. A. Blazhevich and F. S. Abdulin, The Improvement of Methods for Increasing the Capacity of Injection and Petroleum Wells, Report on Subject No. 33, UfNII Productions (1960).
4. V. I. Portnov, Instruments for Use at Depth and the Results of Investigation of the Petroleum Wells of the Bashkirian Field from 1945 to 1960, Report on Subject No. 56-60, UFNII Productions.

LABORATORY STUDIES ON THE CHOICE OF THE MOST PROMISING SURFACTANTS FOR INCREASING PETROLEUM RECOVERY FROM STRATA

I. I. Kravchenko, N. A. Rybinskaya, K. V. Sharina, and V. A. Brovenko

Surfactants added to the water used for flooding strata should satisfy the following requirements [4]:

1) they should be completely soluble in fresh and stratal waters;

2) they should decrease considerably the surface tension at the water–petroleum interface at low concentrations;

3) they should increase the wettability of the rock by water in the presence of petroleum;

4) they should undergo minimum adsorption on the rock surface;

5) they should prevent the formation of a hydrophobic emulsion in the stratum;

6) they should improve the parameters of the displacement of petroleum by water;

7) they should be inexpensive.

Solubility

Aqueous surfactant solutions pumped into the stratum are diluted by the stratal water with which they come into direct contact. The stratal waters of the petroleum deposits in the Urals-Volga oilfield are in most cases hard calcium-chloride waters containing considerable quantities of dissolved salts: NaCl, $CaCl_2$, $MgCl_2$, etc. The solubility of surfactants in these waters is much less than that in fresh water. Moreover, many surfactants may react with the salts in the stratal water to give insoluble precipitates which may reduce the permeability of the stratum. Surfactants used to increase the recovery of petroleum from strata should therefore dissolve completely in fresh and stratal waters.

The solubility of surfactants in these waters and also in kerosene has been determined visually in solutions of different concentration (from 0.01 to 1%) at room temperature.

Anionic surfactants:

a) alkyl sulfates–sulfated esters of aliphatic acids (methyl, butyl, and isobutyl sulforicinoleates), the alkyl sulfates of saturated alcohols–the wetting agents "Novost'" and "Progress";

b) alkyl sulfonates–sulfonate;

c) alkylaryl sulfonates – Sulfonol NP-1, Azolyat A, Azolyat B, and Sodium Detergent DS;

d) petroleum aromatic and cyclic sulfonic acids and their salts–Petrov Contact and NChK.

These surfactants are completely soluble in fresh water, almost insoluble in kerosene, and give flocculent precipitates in stratal water and in mixtures of stratal water with fresh water. These precipitates contain chiefly $CaSO_4$, together with the calcium and magnesium salts of the sulfonic acids present in these surfactants.

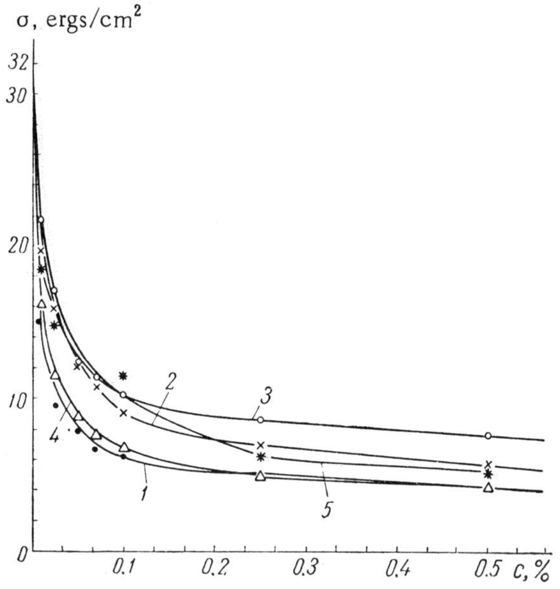

Fig. 1. Isotherms for the surface tension of aqueous solutions of oxyethylated nonionogenic surfactants. 1) OP-10; 2) OP-20; 3) OP-45; 4) OP-7; 5) $KAUFE_{14}$.

Cationic surfactants:

a) amines—Amine C_I, Amine C_{II}, and Amine C—are insoluble in both fresh water and stratal water but dissolve completely in kerosene;

b) 50% solutions of trimethylalkylammonium chlorides and dimethyldialkylammonium chlorides (alkyl 14-18C) in isopropyl alcohol—Arquads S, T, and 2C—dissolve completely in kerosene and in stratal and fresh waters;

c) triethylalkylbenzylammonium chloride—Katamin A, p-alkylbenzylpyridinium chloride (alkyl 16-18C)—Katapin A, the quaternary ammonium salts of the diethylaminomethyl derivatives of the polyethyleneglycol ethers of isooctylphenols—Leveling Agent A—dissolve completely in fresh and stratal waters and are insoluble in kerosene, with the exception of Karbozolin S, which is only partly soluble in stratal water;

d) the quaternary ammonium salts of the diethylaminomethyldiglycolates of a mixture of saturated and unsaturated aliphatic alcohols—Alkamon OS-2 and Alkamon OS-2N—form colloidal solutions in fresh and stratal waters and dissolve completely or partly in kerosene.

Nonionogenic surfactants:

a) oxyethylated phenols and alkylphenols UFE_8, OP-4, OP-7, OP-10, OP-20, OP-45, and $KAUFE_{14}$, with the exception of OP-4, are completely soluble in fresh and stratal waters and insoluble in kerosene. OP-4 is completely soluble in kerosene, insoluble in stratal water,and gives a colloidal solution with fresh water;

b) oxyethylated alcohols: $S_{10}E_{8.3}$, OS-20, $S_{14-16}E_{11}$, $S_{14-16}E_{18.5}$, $KOSE_{15.5}$, $KOSE_{28.6}$, $S_{18}E_{25}$, and $S_{18}P_{27}$—are completely soluble in fresh water and almost insoluble in kerosene. $S_{10}E_{8.3}$, $KOSE_{28.6}$, $C_{18}E_{25}$, and $C_{18}P_{27}$ dissolve completely in stratal water; the others dissolve only partly;

c) oxyethylated aliphatic acids—Stearox-6, OKO, OZhK, and $K_{10-16}E_6$, with the exception of Stearox-6, are partly soluble in fresh and stratal waters and insoluble in kerosene. Stearox-6 is completely soluble in kerosene and almost insoluble in fresh and stratal waters;

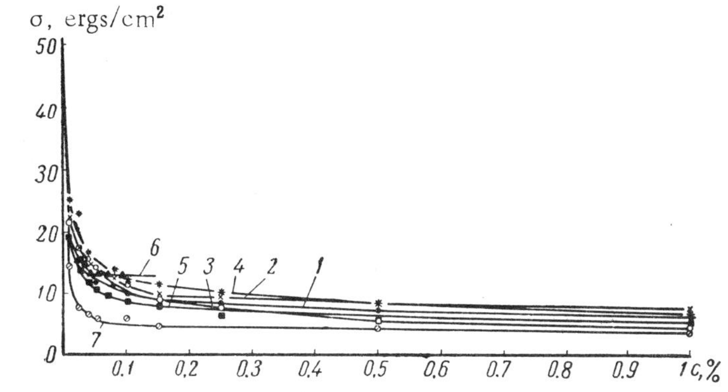

Fig. 2. Isotherms for the surface tension of aqueous solutions of oxyethylated alcohols, Ethomide HT/60, Sulfamide OE-10, and oxyethyl derivatives of xylitol. 1) $S_{15-18}P_{27}$; 2) $S_{15-18}E_{25}$; 3) $S_{10}E_{8\ 3}$; 4) $KOSE_{28.6}$; 5) KS-59; 6) Ethomide HT/60; 7) Sulfamide OE-10.

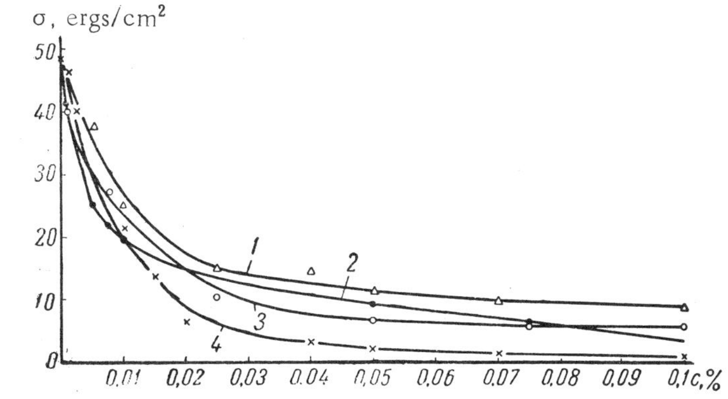

Fig. 3. Isotherms for the surface tension of aqueous solutions of cationic surfactants. 1) Leveling Agent A; 2) Katamin A; 3) Katapin A; 4) Karbozolin O.

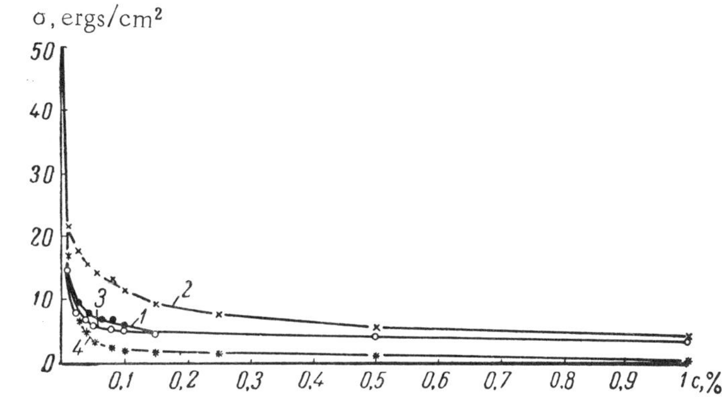

Fig. 4. Isotherms for the surface tension of aqueous solutions. 1) OP-10; 2) $S_{10}E_{8.3}$; 3) Sulfamide OE-10; 4) Karbozolin O.

NOTE: ALL THESE VALUES ARE INTERFACIAL TENSIONS

TABLE 1

Surfactant	Experimental conditions	θ, deg						
		Concentration, %						
		0	0.025	0.05	0.1	0.25	0.5	1.0
OP-10	Medium–kerosene, drop–surfactant solution, surface–quartz	28	28	27	22	21	17	15
Oxamid ST-15		–	29	27	24	22	20	17
KS-59 (OE-10.5)		–	34	33	32	27	22	21
OZhK		–	34	26	23	25	22	20
$KOSE_{28.6}$		–	28	26	19	14	15	15
OKO		–	30	28	25	22.5	17.5	18.5
$S_{10}E_{8.3}$		–	31	29	24	21	17	11
$KAUFE_{14}$		–	29	19	19	17	16	15
UFE_8		–	43	43	25	22	21	19

d) oxyethylated amides–Ethomide HT/60, Oxamid ST-15, and Sulfamide OE-10 – are completely insoluble in kerosene and dissolve in fresh and stratal waters, with the exception of Oxamid ST-15, which is only partly soluble in stratal water;

e) oxyethylated derivates of xylitol–Xylital S-15 and KS-59–are completely insoluble in kerosene. KS-59 is completely soluble in fresh and stratal waters but Xylital S-15 is only partly soluble.

Thus the solubility data indicate that the following surfactants are promising for increasing petroleum recovery from a stratum.

1) Nonionogenic:

a) all oxyethylated phenols and alkylphenols with the exception of OP-4;

b) the oxyethylated alcohols–$S_{10}E_{8.3}$; $KOSE_{28.6}$, $S_{15-18}E_{25}$, and $S_{15-18}P_{27}$ (oxypropylated);

c) the oxyethylated amides–Ethomide HT/60 and Sulfamide OE-10;

d) the oxyethylated derivatives of Xylitol KS-59.

2) Cationic: Katamin A, Katapin A, Leveling Agent A, and Karbozolin O. To these might also be added the three Arquads studied, but these substances are not yet produced in the Soviet Union.

Surface Tension

The surface tension was determined by measuring the volume of drops produced on a micrometric apparatus (the UFNII instrument). The surface tension σ was measured at the interface between kerosene and an aqueous solution of the surfactant. The surface tension of the kerosene at the interface in contact with distilled water was equal to 47 ergs/cm^2. In addition, the surface tension of some aqueous surfactant solutions was measured at the interface between these solutions and Arlanskii petroleum.

Figure 1 gives the isotherms for the surface tension of aqueous solutions of oxyethylated alkylphenols at the interface between these solutions and petroleum, Fig. 2 gives isotherms for the surface tension of aqueous solutions of other types of oxyethylated surfactant at the interface between these solutions and kerosene, and Fig. 3 gives isotherms for aqueous solutions of cationic surfactants at the interface between these solutions and kerosene.

Comparison of the isotherms for the surface tension of the most active surfactants in each group (Fig. 4) showed that the greatest decrease in surface tension at low concentrations is given by OP-10 and Karbozolin O.

TABLE 2

Surfactant	Experimental conditions	θ, deg						
		Concentration, %						
		0	0.025	0.05	0.1	0.25	0.5	1.0
Katapin A	Medium–kerosene,	20	64	73	122	134	129	129
Katamin A	drop–aqueous	–	72	86	78	134	116	114
Karbozolin O	surfactant solution	–	151	146	144	143	142	–
Karbozolin S		–	155	154	163	163	156	150
Leveling Agent A		–	44	41	49	54	61	61

Selective Wetting

The contact angles of wetting θ were determined on an OSK-2 optical bench. The contact angles of wetting for solutions of nonionogenic surfactants on quartz are given in Table 1. The contact angle was measured 1-10 min after the drop had been applied to a grease-free quartz surface. In most cases the wetting angles reached equilibrium values after 10 min.

The contact angles of wetting for solutions of cationic surfactants on a quartz surface are given in Table 2. It can be seen that the wetting of a quartz surface by solutions of nonionogenic surfactants is more effective than that for solutions of cationic surfactants.

The Adsorption of Surfactants from Aqueous Solutions on Solid Adsorbents

The adsorption of surfactants on the solid rock surface plays an important part in the displacement of petroleum by aqueous surfactant solutions.

A surfactant adsorbed from aqueous solutions on solid surfaces gives to these surfaces certain properties which lead to an increase in petroleum recovery. Only part of the adsorbed surfactant is used to increase petroleum recovery, however; adsorption at first plays a positive part by increasing petroleum recovery, but subsequently adsorption continues without an increase in petroleum recovery [1]. Thus the surfactant which is adsorbed after the process of petroleum recovery has become stabilized has no positive action but merely increases the cost of the process.

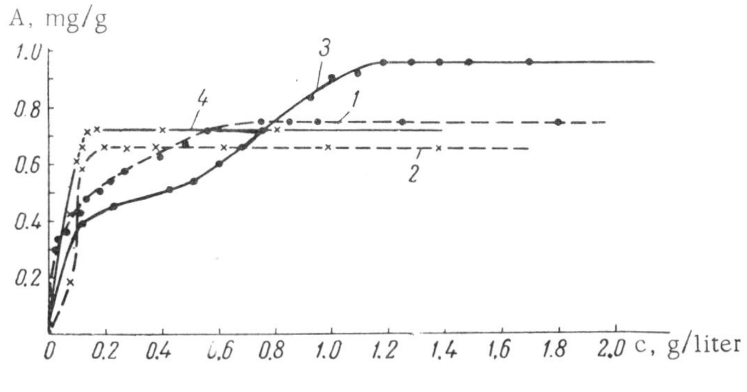

Fig. 5. Isotherms for the adsorption of OP-10 and Oxamid ST-15, obtained gravimetrically and by the surface tension method. 1) OP-10, gravimetric method; 2) OP-10, surface tension method; 3) Oxamid ST-15, gravimetric method; 4) Oxamid ST-15, surface tension method.

TABLE 3

Adsorbent	Specific surface, cm^2/cm^3	Fractional composition, %				
		1-0.5 mm	0.5-0.25 mm	0.25-0.1 mm	0.1-0.06 mm	< 0.06 mm
Arlanskii sandstone	—	0.6	4.0	53.7	25.6	16.1
Quartz sand	7000	—	3.0	57.0	24.5	15.5

Note. The quartz sand fraction < 0.6 gave the following fractions after elutriation:

TABLE 3a* (%)

0.5-0.25	0.25-0.1	0.1-0.05	0.05-0.01	0.01-0.005	0.005-0.001	< 0.001
0.03		0.67	66.2	16.4	11.46	4.7

*From data obtained in the Lithology Laboratory of the UFNII.

Procedure

The solid and liquid phases were taken in the weight ratio 1:3. The series of experiments carried out showed that equilibrium adsorption is reached in 48 h on quartz sand, Arlanskii sandstone, limestone, and dolomite.

In our experiments the surfactant solutions were stirred with the solid adsorbents every day for 3 min.

The adsorption isotherm was constructed on the basis of surface tension isotherms determined before and after adsorption with solutions of known concentration. These isotherms were used to find the equilibrium concentrations for a number of points. The change in the concentrations per unit weight of adsorbent gave the value of the adsorption for each point. The values were used to construct a graph giving the dependence of the adsorption on concentration.

$$A = \frac{(c_1 - c_e)\,V}{p}, \tag{1}$$

where A is the quantity of adsorbed substance for a given concentration, mg/g; c_1 is the concentration before adsorption, g/liter; c_e is the equilibrium concentration after adsorption, g/liter; V is the volume of surfactant solution, ml; and p is the weight of adsorbent, g.

It should be noted that the surface tension method for determining concentrations gives accurate results only with chemically pure compounds. For most nonionogenic and cationic surfactants it gives only approximate results, since these surfactants usually consist of a mixture of more or less homogeneous components and also contain a certain quantity of impurities. For most anionic surfactants this method is quite unsuitable [1].

The determination of concentrations by the surface tension method is also influenced by micelle formation: in the case of micelle formation, the surface tension isotherm will lie above the molecular isotherm.

Although the surface tension method is approximate, it is the least tedious and is easily carried out. In order to estimate the accuracy of this method of determining the concentration of nonionogenic surfactants in solutions before and after adsorption, we carried out experiments on the quantitative analysis of these surfactants by the method involving the formation of the molybdophosphate complex (gravimetric method). Study of a number of nonionogenic surfactants showed that the discrepancy in the determination of the concentrations by the two methods does not exceed 5-10% (Fig. 5).

The fractional composition of Arlanskii sandstone was chosen on the basis of the average data obtained in the granulometric analysis of a large number of core samples from the Arlanskii deposit. The fractional composition of the quartz sand was close to the fractional composition of the Arlanskii sandstaone (Table 3).

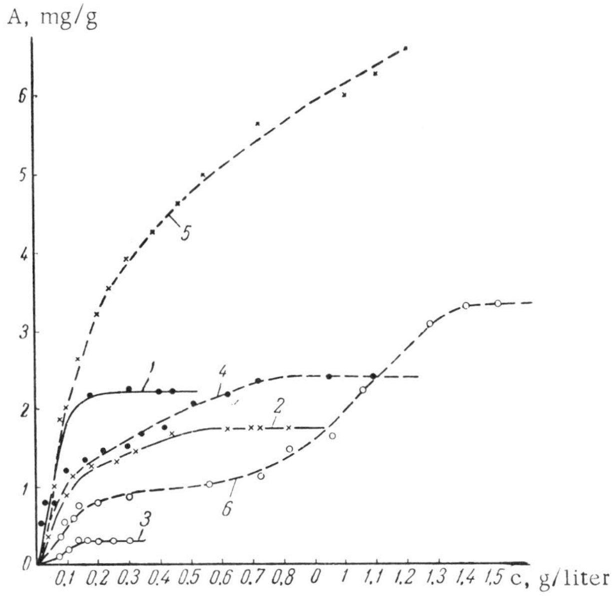

Fig. 6. Isotherms for the adsorption of some nonionogenic surfactants on hydrophilic surfaces and surfaces which had been made hydrophobic. 1) Oxamid ST-15 on a hydrophilic surface; 2) KS-50 on a hydrophilic surface; 3) OP-10 on a hydrophilic surface; 4) Oxamid ST-15 on a hydrophobic surface; 5) SK-59 on a hydrophobic surface; 6) OP-10 on a hydrophobic surface.

The Study of Adsorption on Pure Quartz Sand and on Quartz Sand made Hydrophobic by Arlanskii Petroleum

The sand was made hydrophobic by stirring it with a quantity of Arlanskii petroleum corresponding to 20% of the weight of sand. The optimum quantity of added petroleum was selected experimentally (the maximum quantity of added petroleum at which the sand was completely wetted and no adhesion of the particles took place).

The adsorption isotherms for various surfactants with markely different structures are given in Fig. 6. The adsorption on the surface which had been made hydrophobic was in all cases higher than that on the hydrophilic surface. The results obtained in adsorption experiments under static conditions on hydrophilic and hydrophobic surfaces contradict the data obtained in experiments on the adsorption of OP-10 under dynamic conditions [1], where the least adsorption was observed for the oil-saturated sand and the maximum adsorption was observed for the water-saturated sand. In accordance with G. A. Babalyan and V. M. Baryshev's studies, this can be attributed to the fact that under the filtration conditions the petroleum is only partly eluted, the remaining proportion filling the narrow sections of the pores and thus decreasing the adsorption surface.

Adsorption on Quartz Sand and Arlanskii Sandstone

The experiments on the adsorption on quartz sand and Arlanskii sandstone were carried out with the following surfactants:

1) the oxyethylated alkylphenols OP-10, UFE_8, $KAUFE_{14}$, and VNII NP-58;

2) the oxyethylated alcohols: $S_{10}E_{8.3}$, $S_{15-18}E_{25}$, $S_{15-18}P_{27}$, and oxyethylated alcohols from the Lisichansk Combine 200-240°C fraction, containing 20 moles of ethylene oxide per mole of alcohol;

3) the oxyethylated amides and suloamides of aliphatic acids: Ethomide HT/60, Oxamid ST-15, Oxamid 17-20-15, and Sulfamide OE-10;

4) oxyethylated xylitol derivatives: KS-59, Xylital O-10, and Xylital S-15.

TABLE 4

Surfactant	A_∞, mg/g	$\frac{A_\infty}{c_{sp}}$, g/cm^2	g, ergs/cm^2·mg/g
OP-10	0.72	$2.3\cdot10^{-6}$	0.242
KS-59	3.0	$9.9\cdot10^{-6}$	0.262
Oxamid ST-15	1.86	$6.1\cdot10^{-6}$	0.255
Sulfamide OE-10	0.36	$1.0\cdot10^{-6}$	0.298
$S_{10}E_{8.3}$	–	–	0.415

The isotherms for the adsorption, from their aqueous solutions, of the surfactants in each group which show the least adsorption on quartz sand and Arlanskii sandstone, OP-10, Sulfamide OE-10, Oxamid ST-15, and KS-59 are shown in Fig. 7.

Some of the surfactants whose synthesis requires the use of reagents exhibiting a polymerizing action contain resinous substances. These substances, which exhibit colloidal solubility in the presence of free surfactant, absorb the surfactant and thus distort the surface tension isotherm [1]. Under these conditions the quantitative estimation of the limiting adsorption by the surface tension method becomes impossible. Formula (1) used in the calculations can in fact give only the "apparent" adsorption isotherm based on the apparent difference (c_1-c_e) in the concentrations and not on the actual difference. The "apparent" adsorption isotherm can give only qualitative information on the magnitude of the limiting adsorption [1]. These surfactants include,

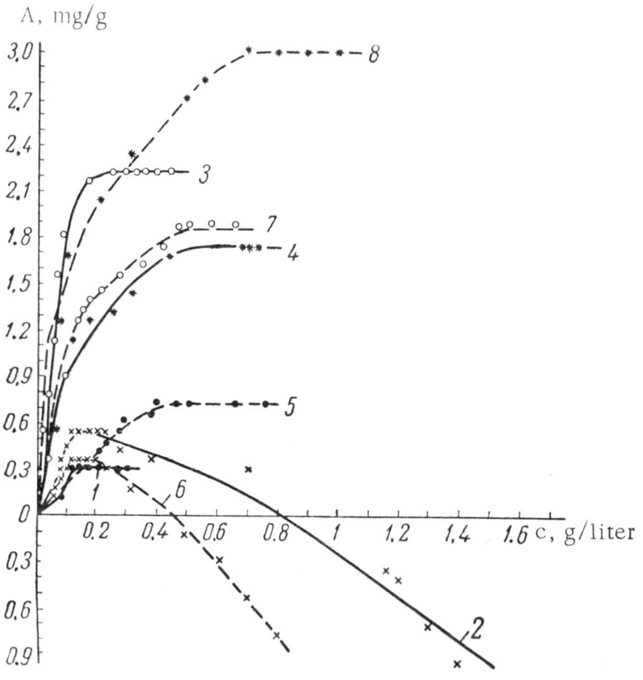

Fig. 7. Isotherms for the adsorption of nonionogenic surfactants on quartz sand and Arlanskii sandstone. 1) OP-10 on quartz sand; 2) Sulfamide OE-10 on quartz sand; 3) Oxamid ST-15 on quartz sand; 4) KS-59 on quartz sand; 5) OP-10 on Arlanskii sandstone; 6) Sulfamide OE-10 on Arlanskii sandstone; 7) Oxamid ST-15 on Arlanskii sandstone; 8) KS-59 on Arlanskii sandstone.

TABLE 5

Surfactant	A_∞, mg/liter
OP-10	0.66
Katapin A	9.12
Mixture of 50% OP-10+50% Katapin A	2.04
Karbozolin O.	5.34
Mixture of 50% OP-10+50% Karbozolin O	2.88

in particular, the nonionogenic surfactant Sulfamide OE-10, whose "apparent" adsorption isotherm begins in the range of positive values, reaches a maximum, and then enters the region of negative values, as can be seen from Fig. 7. In the group of oxyethylated alcohols, the lowest adsorbability is shown by $S_{10}E_{8.3}$, whose adsorption on quartz sand and Arlanskii sandstone is so small that it can be taken as zero. The adsorption isotherm for $S_{10}E_{8.3}$ is not shown in Fig. 7.

Table 4 gives the initial surface activity g for solutions of the least strongly adsorbed surfactants, obtained after adsorption on Arlanskii sandstone. Analysis of Fig. 7 and Table 4 leads to the following conclusions.

1. The oxyethylated alcohol $S_{10}E_{8\ 3}$ undergoes almost no adsorption on quartz sand or Arlanskii sandstone.

2. The adsorption of OP-10 on Arlanskii sandstone is double that of quartz sand.

3. The adsorption of Sulfamide OE-10 and Oxamid ST-15 on Arlanskii sandstone is less than that on quartz sand.

4. The lowest adsorbability on quartz sand (with the exception of $S_{10}E_{8.3}$) is shown by OP-10, whereas the lowest adsorbability on Arlanskii sandstone is shown by Sulfamide OE-10.

5. The greatest initial surface activity on Arlanskii sandstone is shown by $S_{10}E_{8.3}$, followed by Sulfamide OE-10, and the lowest initial surface activity is shown by OP-10.

6. The greatest adsorbability on Arlanskii sandstone and quartz sand is shown by KS-59.

On the basis of these studies, the most effective of the surfactants examined, from the viewpoint of lowest adsorption and greatest decrease in σ, were found to be OP-10, $S_{10}E_{8.3}$, and Sulfamide OE-10. Only OP-10 is produced in large quantities on the industrial scale, so that most of the subsequent studies on the displacement of petroleum by aqueous surfactant solutions were carried out with OP-10.

The Adsorption of Binary Mixtures on Quartz Sand

In many cases, for particular processes, mixtures of different surfactants are more effective than each surfactant taken separately. Thus it has been established that for demulsification purposes, mixtures consisting of three to four surfactants are more effective and more economical. Good results on the displacement of petroleum by water have been obtained by the addition of nonionogenic [1-3] and cationic surfactants [4], the individual cationic surfactants giving better results than the nonionogenic surfactants. The adsorption of cationic surfactants from their solutions on rocks is much greater, however, than that for nonionogenic surfactants and this in most cases prevents the use of the former.

It was therefore of interest to examine the possibility of using mixtures of nonionogenic and cationic surfactants in order to increase petroleum recovery. The first step in this direction was a study of the adsorption of mixtures of surfactants in order to investigate the possibility of decreasing the limiting adsorption of certain cationic surfactants by the addition of nonionogenic surfactants. The experiments were carried out with mixtures of OP-10 (50%)+ Karbozolin O (50%) and OP-10 (50%)+ Katapin A (50%). The limiting adsorption of these surfactants and their mixtures is given in Table 5.

In the case of the 50% mixture of Katapin A and OP-10, the value of A_∞ for Katapin A was decreased by a factor of more than 4, while in the case of the same mixture with Karbozolin O the value was reduced by a factor of 2. This confirms our suggestion that one surfactant has a beneficial action on the other in their mixture.

TABLE 6

Weight of sand, g	Permeability toward air, millidarcys	Porosity, %	Pore space volume, m^3
80	850	29	13.2

Preliminary Experiments on the Adsorption of OP-10 and $S_{10}E_{8.3}$ during Filtration

Since the lowest adsorbability on quartz sand and Arlanskii sandstone is shown by $S_{10}E_{8.3}$, preliminary experiments were carried out on the adsorption of these surfactants on quartz sand during filtration. The porosity characteristics of the medium are given in Table 6.

The filtration was carried out at a pressure drop equal to 400 mm water column. The average volume flow rate amounted to 0.47 ml/min. The concentration of surfactant in the solution amounted to 0.05%. The maximum adsorption of OP-10 amounted to 0.45 mg/g and the adsorption of $S_{10}E_{8.3}$ did not exceed 0.1 mg/g. An interesting feature is that in the initial period of filtration, the adsorption of $S_{10}E_{8.3}$ was so small that it could not be detected by the surface tension method.

On the basis of work which has been carried out in order to increase petroleum recovery, the following surfactants can be recommended for experiments on the study of the displacement of petroleum from a porous medium by surfactant solutions: OP-10, $KAUFE_{14}$, UFE_8, Ethomide HT/60, and $S_{10}E_{8.3}$. The most effective of the surfactants was found to be OP-10.

Literature Cited

1. G. A. Babalyan, I. I. Kravchenko, I. L. Markhasin, and G. V. Rudakov, Physicochemical Principles of the Use of Surface-Active Substances in the Exploitation of Oil-Bearing Strata (Gostoptekhizdat, 1962).
2. G. A. Babalyan, Problems of the Mechanism of Petroleum Recovery (Aznefteizdat, Baku, 1956).
3. G. A. Babalyan, G. V. Rudakov, I. I. Kravchenko, and I. L. Markhasin, "The use of surface-active substances to increase the recovery of petroleum from collectors," Izv. vuz, Neft' i gaz, No. 1 (1961).
4. H. N. Dunning, R. T. Johansen, and L. Hisiao, "The displacement of petroleum by wetting agents," Petrol. Eng., XXX, 1 (1954).

THE EFFECT OF ADDING SURFACTANTS ON THE COEFFICIENT OF COVERAGE OF A POROUS MEDIUM BY WATER

I. L. Markhasin, G. A. Babalyan, and O. I. Tselikovskii

It is known that one of the factors determining the process of displacement of petroleum by water, or more correctly the process of coverage of an oil-saturated porous medium by the displacing water, is capillary impregnation [1]. The roleof capillaryforces in the displacement is particularly important in the case of a heterogeneous stratum with permeability difference boundaries between the zones, or with parts in the form of lenses of differing permeability.

Let us consider the case of a two-layer hydrophilic horizontal stratum. Suppose that the permeability of one layer is k', and that of the other is k", where k' ≪ k". For convenience we will work in terms of the ratio of the rates of movement of the petroleum–water boundary in the two layers:

$$\bar{v} = \frac{v'}{v''}. \tag{1}$$

Suppose that the model stratum under consideration is made up of ideal material. It is clear that in a hydrophilic capillary the rate of meniscus movement under the conditions of an applied drop in pressure may be considered as the sum of two velocities, produced by different causes: the fall in pressure Δp and the capillary pressure p_k.

Equation (1) may therefore be rewritten as:

$$\bar{v} = \frac{\bar{v}'}{\bar{v}''} = \frac{\bar{v}'_{\Delta p} + \bar{v}'_{p_K}}{\bar{v}''_{\Delta p} + \bar{v}''_{p_K}}. \tag{2}$$

1. If the capillary pressure is negligibly small compared with the applied pressure drop, then Eq. (2) has the form:

$$\bar{v} = \frac{\bar{v}'_{\Delta p}}{\bar{v}''_{\Delta p}}. \tag{3}$$

It is known that:

$$v_{\Delta p} = \frac{r^2 (p_1 - p_2)}{8 \mu L}, \tag{4}$$

where r is the capillary radius; $p_1 - p_2$ is the pressure drop; μ is the viscosity of the liquid; L is the capillary length.

Substituting (4) in (3) we have:

$$\bar{v} = \frac{r_1^2}{r_2^2}. \tag{5}$$

According to (2), for an ideal formation medium the mean radius is given by the relation:

$$\bar{r} = \frac{2}{r \cdot 10^3} \sqrt{\frac{k}{m}}, \tag{6}$$

where k is the permeability and m is the porosity.

Substituting the values for r_1 and r_2 from (6) in (5), we have:

$$\bar{v} = \frac{k'}{k''}. \tag{7}$$

Thus, in the case in question, the movement rate of the petroleum–water boundary will be greater in the layer of higher permeability.

2. As $\Delta p \to 0$, Eq. (2) takes the form:

$$\bar{v} = \frac{v'_{p_K}}{v''_{p_K}}. \tag{8}$$

It is known that:

$$p_K = \frac{2\sigma \cos\theta}{r},$$

where σ is the surface tension, θ is the contact angle for selective wetting, and r is the capillary radius. Since $k' \ll k''$, it follows that $r_1 \ll r_2$, and hence:

$$v'_{p_K} > v''_{p_K}. \tag{9}$$

However, if the stratum is very long, the action of viscous forces may be such that the relation (9) does not apply [3].

Let us now consider real porous media.

Suppose that the given stratum element consists of two layers of different permeabilities k' and k", (Fig. 1) where $k' \ll k''$, and that petroleum is displaced along the stratum under the influence of the pressure drop

$$\Delta p = p_1 - p_2.$$

To start with, the petroleum–water boundary in the layer of permeability k" will lead the boundary in the layer of permeability k'. This is because the extent of the stratum will be considerable under real conditions, so that capillary forces will be extinguished by viscous forces [3]. At the same time, the thickness of the layers will not be very great in real strata, and in many cases capillary forces will predominate over viscous forces.

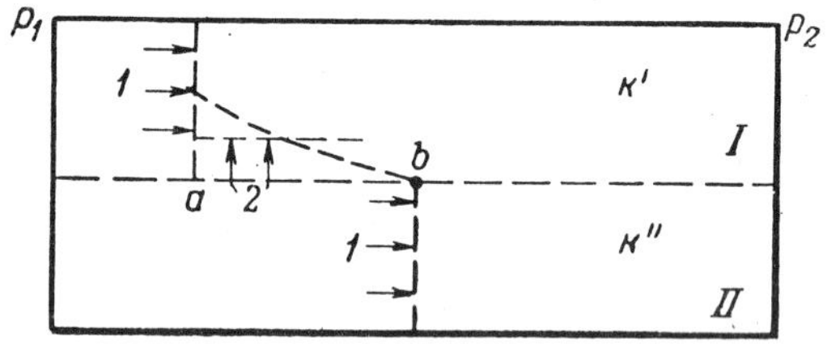

Fig. 1. Diagram of movement of the petroleum–water boundary. 1) Direction of displacement under the action of the pressure drop Δp; 2) direction of displacement under the action of capillary forces.

Thus, in the region ab (Fig. 1) capillary penetration of water from layer II to layer I will commence. It follows that under certain conditions the rate of capillary displacement from layer I may be commensurable with the rate of movement of the petroleum–water boundary in layer II. The above considerations apply only to hydrophilic formations.

In a real stratum these considerations will apply only to the parts containing an appreciable percentage of combined water. It is known [4] that petroleum can

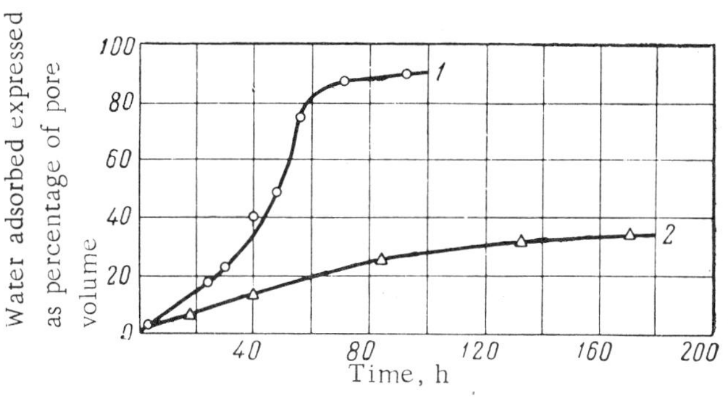

Fig. 2. Relation between volume of water adsorbed by capillary attraction and time, for the penetration of water and aqueous OP-10 solution into petroleum-saturated cores. 1) With 0.05% OP-10 solution; 2) water without OP-10 (no combined water).

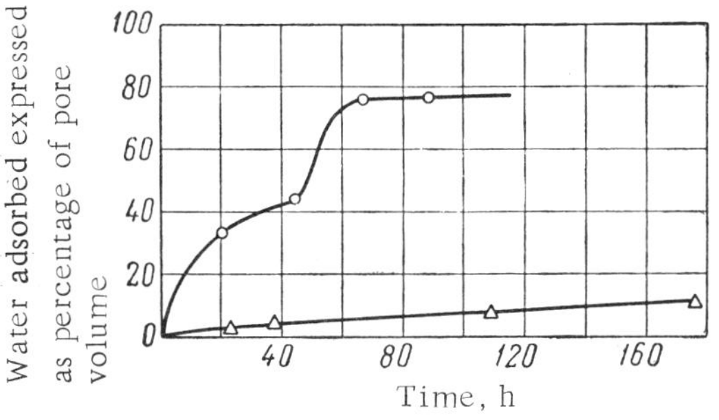

Fig. 3. Relation between volume of water adsorbed by capillary attraction and time, for the penetration of water and aqueous OP-10 solution into petroleum-saturated cores containing buried water. 1) With 0.05% OP-10; 2) without OP-10.

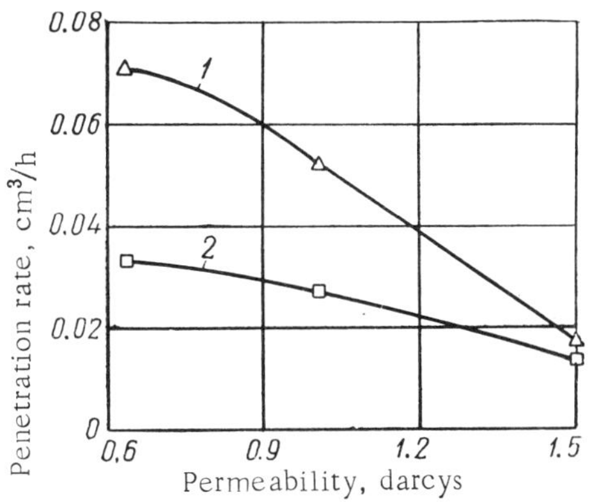

Fig. 4. Relation between the rate of capillary penetration of 0.05% OP-10 solution in water into petroleum-saturated cores and the air permeability of the latter. 1) Mean penetration rate for first 70 h; 2) mean penetration rate for first 10 h (buried water 20%; OP-10 0.05%).

make a porous medium hydrophobic by contact with it. Thus the capillary penetration factor is ineffective or only slightly effective under normal flooding conditions, and this may lead to a small coverage coefficient and, hence, to a relatively low coefficient for extraction of the reserves.

Therefore, to obtain the maximum effectiveness from flooding, it is essential that the water injected into the stratum should be treated so as to increase its rate of capillary penetration into the oil-saturated formation. This can be done by adding surfactant to the water [5].

Experimental

Experiments were carried out to determine the rates of capillary penetration of water, both treated and untreated with surfactant, into a petroleum-saturated porous medium, and also to compare these rates with those for penetration into real strata. The experiments were carried out with cementated samples, prepared by the method of Kosheleva and Kusakov [6]. Combined water in the samples was obtained by the method of Antonov and Berezin [7].

The liquids used in the tests were as follows: petroleum from well 248 of the Arlansk field, degassed without contact with air; formation water from the Kama River, and a solution of 0.05% of the surfactant OP-10 in this water. These liquids were chosen because arrangements are being made to carry out full scale trials in one part of the Arlansk field (round well 248) on addition of the surfactant OP-10 to the injected water, in order to increase the yield of petroleum.

TABLE 1

Sample No.	Sample length, cm	Sample diameter, cm	Porosity as a fraction of unity	Air permeability, darcys
1	10	2	0.35	0.64
2	10	2	0.36	0.65
3	10	2	0.33	1.01
4	10	2	0.33	1.15
5	10	2	0.33	1.38
6	10	2	0.33	1.43
7	10	2	0.33	1.53
8	10	2	0.32	1.78

TABLE 2

Property	Petroleum	Formation water	Underground river water
Density, g/cm^3	0.892	1.171	1
Viscosity, cP	30.06	1.67	–
Surface tension at boundary with petroleum, ergs/cm^2 . .	–	31.2	–
Surface tension at boundary with petroleum, ergs/cm^2 . .	–	–	31.0
Surface tension at boundary with 0.05% OP-10 solution in underground river water	7.8	–	–

TABLE 3

Sample No.	Amount of buried water, %	Amount of water penetrating during treatment time, cm^3	Mean penetration rate, m/year
8	0	5.34	2.89
6	10	3.00	0.86
2	10	8.07	6.90
1	20	7.67	13.00

TABLE 4

Sample No.	Amount of buried water, %	Amount of water penetrating during treatment time, cm^3	Mean penetration rate, m/year
8	0	8.96	11.20
5	10	8.92	13.00
4	10	8.81	22.40
7	20	6.43	4.03
3	20	6.59	14.40
2	20	7.95	17.30

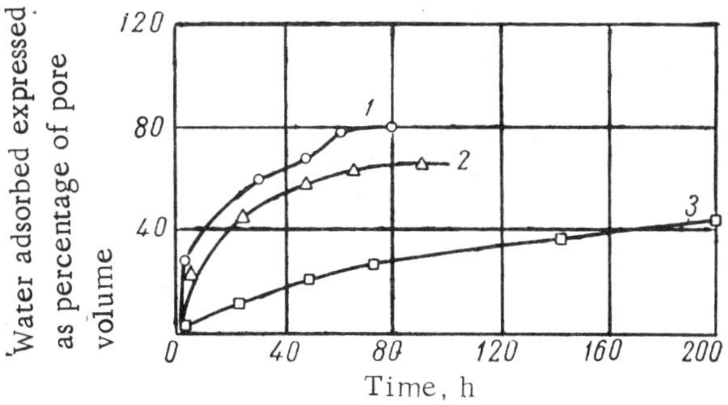

Fig. 5. Change with time of the volume of 0.05% aqueous OP-10 solution adsorbed into cores. 1) k = 0.69 darcy; 2) k = 1.01 darcys; 3) k = 1.53 darcys (buried water 20%; water containing 0.05% of OP-10).

Table 1 shows the properties of the porous media, and Table 2 shows those of the liquids used.

The method of carrying out the experiments was as follows. The cores were extracted and dried to constant weight, and combined water was produced in them. They were then immersed in kerosene in a desiccator, which was highly evacuated for a long time. Completeness of saturation was determined on the basis of the pore volume of the core, its weight with combined water, and its weight with combined water and kerosene. The kerosene-saturated core was packed into a core holder by means of a pneumatic press, and petroleum was filtered through it. Samples of filtrate were taken at intervals to determine their optical densities. It was considered that the core was saturated with petroleum when the optical densities of the filtrate and the petroleum were the same. The core was then transferred to a core holder for capillary penetration tests [5].

Tables 3 and 4 show the results of experiments on the penetration of Kama River water, and of a 0.05% solution of OP-10 in this water, into petroleum-saturated cores.

Figure 2 shows the results of tests on sample No. 8 (Tables 3 and 4). Addition of 0.05% of OP-10 to the water produced a marked increase in the penetration rate. Pentration of OP-10 solution was practically complete after 70 h; the amount of water which had penetrated amounted to 90% of the pore space. In the absence of surfactant, only 38% of the pore space was occupied by the water which had penetrated in 170 h.

Figure 3 shows the results of experiments with 10% of underground water in the rock (from the Arlansk formation). Absorption of OP-10 solution was investigated with sample No. 5 (Table 4), and adsorption of water with sample No. 6 (Table 3). Both samples had almost the same porosity and permeability. Previous investigations [8] had provided a basis for believing that they were also similar with regard to pore space structure. It is clear that the presence of combined water in the porous medium did not reduce the effectiveness of the surfactant. Consequently the views we have expressed relative to the action of surfactant in the absence of combined water are also applicable to the present case, i.e., with 10% of combined water in the pore space.

Figure 4, giving the results of experiments with cores No. 2, 3, and 7 (see Table 4),shows that the air permeability of the porous medium had a pronounced effect on the rate of capillary penetration. The rate of capillary penetration of aqueous OP-10 solution into petroleum-saturated rock decreased with increasing air permeability of the rock, in agreement with relation (9). The same effect was observed with 10% of underground water in the rock (samples No. 4 and 5 of Table 4). In fact, the same was found with the penetration of water which did not contain surfactant (samples No. 2 and 6 of Table 3).

Figure 5 shows curves for the capillary penetration of aqueous OP-10 solution into petroleum-saturated cores, of different air permeabilities, all containing 20% of buried water. Consideration of these curves demonstrates that the air permeability affected the final amount of water adsorbed, i.e., the petroleum yield.

Thus it has been shown experimentally that the capillary penetration of water into a cementated, strongly hydrophilic, porous medium proceeds more rapidly (by a factor of 2 to 3) when the water contains the surfactant

OP-10 than when it does not. It is of interest to compare these experimental results for the rates of capillary penetration with the rates of movement of the water–petroleum boundary under real conditions. According to data [9] obtained from an analysis of the development of the Tuimazinsk petroleum field, the rate of petroleum displacement by water varied from 20 to 150 m/year. Analysis of Table 4 shows that with OP-10 added to the water the rates of capillary penetration were in many cases comparable with the above lower limit for the rate of movement of the petroleum contour.

Returning to the diagram shown in Fig. 1, it should be noted that, according to Eq. (7), under conditions such that the stratum rock is hydrophobic so that there is no capillary penetration, the front of the moving water–petroleum boundary in the layer of higher permeability will be considerably in advance of the front in the layer of lower permeability.

Our experiments have shown that the rate of capillary penetration is greatly increased if surfactant is added to the water so that, for the case shown in Fig. 1, even if the petroleum–water boundary in the layer of low permeability still lags behind that in the layer of high permeability, it will do so to a much lesser extent than when there is no capillary penetration. Thus, addition of surfactant to the injected water, by increasing the rate of capillary penetration, should lead to an increase in the coverage coefficient and hence to an increase in the coefficient of extraction of the reserves.

Literature Cited

1. G. A. Babalyan and Z. F. Rzabekov, Trudy AzNII DN, No. VII.
2. F. I. Kotyakhov, The Fundamentals of the Physics of Petroleum Strata (Gostoptekhizdat, 1956).
3. A. Moore and R. Slobod, Producers Monthly, 20, 10, VIII, 20-30 (1956).
4. V. B. Shnerson, Trudy Instituta Nefti AN SSSR, I, 1 (1949).
5. G. A. Babalyan, I. I. Kravchenko, I. L. Markhasin, and G. V. Rudakov, The Physicochemical Fundamentals of the Use of Surfactants in the Development of Petroleum Strata (Gostoptekhizdat, 1962).
6. I. M. Kosheleva and M. M. Kusakov, Trudy MNI im. I. M. Gubkina, No. 14 (1955).
7. D. A. Antonov and V. M. Berezin, The Technique of Establishing Combined Water, Report No. 21 (UFNII Publications, 1956).
8. M. M. Kusakov and Sh. K. Gimatudinov, Trudy MNI im. I. M. Gubkina, No. 22 (1958).
9. I. G. Permyakov and N. S. Gudok, Neft. Khoz., No. 6 (1961).

A TECHNOLOGICAL SCHEME FOR THE DEVELOPMENT OF THE B_0 STRATUM OF THE POKROVSK FIELD WITH THE USE OF SURFACTANTS DURING FLOODING

V. A. Merkulov, A. T. Morgunov, and V. V. Kukin

The problem of increasing the petroleum yield from productive strata is now perhaps more urgent than ever. Various methods have been proposed for increasing the petroleum yield, but none of them has been developed to the point where it can be reliably recommended for industrial application

Most attention has recently been directed to the possibilities of increasing the effectiveness of flooding by adding surfactant to the water to be injected into the stratum. The results of laboratory investigations usually show that addition of surfactant to the injected water leads to a reduction in the residual petroleum content of a productive stratum. The advantages of using surfactant in flooding has also been confirmed abroad by field tests.

It should be noted, however, that the results obtained by different investigators show little agreement, and that the views of the investigators are often contradictory. The primary reason for this is that no satisfactory model for investigating the process has yet been developed. A well-defined hydromechanical criterion for similarity has not been developed, even for the simple case of flow of unmixed liquids in a porous medium, and, moreover, the need for such a criterion is often ignored in experimental work. A theoretical basis for such a model is generally lacking for the more complicated cases, where various physicochemical processes can occur in the porous medium. Thus the development of a system of physicochemical models is now one of the most important tasks.

This need is aggravated because the results obtained under laboratory conditions cannot be applied completely to a natural formation. It has been shown in papers published by the UFNII that, under laboratory conditions, the use of OP-10 increases the petroleum yield and reduces the displacement time. This Institute has a plan for the development of an experimental section of the Arlansk field, where it is believed that the petroleum yield will be increased by about 10%. The first experimental results obtained by the UFNII should clearly serve as a basis for the choice of surfactant and of the objectives for carrying out this process.

(In 1000 m^3)

Part	Rock volume	Volume of porous medium	Volume of buried water	Volume of petroleum under formation conditions	Volume of petroleum under surface conditions
Northern	11375.0	2343.25	281.19	2062.06	1891.8
Central	4318.75	899.66	106.76	782.90	718.26
Southern	7850.0	1617.10	194.05	1423.05	1305.55

For field trials it is first of all desirable to select strata or zones where it will be possible to make a precise comparison of the results with and without the use of surfactant. The petroleum deposits of the B_0 stratum of the Pokrovsk field in the Kuibyshev region are interesting in this respect. The shape of the formation creates favorable conditions for carrying out experiments on the injection of aqueous surfactant solutions into the stratum.

The deposits are confined to a sandstone stratum of the Tul'sk horizon, which is shaped like a ribbon 500 to 700 m wide and extending for 11 km. It lies at a depth of 1600 m. Its cross section is in the form of a lens, embedded in clay. Lithologically it is a fine-grain sandstone, weakly cementated with clay. The stratum has a mean porosity of 20.6%, a mean permeability of 700 millidarcys, and an initial water saturation of 12%.

Along its axis the stratum is relatively homogeneous with regard to trapping properties. Its thickness is also relatively constant along the axis, with a mean value of 6 to 9 m. An exception to this is the part round wells 184, 183, 118, and 632, where the thickness is only 2 m and where there is marked deterioration of the trapping properties. Thus the stratum is separated naturally into two parts, with a restricted hydrodynamic link between them.

Main Physicochemical Properties of Petroleum Stratum B_0

Initial formation pressure, atm	175.0
Formation temperature, °C	45.0
Saturation pressure at 45°C, atm	60.0
Weighted mean stratum pressure on April 1, 1962, atm	144.0
Phase factor, m^3/m^3	33.4
Density of petroleum, g/cm^3:	
Under formation conditions	0.804
Under atmospheric conditions	0.848
Petroleum shrinkage, %	7.8
Volumetric coefficient	1.09
Petroleum viscosity, cP:	
Under formation conditions	2.8
Under atmospheric conditions	8.12
Content, %:	
Sulfur	0.94
Silica gel resins	6.23
Asphaltenes	0.63
Paraffin	6.38

The formation was brought into production in 1955. Flooding was started in 1958.

The formation can be divided into three parts, in accordance with the development conditions: northern, between injection wells 403 and 82; central, between injection well 403 and the low permeability zone; and southern, between the poor permeability zone and injection well 400. The reserves in these parts are shown in the table on the preceding page.

The central part has hardly been developed at all up to the present. Analysis of the previous development of the formation shows that the central part of the stratum B_0 should be suitable for carrying out field trials on the use of surfactants in flooding. Therefore, in a preliminary technological plan for development of the B_0 stratum, allowing for experiments to be carried out, the stratum has been divided up into the above three separate parts. The central part, between injection wells 82 and 630 (the latter will be converted for discharge of what is injected when the experiment starts), and the southern part, between injection well 400 and the zone of low permeability, will be developed with normal flooding by fresh water. The proposed plan for separating the stratum into separate parts has the advantage that it makes it possible to determine, with adequate precision, the results of flooding with and without surfactant in the separate parts, and hence to obtain definite proof as to whether it is advantageous to introduce surfactant in flooding to increase the petroleum yield.

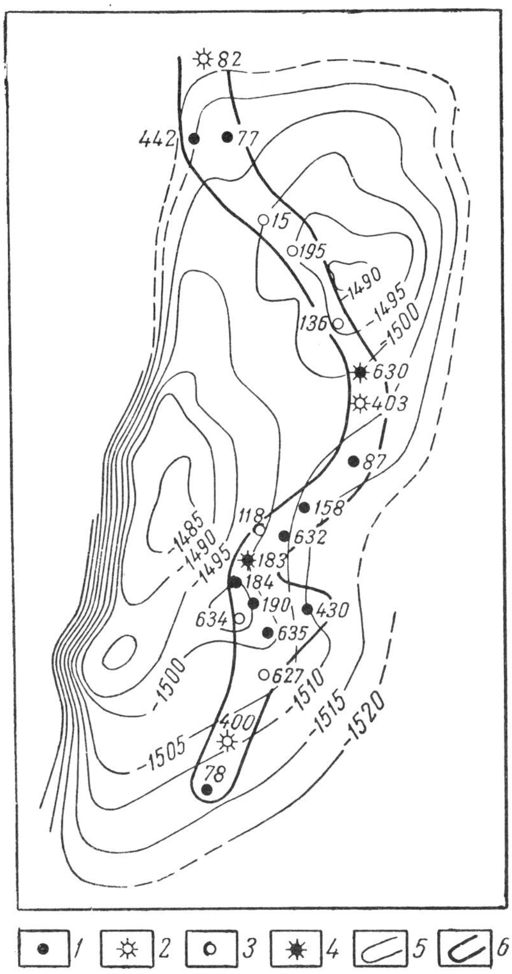

1) Exploitation wells; 2) injection wells; 3) observation wells; 4) exploitation-injection wells; 5) isohypses for the bottom of the stratum B_0; 6) boundary limits.

The stratum B_0 of the Pokrovsk field is well-suited for this experimental work, because the special features of the stratum structure should make it possible to determine the effectivness of surfactant for increasing petroleum yield in a relatively short time and at relatively low cost.

The choice of surfactant and of the amount to use still presents some difficulty. For a number of reasons it has not yet been possible to carry out laboratory work on the displacement of petroleum by surfactant solution directly on a model of the B_0 stratum. The choice will therefore be made on the basis of the UFNII recommendations that the nonionic surfactant OP-10 should be used for experimental work. It is planned to inject into the experimental part a 0.05% aqueous solution of OP-10, the volume to be twice that of the oil-saturated part of the pore space, and to follow this up with fresh water, whose volume will be equal to that of the oil saturated part. Fresh water, equal in volume to three times that of the oil-saturated part of the pore space, will be injected into the northern and southern parts of the formation.

In a previous report of the NPU of Chapaevskneft' the project for development of the B_0 stratum was based on a petroleum yield of 0.5. On the basis of the UFNII data, and of analysis of the physicochemical properties of the B_0 stratum, it is hoped to increase the petroleum yield in the experimental part by 10%.

The B_0 formation has not yet been adequately investigated. In particular, the accepted values for the porosity, permeability, initial petroleum saturation, and expected petroleum yield should be known more precisely. However, there is no reason for doubting that, from the point of view of lithological characteristics and petroleum saturation, the B_0 formation satisfies the requirements for carrying out the experiment. The parameters of the stratum will be determined more precisely by geophysical methods, field trials, and laboratory tests. Moreover, it is planned, in the near future, to drill four more wells (636, 637, 638, 639) in the Pokrovsk field, which will provide continuous cores from the B_0 formation. These cores will be used to study the effects of surfactant on the petroleum yield, under conditions modelled on those of the B_0 formation, in the laboratory for the physics and chemistry of strata of the Kuibyshev NII NP. The results should provide additional material for more precise technological planning of the development of the B_0 formation, and the requirements for carrying out the experiments of injection of surfactant solution.

There should be no objection to the experiments from the point of view of the properties of the B_0 formation petroleum, since these are very typical of Povolzh' petroleum. Another advantage of carrying out this experiment at the Pokrovsk field is that there is no need to drill additional wells for the experimental work.

Approximate calculations indicate that the total duration of the experiments should be 10 to 12 years. The total quantity of OP-10 required will be 800 tons. The experiment will be rather prolonged and expensive because of the big volume of the part selected. Although this is a disadvantage, there are compensating advantages. The fact that very large volumes have been selected for the experiment means that the probability of error in calculating the increase in petroleum yield is much less than in the case of small volumes. This is such an important factor that it was decided to accept the disadvantage of the long time required and to carry out the prolonged tests which would give fully reliable results. However, it will not be necessary to wait so long before any results are available. Information will be obtained gradually, as the experiment proceeds, from comparison of acceptance profiles of injection wells, comparison of yield rates, etc.

Values for the anhydrous yield of the formation should be available after two to three years. Some definite conclusions should be obtained in three to four years, after the injection of one volume of water.

The proposed technological scheme should be considered as a preliminary to a fully worked out plan for development of the B_0 formation of the Pokrovsk field, in accordance with the requirements established by the experiments.

THE USE OF SURFACTANTS TO INCREASE THE RECOVERY OF PETROLEUM FROM STRATA

K. F. Oshchepkov

Low recovery of petroleum from strata is due to the fact that the petroleum adheres to the walls of the pores and is not washed out by water, and at small pore volumes the quantity of petroleum forming the film is extremely large. In order to increase the recovery of petroleum, therefore, it is necessary to improve the flushing properties of water by extracting the surface-active components from the petroleum in contact with it to form new water-soluble surface-active compounds based on these components, or by appropriate treatment of the injected waters by various surfactants.

In its passage, the injected water does not enter all the pores but tries to find a path through the largest pores, so that some of the oil-saturated pores are bypassed by the water. Thus the water used for flooding should exhibit good penetration into the pores, irrespective of their dimensions, the type of minerals making up the rock, or the degree of wetting of the rock by petroleum. This, in turn, would lead finally to extensive coverage of the pores, irrespective of their size and the nature of the deposits. Thus in order to achieve the maximum recovery of petroleum by the flooding of oil deposits the choice and preparation of the water to be used for flooding should satisfy these requirements to the maximum extent.

The principal method of satisfying the necessary conditions of the waters forced into the stratum during flooding is by treatment with surfactants. The aim of this treatment is to decrease the surface tension at the boundary with the oil to the maximum extent. This is brought about since the surfactant molecules at the oil–water interface direct their hydrophilic (i.e., polar) end towards the water, and their hydrophobic (i.e., nonpolar) end towards the petroleum. When surfactant additives are used, the water moving into the stratum not only displaces the free volume of the oil but also decreases the thickness of the oil films adhering to the walls of the pore, or breaks down these films with a resultant increase in petroleum recovery. Moreover, the addition of surfactants to the water improves the penetration of the latter into the fine pores and channels of low cross section.

Surfactants should be of particular importance for flooding oil deposits in regions where the stratal waters and surface waters (infrabed or river waters) are not alkaline, e.g., the waters from the deposits in the Volgograd region, where the stratal waters are of the calcium-chloride type and are characterized by high hardness and low pH (pH = 6.8), while the infrabed waters (in the region of the town of Zhirnovsk) are of the sodium-sulfate type and, although they are of a type intermediate between hard and alkaline waters, they are characterized by an extremely low alkalinity (pH = 7.8).

The task of increasing the recovery of petroleum from strata by flooding is always associated with the problem of utilizing stratal waters. It is rational to inject these waters back into the stratum after appropriate treatment (purification from mechanical impurities, iron compounds, petroleum products, etc.).

In 1960-1961, laboratory experiments were carried out to evaluate the petroleum-flushing properties of natural stratal water (from the Zhirnovskii deposit) and infrabed waters from the Medveditsa River, and also to examine the influence of treatment of these waters by various surfactants on their oil-flushing properties.

These experiments involve the displacement of petroleum from oil-saturated soil by pure water and water treated with surfactants. The experiments were carried out under comparable conditions by the following method.

The model soils were prepared from the following quartz sand fractions: < 0.25 mm 75% and 0.5–0.25 mm 15%. The fraction < 0.25 contained 2.49% clay particles. In addition, 10% (by weight) of limestone with particle size < 0.25 mm (ground cores of crystalline limestone) was added to the sand.

The sand and limestone were mixed thoroughly until a homogeneous mixture was obtained. The mixture was then placed in tubes with diameter of approximately 32 mm and length approximately 800 mm. The porosity of the mixture was approximately 30% and the average water permeability approximately 0.900 darcy.

The contents of the tube were saturated under vacuum with stratal water from the Bashkirskii level of the Zhirnovskii deposit, with the following composition:

	g/cm^3
Cl	76325.00
SO_4	37.04
HCO_3	54.90
Ca	8517.0
Mg	2874.96
SiO_2	16.00

The density of this water is 1.092 g/cm^3 and its pH = 6.8. The stratal water was then displaced from the model soil by petroleum from the Verkhnebashkirskii horizon of the Zhirnovskii deposit, with the following characteristics.

Density at 20°C, g/cm^3	0.880
Viscosity, cP	38.0
Concentration, %:	
Paraffin wax at 50°C	2.41
Sulfur	0.76
Sulfuric acid tars	22.0
Silica gel tars	9.60
Asphaltenes	1.25
Acid number, mg KOH	0.082

The water was displaced by petroleum until the oil leaving the model soil contained no trace of water. The remaining water was regarded as connate water and its concentration in the sand amounted on the average to approximately 30%, while the oil saturation amounted to approximately 70%.

The petroleum was displaced from the model soils by different waters with and without the addition of surfactants at a pressure drop of 100 cm of water column, corresponding to the rates of movement of the VNK, which with rare exceptions do not exceed 25 m/sec.

The infrabed water used together with the stratal water had the following composition:

	mg/liter
Cl	46.15
SO_4	139.91
HCO_3	274.50
Ca	100.20
Mg	31.45
SiO_2	26.00

The density of this water is 1.002 g/cm^3 and its pH = 7.8.

During the experiment, a fourfold volume (relative to the absolute volume of the pores) of displacing water filtered through the porous medium, so that at the end of the experiment the current water factor reached approximately 100%.

	Petroleum recovery, %					Time, h	
	Water-free	With a total water factor of				Water-free period	Water-containing period
		1.0	2.0	3.0	final 8.0		
Pure stratal water	51.0	58.3	61.64	63.1	66.9	20.5	54.6
Stratal water containing 0.2% Nekal	62.1	65.2	66.6	67.5	69.8	20.5	89.0
Pure infrabed water	55.0	60.2	65.7	68.5	74.5	28.5	98.0
Infrabed water containing 0.05% secondary oxyethylated alcohol	60.8	62.8	64.4	65.1	68.3	28.5	170.0

The displacement coefficients, expressed as percentages (for the sake of brevity these figures will subsequently be referred to as the petroleum recovery), were determined for the two waters and for aqueous solutions of different surfactants. The data obtained in experiments on the displacement of petroleum by pure stratal and infrabed waters and by the same waters treated with surfactants show that in all cases the water-free petroleum recovery for displacement of the petroleum by stratal waters treated with different surfactants is considerably greater than the corresponding petroleum recovery for displacement of the petroleum by pure stratal waters, and in the great majority of cases is greater than the water-free petroleum recovery obtained by displacement with pure infrabed waters. The highest final petroleum recovery (taken for an overall water factor of 8.0), however, was achieved by displacing the petroleum with pure infrabed water. It should be noted that none of the experiments in which surfactants were used showed an increase in the rates of displacement; on the contrary, in all cases the total time required for displacement of the petroleum (water-free and water-containing periods) increased considerably (by factors of 1.5 and above) compared with the time required for the displacement of petroleum by pure waters.

The highest rate of displacement of petroleum in the water-free and water-containing periods for a given pressure drop is shown by the pure stratal water. Considerable adsorption of surfactants by the rocks was observed during the experiments. When four to five pore volumes of surfactant solution were pumped through the porous medium, the values of the surface tension of the solutions at the outlet were in all cases greater than those of the original solutions (at the inlet).

Special experiments showed that the surfactants are strongly adsorbed on the clay particles of the rock and that very little adsorption takes place on quartz sand. The carbonate rocks dolomite and limestone occupy an intermediate position between the clays and the quartz sand. It was also observed that the higher the solubility of the surfactants in water, the lower their adsorption from a given solution on rock.

The adsorption of surfactants increases on clays, limestone, and dolomite which have been wet with petroleum. These preliminary data require further confirmation and analysis.

The petroleum recovery in the water-free period deserves particular attention. The best results in this case are obtained when the oil is displaced by stratal waters containing 0.2% of Nekal and by infrabed waters containing 0.05% of secondary oxyethylated alcohols,* containing 18 moles of ethylene oxide. The table gives the chief parameters of the displacement of petroleum by pure stratal and infrabed waters and by the same waters treated by the surfactants indicated.

The lowest water-free (51.0%) and final (66.9%) petroleum recovery is obtained by displacing the petroleum with pure stratal water. The time required for displacement is also a minimum (20.5 h for the water-free period and 54.6 h for the water-containing period). When 0.2% of the surfactant Nekal is added to the stratal water, the water-free petroleum recovery reaches a maximum (62.1%) in the time (20.5 h) required for

* Secondary oxyethylated alcohols are the products obtained by the oxyethylation of a mixture of various alcohols containing 10-20 carbon atoms, by 18 moles of ethylene oxide.

displacement of the petroleum by pure stratal water. The increase in the quantity of petroleum in the water-containing period for displacement by pure stratal water amounts to 15.9%, while that for displacement by Nekal solution amounts to only 7.7% and requires a longer period (89.0 h).

When 0.05% of secondary oxyethylated alcohols is added to the infrabed water, the highest petroleum recovery in the water-free period (60.8%) is observed in the time required for displacement by pure infrabed water—28.5 h. The petroleum recovery in the water-containing period is much lower than that for displacement by pure infrabed water (68.3% compared with 74.5%). The time required for displacement is also considerably greater (170 h compared with 98 h).

On the basis of the results obtained it may be recommended that in the first stage of development flooding should be carried out by stratal waters containing 0.2% Nekal in order to extract the maximum quantity of water-free petroleum in the minimum time. When no stratal water or insufficient stratal water is available, the use of infrabed water containing 0.05% of secondary oxyethylated alcohols is recommended.

The data obtained in the experiments correspond to the displacement of petroleum in the water-free and water-containing periods under constant conditions at constant pressure gradient. In practice, from the start of the flooding of the operational wells by approaching waters, the conditions of removal of the liquid (petroleum and water) from the stratum will change to forced conditions, i.e., there will be a considerable increase in the volumes removed and hence in the volumes forced into the stratum. At this stage of development it is more advantageous to inject pure infrabed water into the stratum, since flooding by this water gives maximum increase in petroleum recovery during the water-containing period. The duration of the water-containing period will here be determined by the degree of forcing of liquid from the stratum.

Thus in order to achieve maximum petroleum recovery and maximum reduction of the time required for development in the case under consideration, two main lines of approach may be adopted.

1. When a sufficient quantity of stratal water is available, it is most advantageous in the first stage to use area flooding or inner-contour flooding of the deposit by these waters with the addition of 0.2% Nekal. These waters are injected until they approach the operational wells, after which pure infrabed water is injected into the deposit and forced removal of liquid from the wells is adopted.

2. When there are no stratal waters or insufficient stratal waters, the most effective means of increasing petroleum recovery in the first stage (in the water-free period) of development is by area flooding or inner-contour flooding of the deposit by infrabed waters containing 0.05% of secondary oxyethylated alcohols. When the operational wells begin to be flooded, pure infrabed water is injected into the stratum and forced removal of liquid from the wells is adopted.

Under the specific conditions of the Volgograd region it is recommended that these techniques be adopted as an experiment in the development of restricted sections of the Verkhnebashkirskii oil-bearing stratum of the Zhirnovskii and Bakhmet'evskii deposits, which are characterized by comparatively low collector properties and high variations in area and thickness. The greatest efficiency in this case may be achieved by area flooding in an early stage of development of the deposit. The low permeability of the strata of the deposit (of the order of 1-100 millidarcys) provides no obstacle to this method. Experimental work in other countries has shown that with a suitable choice of water, the successful flooding of sandstones with permeabilities of a fraction of a millidarcy is possible.

EXPERIENCE IN THE USE OF SURFACTANTS IN THE ESTABLISHMENT OF WELLS IN SIBERIA

G. T. Ovnatanov

In Siberia, in the Irkutsk amphitheatre in particular, oil- and gas-bearing strata are encountered in Cambrian deposits. Although the total quantity of petroleum recovered from tertiary deposits over the whole world amounts to 60%, the quantity recovered from Cambrian deposits is only 1%. Thus,in the world petroleum recovery balance, Cambrian deposits have not yet occupied a significant place.

Finally, although up to the present petroleum has been recovered from granular rocks, the petroleum in Cambrian deposits is found chiefly in fissured rocks. It is necessary first of all to clarify the concept of fissured rock.

The term granular rocks is taken to mean rocks whose capacity is provided by pores. The latter also act as channels leading to the well. In fissured rocks the capacity is provided by pores, pockets, and other types of cavity. In this case, however, the channels leading to the wells are chiefly microcracks which at low depths have openings of not more than 100-150 μ.

The Yarega stratum in Ukhta lies at a depth of 200-220 m. Here, in actual fact, the cracks have larger openings as a result of the shallow depth at which the rocks are encountered, so that the opening of the cracks cannot be transferred mechanically to lower depths.

On the basis of an analysis of the conditions of liquid flow into wells drilled in fissured reservoirs, it has been concluded that in order to understand the mechanism of the movement of a liquid in fissured reservoirs, it is necessary to make use of branches of science such as mining engineering.

The physical chemistry of surface phenomena has been developed in the well-known work by Academician P. A. Rebinder and his school, and has been applied to petroleum recovery processes by Professor G. A. Babalyan.

It is impossible to establish the presence of fissures at low depths without detailed petrographic studies.

A petrographic emphasis in experimental work is extremely important and has been developed by Professor E. M. Smekhov and his co-workers.

This emphasis has also been developed in our studies in Siberia.

The studies covered the Irkutsk amphitheatre (Cambrian deposits), the Minusinsk basin (Middle Devonian) with fissured rocks, the Kemerovskii Oblast', and Tyumen' (Mesozoic). A number of studies were also carried out in Byelorussia, Turkmenia, and Armenia.

We shall deal with some of the main conclusions reached on the basis of many years' work in Siberia.

In fissured rocks the construction of the well filter should differ from that of ordinary wells.

Petrographic study of rocks has shown that Cambrian deposits usually contain 40-70 fissures per meter of thickness of the stratum. Sometimes the number of fissures decreases to 5-10. It is evident that if the usual well construction is adopted the possibility that the perforations will lie opposite the fissures will be left absolutely to chance.

1. In fissured rocks it is preferable to have an open end face. An open end face, however, makes it difficult to treat the end face zone with various reagents. Successful treatment requires a planned separating interval of treatment away from the shaft of the well.

2. During the opening of the stratum, annular stresses are produced in the neighborhood of the end face, which are approximately double the stresses produced under the influence of the rock pressure.

Under the influence of the annular stresses the vertical fissures close up, while the horizontal fissures remain open. Thus to ensure connection with the stratum it is necessary to open the fissure and to prevent its subsequent closure. If this is not done it is impossible to obtain a connection to the stratum.

During the drilling of three wells in the Atovskii area we were able to establish that the pressure at which the fissures open amounts to 230 atm. In the Novo-Mikhailovskii area this pressure amounts to 380 atm.

We were able to prove theoretically and practically that the annular stresses extend to a distance of approximately one meter. It is therefore necessary to open up the fissure and keep it open for this distance.

We have also been able to calculate the rate at which the walls of the fissure close.

In particular, the average rate at which the walls of a fissure close in the Minusinsk basin amounts to 15-30 A/sec. The walls of the fissures require approximately 20-25 h for closure.

Surfactants play an extremely important part in the opening and closing of fissures. In the Atovskii area we first obtained 1-5 liter of petroleum per day, after which the flow of oil stopped. This area lies at a distance of 300 km to the north of Irkutsk (in the midst of the taiga).

The stratum was opened in well No. 2 at a depth of 1983 m and the first 50 liters of oil were obtained at very high depressions at the end face.

After the second acid treatment using surfactants, we obtained the first industrial oil—5 tons per day. This was a considerable achievement. An interesting feature is that the stratal pressure in January 1960 amounted to 220 atm, while by September 1961 it had increased to 240 atm. The stratal pressure usually decreases with time but here it began to increase. Because of the lack of petroleum-receiving assemblies the well operated under unstable conditions. The stratal pressure increased because the Osinskii horizon, which is made up of dolomite (thickness 50 m), is covered by a 600 m layer of rock salt. Because of its (plastic) deformation, the rock pressure began to be transferred to the liquid. Observation shows that the stratal pressures continue to increase and do not decrease.

Quartz sand is used everywhere for the hydraulic rupture of the stratum. It serves to fill the fissures. The quantity of sand used reaches 30 tons. Quartz sand, however, has a low mechanical strength of 300 kg/cm^2. It has been found that a large quantity of sand cannot be distributed in fissured rocks and that a small quantity does not withstand the rock pressure. Here we require sands with a strength of not less than 2-2.5 thousand kg/cm^2. Small quantities of sand are used at a thickness of 300 m in the Minusinsk basin, since the volume of the fissures amounts to only 1.5 m^3. The maximum opening of the fissures can increase their volume by a factor of not more than 2. In the Atovskii area the volume of the fissures is even smaller—0.48 m^3 per 50 m of horizon thickness.

In this connection we carried out work on the preparation of a new reinforcing material consisting of small glass spheres with strengths of 1.5-3 thousand kg/cm^2. This material was prepared by the Glass Institute in accordance with a resolution of the State Committee for the Fuel Industry, USSR Gosplan.

Several wells were drilled in the Novo-Mikhailovskii area and well No. 1 was drilled using a solution of specific gravity 1.6 (treated with baryta). A vigorous gusher was obtained in spite of the increase in specific gravity.

Well No. 8 was drilled at a distance of 50-60 m from well No. 1, using an aqueous solution of OP-10 (specific gravity 1). The well was taken to the planned depth. At high end face pressures, the cracks open up, flow from the stratum is initiated, and a gusher is obtained, whereas the use of an aqueous OP-10 solution did not have these consequences and the well was drilled without complications.

The flushing liquid used to open up fissured rocks must be highly pure. If a suspension enters the fissures it subsequently sticks when the fissure closes. The results of our work show that petroleum can be obtained from Cambrian deposits.

From G. A. Babalyan's studies it is known that an emulsion is formed when water penetrates into and is displaced from the end face zone. In Nazino (Western Siberia) it has been possible to extract from the end face of the well an emulsion flowing directly from the end face zone. The emulsion was formed as a result of the penetration of technical waters into the stratum.

S. A. Khristianovich has put forward the theory that when the stratum is opened the rock pressure is relieved as a result of the formation of an arch of plastic rock forming the roof of the productive horizon.

This pressure relief has been detected in the Plotnikovskii area (Kuzbass). Five hydraulic breakthroughs were made in the section of well No. 4. In three cases the rock broke at pressures of 300 atm and in two cases the stratum could not be broken at a pressure of more than 500 atm. The pressure relief corresponded to the calculated data obtained by S. A. Khristianovich and lay in the range 200-250 atm.

In conclusion it should be noted that in addition to OP-10, the cationogenic surfactant Katapin, which has excellent inhibiting properties, has also been used successfully. Acid treatment of wells using a mixture of Katapin and OP-10 gives exceptionally good results.

THE DOSAGE OF SURFACTANTS REQUIRED FOR THE TREATMENT OF FACIAL ZONES

E. G. Khalilov, I. I. Kravchenko, and G. A. Babalyan

Surfactants have recently been used extensively in the petroleum industry, both in this country and abroad (mainly USA). For example, they are widely used in the petroleum fields of Apsheron for acid treatment, opening up and washing of petroleum and injection wells, to combat the formation of sandstone blockages, etc.

However, the dosage of surfactant is formulated without allowing for the characteristics of the facial zone rock and its capacity to adsorb surfactant. In order to establish the required dosage, it is necessary in each case to determine the equilibrium value of the adsorption of the surfactant used in practice by the characteristic rocks (sands) of the various petroleum traps.

Our work was restricted to the adsorption of the nonionic surfactants OP-10 and KOS$E_{28.6}$ by pure quartz sand, sands from the Kirmakinsk series (KS) of the Balakhansk field, the super-Kirmakinsk series and (SKS) from the Karadagsk region, and the Apsheron formation from the Bibi-Eibatsk field.

OP-10 is the reaction product of 10 moles of ethylene oxide with a mixture of mono- and di-alkylphenols, where the alkyl groups contain 8 to 10 carbon atoms. A laboratory sample from VNII Neftekhim was used.

KOS$E_{28.6}$ is a condensation product of alcohols, obtained by Oxo-synthesis and containing 12 to 16 carbon atoms, with 28.6 moles of ethylene oxide.

The sands used in the experiments were first cleaned as follows. They were washed with fresh (Shollar) water, extracted with a 1:3 mixture of alcohol and benzene to remove all the petroleum and asphalt resins, washed again with distilled water, and dried to constant weight in an oven at 100 to 105°C.

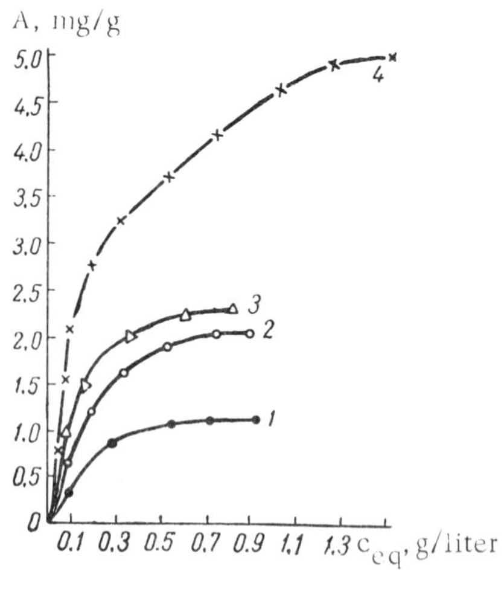

Fig. 1. Isotherms for the adsorption of OP-10 from aqueous solution on to various sands. 1) Quartz sand; 2) SKS; 3) Apsheron formation; 4) KS.

The table shows the characteristics of the sands investigated.

The experiments were carried out as follows. The surface tension at the interface with purified kerosene ($\sigma = 48.1$ ergs/cm^2) was determined for surfactant solutions of different concentrations. A given weight of sand (10 g) was then introduced into a surfactant solution of known concentration, time was allowed to establish adsorption equilibrium (this took two days), and the surface tension was measured again.

The limiting adsorption of surfactant was calculated from the equation:

$$A = \frac{(c_1 - c_{eq})\,V}{p},$$

where A is the amount of surfactant adsorbed from the solution of given concentration, in mg/g; c_1 is the concentration before adsorption, in g/liter; c_{eq} is the equilibrium concentration after ad-

Sand	Size distribution, %					< 0.075 mm	Density, g/cm³	Specific surface, cm²/cm³	Porosity, %	Carbonate content, %
	> 0.5 mm	0.5-0.25 mm	0.25-0.15 mm	0.15-0.1 mm	0.1-0.075 mm					
Quartz sand	–	7.65	74	17.80	0.8	0.2	2.62	–	32.3	–
SKS sand	–	24.6	37.75	7.30	26.7	3.6	2.68	–	35.8	1.6
KS sand	–	8.0	29.9	50.6	9.5	1.9	2.65	–	35.7	4.8
Apsheron formation sand	5.24	50.8	27.9	11.6	2.6	1.8	2.67	134.8	27.7	35

sorption, in g/liter; V is the volume of surfactant solution, in ml (in our experiments V = 30 ml); p is the weight of adsorbent, in g.

Surface tensions were determined by the method of measuring the volume of a drop, squeezed out by the micrometer control of a UFNII stalagmometer.

Figure 1 shows adsorption isotherms $A = f(c)$ for OP-10 with the various sands.

The adsorption was greatest on KS sand(45 mg/g) and least on quartz sand. Adsorption of OP-10 was practically the same on SKS and Apsheron formation sands.

Figure 2 shows that $KOSE_{28.6}$ was also more strongly adsorbed on KS sand (9.3 mg/g) than on quartz sand (2.15 mg/g). Figure 3 compares the adsorption of OP-10 and $KOSE_{28.6}$ on quartz and KS sands. OP-10 was less strongly adsorbed than $KOSE_{28.6}$ on all the sands. In addition, for the same concentration in distilled water, OP-10 reduced the surface tension at the interface with kerosene more than $KOSE_{28.6}$ did. The above results indicate that the adsorbtion of surfactant by various sands depends on the physical properties of the rock and on the characteristics of the surfactant.

It is, therefore, necessary that the surfactant dosage for treating the facial zone of a well should be determined by calculation from the limiting value of possible surfactant adsorption. The requisite dosage for OP-10 should be less than that for $KOSE_{28.6}$. The amount of surfactant required to treat a well in KS should be almost twice that required to treat a well in the Apsheron formation or in SKS.

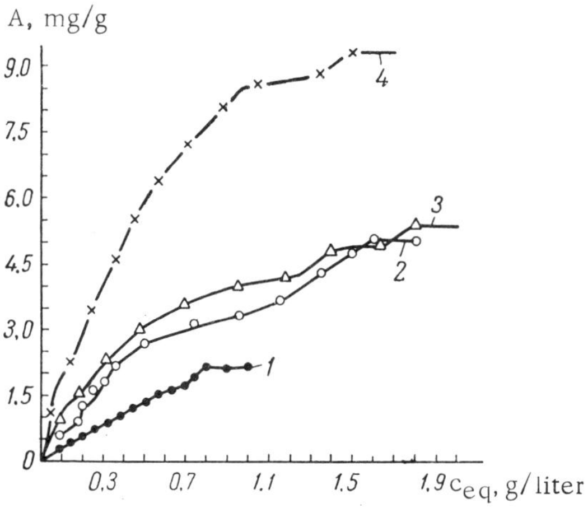

Fig. 2. Isotherms for the adsorption of $KOSE_{28.6}$ from aqueous solution onto the various sands. 1) Quartz sand; 2) Apsheron formation; 3) SKS; 4) KS.

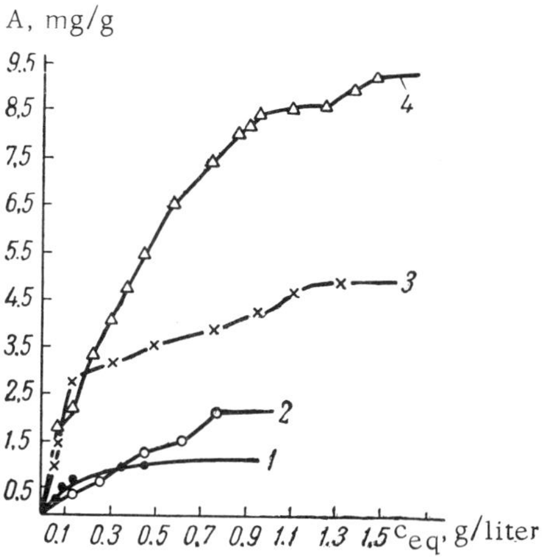

Fig. 3. Isotherms for the adsorption of OP-10 and $KOSE_{28.6}$ from aqueous solution onto quartz and KS sands. 1) OP-10 on quartz sand; 2) $KOSE_{28.6}$ on quartz sand; 3) OP-10 on KS; 4) $KOSE_{28.6}$ on KS.

Suppose that aqueous solutions of the surfactants OP-10 and $KOSE_{28.6}$ are to be injected, so that they penetrate to a radius R, equal to a depth of 1 m in a KS stratum, and that the specific surface of the rock is the same as that of the KS investigated in the laboratory. Assume that the thickness of the stratum is h = 1 m, the specific gravity of the sand KS is $\gamma = 2.65$, and the porosity m = 35.7%.

The quantity of surfactant Q required to achieve the limiting adsorption A_m in the part of the stratum under consideration may be calculated from the following equation:

$$Q = A_m \pi R^2 h \gamma (1 - m) + \pi r^2 h m c_{eq}.$$

The first term on the right hand side of the equation is the amount of surfactant adsorbed and the second term is the amount of surfactant in the solution in the pore space. Here we have assumed that the penetration rate of the limiting adsorption front is greater than the filtration rate. If the reverse is true, then the depth of penetration of the limiting adsorption front will be less and the concentration of surfactant in the solution in the pore space will be greater. Calculations show that for the case considered the requirements of OP-10 and $KOSE_{28.6}$ will be 28 and 51 kg respectively.

Experiments have shown that the equilibrium adsorption for filtration of solution through a sand saturated with petroleum is about half that for filtration through a sand saturated with water [1].

Therefore, in calculating the dosage required for an anhydrous well, the limiting adsorption found under laboratory conditions should be reduced by a factor of 1.5 to 2. However, if the well has been flooded extensively, the value corresponding to the laboratory data should be used.

Literature Cited

1. G. A. Babalyan, I. I. Kravchenko, I. L. Markhasin, and G. V. Rudakov, Physicochemical Fundamentals of the Use of Surfactants in the Development of Petroleum Strata (Gostoptekhizdat, 1962).
2. G. A. Babalyan, G. V. Rudakov, I. L. Markhasin, and I. I. Kravchenko, "The use of surfactants to increase the petroleum yields from strata", Trudy UFNII, No. VII (1961).
3. G. A. Babalyan, "The theory of surfactant action in the process of well development", Trudy UFNII, No. VI (1960).
4. P. A. Rebinder, The Use of Surfactants in the Petroleum Industry (Gostoptekhizdat, 1961).

CELLULOSE SULFATES AND SULFONATED CELLULOSE ETHERS–EFFECTIVE STABILIZERS OF AQUEOUS SUSPENSIONS OF NONCLAY ROCKS

S. N. Yatrov, I. E. Gol'dshtein, I. E. Shevaldin,
N. F. Semenko, and T. S. Osmolovskaya

Synthetic surfactants based on cellulose ethers and esters are being used on an ever-increasing scale for the stabilization of aqueous mineral suspensions and in a number of other technological processes. Extensive use has been made of carboxymethylcellulose (KMTs)–the product of the reaction between alkali cellulose and sodium monochloroacetate. In other countries use is made of carboxyethylcellulose – the product obtained by treating cellulose with acrylonitrile in the presence of alkali. Studies on the development and use of these reagents for drilling and in other branches of industry in the USSR have been carried out by K. F. Zhigach and co-workers.

Among the materials which appear promising for use as stabilizers of mineral suspensions are the sodium salts of cellulose sulfate and sulfoethylcellulose. Cellulose sulfate (STs) is obtained as a white or pale yellow powdery or fibrous mass. In our work we used a product synthesized in the laboratory under the direction of A. B. Gershenovich and M. A. Rabinovich. The esterification of cellulose with sulfuric acid and the subsequent neutralization of the acid sulfate with sodium hydroxide take place according to the reactions:

$$C_6H_9O_4OH + H_2SO_4 \rightarrow \underset{\text{cellulose hydrogen sulfate}}{C_6H_9O_4OSO_3H} + H_2O$$

$$C_6H_9O_4OSO_3H + NaOH \rightarrow \underset{\text{sodium cellulose sulfate}}{C_6H_9O_4OSO_3Na} + H_2O.$$

The esterification is carried out under mild conditions with miniumum degradation to give a product which is fairly soluble in water and is stable to the action of salts causing hardness. The viscosity of STs solutions increases considerably with increase in concentration.

Sodium sulfoethylcellulose (ESTs) is obtained as a friable yellowish powder. The synthesis is carried out as follows:

$$ClCH_2CH_2Cl + Na_2SO_3 \rightarrow \underset{\text{Sodium Chloroethanesulfonate}}{ClCH_2CH_2SO_3Na} + NaCl$$

$$C_6H_9O_4OH + NaOH \rightleftarrows \underset{\text{Alkali cellulose}}{C_6H_9O_4ONa} + H_2O$$

$$C_6H_9O_4ONa + ClCH_2CH_2SO_3Na \rightarrow \underset{\text{Sodium sulfoethylcellulose}}{C_6H_9O_4CH_2CH_2SO_3Na} + NaCl,$$

The experimental studies of the stabilizing adsorptive action of cellulose sulfate and sodium sulfoethylcellulose were carried out with aqueous suspensions of nonclay rocks. We studied the influence of the addition of these surfactants on the sedimentational stability and water loss for suspensions of carbonate and sulfate rocks, aleurolites, and argillites. We studied the change in other characteristics (specific gravity, viscosity, and limit-

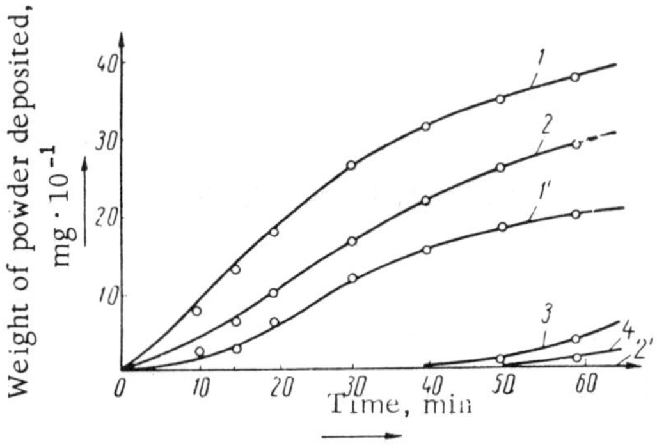

Fig. 1. Influence of the addition of cellulose sulfate on the stability of suspensions of anhydrite powder with particle size 0.2-0.6 μ in fresh and mineralized waters. Concentration of STs in the fresh-water suspension: 1) 0.25%; 2) 0.5%; 3) 1.0%; 4) 1.5%. Concentration of STs in the mineralized-water suspension (10% NaCl): 1') 0.25%; 2') 0.5%.

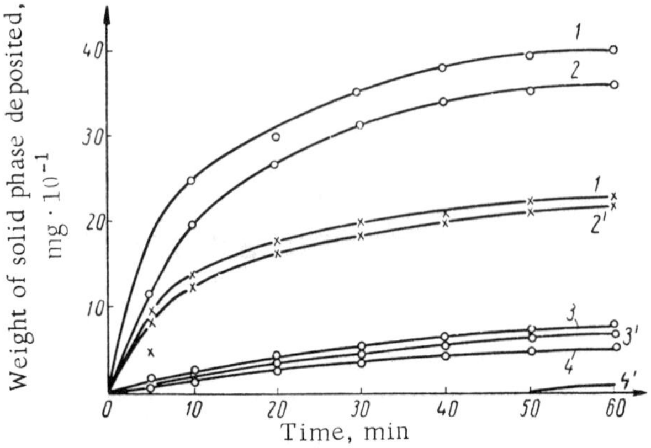

Fig. 2. Influence of the addition of cellulose sulfate on the stability of suspensions of anhydrite powder with particle size 0.6-1.4 μ in fresh and mineralized waters. Concentration of STs in the fresh-water suspensions: 1) 0.25%; 2) 0.5%; 3) 1.0%; 4) 1.5%. Concentration of STs in the mineralized water suspension: 1') 0.25%; 2') 0.5%; 3') 1.0%; 4') 1.5%.

ing shearing stress) of these suspensions with change in the concentration of cellulose sulfate and sodium sulfoethylcellulose. We studied suspensions formed by fresh water and by mineralized waters with different degrees of mineralization. The sedimentational stability was studied by weighing, on a torsion balance, the quantity of solid phase deposited by the suspension with time for different reagent concentrations; we also determined the settling of the dispersion medium and the deposit of disperse phase. The influence of added cellulose sulfate on the stability of suspensions of anhydrite powder in fresh and mineralized waters is shown in Fig. 1; the particle size of the powder was 0.2-0.6 μ. The influence of the addition of cellulose sulfate on the stability of suspensions of anhydrite powder with particle size 0.6-1.4 μ in fresh and mineralized waters is shown in Fig. 2.

It can be seen from the figures that cellulose sulfate is an active stabilizer for the anhydrite suspensions: the quantity of solid phase undergoing sedimentation decreases sharply with increase in the concentration of

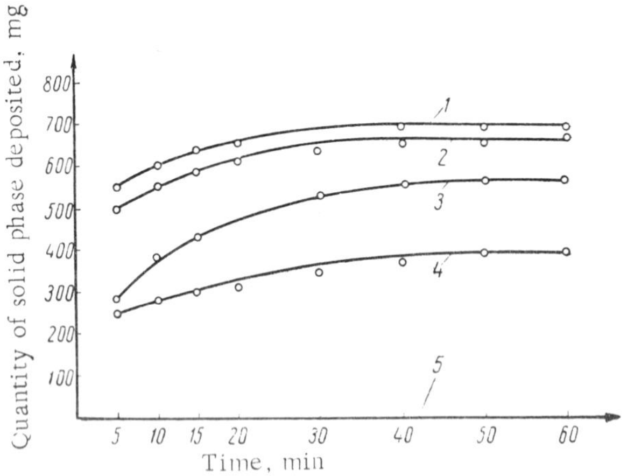

Fig. 3. Curves for the sedimentation of particles of solid phase for an anhydrite suspension at different sodium sulfoethylcellulose concentrations. ESTs concentration: 1) 0.1%; 2) 0.5%; 3) 1.0%; 4) 2.0%; 5) 3.0%.

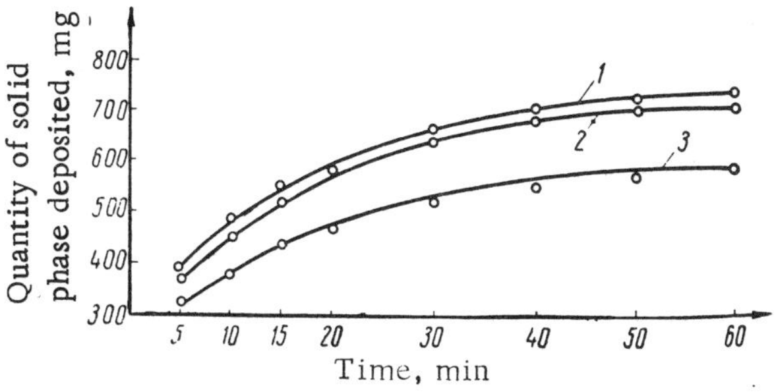

Fig. 4. Curves for the sedimentation of particles of solid phase for an anhydrite suspension at different degrees of mineralization. NaCl concentration: 1) 0%; 2) 5%; 3) 15%.

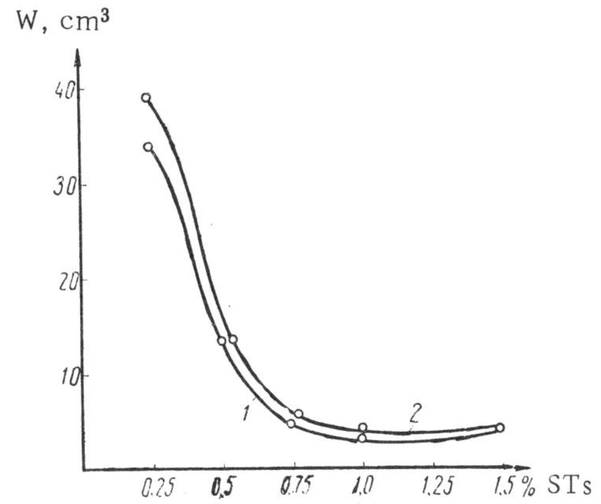

Fig. 5. Change in the water loss for argillite suspensions in fresh and mineralized water at different cellulose sulfate concentrations. 1) 10% suspension of argillite in fresh water; 2) with the addition of 5% $CaCl_2$.

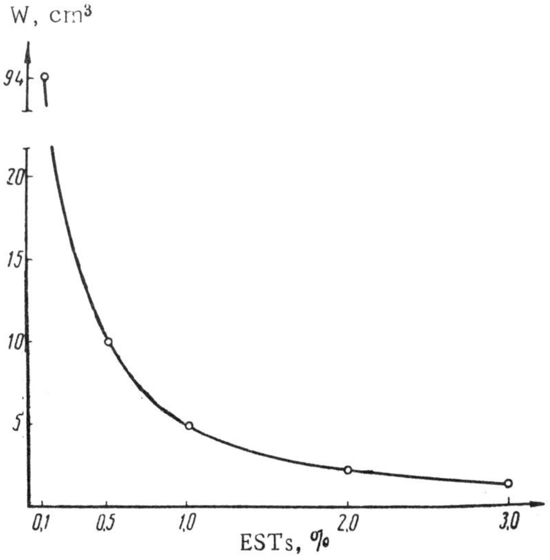

Fig. 6. Change in the water loss for a suspension of anhydrite in fresh water with change in the sodium sulfoethylcellulose concentration.

the reagent, and the adsorptive action increases in the presence of NaCl; the addition of small quantities of STs leads to almost complete stabilization of the system, and sedimentation of the solid phase stops. This state is achieved in a mineralized suspension from powders with particle size 0.2-0.6 μ in the presence of 0.5% STs (2' in Fig. 1); in the case of powders with particle size 0.6-1.4 μ this state is reached in the presence of 1.5-1.75% STs (4' in Fig. 2).

Curves for the sedimentation of particles of the disperse phase of suspensions treated with sodium sulfoethylcellulose are shown in Figs. 3 and 4. It can be seen from the figures that this reagent is also an extremely effective stabilizer for suspensions of nonclay rocks in fresh and mineralized waters; the addition of comparatively small quantities leads to almost complete stabilization of the system. Sedimentation of the particles of solid phase in a fresh-water suspension is stopped by the addition of 2.5-3.0% ESTs (Fig. 3); the activity of ESTs as a stabilizer increases with increase in the degree of mineralization (Fig. 4).

The decrease in water loss is one of the main characteristics of the adsorptive action of surfactants and the stabilization of aqueous mineral suspensions. The change in the water loss for argillite suspensions in fresh water and water mineralized with $CaCl_2$ at different concentrations of cellulose sulfate is shown in Fig. 5. The change in the water loss for a suspension of anhydrite in fresh water at different sodium sulfoethylcellulose concentrations is shown in Fig. 6, and the change in water loss for this suspension at different degrees of mineralization is shown in Fig. 7.

The curves in Figs. 5-7 confirm the high stabilizing power of cellulose sulfate and sodium sulfoethylcellulose. With increase in the concen-

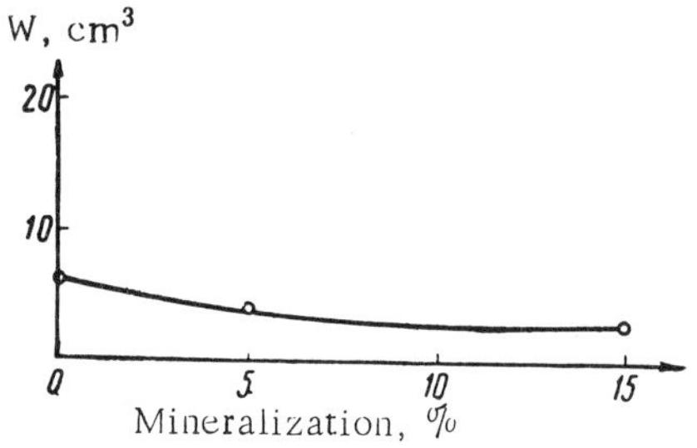

Fig. 7. Change in the water loss for an anhydrite suspension treated with sodium sulfoethylcellulose for different degrees of mineralization.

tration of these stabilizers, the water loss for the suspensions decreases sharply, reaching minimum values of 3-4 cm^3 for 1.25-1.5% STs and 2.5-1.7 cm^3 for 2-3% ESTs. With increase in the degree of mineralization, the water loss continues to decrease (Fig. 7), indicating an intensification of the surface activity.

Thus in the cases under consideration, inorganic electrolytes (NaCl and $CaCl_2$) are active agents which increase the stabilizing action of the main stabilizers—cellulose sulfate and sodium sulfoethylcellulose.

In the determination of the changes in the settling of dispersion medium and deposit of solid disperse phase with change in the concentration of cellulose sulfate and sodium sulfoethylcellulose, it was established that sedimentational stability of the suspension is achieved for fresh-water suspensions when 2.0-2.5% of cellulose sulfate or 2.5-3.0% of sodium sulfoethylcellulose are added, and for mineralized suspensions when 1.5-2.0% cellulose sulfate or 2.0-2.5% sodium sulfoethylcellulose are added.

The adsorptive activity of these stabilizers depends on the method by which they are produced and their degree of esterification or sulfonation and polymerization (this question will be discussed more fully).

Studies were made of the comparative activity of the adsorptive action on unstable mineral suspensions of nonclay rocks for the above surfactant stabilizers and for the most active chemical reagents used in drilling. These studies confirm that cellulose sulfate and sodium sulfoethylcellulose exhibit considerable advantages. For example, when 0.25-1.0% of carboxymethylcellulose was added to a mineralized suspension, the water loss corresponded to complete filtration and decreased to 27.5 cm^3 in the presence of 1.5% KMTs; when the same suspension was treated with 1-1.5% of cellulose sulfate, the water loss reached a minimum value of 4-5 cm^3.

It was of interest to examine the ability of cellulose sulfate and sodium sulfoethylcellulose to retain their stabilizing properties at increased temperatures. The thermal stability of stabilizers used for drilling deep wells is now of paramount importance. In view of the marked increase in the depth of wells already noted and, in particular, the proposed drilling of the sedimentary cover, this property of surfactants will become the chief characteristic determining the possibility of using these surfactants to stabilize drilling liquids.

Aqueous chalk suspensions were used in the study of the thermal stability of STs and ESTs. We determined the sedimentational stability, water loss, and viscosity of these suspensions treated with STs and ESTs at normal temperature (20°C). We then determined the same parameters for suspensions heated to 60, 100, 150, and 200°C. It can be seen from Figs. 8-10 that cellulose sulfate and sodium sulfoethylcellulose retain their

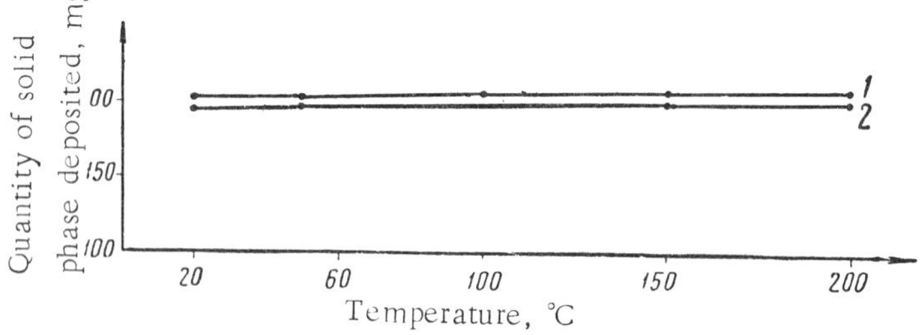

Fig. 8. Sedimentational stability of chalk suspensions treated with 1% STs (1) and 1% ESTs (2) at different temperatures.

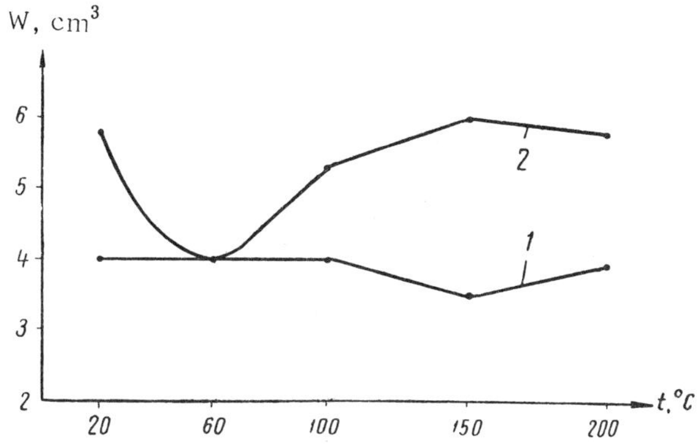

Fig. 9. Water loss for chalk suspensions treated with 1% STs (1) and 1% ESTs (2) at different temperatures.

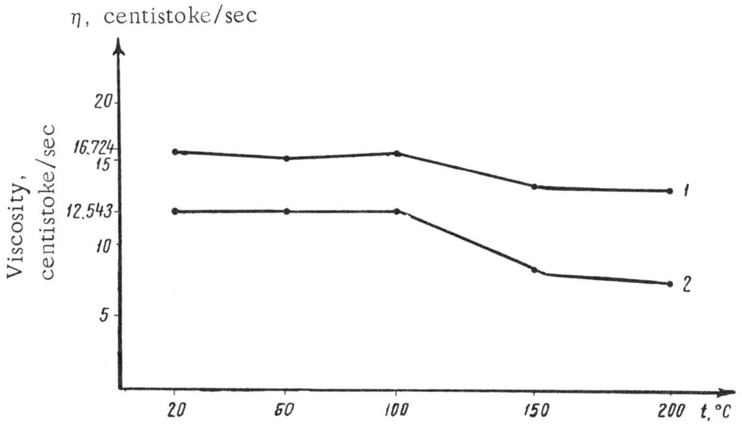

Fig. 10. Viscosity of chalk suspensions treated with 1% STs (1) and 1% ESTs (2) at different temperatures.

stabilizing properties in the temperature range studied. Thermal treatment of the reagents under the conditions described did not change their surface activity; the sedimentational stability (Fig. 8) and water loss (Fig. 9) remained for the most part at the same level. These data enable cellulose sulfate and sodium sulfoethylcellulose to be classified as heat-resistant surfactants and this increases even further their values as stabilizers of aqueous mineral suspensions.

Summary

New synthetic surfactants—the sodium salts of cellulose sulfate and sulfoethylcellulose—are extremely effective stabilizers for mineral suspensions of nonclay rocks in fresh or mineralized waters. These surfactants also retain their stabilizing properties at increased temperatures.

THE USE OF SURFACE-ACTIVE COLLOIDS IN THE PETROLEUM INDUSTRY

K. F. Zhigach

Academician P. A. Rebinder has described in detail the theoretical principles of the application of surfactants and their importance in the national economy. Four types of surfactant are at present used in industry.

1. Potential-determining surface-inactive substances, i.e., substances which are not adsorbed at a water–air boundary. These include first of all the ordinary electrolytes: sodium chloride, potassium chloride, aluminum chloride, sodium carbonate, alkali, etc. These substances are also effective as reagents for decreasing hardness during drilling at low concentrations, and in other processes involving control of the properties of dispersed systems. Their action is due to the development of an electrical double layer at the surface of the particles.

2. Organic surface-active electrolytes or, as they are called, ionogenic surfactants with an active organic cation or anion.

3. Nonelectrolytes, nonionogenic surfactants.

4. Surface-active colloids and hydrophilic macromolecular compounds; it is with these that the present paper is concerned.

Whereas electrolytes act by controlling the properties of the electrical double layer, in the case of surfactants of the second and third types (organic electrolytes and nonelectrolytes) the actual decrease in surface energy accompanying the adsorption of surfactant on the phase boundary is naturally of considerable importance [2].

Surface-active colloids are of two types: electrolytes and nonelectrolytes. Their action involves a decrease in the surface energy at the boundary of separation between the phases with the formation, on the surface of separation, of a protective structured film which may produce a marked change in the properties of various disperse systems.

In recent years, surface-active colloids have begun to acquire ever-increasing importance in the petroleum industry and in other branches of industry. These materials include natural plant colloids, sulfite-alcohol waste, which has for long been used in the petroleum industry, starch and its derivatives, alkaline extracts of peat, brown coal, water-plant extracts, etc. All these products are widely used as protective colloids during drilling and in a number of other branches of industry—the textile industry, the manufacture of adhesives, etc.

In recent years the demand for these protective colloids has continued to increase. At the same time there has been a corresponding increase in the quality required. This has raised the question of the utilization of industrial synthetic macromolecular compounds with the required properties, in addition to natural plant colloids. Effective synthetic compounds include the derivatives of water-soluble and alkali-soluble cellulose ethers and esters. The production of these materials increases annually. This is particularly true of KMTs (carboxymethylcellulose); the technology of the application of this material was developed in the MINKh i GP. Carboxymethylcellulose is a product obtained by condensing alkali cellulose with sodium monochloroacetate. It was the first (and is possibly at present the most effective) protective colloid.

A very promising group of materials consists of the cellulose sulfates, prepared by treatment of cellulose with sulfuric acid in the presence of various alcohols which moderate the sulfonation process. Another is carboxyethylcellulose, which is prepared by treating cellulose with acrylonitrile in the presence of alkali. Carboxyethylcellulose is widely used abroad as a filler in the textile and food industries and a number of other fields. The preparation and use of these materials have been described in detail earlier [1].

Interesting work on the preparation of surface-active reagents based on oxidized lignin (lignin is a material in plentiful supply) is being carried out in the VNIIBT under the direction of I. B. Adel' with the cooperation of N. N. Shorygina (Organic Chemistry Institute, USSR Academy of Sciences). Treatment with a dilute solution of nitric acid or with chlorine, followed by treatment with alkali, gives a very effective reagent for decreasing the viscosity of clay suspensions.

At the present time, a number of products are being synthesized from phenols obtained from wood and coal tars—the so-called synthanes.

Thus there are now available many extremely effective surface-active stabilizers for the different branches of the petroleum industry.

Let us consider KMTs, which is most widely used in the petroleum-recovery industry.

Three grades of KMTs have now been developed and are being produced industrially. These include the high-viscosity material known as GETs which is added to clay-free and silicate-salt liquids during drilling, the intermediate-viscosity material which is used in drilling, and the low-viscosity KMTs, which is a very effective reagent for decreasing viscosity in structured systems. Thus different products are obtained depending on the conditions of KMTs production—temperature, duration of aging of the alkali cellulose, etc. The properties of these materials can be controlled in two directions: by degradation of the cellulose, in which its large molecules are broken down to smaller fragments, or by increasing the number of substituent functional groups. Even when we obtain a product with definite chemical properties, these products may nevertheless differ from one another, since a KMTs solution is a colloidal polydisperse system. Fractionation of an ordinary KMTs specimen with a given chemical composition gives a number of fractions with different viscosities and with different stabilizing and peptizing properties.

When 0.25% of a high-molecular KMTs fraction and 0.25% of a low-molecular fraction are added to a suspension, the viscosity of the system decreases sharply, although the total concentration increases, i.e., the high-molecular fractions combine with the low-molecular fractions and block them, and the structure breaks down. On the basis of chemical composition, by controling the degree of dispersion, it is thus possible to obtain a whole range of products with different viscosity and molecular weight, whose fractions react actively with one another. We have carried out work together with the VNIIBT on the use of KMTs in drilling, and together with the Fat Industry Institute on the use of KMTs in the production of wetting agents.

The function of the KMTs stabilizer in wetting agents is the same as that in drilling, and involves the stabilization of dispersed particles. It is known that it is possible to launder a fabric with synthetic products in the absence of KMTs, but if the fabric was white before being dirtied, it appears grey after laundering. This is due to the fact that the very fine particles of dirt which have been washed out are not stabilized by the other components of the wetting agent and are redeposited on the fabric and adhere to it. To keep these particles in the suspended state, the suspension must be effectively stabilized. KMTs and a number of its products act as very effective stabilizers.

It is necessary to select an appropriate product with a degree of polymerization and etherification giving maximum activity.

In a number of cases a low-quality product is obtained because the necessary technological conditions have not been ensured. The synthesis of these products is fairly complex. The establishment of their production requires an intensive study of their chemical and colloidal properties, etc.

The use of KMTs is particularly effective in aggressive media when the suspension of drilled rock contains a large quantity of salts.

In these cases the addition of cellulose products can give effectively stabilized suspensions with a high concentration of salts. KMTs is not a particularly effective stabilizer in the presence of calcium chloride, however, so that most attention is being paid to the production of cellulose sulfates or sulfonated cellulose ethers. These are effective reagents which are stable in aggressive media containing calcium, and are moreover obtained from a more plentiful raw material.

KMTs is more effective in sodium chloride solution. Calcium chloride has almost no influence on the properties of clay suspensions but it reduces the efficiency of action of KMTs.

The same is true of the static shearing stress. As the suspension becomes contaminated, in the presence or absence of KMTs, the limiting shearing stress increases, i.e., the suspension become thicker. In the presence of sulfonated cellulose, however, almost no thickening takes place. It is therefore necessary to begin the large scale production of sulfonated celluloses in addition to KMTs.

Surface-active colloids are used as active stabilizers of dispersed systems. They effectively prevent coagulation in a variety of aggressive coagulating media.

P. A. Rebinder has spoken of the influence of various materials which act as reagents decreasing hardness. Let us examine the question of whether surface-active colloids decrease hardness and how they influence the mechanical rate of drilling.

First of all, they produce high-quality flushing liquids and thus ensure accident-free drilling. Observations show that their use gives no marked increase in the mechanical rate of drilling.

P. A. Rebinder has pointed out that use is now being made of flushing liquids containing different substances which decrease hardness. It may be added that any surfactant decreases the hardness of rock during drilling but also decreases the hardness of the tool. The total effect obtained may be positive (i.e., correspond to an increase in the mechanical rate of drilling) or negative, depending on which of the above factors predominates.

A study has been made of the influence of various additives in flushing liquids on tool wear. As a result of a study of wear in hardened steel during friction against quartz, it was established that typical potential-forming substances (such as electrolytes and the alkali Na_2CO_3) increase wear at low concentrations. When KMTs is used, the tool wear decreases by a factor of almost 4.

This has been established not only by laboratory experiments but also by data obtained during drilling in a number of regions. In turbine drilling with KMTs additives, for example, increases were noted in the turbodrill's period of operation between overhauls, the distance drilled per bit, etc.

Whereas 20 years ago the problem was to find reagents (for decreasing hardness) to increase the mechanical rate of drilling (at the present time reagents which decrease hardness are always used during drilling), a new problem has now arisen which is perhaps more important: it is necessary, by adding surface-active substances to flushing solutions, to ensure a high stability in the cutting tool in order to increase the distance cut per drill bit. This is particularly important during deep drilling, since the lifting and lowering operations take up a great deal of time. Reagents for decreasing the hardness of rock have now been introduced and are giving the required effect.

A search must be made to find additives which increase the wear-resistance of the drill by improving the lubricating and cooling action as a result of the formation of a colloidal film on the metal surface in order to ensure the maximum possible distance traveled per bit by the use of surfactants. For this purpose it is necessary, in the evaluation of different flushing liquids, to introduce the criterion of metal wear-resistance in addition to the criteria of viscosity, filtration, and the other usual parameters.

In recent years, studies have revealed another important field of application of surfactants, in this case water-soluble sulfates or sulfonated ethers. P. A. Rebinder and co-workers on the basis of studies on the theory of setting of mineral binding agents, have developed theories which are now generally accepted) as to the influence of added surfactants in cements. The materials discussed here can also be used as additives in cement mixes and concretes, and this is of particular importance in the cementation of deep wells under conditions of cement hardening at high temperatures. The additives increase the strength of the cement block and decrease its permeability. In this case, cellulose sulfates or sulfonated cellulose ethers are particularly effective.

We have prepared artificial cores of cement block and studied their strength and permeability by different methods under pressure. The filtration of water through ordinary cement block in 1 h amounted to 30 cm^3 but this decreased to zero when 0.1% of sulfonated cellulose was added. It has thus been possible to prepare practically water-impermeable and petroleum-impermeable cement block.

A number of American patents have recently appeared recommending the use of sulfonated celluloses as cement additives.

We shall not consider in detail the uses in other branches of industry but merely point out that KMTs, particularly the low-viscosity material, has now proved to be an active flotation reagent. In the textile and other industries these reagents are used as thickening agents in print dyeing and for other purposes.

Nonaqueous flushing liquids have also been developed in the MINKh i GP. These are complex nonaqueous hydrocarbon systems in which the liquid phase is diesel fuel with a certain concentration of aromatic and other hydrocarbons. Oxidized bitumen is used as colloidal structure-forming phase; during filtration this blocks the pores of the stratum, preventing the penetration of flushing liquid, and this is very important. Oxidized paraffins with a definite degree of oxidation are also used as products facilitating structure formation. These systems are designed for drilling under extremely complicated conditions where ordinary flushing liquid is ineffective; they are also used in the case of well collapse and in the opening of oil-bearing strata, to prevent penetration of water into the stratum.

Laboratory studies and industrial tests show that in the opening of strata the productivity of wells is increased considerably by the use of nonaqueous liquids. The following data, in particular, were obtained in the Tatar Autonomous Republic. The quantity of petroleum obtained in four years from well No. 362 was greater by 15 thousand tons than that obtained from neighboring wells drilled in better sections of the stratum. Thus the use of nonaqueous systems appears extremely promising. These liquids have found extensive use in the USA.

In order to facilitate their production, we have developed a formulation for the preparation of a bitumen powder containing added reagents. This powder can be dissolved and used for the rapid production of a high-quality liquid, e.g., a mixture of bitumen and quicklime.

NChK is also a very effective structure-forming additive. Comparisons have been made of OP-10, OP-20, and various American materials containing NChK, which are of course less satisfactory since they contain not more than 10% of the active constituent. The addition of Georgian NChK to Tuimazinskii petroleum, however, gives a more stable emulsion. This is quite understandable since this NChK contains only 2% of the active constituent, the rest being asphaltenes which are not demulsifiers but are emulsion stabilizers. Before NChK is rejected, industry must be made to produce NChK to a strict standard and improve the technology of its production.

Literature Cited

1. K. F. Zhigach, M. Z. Finkel'shtein, E. M. Mogilevskii, and I. M. Timokhin, Khimicheskaya nauka i promyshlennost', 4, 6, 718 (1959).
2. P. A. Rebinder, L. A. Shreiner, and K. F. Zhigach, Reagents for Decreasing Hardness During Drilling (Izd. AN SSSR, Moscow–Leningrad, 1944).

THE STABILIZATION OF HIGHLY MINERALIZED SUSPENSIONS OF DRILLED NONCLAY ROCKS (From Data Obtained in Prospecting Regions in Eastern Siberia)

V. A. Glebov and S. N. Yatrov

The problem of the stabilization of highly mineralized suspensions of nonclay rocks is of considerable theoretical and practical importance. Studies in recent years have shown that the chemical reagents widely used in modern drilling technology to increase the aggregative stability of drilling fluids belong to the class of surfactant stabilizers [1, 2, 3] whose adsorptive action is the chief factor in the stabilization of aqueous mineral suspensions. Inorganic electrolytes (in this case NaCl) are active additives which selectively increase the surface activity of the stabilizer reagents [3, 4, 5]. The combined use of surfactant stabilizers and mixtures of surfactants with active additives can also increase selectively the stabilization of highly mineralized suspensions of drilled nonclay rocks [1, 3, 5].

On the basis of these considerations, experiments were designed to study the stabilization of artificial highly mineralized suspensions and suspensions with a high degree of mineralization formed during the drilling of wells under the conditions obtaining in prospecting regions in Eastern Siberia.

In the stratigraphic section of the areas drilled in this region, carbonate rocks (dolomites, limestones), halide rocks (rock salt), argillites, and aleurolites predominate; clay rocks are almost completely absent. Because of the thick deposits of rock salt, the degree of mineralization of the suspensions is very high; in the absence of complications the drilling fluid consists of a suspension supersaturated with sodium chloride. In a large part of the section, particles of rock salt are present in the disperse phase of the drilling fluid during drilling.

The stabilization and control of the properties of clay suspensions prepared from imported clays under conditions of such high mineralization involve considerable difficulties and up to the present they represent an

TABLE 1

Parameters of original liquid			Quantity of reagents added					Parameters of resulting fluid		
ρ, g/cm^3	T, sec	W, cm^3/30 min	Starch	SSB	KMTs	NaOH	Bentonite	ρ, g/cm^3	T, sec	W, cm^3/30 min
1.20	15	100	2	–	–	2	–	1.20	20	15
1.20	15	100	4	–	–	2	–	1.20	35	10
1.18	15	100	0.6	1	1	1.2	0.5	1.18	26	15
1.18	15	100	0.3	1	2	1	0.5	1.19	48	12
1.18	15	100	0.6	1	2	0.7	0.5	1.18	51	12
1.18	15	100	0.3	1	2	0.5	0.5	1.19	60	7

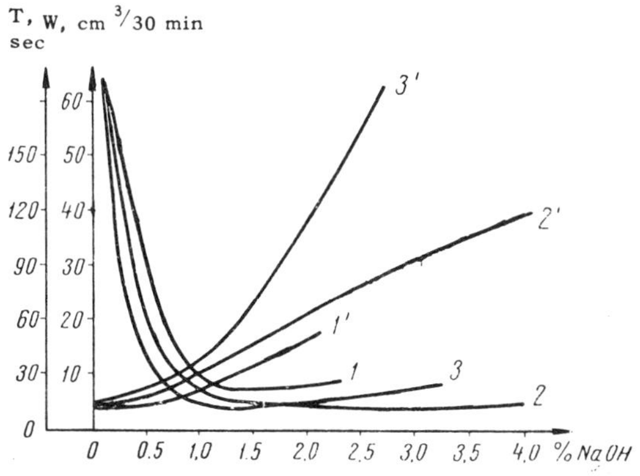

Fig. 1. Change in the parameters of a highly-mineralized carbonate suspension taken from well 3-R. 1 and 1') Water loss and viscosity respectively for the addition of 20% KShchR (2 : 6) to the suspension; 2 and 2') water loss and viscosity respectively for the addition of 25% KShchR (3 : 6) to the suspension; 3 and 3') water loss and viscosity respectively for the addition of 30% KShchR (3 : 6) to the suspension.

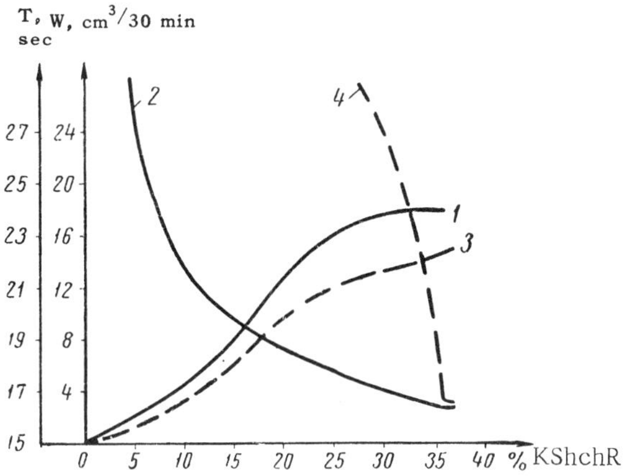

Fig. 2. Change in the parameters of a carbonate-halide suspension taken from well 2-R. 1 and 2) Viscosity and water loss respectively for a suspension treated with KShchR and 1% NaOH; 3 and 4) viscosity and water loss respectively for a suspension treated with KShchR + 10% NaOH + 1% SSB.

unsolved problem. As a rule, the properties of clay suspensions deteriorate rapidly with increase in the mineralization of the dispersion medium; the system separates into layers and coagulation takes place. Moreover, the preparation of clays and clay suspensions in Eastern Siberian regions is very tedious and uses up much time and material. The clay often has to be transported over distances of up to 400-500 km under difficult conditions.

The possibility of producing and using stable drilling fluids with a disperse phase consisting of a mixture of drilled nonclay rocks from these regions is naturally of particular importance. In our experimental studies of highly-mineralized suspensions of drilled nonclay rocks, we examined the stabilizing action of the following: a) individual surfactant stabilizers from the chemical reagents used in drilling; b) different combinations of these surfactants; and c) mixtures of surfactant stabilizers with active additives—inorganic electrolytes.

The surfactant stabilizers studied included sulfite-alcohol waste, carboxymethylcellulose, and starch; the active additives studied were sodium hydroxide, sodium silicate (water glass), and the rock salt present in the fluid. A small percentage of bentonite was also used as active additive.

The effect of different combinations of these surfactants and active additives on a mineralized suspension is shown in Table 1. In the experiments rock salt was dissolved in fresh water until a completely saturated solution was obtained, after which the surfactants and active additives were introduced. It can be seen from Table 1 that in the presence of small quantities of starch, SSB, and KMTs, in different combinations and in mixtures with active additives (NaOH and bentonite), the water loss from the brine decreases sharply (from complete filtration to 15-7 cm^3), indicating stabilization of these highly mineralized suspensions.

The change in the parameters of a natural highly-mineralized carbonate suspension taken from exploratory well 3-R (Sheloninskaya area) during treatment with starch-alkali reagent (6% starch and 2-3% NaOH) with the further addition of NaOH, is shown in Fig. 1. The original parameters of the suspension were as follows: $\rho = 1.20\ g/cm^3$, relative viscosity (relative to SPV-5) T = 16 sec, and water loss W > 100 cm^3 in 30 min.

The change in the parameters of a carbonate-halide suspension, taken from well 2-R at a depth of 2400 m, during treatment with starch-alkali reagent containing added NaOH and SSB, is shown in Fig. 2.

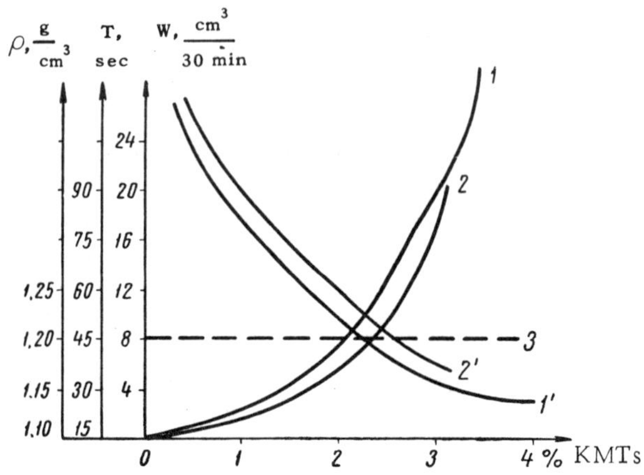

Fig. 3. Change in the parameters of a highly mineralized suspension for treatment with KMTs and added bentonite. 1 and 1') Viscosity and water loss respectively for a suspension containing 0.5% bentonite; 2 and 2') viscosity and water loss respectively for a suspension containing 1% bentonite; 3) density.

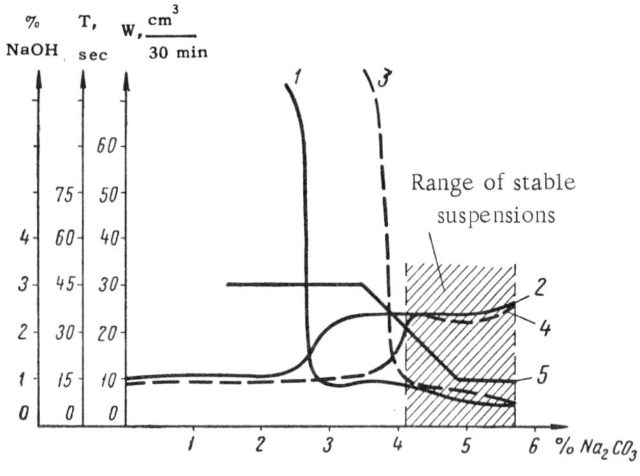

Fig. 4. Change in the properties of a highly mineralized natural carbonate-halide suspension for treatment with KShchR (2 : 6), SSB, NaOH, and Na_2CO_3. 1 and 2) Water loss and viscosity respectively for a suspension treated with 20% KShchR (2 : 6), NaOH, and Na_2CO_3; 3 and 4) the same as in 1 and 2, + 1% SSB; 5) quantity of NaOH added to the suspension during complex treatment.

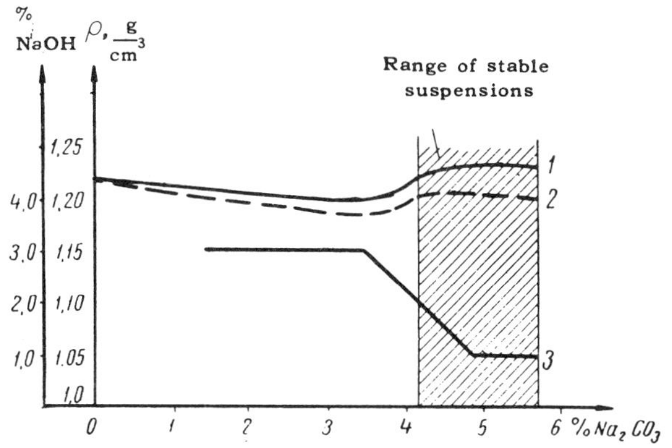

Fig. 5. Change in the specific gravity of a highly mineralized suspension during treatment with KShchR (2 : 6), SSB, NaOH, Na_2CO_3. 1) Density of a suspension treated with 20% KShchR (2 : 6), NaOH and Na_2CO_3; 2) the same as in 1 + 1% SSB; 3) quantity of NaOH added to the suspension during complex treatment.

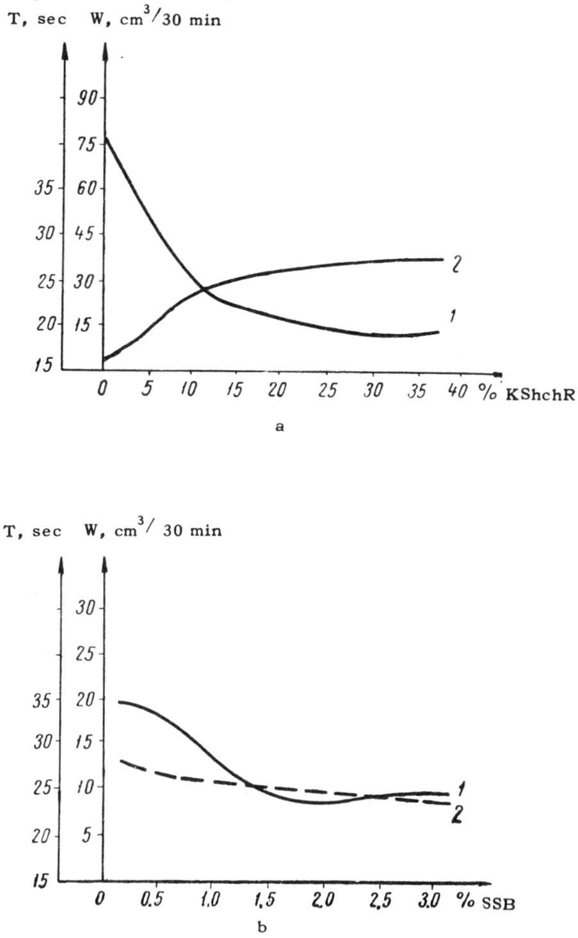

Fig. 6. Change in the parameters of a natural carbonate-halide suspension (parameters of the original suspension: $\rho = 1.34$ g/cm^3; T = 16 sec; W = 75 cm^3/30 min) with the addition of: a) 1.5% SSB and KShchR; b) 30% KShchR and SSB. 1 and 2) Water loss and viscosity respectively.

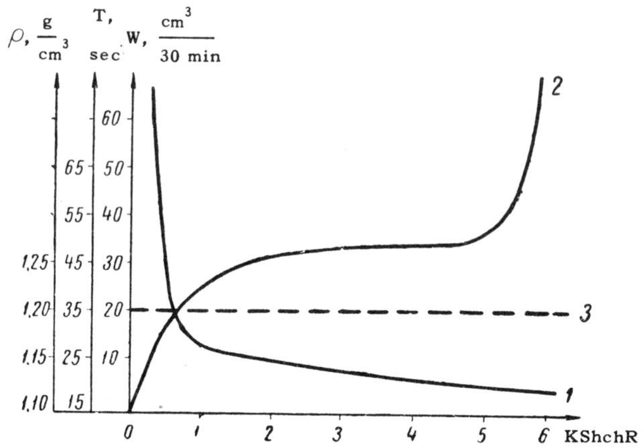

Fig. 7. Change in the parameters of a highly mineralized suspension from well 3-R (parameters of original suspension: $\rho = 1.20$ g/cm^3; T = 15 sec; W > 100 cm^3/30 min) during complex treatment with starch (1.5%), SSB (1%), KMTs and NaOH (1.2%). 1) Water loss; 2) viscosity; 3) density.

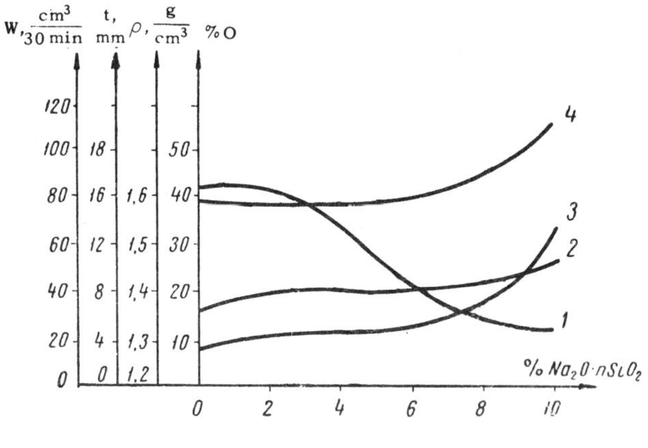

Fig. 8. Change in the properties of a natural highly mineralized suspension from well 2-R (parameters of original suspension: $\rho = 1.34$ g/cm^3; T = 16 sec; W = 75 cm^3/30 min) during treatment with sodium silicate. 1) Amount of settling per day; 2) density; 3) thickness of filtration crust; 4) water loss.

TABLE 2

Parameters of original suspension				Quantity of reagent added		Parameters of resulting suspension			
ρ, g/cm³	T, sec	W, cm³/30 min	θ_1, mg/cm²	Na_2O and SiO_2	NaOH	ρ, g/cm³	T, sec	W, cm³/30 min	θ_1, mg/cm²
1.18	19	6	0	0.5	–	1.16	16	30	0
1.18	19	6	0	1.0	0.5	1.18	16	10	0
1.18	19	6	0	1.5	0.75	1.18	16	10	0
1.18	19	6	0	2.0	1.0	1.18	16	10	0
1.18	19	6	0	3.0	1.5	1.18	17	12	0
1.18	19	6	0	3.0	1.5	1.18	17	12	

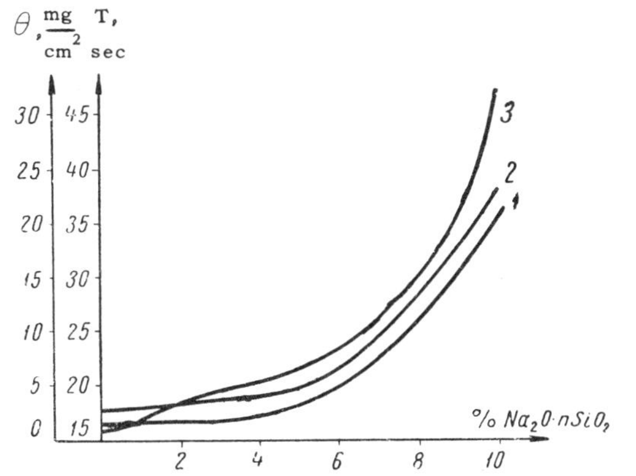

Fig. 9. Influence of sodium silicate on the structureal-mechanical properties of a highly-mineralized suspension. In the presence of 10% $Na_2O \cdot nSiO_2$, coagulation of the solid phase takes place (a large quantity of sludge in coarse aggregates forms on the walls of the viscometer). 1) Limiting shearing stress θ_1; 2) limiting shearing stress θ_{10}; 3) viscosity.

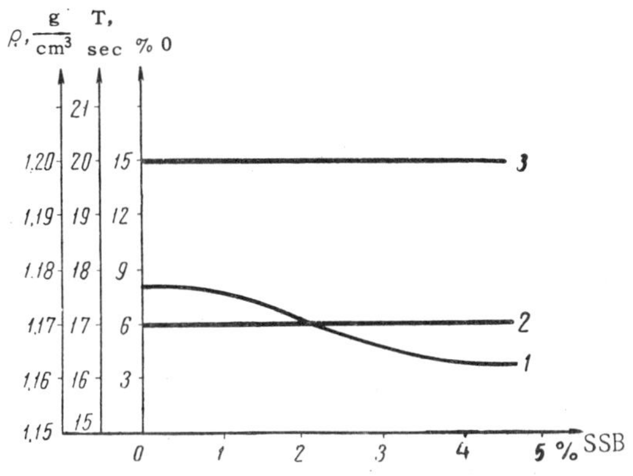

Fig. 10. Influence of SSB on the parameters of a highly mineralized suspension. 1) Water loss; 2) density; 3) viscosity.

It can be seen from Fig. 1 that the starch-alkali reagent stabilizes these suspensions. With increase in the alkalinity and hence increase in pH, the water loss decreases more intensively (curves 1 and 2 in Fig. 1) but only as far as a certain limit (curve 3). As is well known, starch is an effective reagent for decreasing the water loss from clay suspensions with fairly high pH, but depending on their method of preparation, not all starch specimens are effective for mineralized suspensions [6]. The addition of SSB leads to a more intensive decrease in the water loss (curve 4 in Fig. 2). The change in the parameters of a highly mineralized suspension during treatment with KMTs and added bentonite is shown in Fig. 3. This combination of surfactant and additive also decreased the water loss to minimum values. The stability of suspensions treated with stabilizer alone, in this case the starch-alkali reagent (Fig. 1) and KMTs (Fig. 3), is not permanent; the water loss properties deteriorate in one to three days.

With a considerable increase in the concentration of the starch reagent (up to 20%) and the mixture of active electrolyte additives (NaOH and Na_2CO_3) in optimum proportions, stable suspensions are obtained whose stability is permanent (Figs. 4 and 5).

Treatment of the unstable suspensions taken from the wells with starch reagent and SSB, starch reagent SSB alone, and KMTs also gave stable suspensions (Figs. 6 and 7). An increase in the stability of the stabilized suspensions with increase in the quantity of the components was observed. Thus a suspension treated with starch reagent, SSB, and KMTs showed almost no change in properties over a period of 90 days. The stabilization of unstable suspensions by the simultaneous action of starch reagent, SSB, and KMTs, was used in the drilling of exploratory wells in the Sheloninskii area of IrkutskOblast'. Good results were obtained: in a suspension with initial parameters $\rho = 1.20$ g/cm³, T = 15 sec, and W > 100 cm³, the water loss after addition of the reagents decreased to 6-7 cm³ in 30 min; the relative viscosity increased slightly to 22-24 sec.

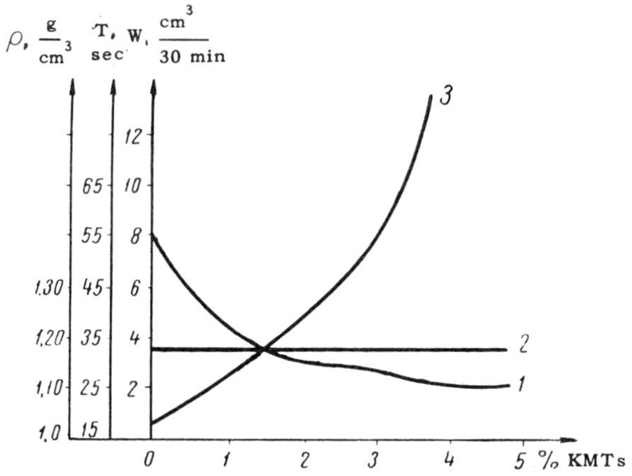

Fig. 11. Influence of KMTs on the parameters of a highly-mineralized suspension. 1) Water loss; 2) density; 3) viscosity.

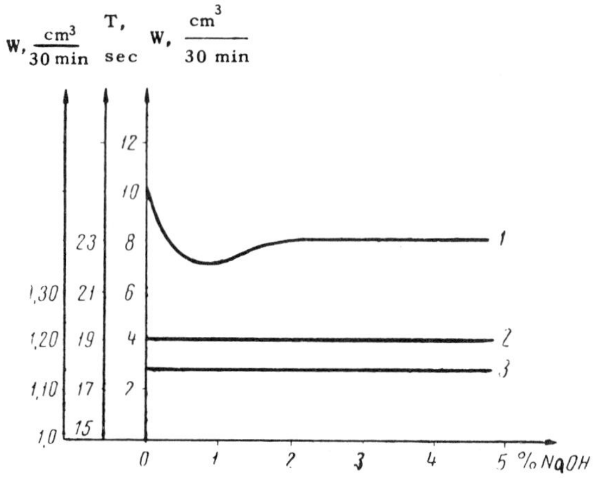

Fig. 12. Influence of NaOH on the parameters of a highly mineralized suspension. 1) Water loss; 2) density; 3) viscosity.

We studied the influence of active additives—inorganic electrolytes—on unstable, highly mineralized suspensions of nonclay rocks. The change in the rheological and structural-mechanical properties of this suspension as sodium silicate ($Na_2O \cdot nSiO_2$) is added as shown in (Figs. 8 and 9. It can be concluded from these curves that active additives also have a stabilizing action on unstable suspensions: increase in the $Na_2O \cdot nSiO_2$ content leads to a considerable decrease in the amount of settling per day and increases the density (Fig. 8). The mechanism of action of the active additives differs from that of the surfactant stabilizers, however, since although the sedimentational stability increases, the filtration properties remain unchanged or even deteriorate (Fig. 8). This indicates that the active additives influence chiefly the electrokinetic forces at the surface of separation between the phases and do not create adsorption layers with the required structural-mechanical properties or orientated ("combined") water molecules. This mechanism of the action of $Na_2O \cdot nSiO_2$ also explains the increase in the structural-mechanical properties of the suspension with increase in the concentration of this structure-forming electrolyte (Fig. 9).

The stabilized, highly mineralized suspensions in the well during drilling were subjected to further treatment with surfactant stabilizers. Their filtration properties improved considerably and the water loss decreased to minimum values of 3-2 cm^3. The influence of further additions of SSB on the properties of a suspension stabilized with starch, KMTs, and SSB (a suspension taken from well 3-R at a depth of 2650 m) is shown in Fig. 10, and Fig. 11 shows the influence of the addition of KMTs on the properties of a suspension treated with starch, NaOH, and Na_2CO_3 (a suspension taken from well 2-R at a depth of 2700 m). Further addition of electrolytes to the stabilized suspension does not lead to an improvement in its filtration properties: the water loss remains at its previous value (Fig. 12) or even increases (Table 2).

The structural-mechanical properties of highly mineralized suspensions are not improved by the addition of electrolytes under these conditions (Table 2). Further studies are being made of these problems.

Literature Cited

1. P. A. Rebinder, "Surface-active substances and their uses," Khimicheskaya nauka i promyshlennost', IV, 5 (1959).
2. Chiang Lung, N. N. Serb-Serbina, and P. A. Rebinder, "A study of structure formation and the rheological properties of aqueous dispersions of bentonite clays in their reaction with humates," Kolloidnyi zhurnal, 23, 3 (1961).
3. S. N. Yatrov, Natural Flushing Liquids in Drilling (Liquids Based on Drilled Rock), (Gostoptekhizdat, 1963).
4. A. B. Taubman, "The physicochemical basis of the wetting action of surface-active substances," Khimicheskaya nauka i promyshlennost', IV, 5 (1959).

5. A. M. Schwartz, J. W. Perry, and J. Berch, Surface-Active Agents and Detergents [Russian translation, edited by A. B. Taubman] (IL, 1960).
6. W. F. Rogers, Composition and Properties of Oil-Well Drilling Fluids [Russian translation],(Gostoptekhizdat, 1960).

ADSORPTION PROCESSES DURING THE FORMATION AND CONTROL OF THE PROPERTIES OF FLUSHING LIQUIDS FROM NONCLAY ROCKS

S. N. Yatrov

High hydrophilic properties (the ability to undergo considerable swelling and spontaneous dispersion in water to produce colloidal particles), characterizing a high degree of molecular interaction (affinity), are now necessary in materials used as the disperse phase in drilling fluids. These requirements are satisfied by colloidal (sodium) clays. Drilling fluids from sodium clays are sufficiently stable and their rheological and structural-mechanical properties can be controled over a wide range. Under straightforward drilling conditions it is possible to use two-component systems consisting of clay and water without the addition of surfactant stabilizers and electrolytes.

Depending on the degree of their physicochemical interaction with the aqueous medium, however, clays provide an extremely varied range of materials for the production of drilling fluids.

Calcium clays, for example, do not exhibit such favorable properties: they are less extensively hydrated and their dispersion and swelling in water is poor. The least active in water are kaolinite clays; practical experience has shown that drilling fluids of suitable quality cannot be obtained from kaolinite clays without the addition of surfactant stabilizers.

Electron-microscope photographs of montmorillonite, halloysite, and kaolinite are shown in Figs. 1, 2, and 3 [1]. Particles of montomorillonite and bentonite clays, in which this mineral predominates, from long fine flakes with indistinct edges which have undergone considerable swelling and have lost their shape to some extent (Fig. 1). This characteristic feature of particles of montmorillonite in water is due to the high degree of interaction at the surfaces of separation between the phases. Particles of halloysite (Fig. 2) have a very elongated (acicular) shape. According to P. A. Rebinder's views, elongated shapes in particles (Figs. 1 and 2) facilitate structure formation and adhesion of particles at regions with a high radius of curvature under the influence of active electrolyte additives, and produce a structured carcass framework in the system [2]. Particles of kaolinite in water form clearly-defined distinct hexagonal platelets (Fig. 3), indicating insufficient interaction (affinity) between this type of clay and the aqueous medium.

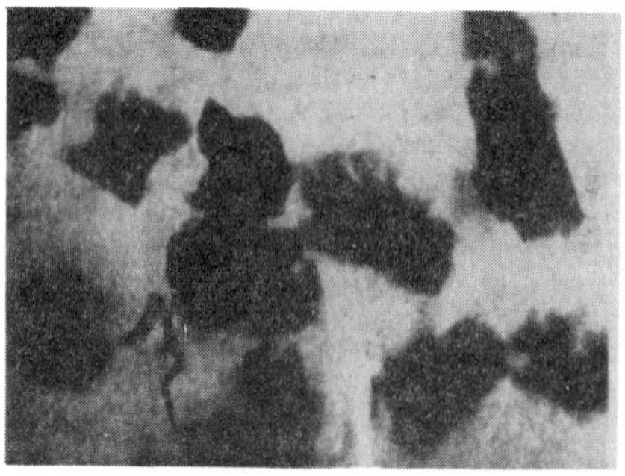

Fig. 1. Photomicrograph of montmorillonite.

Fig. 2. Photomicrograph of halloysite.

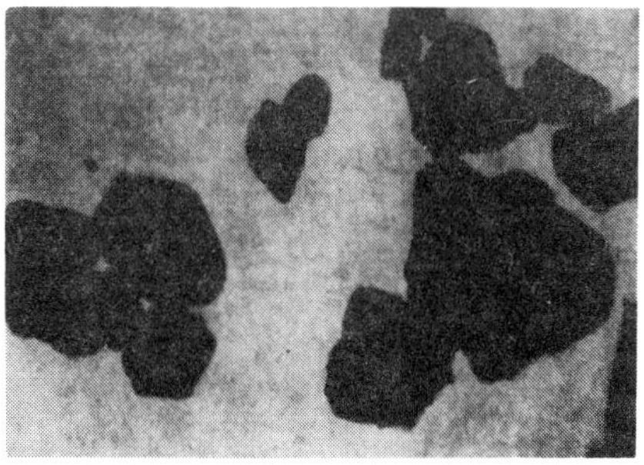

Fig. 3. Photomicrograph of kaolinite.

The physicochemical characteristics of nonclay rocks, which are being used on an ever-increasing scale as the basis of the solid disperse phase in flushing liquids, indicate that the interaction which they undergo with the aqueous dispersion medium is insufficient for the production of stable flushing liquids. These rocks require forced dispersion. They are incapable of undergoing spontaneous dispersion in water, their degree of swelling is extremely low, and they contain no ion-exchange complex. The adsorption layers formed when the particles of the disperse phase of nonclay rocks are hydrated do not exhibit the structural-mechanical properties necessary for stabilization of the system. The natural aqueous suspensions formed in a well during the drilling of carbonate, sulfate, and other nonclayey rocks form dilute disperse systems; in a state of rest, the disperse phase and the dispersion medium separate in a few minutes. Stabilization of such systems can be achieved only by the addition of surfactant stabilizers which create adsorption layers with the necessary structural-mechanical properties: viscosity, elasticity, and shear strength. Under the electron microscope, particles of nonclay rocks show clearly defined distinct outlines, indicating a comparatively low degree of molecular interaction and inability to undergo spontaneous dispersion and swelling. Electron-microscope photographs of anhydrite, argillite, and aleurolite are shown in Figs. 4, 5, and 6. Anhydrite forms ellipsoidal particles with radial projections (Fig. 4). The shape of argillite and aleurolite particles also approximates to that of elongated polyhedra (Figs. 5 and 6). The particles of these rocks do not form such markedly elongated platelets with high radii of curvature as those formed by montmorillonite and, more particularly, halloysite.

Thus electron-microscope studies of the disperse phase of aqueous clay and nonclay colloidal suspension systems confirm the physicochemical surface properties of the particles, the difference in the degree of molecular interaction (affinity) with the dispersion medium, and the differences in shape (structure).

These interaction characteristics are factors exerting a decisive influence on the physicochemical properties of drilling fluids. The characteristic features of the surface adsorption interaction are responsible for the inherent disadvantages associated with clay suspensions—the marked tendency to undergo considerable thickening and the loss in quality under the influence of mineralized stratal waters and water-soluble rocks. For example, the tendency to undergo spontaneous dispersion leads to an increase in the quantity of particles of the disperse phase in unit volume and hence to an increase in the structural-mechanical properties.

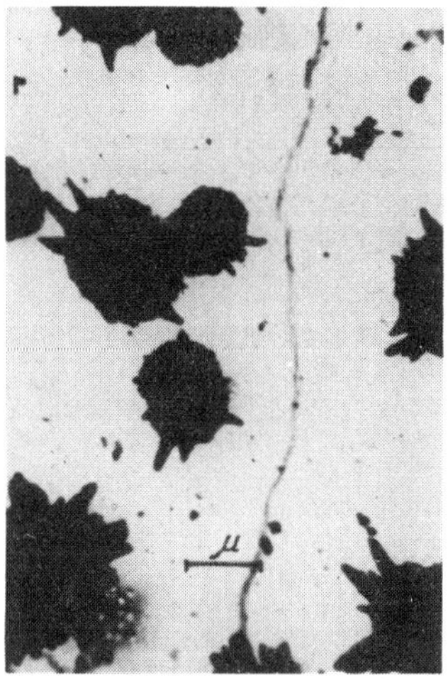

Fig. 4. Particles of anhydrite under the electron microscope.

Fig. 5. Particles of argillite under the electron microscope.

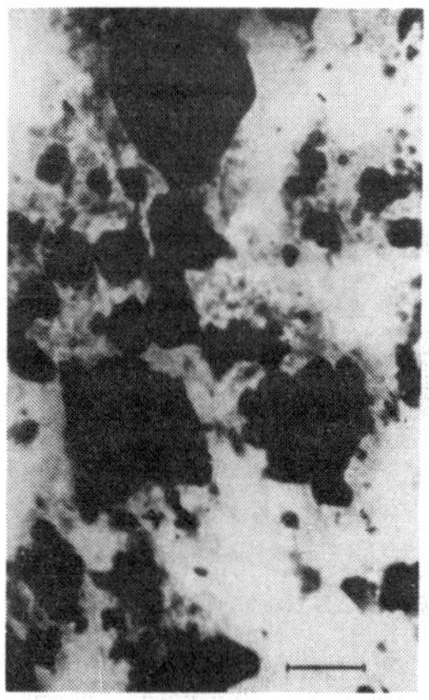

Fig. 6. Particles of aleurolite under the electron microscope.

The chief technological advantage of flushing liquids from nonclay rocks, which distinguishes them from clay suspensions, is their optimum physicomechanical properties and their ability to retain high mobility and low viscosity at high concentrations of the solid phase (sufficiently high specific gravity) and simultaneously to exhibit good filtration characteristics—low water loss with a thin, compact, and nonadhesive filtration crust. As a rule, suspensions of nonclay rocks are thixotropic but have low values of the limiting shearing stress (from standard measurements). These suspensions do not undergo unnecessary thickening and their limiting shearing stress does not increase to extremely high values. Practical experience in drilling wells with natural flushing liquids from nonclay rocks shows that the action of mineralized stratal waters and water-soluble rocks on these suspensions is not so harmful ("aggressive") as that on clay suspensions.

The above properties have a favorable influence in well drilling and lead to an improvement in the technical and economic characteristics by increasing the rate of drilling and reducing its cost.

Thus the characteristic physicochemical features of the interaction between the particles of nonclay rocks and an aqueous dispersion medium (the low degree of hydrophilic character and the characteristic features of the action of ions) in this case are beneficial factors leading to the production of flushing liquids with optimum rheological and structural-mechanical properties.

These results indicate the considerable advantages of nonclay rocks, not only in drilling with natural flushing liquids but also when it is necessary to use imported nonclay rocks in place of clay rocks.

According to modern theories of the physical chemistry of disperse systems, the chief factor in the stabilization of heterogeneous hydrophobic systems, including aqueous suspensions of nonclay rocks, is the adsorption of surfactants from solutions on the surface of separation between the solid particles and the liquid. Studies have been made of the adsorptive action of alkali-carbonate reagent, sulfite-alcohol waste, cellulose sulfate, sulfoethylcellulose, and shale tar sulfonates—reagents used to stabilize aqueous suspensions of drilled nonclay rocks. The adsorption properties were studied with limestone and anhydrite powders, the particle size of the main fractions being 1-2 μ. The studies showed that in spite of the difference in their chemical character, the above reagents, in aqueous suspensions of drilled nonclay rocks, behave as surfactants and are adsorbed on the particles of the disperse phase, decreasing the interfacial surface tension. The adsorption layers

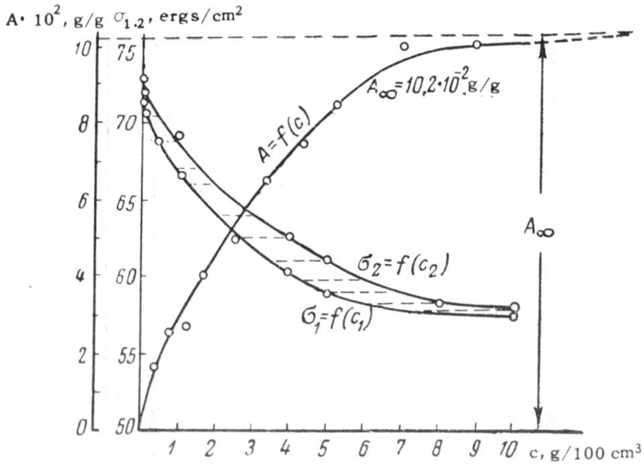

Fig. 7. Surface tension isotherms (aqueous SSB solutions and air) and the adsorption of SSB on limestone particles.

Concentration of ESTs in solution, %									
Before adsorption	2	1.75	1.5	1.25	1.0	0.75	0.5	0.25	0.1
After adsorption	1.87	1.49	1.30	1.03	0.83	0.54	0.41	0.22	0.05

formed in the process exhibit definite structural-mechanical properties which are the chief factor in the stabilization of suspensions, bringing the latter into a study of relative equilibrium and giving them aggregative stability.

Curves showing the change in the surface tension of aqueous solutions of sulfite-alcohol waste (SSB) and shale tar sulfonates (SNS) at the interface in contact with air with change in the concentration of these surfactant stabilizers before and after adsorption, are given in Figs. 7 and 8 respectively; Fig. 9 shows the analogous dependence of σ on c for aqueous solutions of cellulose sulfate (STs) at the interface in contact with heptane. The same figures give the isotherms for the adsorption of SSB on limestone particles and for the adsorption of SNS and STs on anhydrite particles.

The $\sigma = f(c)$ curves show that σ decreases regularly with increase in the concentration of the stabilizer solutions; σ decreases more intensively at low values of c, after which the curves flatten out, corresponding to gradual saturation of the adsorption layer. The more active adsorption at low concentrations of stabilizer and the gradual saturation of the adsorption layer with increase in c are confirmed by the adsorption isotherms [$A = f(c)$]. The limiting adsorption A_∞ (g/g) is determined by the adsorptive properties of the stabilizer under the given conditions and by the degree of dispersion of the powder.

The change in the concentration of sulfoethylcellulose (ESTs) after adsorption on anhydrite powder is shown in the table. The quantity c was determined colorimetrically.

The table also confirms the above description of adsorption processes in the stabilization of mineral suspensions of nonclay rocks.

Experimental studies and practical experience in the use of natural flushing liquids have shown that in most cases it is advantageous to use surfactant stabilizers for mineral suspensions in mixtures with other surfactants exhibiting stabilizing properties, and also in conjunction with substances, in this case surface-active substances, which behave as active additives, increasing the adsorption properties of the main stabilizers and improving the conditions for the control of the physicomechanical properties of drilling fluids. The active additives can, in principle, be inorganic or organic compounds, the former being of greater importance [3]. In our work on the stabilization and control of the properties of mineral suspensions of nonclay rocks, substances of different chemical character, chiefly inorganic electrolytes, were used as active additives [4].

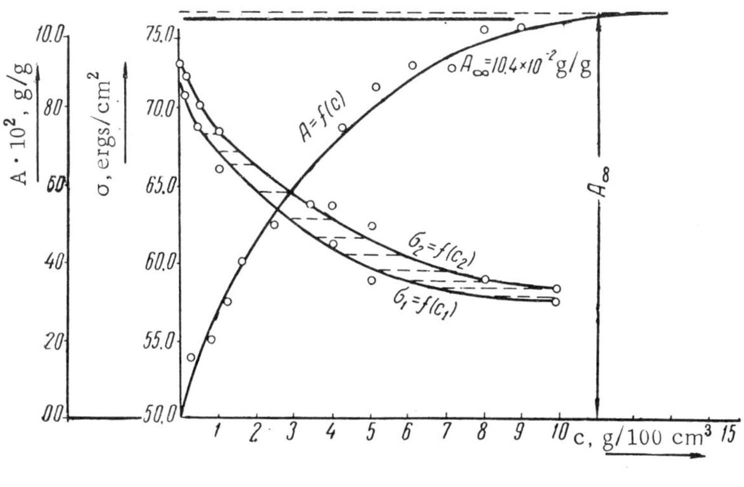

Fig. 8. Surface tension isotherms (aqueous SNS solutions and air) and the adsorption of SNS on anhydrite particles.

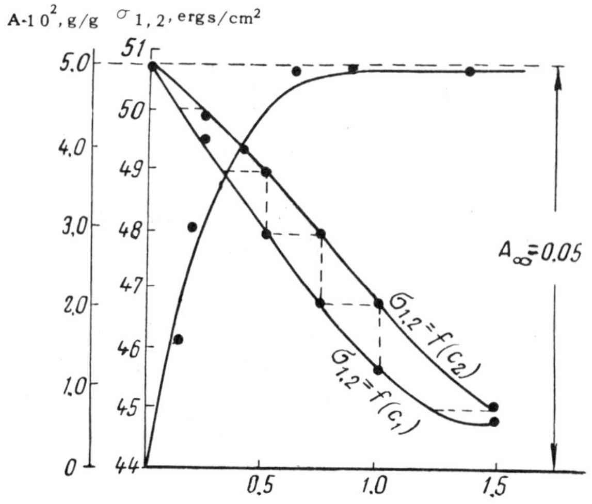

Fig. 9. Surface tension isotherms (cellulose sulfate and heptane) and the adsorption of cellulose sulfate on anhydrite particles.

The increase in the stabilizing power of UShchR and SSB when used together is shown in Fig. 10.

The increase in the stabilizing power of UShchR and SSB when used together with the active additive sodium silicate (water glass) is shown in Fig. 11.

Isotherms for the surface tension at the interface between air and aqueous solutions of shale tar sulfonates (SNS) and the same solutions containing added starch and starch reagent are shown in Fig. 12. It can be seen from Fig. 12 that in the presence of small quantities of the second stabilizer, the adsorption increases and the stabilizing power of the surfactants is raised.

The influence of NaCl on the sedimentational stability of an anhydrite suspension treated with SSB and starch reagent is shown in Fig. 13, and Fig. 14 shows the influence of NaCl on the stability of a suspension treated with cellulose sulfate. In this case, the inorganic electrolyte functions as an active additive, leading

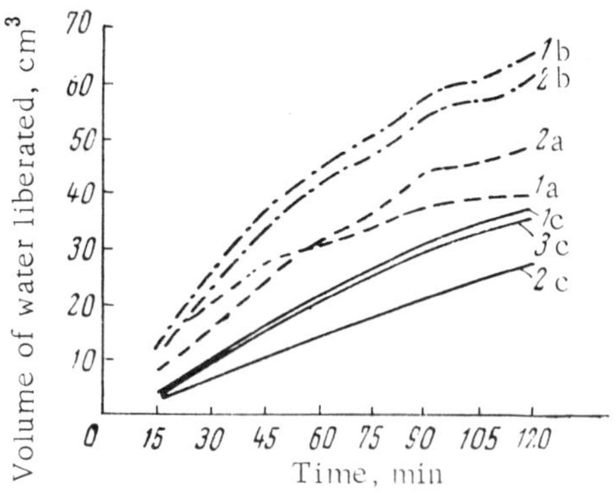

Fig. 10. Additives: 1a) 5% UShchR; 2a) 10% UShchR; 1b) 5% SSB; 2b) 10% SSB; 1c) mixture of 5% UShchR and 5% SSB; 2c) mixture of 10% UShchR and 5% SSB; 3c) mixture of 5% UShchR and 10% SSB.

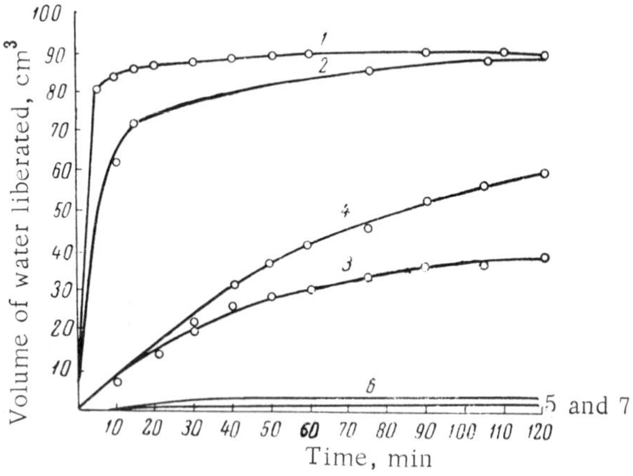

Fig. 11. 1 and 2) Limestone suspension from well; 3) the same with the addition of 10% UShchR; 4) the same with the addition of 10% SSB; 5) the same with the addition of 5% UShchR + 2% water glass; 6) the same with the addition of 5% SSB + 2% water glass; 7) the same with the addition of 2% water glass.

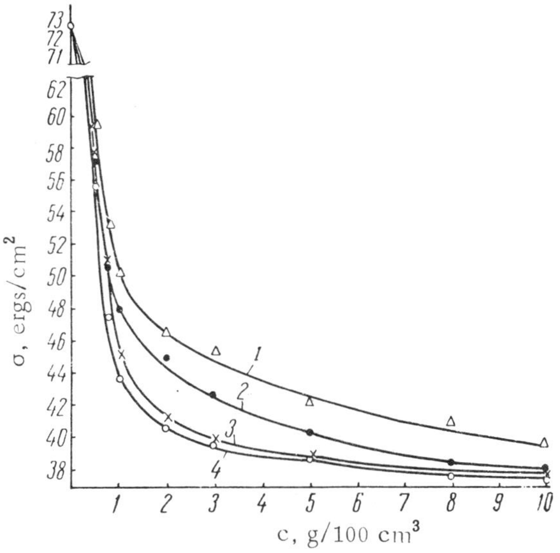

Fig. 12. 1) SNS solution before adsorption; 2) SNS solution after adsorption on anhydrite particles; 3) SNS solution with added starch after adsorption; 4) SNS solution with added starch reagent.

to an increase in the stabilization of the systems. The unique character of the action of inorganic electrolytes—mild coagulants of the water-glass type—on mineral suspensions of nonclay rock is noteworthy. This reagent leads to active stabilization of the suspensions at low concentrations, as can be seen from Fig. 11. With increase in the water glass content the suspension is converted to a sol-like system with regular distribution of the solid phase (Fig. 15). While stabilizing the suspension, in the sense of increasing the sedimentational stability, however, water glass has almost no influence on its filtration properties, and the water loss from the system remains high. Complete filtration of water takes place in a carbonate suspension in the presence of 10% water glass solution, and at 25% the water loss reaches 60 cm^3. This indicates that the electrolytes influence chiefly the electrokinetic forces at the surface of separation between the phases, and also that maximum stability of the systems requires the addition of surfactant stabilizers which produce adsorption layers with viscous and elastic strength properties [5].

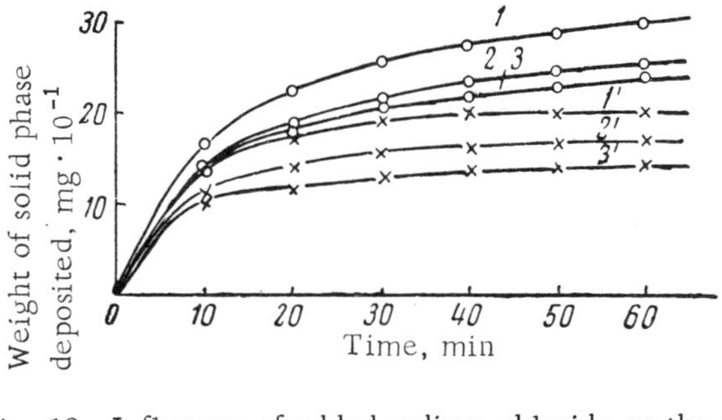

Fig. 13. Influence of added sodium chloride on the stability of an anhydrite suspension from powder No. 1, treated with SSB and starch-reagent. SSB solutions with the addition of 1% starch: 1, 2, 3) SSB concentrations 0.25, 1, and 5% respectively; 1', 2', 3') the same with 10% NaCl in the suspension.

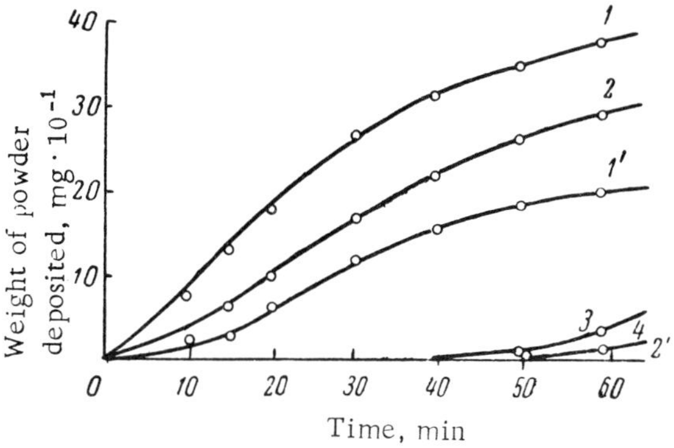

Fig. 14. Influence of cellulose sulfate and added NaCl on the stability of a suspension of anhydrite powder No. 1. Cellulose sulfate solutions: 1, 2, 3, 4) 0.25, 0.5, 1, and 1.5% respectively; 1', 2') the same as 1 and 2 with the addition of 10% NaCl.

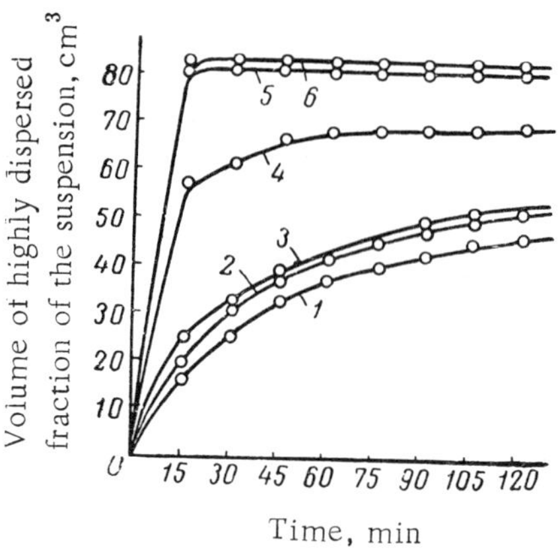

Fig. 15. Change in the volume of the highly dispersed fraction of the suspension with change in the water glass concentration. Water glass concentration: 1) 2%; 2) 4%; 3) 5%; 4) 10%; 5) 15%; 6) 20%.

The degree of subdivision of the solid phase is of considerable importance for the successful stabilization of aqueous suspensions of nonclay rocks. Since these rocks are unable to undergo spontaneous dispersion and swelling in water, this factor must always be taken into account in work on the use of flushing liquids from nonclay rocks. The smaller the particle size of the disperse phase, the greater the stability of the system attained for stabilizers of given quality and quantity.

The influence of the degree of dispersion of the solid phase and the concentration of UShchR on the sedimentational stability of an argillite suspension is shown in Fig. 16. The studies were carried out with a suspension of powder No. 1 with particle size 0.2-0.6 μ and powder No. 2 with particle size 0.6-1.4 μ.

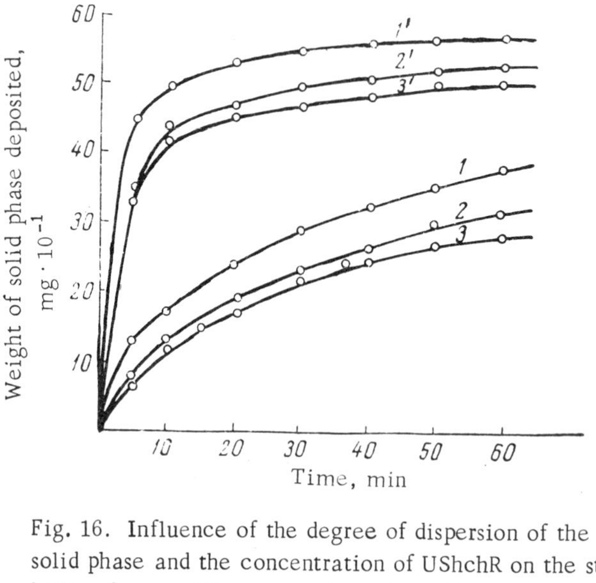

Fig. 16. Influence of the degree of dispersion of the solid phase and the concentration of UShchR on the stability of an argillite suspension. 1, 2, 3) Suspensions of powder No. 1 (UShchR concentration 2.5, 7.5, and 15% respectively); 1', 2', 3') the same for a suspension of powder No. 2.

Summary

1. The characteristic physicochemical features of the interaction between the particles of the disperse phase in suspensions of nonclay rocks and the aqueous dispersion medium lead to the production of flushing liquids with optimum rheological and structural-mechanical properties. The thickening of these flushing liquids is not so great as that of clay suspensions and the flushing liquids retain their mobility and low viscosity at high concentrations of solid phase compact, nonadhesive filtration crusts. They exhibit thixotropic properties with low-limited shearing stress values and are stable to the action of salts from mineralized stratal waters and water-soluble rocks. These qualities indicate that it is advantageous to use nonclay rocks for drilling with natural flushing liquids and also, in a number of cases, to use these rocks instead of clay rocks for the preparation of drilling fluids from imported materials.

2. Stable flushing liquids from nonclay rocks can be obtained only by using surfactant stabilizers. The chemical reagents used to stabilize suspensions of nonclay rocks act as surfactants and are adsorbed on the particles of the disperse phase, reducing the surface tension and forming adsorption layers whose structural-mechanical properties are the chief factor in the stabilization of the system.

3. In general, it is advantageous to use mixtures of surfactant stabilizers or surfactant stabilizers combined with active additives, for the most part inorganic electrolytes; both types of material have a selective action and increase the efficiency of the stabilization of suspensions of nonargillaceous rocks.

Literature Cited

1. F. D. Ovcharenko, The Hydrophilic Properties of Clays and Clay Minerals (Izd. AN USSR, Kiev, 1961).
2. P. A. Rebinder, "The structural and mechanical properties of clay rocks and modern theories of the physical chemistry of colloids," Proceedings of the Conference on the Geological Engineering Properties of Clay Rocks, Vol. 1 (Moscow, 1956).
3. A. M. Schwartz, J. W. Perry, and J. Berch, Surface-Active Agents and Detergents [Russian translation, edited by A. B. Taubman], (IL, 1960).
4. S. N. Yatrov, Natural Flushing Liquids in Drilling (Liquids Based on Drilled Rock), (Gostoptekhizdat, 1963).
5. P. A. Rebinder and A. B. Taubman, "Notes on the problem of the aggregative stability of disperse systems," Kolloidnyi zhurnal, 23, 3 (1961).

A STUDY OF SURFACE-ACTIVE ADDITIVES IN THE FLUSHING LIQUIDS USED FOR OPENING UP STRATA

N. I. Skomorovskaya and E. G. Ivanova

During the opening of a productive stratum the flushing liquid generally penetrates into the zone next to the end face of the well.

When water is used as flushing liquid and penetrates into the zone next to the end face, it displaces petroleum from this zone into the depth of the stratum and creates a zone of decreased oil permeability at a distance from the well.

A decisive part in the processes involved in the mutual displacement of petroleum and water is played by the molecular surface phenomena in the porous medium [6-8].

In recent years considerable attention has been paid to the use of surfactants to decrease the undesirable influence of the flushing liquid on the permeability of the zone of the productive stratum next to the end face.

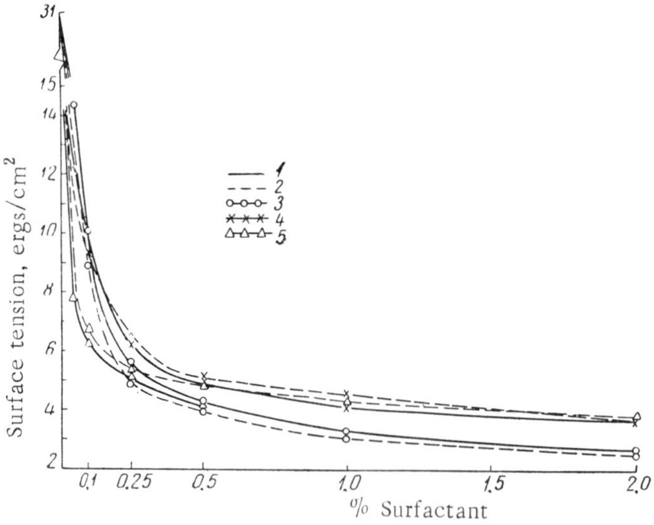

Fig. 1. Dependence of the surface tension at the aqueous solution–petroleum boundary on the surfactant concentration. 1) Surface tension in solutions prepared from distilled water; 2) surface tension in model drilling water; 3) solution containing UFE_8; 4) solution containing $KAUFE_{14}$; 5) solution containing OP-10.

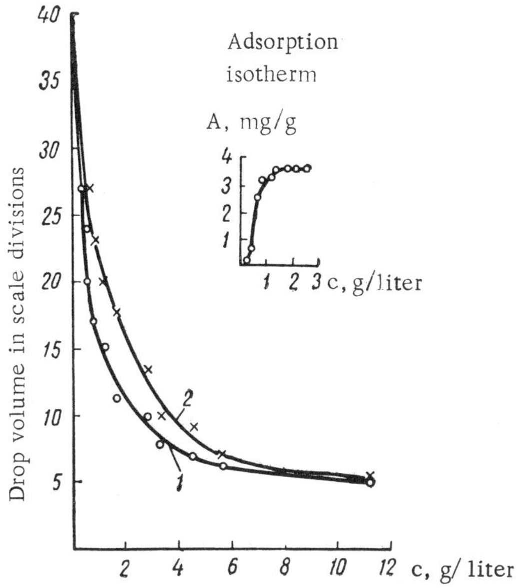

Fig. 2. Adsorption of UFE_8 from distilled water by finely ground Devonian sandstone (weight of sandstone 10 g, volume of solution 30 cm^3). Surface tension of solutions with initial concentrations: 1) before adsorption; 2) after adsorption.

c, g/liter	Δc, g/liter	A, mg/g
0.25	0.07	0.21
0.35	0.15	0.45
0.45	0.20	0.60
0.50	0.50	1.50
0.55	0.70	2.10
0.65	0.85	2.55
0.93	1.07	3.21
1.15	1.10	3.30
1.33	1.17	3.51
1.80	1.20	3.60
2.05	1.20	3.60
2.30	1.20	3.60
2.55	1.20	3.60

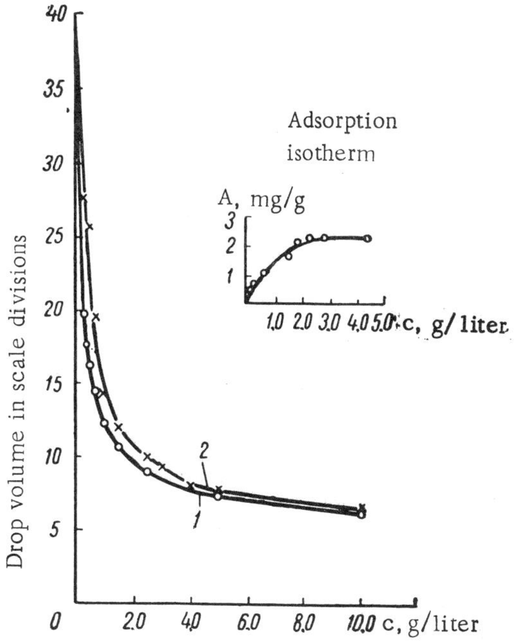

Fig. 3. Adsorption of $KAUFE_{14}$ from distilled water by finely ground Devonian sandstone (weight of sandstone 10 g, volume of solution 30 cm^3). Surface tension of solutions with initial concentration: 1) before adsorption; 2) after adsorption.

c, g/ liter	Δc, g/ liter	A, mg/g
0.23	0.17	0.51
0.28	0.22	0.66
0.43	0.27	0.72
0.65	0.35	1.05
1.05	0.45	1.35
1.45	0.55	1.65
1.80	0.75	2.10
2.25	0.75	2.25
2.75	0.75	2.25
3.30	0.75	2.10
3.85	0.75	2.25
4.25	0.75	2.25

In view of the important part played by surface tension and selective wetting in the processes involved in the mutual displacement of petroleum and water, the most important property of a surfactant used as additive in a flushing liquid for the opening of a productive stratum is a high surface activity at the petroleum–water boundary. Of particular importance are the change in the wettability of the rock under the influence of the surfactant and the retention of surface-active properties in the reaction of mineralized stratal water with adsorbent rock particles. Moreover, the surfactants should not form adsorption layers with a gel-like structure at the boundary of separation between the phases, since this produces high hydraulic resistance.

On the basis of these requirements the most effective materials for this particular application may be the nonionogenic surfactants.

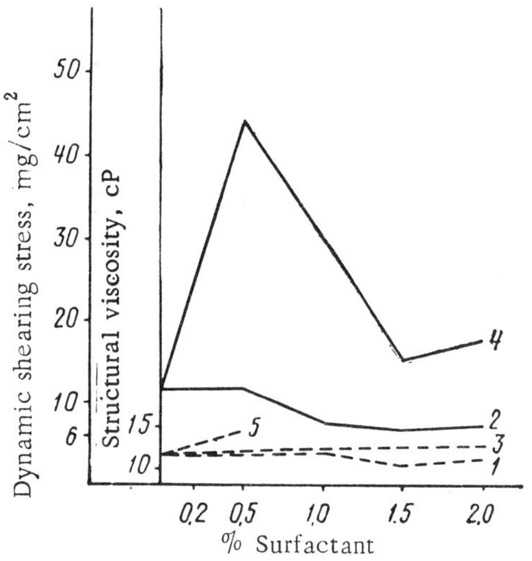

Fig. 4. Influence of added surfactant on the structural viscosity and limiting dynamic shearing stess of a clay suspension (stabilized). 1, 2) Solutions containing UFE_8; 3, 4) solutions containing $KAUFE_{14}$; 5) solutions containing OP-10. Solid lines) limiting dynamic shearing stress; broken lines) structural viscosity.

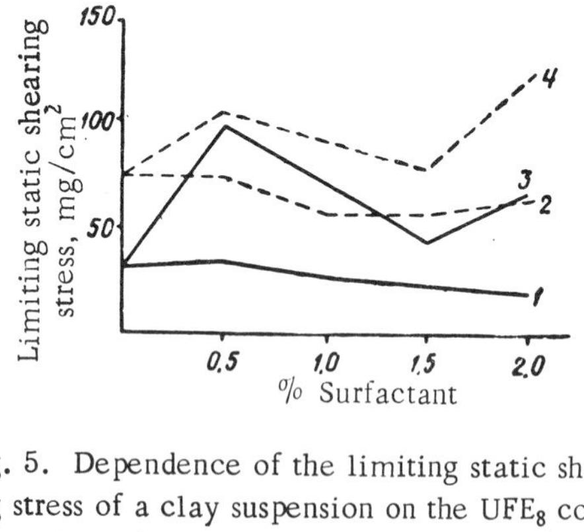

Fig. 5. Dependence of the limiting static shearing stress of a clay suspension on the UFE_8 concentration. Solid lines) limiting static shearing stress after 1 min; broken lines) limiting static shearing stress after 10 min. 1, 2) Solution containing UFE_8; 3, 4) solution containing $KAUFE_{14}$.

In a study of a number of nonionogenic surfactants synthesized by the VNII Neftekhim, it was established that these materials decrease to a considerable extent the surface tension at a petroleum—water boundary; the highest surface activity was shown by UFE_8—a product obtained by condensing a coal phenol fraction with ethylene oxide, and $KAUFE_{14}$—a product obtained by condensing ethylene oxide with phenols alkylated by a cracking-gasoline fraction.

These products were selected for further studies because of their high surface-active properties, high solubility in water, and minimum foam-forming power compared with other nonionogenic surfactants studied.

We determined the surface tension of aqueous solutions of the surfactants at the boundary with Tuimazinskii petroleum in solutions prepared from distilled water and in solutions prepared from mineralized water. The latter consisted of a drilling-water model containing 23% of stratal Devonian water from Tuimazinskii deposit and 77% of Ufa mains water.

The surface tension isotherms $\sigma = f(c)$ for solutions of UFE_8 and $KAUFE_{14}$ in distilled water and the model drilling water are shown in Fig. 1.

It can be seen that the surface tension of solutions of UFE_8 and $KAUFE_{14}$ in distilled water at concentrations of 0.25% decreases from 30 to 5.6 ergs/cm^2 for UFE_8 and to 6.2 ergs/cm^2 for $KAUFE_{14}$; at concentrations of these surfactants equal to 1% the surface tension decreases to 3.3 and 4.1 ergs/cm^2 respectively.

The surface tension of solutions of $KAUFE_{14}$ in mineralized water remains practically unchanged. The surface tension of a solution of UFE_8 in mineralized water at all concentrations is slightly lower than that in distilled water.

The ability of nonionogenic surfactants to preserve their surface activity in the presence of inorganic ions has been noted by a number of workers [1, 3, 5].

This property of nonionogenic surfactants gives them an advantage as additives to flushing liquids for opening the productive stratum, since the flushing liquid usually contains a large quantity of mineral salts, and contact between the flushing liquid and connate water in the pore space of the stratum is always unavoidable.

The change in the surface-active properties of the liquid penetrating into the pores of the productive stratum will depend to a considerable extent on the adsorption of surfactant by the rock.

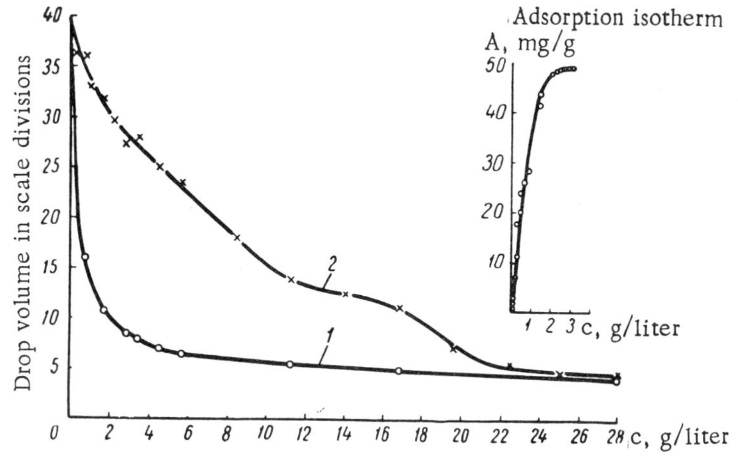

Fig. 6. Adsorption of UFE_8 from distilled water by clay particles after 7 days (weight of clay 10 g, volume of solution 30 cm^3). Surface tension of solutions with initial concentration: 1) before adsorption; 2) after adsorption.

c, g/liter	Δc, g/liter	A, mg/g	c, g/liter	Δc, g/liter	A, mg/g
0.05	0.35	1.05	1.35	13.90	41.70
0.07	0.58	1.74	1.45	14.55	43.65
0.10	0.90	2.70	1.65	15.35	46.05
0.20	2.40	7.20	2.10	15.90	47.70
0.25	3.75	11.25	2.25	16.00	48.00
0.30	5.95	17.85	2.40	16.10	48.30
0.40	6.40	20.1	2.60	16.15	48.45
0.50	8.00	24	2.70	16.20	48.60
0.60	8.75	26.25	2.85	16.15	48.45
0.75	9.25	27.75	2.95	18.2	48.60
0.85	9.42	28.26			

Figures 2 and 3 give curves showing the dependence of the surface tension (drop volume in scale readings) on the concentration of the solutions before and after adsorption, together with the adsorption isotherms for UFE_8 and $KAUFE_{14}$, determined for the adsorption of these surfactants by finely ground Devonian sandstone from the Tuimazinskii deposit (fractional composition of sandstone: 0.5-0.25 mm fraction–20.7%, 0.25-0.1 mm fraction–68.6%, fraction below 0.1 mm–10.7%.

The maximum adsorption of UFE_8 from solutions in distilled water amounts to 3.6 mg/g of sandstone, and the maximum adsorption of $KAUFE_{14}$ amounts to 2.25 mg/g of sandstone.

We also determined the adsorption of UFE_8 by Devonian sandstone from the model drilling water and from a clay suspension filtrate; the values showed a decrease in adsorption with increase in the degree of mineralization of the aqueous medium.

The maximum adsorption from the clay suspension filtrate amounted to 2.85 mg/g of sandstone, while that from the model drilling water amounted to 2.25 mg/g of sandstone.

The interfacial surface tension after adsorption by Devonian sandstone was maintained within the range of minimum values before adsorption for UFE_8 concentrations of approximately 0.6-0.8%.

The flushing liquid (water or clay suspension) for opening a productive stratum is selected on the basis of the specific geological condition of the petroleum deposit.

The permeability of the zone next to the end face also decreases when a clay suspension is used as flushing liquid.

TABLE 1. Influence of Added Surfactants on the Physicochemical and Rheological Properties of Clay Suspensions

Added surfactant	Surfactant concentration, %	Viscosity according to SPV-5, sec	Water loss after 30 min, cm^3	Thickness of clay crust, mm	Specific gravity, g/cm^3	Stability, g/cm^3	Limiting static shearing stress, mg/cm^2		Limiting dynamic shearing stress, mg/cm^2	Structural viscosity, cP	Surface tension of clay suspension filtrate, dyn/cm	pH of clay suspension filtrate
							After 1 min	After 10 min				
0	0	30.0	7.5	2.0	1.19	0.00	30.6	74.8	12.7	11.57	24.70	9.3
UFE_8	0.5	31.0	7.0	2.0	1.19	0.00	34.0	74.8	11.7	11.97	12.03	9.1
UFE_8	1.0	30.0	6.0	2.0	1.19	0.00	27.2	57.8	7.6	11.99	9.66	9.2
UFE_8	1.5	28.0	6.0	2.0	1.19	0.00	23.8	57.8	6.8	10.60	4.14	9.2
UFE_8	2.0	30.0	7.0	2.0	1.19	0.00	20.4	64.6	7.4	11.33	3.15	9.3
$KAUFE_{14}$	0.5	35.0	7.5	2.0	1.19	0.00	98.6	105.4	44.2	12.20	—	—
$KAUFE_{14}$	1.5	35.0	6.5	2.0	1.19	0.00	44.2	78.2	15.3	12.67	5.36	9.3
$KAUFE_{14}$	2.0	36.0	7.0	2.0	1.19	0.00	68.0	122.4	17.7	12.80	4.48	9.3

TABLE 2. Establishment and Productivity of Wells Opened with Water Containing Added UFE_8 and Wells Opened with Clay Suspensions

Well No.	Thickness of opened section of stratum	Type of flushing liquid	Initial pressure, atm		Number of shots and type of perforations	Net time required to establish well, h	Initial productivity coefficient, tons/day · atm	Start of operation	Type of operation	Average output of petroleum during first month of operation	Porosity coefficient, %	Oil-saturation coefficient, %
			$P_{stratal}$	Pend-face								
1452	6.8	Water + UFE_8	149.5	136.8	220 VPK -103	13.5	3.9	9-10-59	Gusher	49.6	19.4	90.9
937	7.0	Clay suspension	148.2	134.2	210 VPK -103	—	3.6	6-58	"	33.8	22.1	89.5
1296	1.5	" "	131.7	124.1	45 VPK -103	144	2.9	7-1-56	"	20.7	17.2	84.5
944	5.7	" "	119.4	111.1	170 VPK -103	56	4.2	9-7-57	"	38.11	21.2	88.2
333	3.0	" "	139.1	128.8	100 VPK -103	41	2.1	10-11-58	"	20.6	19.8	88.5

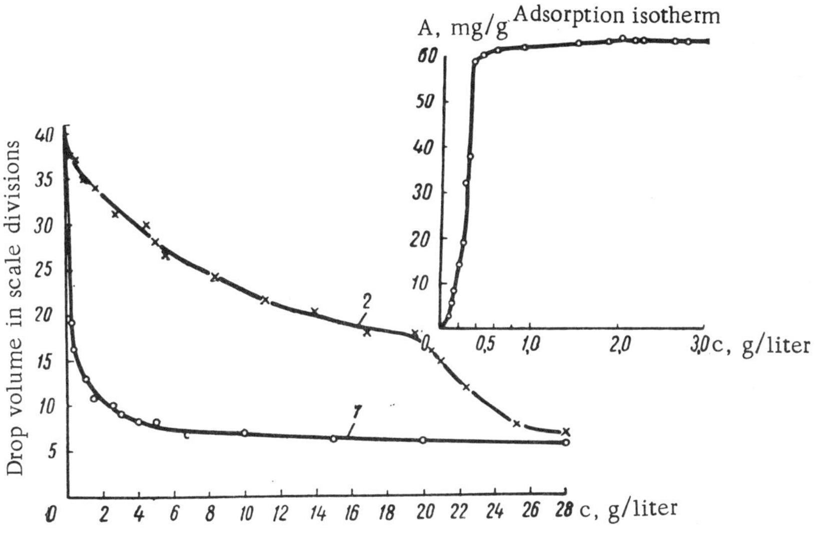

Fig. 7. Adsorption of $KAUFE_{14}$ from distilled water by clay particles after 7 days (weight of clay 10 g, volume of solution 30 cm^3). Surface tension of solutions with initial concentration: 1) before adsorption; 2) after adsorption.

c, g/ liter	Δc, g/ liter	A, mg/ g	c, g/ liter	Δc, g/liter	A, mg/g
0.05	0.45	1.35	0.90	20.60	61.80
0.08	0.88	2.64	1.35	20.65	61.95
0.10	0.90	2.70	1.50	21.00	63.00
0.13	1.87	5.61	1.85	21.15	63.45
0.15	2.85	8.55	2.00	21.25	63.75
0.20	4.80	14.4	2.15	21.25	63.75
0.25	6.75	19.25	2.25	21.25	63.75
0.27	10.73	32.19	2.35	21.25	63.75
0.30	12.70	38.10	2.60	21.25	63.75
0.35	19.65	58.95	2.75	21.25	63.75
0.45	20.05	60.15	3.00	21.25	63.75
0.60	20.40	61.20			

In this case the influence of the clay suspension is determined not only by the surface properties of the filtrate but also by the physicochemical and – more particularly, by the rheological – properties of the clay suspension itself.

The clay suspensions used to open up productive strata can be improved considerably by the addition of surfactants.

The influence of surfactants on the properties of mineral particles and clay suspensions in the processes aimed at intensifying petroleum recovery and improving the establishment of oil wells has been studied by a number of workers in the Soviet Union and abroad [4, 7, 9].

The chief advantages of clay suspensions containing surfactants are the decrease in the undesirable influence of the filtrate and clay particles on the permeability of the zone next to the end face, the possibility of controlling the rheological properties of the suspension, and the chemical and thermal stability of the solution. These advantages are due to a considerable extent to the flocculating effect produced by the addition of appropriate surfactants.

Work by P. A. Rebinder [6-8] has shown that in the adsorption of certain surfactants on the surface of hydrophilic minerals, the surfactant molecules are attached by chemosorption, with their polar groups directed

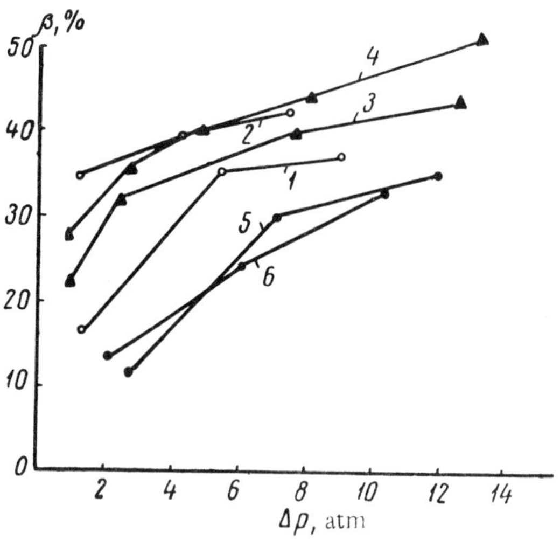

Fig. 8. Restoration of the permeability of Devonian sandstone cores (length of specimens with packing 19.6 cm). 1) β for filtration: petroleum—distilled water—petroleum ($k_{for\ air}$ = 0.220 darcy); 2) β for filtration: petroleum—1.5% solution of UFE_8 in distilled water—petroleum ($k_{for\ air}$ = 0.26 darcy); 3 β for filtration: petroleum—model drilling water—petroleum ($k_{for\ air}$ = 0.490 darcy); 4) β for filtration: petroleum—1.5% solution of UFE_8 in drilling water—petroleum ($k_{for\ air}$ = 0.510 darcy); 5) β for filtration: petroleum—clay suspension filtrate—petroleum ($k_{for\ air}$ = 0.220 darcy); 6) β for filtration: petroleum—distilled water—petroleum ($k_{for\ air}$ = 0.206 darcy).

toward the mineral and their hydrocarbon chains directed toward the aqueous medium; this produces hydrophobic properties on the mineral surface.

In aqueous suspensions this production of hydrophobic properties brings about loose adhesion of the particles by their outward-pointing hydrocarbon chains, i.e., to flocculation [6-8].

According to Burdin and Viner [2], nonionogenic surfactants which are readily soluble in water act as effective flocculating agents for clay particles. Their flocculating action leads to the formation of aggregates of particles suspended in the dispersion medium and produces a decrease in the limiting static shearing stress, structural viscosity, and limiting dynamic shearing stress, i.e., it improves the rheological characteristics of the suspension.

The rheological parameters—the structural viscosity and limiting dynamic shearing stress of the clay solutions—were determined on rotational viscometers of the MacMichael type, developed in the UFNII.

Studies were made of clay suspensions prepared from Talalaevskii clay with a concentration of 40% and containing calcined soda in quantities amounting to 2% of the weight of the clay.

Table 1 gives data on the influence of UFE_8 and $KAUFE_{14}$ on the physicochemical and rheological properties of the clay suspensions.

The clay suspensions containing added UFE_8 and $KAUFE_{14}$ remain stable at all concentrations studied. The water loss, thickness of the clay crust, and specific gravity of the solution remain practically unchanged.

The most interesting data on the dependence of the structural viscosity and dynamic shearing stress on the concentration of surfactant in the solution are given in Fig. 4.

In suspensions containing added $KAUFE_{14}$ at a concentration of 0.5%, a considerable increase in the strength of the structure is observed; this is shown by the increase in the limiting static shearing stress and by the increase in the rheological parameters. At a concentration of 1.5% $KAUFE_{14}$, the strength of the structure decreases but the values of the above quantities remain higher than their values for the original solution. Further increase in the $KAUFE_{14}$ concentration leads to a further increase in these quantities.

It can be seen that the structural viscosity and limiting dynamic shearing stress curves for the solution containing UFE_8 show a clearly defined minimum at a concentration of 1.0-1.5% UFE_8. The curves giving the dependence of the limiting static shearing stress of the suspension on the UFE_8 concentration are given in Fig. 5 and are similar in character.

The nature of the influence of UFE_8 on the structural-mechanical and rheological properties of the clay suspensions indicates that they have a flocculating action on the clay particles and that it is possible to obtain clay suspensions with controled rheological parameters.

The flocculating action of nonionogenic surfactants of the oxyethylated nonylphenol type on clay suspensions is due to their adsorption on the clay particles, as has been shown in [2].

The efficiency of action of the surfactants depends on their molecular structure and the balance between the polar and nonpolar properties.

We determined the adsorption of UFE_8 and $KAUFE_{14}$ by clay particles from a suspension of Talalaevskii clay in distilled water.

The adsorption isotherms for UFE_8 and $KAUFE_{14}$ are shown in Figs. 6 and 7. The magnitude of the maximum adsorption amounts to 48.6 mg/g of clay for UFE_8 and 63.7 mg/g of clay for $KAUFE_{14}$.

The influence of UFE_8 and $KAUFE_{14}$ on the surface tension and pH of the clay suspension filtrates is shown in Table 1, from which it can be seen that the surface tension of the clay suspension filtrates containing UFE_8 is lower than that of the suspension filtrates containing $KAUFE_{14}$ at the same concentrations. The smallest surface tension value is observed at a concentration of 2% UFE_8.

The influence of surfactants on the oil permeability of a productive sandstone can be estimated from the change in the oil permeability of specimens of the sandstone during the displacement of water and aqueous surfactant solutions from the pore space of the specimen by petroleum.

The influence of UFE_8 can be expressed most clearly by the coefficient for decrease in permeability β (in %), proposed by K. F. Zhigach for the estimation of the influence of flushing liquids. This coefficient characterizes the ratio of the oil permeability of the specimens during repeated filtration to the permeability during initial filtration.

The dependence of the coefficients for the decrease in permeability during filtration of different liquids on the pressure drop (Δp) for specimens of length 19.6 cm with packing is illustrated by the curves in Fig. 8, from which it can be seen that for all experimental conditions, the addition of UFE_8 to water increases the coefficient for the decrease in the oil permeability of the specimens.

In the experiments with distilled water and with specimens without residual water, β increases by a factor of 3.5-5, while in experiments with model drilling water and with specimens containing residual water, β increases by a factor of 1.2-1.4.

With increase in the pressure drop from 3 to 15 atm, the influence of UFE_8 is less marked, but the curves giving the restoration of the oil permeability of the specimens during filtration after UFE_8 solution in this case also lie above the curves giving the restoration of the oil permeability of the specimens for filtration after water.

In the filtration of a 1.5% solution of UFE_8 in model drilling water, we determined its adsorption by Devonian sandstone saturated with petroleum.

The adsorption of UFE_8 during filtration until its concentrations at the specimen inlet and outlet were equal, amounts to 1.75 mg/g or 4.8 kg/m^3 of rock.

In 1959 the Surface-Active Substances Laboratory and the Flushing Liquids Laboratory of the UFNII, together with the Tuimazaburneft' Trust, carried out industrial tests on the opening of a productive stratum by water containing UFE_8 in well No. 1452 of the Tuimazinskii deposit and well No. 757 of the Shkapovskii deposit.

The UFE_8 was added to the technical water before the productive horizon (Pashiiskii layers) in quantities amounting to 1.2-1.5%.

The concentration of UFE_8 decreased during drilling as a result of its adsorption by drilled rock and by replenishment of the volume of circulating liquid by water containing no UFE_8 to compensate for the adsorption of flushing liquid.

The concentration of UFE_8 during drilling in the productive strata was 0.5-0.55% for well No. 1452 and 0.20-0.25% for well No. 757. Both wells were established rapidly and gave petroleum in the form of a gusher.

The initial output of petroleum from well No. 757 amounted to 90 tons per day compared with 50 tons per day in well No. 753, situated alongside the former and opened with high-quality clay suspension.

Comparative data on the establishment and productivity of well No. 1452 and wells situated alongside it and drilled with high-quality clay suspension are given in Table 2.

The following features can be seen from Table 2: the time required to establish well No. 1452, whose productive stratum was opened up with water containing UFE_8, was much less than the time required to establish wells opened with a high-quality clay suspension (13.5 h for well No. 1452 compared with 40-144 h for wells No. 1296, 944, and 333); the average daily output from well No. 1452 during the first month of operation was 40-50% greater than the average daily output from four comparable wells in the same period.

An industrial research team from the UFNII made a study of gusher wells No. 1452 and 937 in order to estimate quantitatively the change in the properties of the zone next to the end face in the productive stratum during its opening with water containing UFE_8.

It can be seen from Table 2 that the productivity coefficient for well No. 1452 is slightly higher than that for well No. 937, and that the conductivity of the stratum in the region of well No. 1452 was equal to that of the drainage zone stratum in well No. 937.

The ratio of the permeability coefficient of the zone next to the end face k_1 to the permeability coefficient of the distant zone k for well No. 937 is less than unity, however, indicating a decrease in the permeability of the stratum in the zone next to the end face.

In well No. 1452, opened up with water containing UFE_8, the ratio $k_1/k = 1$ (i.e., no decrease in permeability in the zone next to the end face) is observed.

Thus the industrial tests gave the following results.

1. The time required to establish wells whose productive stratum was opened up with water containing UFE_8 was much less than that required to establish wells drilled with clay suspension.

2. The addition of UFE_8 to the water during the opening up of the productive stratum made it possible to preserve the natural permeability of the zone of the stratum next to the end face and to obtain higher initial outputs than in wells opened up with high-quality clay suspension.

3. As a result of the favorable influence of UFE_8 on the collector properties of the stratum, the time required to drill oil wells can be decreased by drilling the productive horizon with water containing added UFE_8 as flushing liquid.

4. Turbine drilling with water containing added UFE_8 as flushing liquid took place normally and no complications due to foam formation were observed.

It should be noted that technical coal phenol fractions are used for the production of UFE_8, so that the cost of this material is lower than that of other surfactants of the oxyethylated phenol type (OP-10, OP-7, $KAUFE_{14}$).

In 1960 more extensive industrial tests on the opening up of a productive stratum by means of water containing added UFE_8 are to be carried out; these tests will require 50-60 tons of the nonionogenic surfactant UFE_8.

Literature Cited

1. S. F. Bezuglyi, L. Z. Osityanskaya, and B. A. Akimov, "Surface-active substances used in the production of chemicals for plant protection," Khimicheskaya nauka i promyshlennost', 4, 5 (1959).
2. R. F. Burdin, "A new flushing liquid for drilling wells with high end-face temperatures" [Russian translation No. 98/576] (1957).
3. R. A. Guereca, "The selection of surface-active substances according to their oil-displacing power" [Russian translation No. 171/55] (1956).
4. M. A. Geiman, A. N. Larin, V. B. Shneerson, and R. A. Fridman, "Salts of aromatic sulfonic acids (DS) used in petroleum recovery," Tr. Instituta Nefti AN SSSR, Vol. 6 (1955).
5. K. G. Mizuch and R. A. Lapina, "Surface-active substances based on ethylene oxide," Khimicheskaya nauka i promyshlennost', 4, 5 (1959).
6. P. A. Rebinder, The Physical Chemistry of Flotation Processes (Metallurgizdat, Moscow–Leningrad, 1933).

7. P. A. Rebinder, Studies in the Applied Physical Chemistry of Surface Phenomena (ONTI, Moscow–Leningrad, 1936).
8. P. A. Rebinder, The Part Played by Gases and Reagents in Flotation Processes (Izd. AN SSSR, Moscow–Leningrad, 1950).
9. V. B. Shneerson, "A new method for studying the production of hydrophobic properties in rocks under the influence of petroleums," Tr. Instituta Nefti AN SSSR (1949).

THE USE OF SOLUTIONS OF NONIONOGENIC SURFACTANTS FOR OPENING UP OIL-BEARING STRATA

E. G. Ivanova, N. I. Skomorovskaya, and V. I. Tokunov

It is well known [1, 2, 3, 4] that the penetration of flushing liquid into the end-face zone in many cases leads to a sharp decrease in the productivity of wells and is one reason for increase in the time taken to recover the petroleum.

This undesirable effect on productivity is particularly marked when the producing horizon is opened up using water as flushing liquid. Thus, in spite of the increase in the rate of drilling when water is used as flushing liquid, its use is restricted to the region before the producing horizon.

Work which has been carried out in a number of Scientific Research Institutes has shown the high efficiency of the addition of surfactants to water in order to decrease its undesirable influence on the productivity of wells [5, 6, 7, 8, 9].

In this case, the choice of surfactant is determined not only by its physicochemical properties but also by the physicochemical characteristics of the reservoir and by the geological conditions of the deposit being drilled.

TABLE 1. Productivity of Wells Opened Up with Different Flushing Liquids

Characteristics of the flushing liquid used to open up the productive stratum	Arlanskii area		Nagaevskii area	
	Productivity coefficient, tons/day · atm	Specific initial output per meter of effective thickness, tons/day	Productivity coefficient, tons/day · atm	Specific initial output per meter of effective thickness, tons/day
Clay suspension with water loss less than 20 cm^3	4.8	7.2	4.4	4.4
Clay suspension with water loss greater than 20 cm^3.	3.4	5.4	3.0	2.6
Water containing added Sulfonol	2.4	3.8	2.6	2.4
Water	–	–	1.5	0.94

TABLE 2. Influence of Stratal Waters on the Properties of Surfactant Solutions

Characteristics of the medium	Surface tension of a 0.5% solution of the surfactant, ergs/cm^2 — Concentration of Carboniferous stratal water, %				Characteristics of surfactant solutions after reaction with stratal water
	0	10	30	50	
			OP-10		
Artinskii water	3.48	3.68	3.48	3.25	All solutions transparent
			UFE_8		
The same	2.37	2.42	2.36	2.30	The same
			Sulfonol		
The same	7.30	12.60	14.40	16.10	Voluminous precipitates obtained in all solutions. Turbid filtrates obtained after separation of precipitates

Laboratory and industrial tests carried out in the UFNIII over a period of years have established [6] the efficiency of nonionogenic surfactant additives of the oxyethylated phenol type for opening up productive strata, using water as flushing liquid, particularly when horizons with high-pressure mineralized waters are encountered in the section.

One such deposit is the Arlanskii deposit in the northwest of the Bashkir Republic. The waters of the Sakmaro-Artinskii layer in this deposit have a high stratal pressure and are highly mineralized and saturated with hydrogen sulfide. Stratal waters are found in a number of other horizons.

Until recently the opening up of the productive strata of the carboniferous formation in wells of the Arlanskii deposit was carried out using a clay suspension. As a result of the salinity of the stratal waters, the properties of the suspension deteriorate rapidly and wells opened up using suspensions with high water loss generally show reduced output.

In a number of wells in this deposit the productive stratum was opened up using water along and water containing added Sulfonol.

Comparative analysis of the drilling results and the output obtained from 30 wells situated in comparable conditions showed that the opening up of a productive stratum using water alone or water with added Sulfonol leads to an even greater decrease in the productivity of the wells.

Table 1 gives values for the average productivity coefficients and specific initial output per meter of effective stratum thickness for a group of wells opened up using different flushing liquids in the Nagaevskii and Arlanskii areas of the Arlanskii deposit.

It can be seen from Table 1 that the highest values of the productivity coefficients and specific initial outputs are shown by wells opened up with a clay suspension with a water loss less than 20 cm^3. Values of the water loss for the clay suspensions in the range 10-16 cm^3 were observed in wells opened up with clay suspensions stabilized by condensed sulfite-alcohol waste (KSSB) and sodium hexametaphosphate (GMPhN) according to the formula developed by the UFNII [10].

TABLE 3. Characteristics of the Drilling of Wells using Aqueous OP-10 Solution and Clay Suspension

Well No.	Flushing liquid	Effective thickness of section of stratum opened, m	Time required to establish well, h	Exploitation method	Initial output, tons/day	Diameter of connecting pipe, mm
182	Water containing OP-10 . .	12.6	0.0	Gusher	148.0	13.0
183	Clay suspension	15.4	3.5	"	73.0	7.0
344	Water containing OP-10 . .	14.5	9.0	"	79.0	9.0
345	Clay suspension	18.6	55.0	"	80.0	9.0
111	Water containing OP-10 . .	5.2	—	Deep pumping	9.0	—
269	Clay suspension	4.8	—	Ditto	10.0	—
360	Water containing OP-10 . .	6.6	—	"	18.0	—
464	Clay suspension OP-10 . . .	6.0	—	"	16.0	—

Since in the Arlanskii deposit the greater part of the well section consists of hard rocks, and drilling to the roof of the producing horizon is carried out using water as flushing liquid, the drilling of wells to the projected depth without changing to a clay suspension is of considerable interest and can be carried out using water containing added surfactants.

When anionogenic surfactants, including Sulfonol, are used in contact with mineralized waters, they react with the salts and coagulate.

Studies which we have carried out have shown that in solution in Artinskii water and in mixtures with the stratal water from the Carboniferous layer of the Arlanskii deposit, Sulfonol coagulates with the formation of voluminous flocculent precipitates.

Table 2 gives the values of the surface tension of a 0.5% solution of Sulfonol in Artinskii water, together with data for its reaction with 10, 30, and 50% Carboniferous stratal water, and also gives for comparison the values of the surface tension for solutions of the nonionogenic surfactants UFE_8 and OP-10.

It can be seen from Table 2 that in mineralized solutions Sulfonol not only forms precipitates but also partially loses its surface activity, whereas the high surface activity of UFE_8 and OP-10 is completely preserved in mineralized media. This type of interaction between Sulfonol and mineralized stratal waters is evidently responsible for the unsatisfactory results obtained in the opening up of the produtive stratum using water with added Sulfonol in wells in the Arlanskii deposit.

Experience in the use of nonionogenic surfactants in water for drilling wells to the projected depth has shown that satisfactory results can be obtained by using these surfactants under conditions where the mineralization of the drilling water is increased by stratal waters.

The productive stratum was opened up using an aqueous solution of OP-10 in four wells (182, 344, 111, and 360) of the Arlanskii deposit. Wells No. 183, 345, 269, and 464 were drilled simultaneously, but in these wells the productive stratum was opened up using a high-quality clay suspension.

The wells in both groups were drilled under analogous conditions and the parameters of the productive strata opened up by these wells are comparable.

0.5% OP-10 was added to the drilling water before the roof of the producing horizon.

During the drilling of the producing horizon, the concentration of surfactant amounted to 0.3-0.57%. The average quantity of OP-10 used in one well with a depth of 1350 m amounted to 2.4 tons.

All the wells opened up using aqueous OP-10 solution were established and gave a petroleum output suitable for industrial purposes.

The characteristics of the drilling and the initial outputs of wells opened up using aqueous OP-10 solution and high-quality clay suspension are given in Table 3 and show that the use of aqueous OP-10 solution gave initial outputs which were practically identical with those from wells opened up with the high-quality clay suspension.

Industrial tests on the opening up of the productive stratum with water containing the nonionogenic surfactant UFE_8, carried out in wells of the Tuimazinskii and Shkapovskii deposits in the Bashkir Republic, showed that the addition of 0.5-0.8% UFE_8 leads to rapid establishment of the wells and retention of the natural oil permeability of the end-face zone.

The initial output from wells opened up with water and aqueous UFE_8 solution was found to be 40-50% higher than the initial output from wells opened up using a high-quality clay suspension. The current outputs from these wells remained at a constant high level for a prolonged period.

An interesting feature is that the operational characteristics of the drilling bits, obtained in the period when drilling was carried out using OP-10 solution as flushing liquid, were higher than those obtained during drilling when the end face was flushed with clay suspension or with water.

In addition to the general improvement in the operational characteristics of the drilling bits during drilling when aqueous OP-10 solution was used, an increase in the duration of a single drilling by an average of 13% was observed, compared with the duration of a single drilling when water was used, indicating that the efficiency of the drill bit bearings is increased.

For a complete well-drilling operation, as a result of the saving in the preparation of the clay suspension, the decrease in the time required for lowering and lifting operations, the decrease in the time required for mechanical drilling, and the decrease in the consumption of drill bits, there is a decrease in the cost of the drilling operations with favorable economic results.

The satisfactory results obtained in the drilling and opening of productive strata using aqueous solutions of nonionogenic surfactants make it possible to recommend these materials for extensive introduction into drilling practice.

In order to study the influence of the addition of surfactants to water forced into the stratum in order to increase petroleum recovery and the rate of development of the oil deposits in the Arlanskii oil-field, it is proposed to organize an experimental section containing three flooding centers [11].

In the three flooding centers, which are situated on the Nagaevskii dome, twelve more wells are to be drilled in addition to wells No. 29, 248, and 380, already drilled. The distance between the operational wells and the driving wells will be 100 m.

From the economic viewpoint, for this distance between wells it would have been advantageous to use group drilling. The drilling of wells with many open end faces and with branching from the main shaft is excluded, however, since the geological conditions of the Arlanskii deposit make it necessary to reinforce the shaft of the wells with lining tubes.

Two-shaft drilling is also impossible since the light BU-75Br assemblies at present in use in Arlan are unsuitable for this purpose.

The use of other drilling assemblies for two shafts with one assembly at the well depth of 1350 m in Arlan and a distance of 100 m between the wells, and the cutting of the sloping section of the shaft at a depth of 30-50 m, require the selection and maintenance of a curvature of less than 5°, and with this choice of curvature it is difficult to control the azimuth with existing geophysical equipment.

In the experimental section, therefore, it is more convenient to drill vertical wells with BU-75Br assemblies, using the drilling technique generally adopted in Arlan for drilling small-diameter wells.

It is recommended that in a number of wells the flushing liquids should be clay suspensions with combined stabilization by KSSB, GMPhN, and starch reagent and also aqueous solutions of the nonionogenic surfactants UFE_8 or OP-10, according to the formula developed by the UFNII [6, 10].

In two wells of the experimental section, test cores with a natural concentration of residual water will be obtained by using oil-base suspensions, based on the formula developed by the Gubkin MINKh i GP [12], as flushing liquid in the drilling of the productive stratum.

When drilling is complete at each well in the experimental section, a range of studies will be carried out to estimate the collector properties of the productive stratum and the change in the oil permeability of the end-face zone during the opening of productive strata by different flushing liquids.

Literature Cited

1. F. I. Kotyakhov, Fundamental Physics of the Oil-Bearing Stratum (Gostoptekhizdat, 1956).
2. G. A. Babalyan, Problems of the Mechanism of Petroleum Recovery (Aznefteizdat, 1956).
3. K. F. Zhigach and K. F. Paus, "The influence of flushing liquids on the permeability of cores," Neft. khoz., No. 11 (1957).
4. G. A. Babalyan, "The physicochemical principles of the use of surface-active substances in the drilling of wells in order to intensify the recovery of petroleum and increase the petroleum output from strata," Proceedings of the First All-Union Conference on the Use of Surface-Active Substances in the Petroleum Industry (Gostoptekhizdat, Moscow, 1961).
5. K. A. Burdyn' and L. V. Lyutin, "Comparative studies of surfactant materials for restoring the permeability of the end-face zone," Proceedings of the First All-Union Conference on the Use of Surface-Active Substances in the Petroleum Industry (Gostoptekhizdat, Moscow, 1961).
6. N. I. Skomorovskaya and E. G. Ivanova, "The use of surfactants in the opening up of the productive stratum of oil wells to preserve the natural oil permeability of the end-face zone of the stratum," Proceedings of the First All-Union Conference on the Use of Surface-Active Substances in the Petroleum Industry (Gostoptekhizdat, Moscow, 1961).
7. A. G. Kochmarev, "Experience in the use of surfactants in the opening up of productive strata and the establishment of oil wells," Proceedings of the First All-Union Conference on the Use of Surfactants in the Petroleum Industry (Gostoptekhizdat, Moscow, 1961).
8. V. P. Bobelyuk, "The treatment of oil wells with surfactant solutions. From data obtained in the Romanian People's Republic and in the USA," Collection of Scientific and Technical Papers on Petroleum Recovery (Gostoptekhizdat, 1960).
9. G. I. Kiselev, "The influence of flushing liquids and added surfactants on the productivity of wells," Oil and Gas Technology News, Neftepromyslovoe delo, No. 6 (1961).
10. E. G. Ivanova and N. I. Skomorovskaya, "The combined stabilization of clay suspensions under complicated conditions in the Arlanskii deposit," Oil and Gas Technology News, Neftepromyslovoe delo, No. 4 (1962).
11. K. F. Zhigach, B. I. Adel', V. M. Mukhin, V. N. Demeshev, and N. N. Goncharov, "Oil-based flushing liquids for the opening up of productive strata in drilling under complicated conditions," Neft. khoz., No. 8 (1956).
12. G. A. Babalyan, G. M. Ovanesov, A. M. Baikov, and I. L. Markhasin, "The development of new methods for increasing the recovery of petroleum from strata by using surface-active substances," Report on Theme No. 24-61, UFNII Fund (1962).

THE CHOICE OF SURFACTANT TO RENDER THE FACIAL ZONE HYDROPHOBIC IN ORDER TO KEEP OUT BOTTOM WATERS

E. N. Umrikhina

Injection into a stratum of a reagent which renders the rock hydrophobic can provide an effective method for increasing the productivity of an exploitation well [1]. Rendering the rock hydrophobic can prevent breakthrough of bottom water or reduce the content of the latter in the petroleum recovered.

A surfactant adsorbed on the surface of a solid body can markedly affect its molecular nature. Long hydrocarbon chains, directed outwards, prevent wetting of a solid surface by water [2].

Under these conditions, when the pore space (channels) has been filled with a hydrocarbon liquid (kerosene or petroleum), the capillary forces at the boundary between the petroleum and the formation water will act against the water and may cause considerable resistance to the movement of water into the facial zone. The choice of an effective surfactant for rendering the rock hydrophobic depends on the chemical nature of the solid mineral surface and its ability to combine with the polar groups of the surfactant.

A siliceous surface, such as that of a quartz sand, is made hydrophobic by a cationic surfactant, but a carbonate rock, such as dolomite, requires an anionic surfactant [2].

No core materials were available for experiments on the choice of surfactants for studies on their capacity to make rocks hydrophobic, so the tests were carried out with artificial sandstones. Cylindrical samples, 4 cm in diameter and 5 cm high, were prepared by the method of M. M. Kusakov, i.e., by bonding quartz sand with sodium tetraethylorthosilicate. The contact angle for wetting between a drop of distilled water and air on a glass plate treated with tetraethylorthosilicate was 38.5°, while that between purified kerosene and air was 70.5° [3].

Brief Account of Characteristics of Surfactants Investigated

Tests were carried out with cationic, anionic, and nonionic surfactants.

Cationic Surfactants

1. Alkamon OS-2. This is a quaternary salt of the diethylaminomethyldiglycolic ether of high fatty alcohols (saturated alcohol):

$$[R{-}OCH_2CH_2OCH_2CH_2OCH_2\underset{\displaystyle \overset{|}{CH_3}}{N}(C_2H_5)_2]^+ C_2H_5SO_3^-,$$

where $R = C_{14-18}H_{29-37}$. It does not dissolve in kerosene and forms a colloidal solution in water.

2. Alkamon OS-2N. This is a similar salt, except that the alcohols are unsaturated. It forms a colloidal solution in water and dissolves readily in kerosene.

3. Arquad 2C. This is a dialkyldimethylammonium salt:

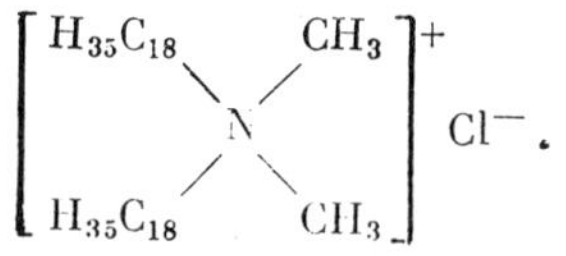

It dissolves readily in kerosene, but not so well in water.

4. Arquad S-50 and Arquad T-50. These are derivatives of the monoalkyltrimethylammonium chloride:

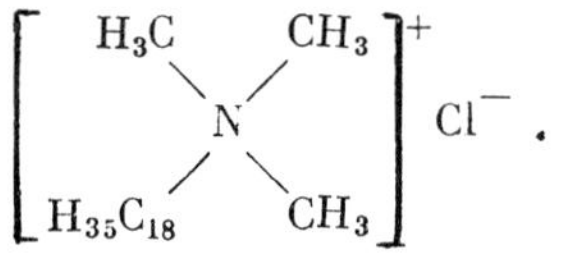

where S is soya bean oil and T is tallol. Both Arquads are readily soluble in water and in kerosene.

5. Karbozolin O. This is a quarternary salt of 1-hydroxyethyl-2-heptadecenylimidazoline:

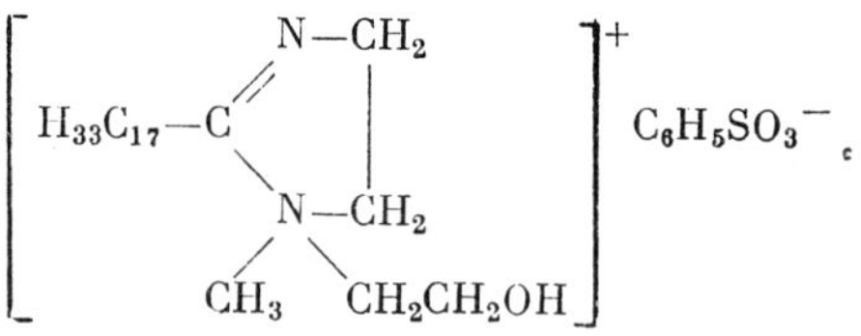

It is readily soluble in water but is insoluble in kerosene.

6. Karbozolin S. This is a quaternary salt of 1-hydroxyethyl-2-heptadecylimidazoline:

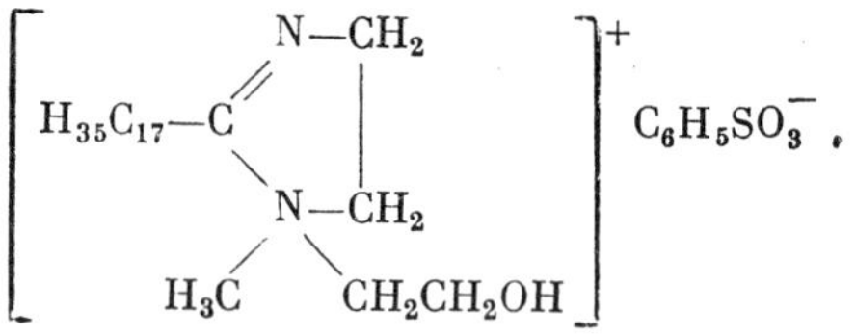

It also dissolves readily in water but is insoluble in kerosene.

The Alkamons and Karboxolins were synthesized by the NIOPiK, and the Arquads by the VNIIZh.

7. Katapin A. This is a para-alkylbenzylpyridinium chloride:

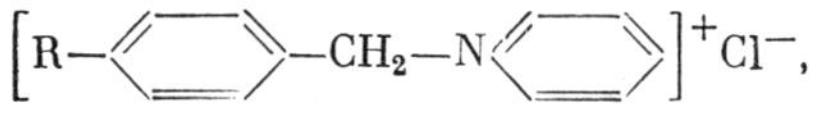

where R is an alkyl group containing 12 to 18 carbon atoms. It dissolves readily in water but does not dissolve in kerosene.

8. Katamin A. This is a para-alkylbenzyltriethylammonium chloride:

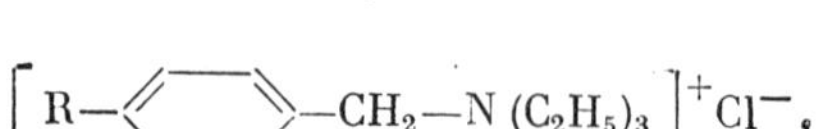

where R is an alkyl group containing 12 to 18 carbon atoms. It is readily soluble in water, but not in kerosene.

9. Diamine dioleate. This has the structure:

$$[R-NH_2-C_3H_6-NH_3]^{2+}[C_{17}H_{35}COO]_2^- .$$

It dissolves in kerosene, but not in water.

10. Amines C_I and C_{II}. These are high molecular aliphatic amines, with normal alkyl chains containing 8 to 18 carbon atoms. Coconut oil is used to prepare Amine C_{II} and stearic acid to prepare Amine C_I.

Diamine dioleate and Amines C_I and C_{II} were products of laboratory synthesis from the Moscow branch of the VNIIZh.

Anionic Surfactants

1. Mylonaft. This is a mixture of the sodium salts of naphthenic acids (43%), water, and small amounts of mineral oil and mineral salts. It is obtained by treating the waste products from the purification of kerosene, gas oil, solar distillate, and light oil fractions of petroleum. We used material from the Bakinsk plant. Mylonaft is soluble in water.

2. Copper salts of naphthenic acids, obtained by the interaction of copper salts with Mylonaft. These are soluble in kerosene, but not in water. Samples were obtained from the laboratory of UFNII.

3. SMS. This is a copper sulfonate, obtained by reacting sodium sulfonates (SNS) with $CuSO_4 \cdot 5H_2O$. It is insoluble in water and only sparingly soluble in kerosene. A sample of SMS was obtained from KuibyshevNII.

4. Substance P. This is a water-soluble alkyl sulfonate, synthetized by the Moscow branch of the VNIIZh.

5. Alcohols from sperm whale fat. These are mixtures containing 35 to 40% of oleyl alcohol, 25 to 30% of cetyl alcohol, up to 10% of unsaturated alcohols above C_{18}, some tetradecyl and octadecyl alcohols, and cholesterol. They are soluble in kerosene. A sample was obtained from the Vakhitov Fat Combine, Kazan.

Nonionic Surfactants

1. OP-10. This is the reaction product from 10 moles of ethylene oxide and a mixture of mono- and di-alkylphenols, where the alkyl groups are C_8 to C_{10}:

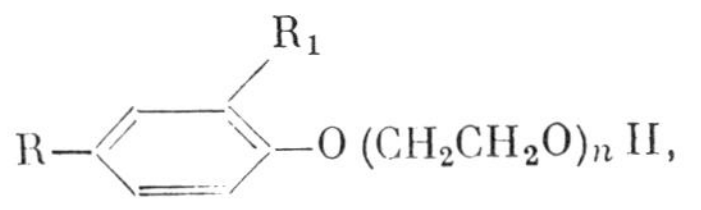

where n = 10. It is readily soluble in water.

2. OP-4. This is the reaction product from 4 moles of ethylene oxide and a mixture of mono- and di-alkylphenols, where the alkyl groups are C_8 to C_{10}. It dissolves readily in kerosene, oils, and organic solvents.

3. Sulfamide OE-10. This is an ethylene oxide sulfamide:

$$R{-}SO_2N\begin{cases}(CH_2{-}CH_2{-}O)_{n_1}H \\ (CH_2{-}CH_2{-}O)_{n_2}H\end{cases},$$

where $n_1 + n_2 = 10$, and R is an acid residue obtained by the oxidation of soft paraffins. This dissolves in water, but not in kerosene. It was synthetized by A. I. Gershenovich. For Sulfamide OE-5, $n_1 + n_2 = 5$.

The only silico-organic compound tested was GKZh-94, a liquid which renders surfaces hydrophobic, obtained by hydrolysis of ethyldichlorosilane. It readily emulsifies in water, to give a stable emulsion, and is readily soluble in kerosene.

Capillary Penetration

Capillary penetration of samples treated with surfactant is very useful for determining the extent to which a sandstone has been rendered hydrophobic.

The experiments on capillary penetration were carried out by the following method. The porosity and air permeability of the sandstone were first determined, and the sample was then saturated with water under

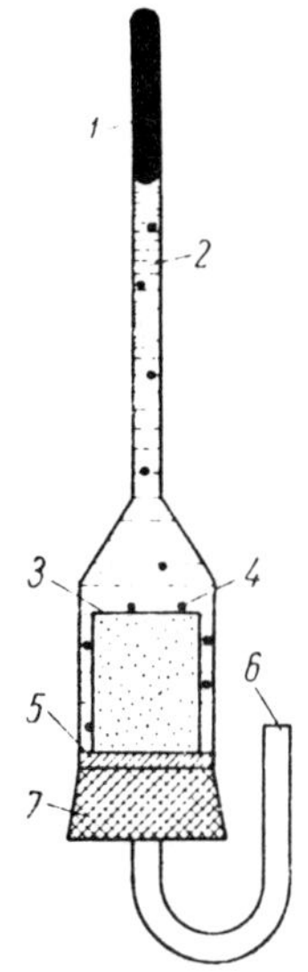

Fig. 1. Instrument for measuring the rate of capillary penetration. 1) Kerosene; 2) water; 3) core; 4) displaced kerosene; 5) filter; 6) tube; 7) rubber bung.

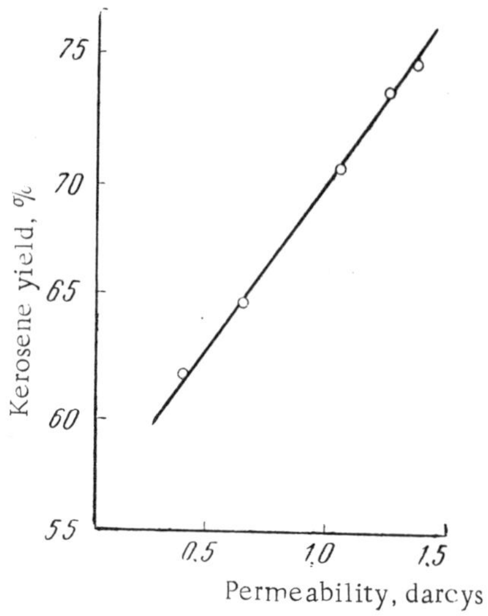

Fig. 2. The effect of the permeability of the sample on the kerosene yield from capillary penetration.

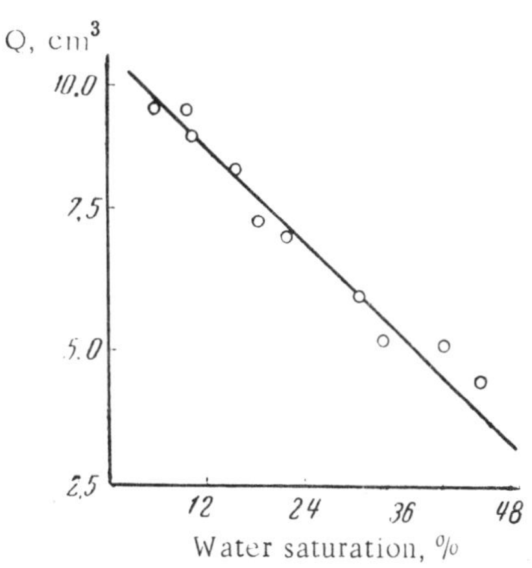

Fig. 3. The effect of the water saturation of the sample on capillary penetration.

vacuum and its water permeability was determined. A kerosene or aqueous solution of surfactant (or pure kerosene) was next filtered through, and the sample was allowed to stand for 18 to 20 h to allow complete adsorption of surfactant onto the sand surface.

If an aqueous solution of surfactant had been used, it was then necessary to filter kerosene through the sample. The kerosene used in all the experiments had the following characteristics: viscosity 1.6 cP at 20°C, surface tension at interface with distilled water 42.5 ergs/cm^2. Finally the sample was placed in the apparatus for capillary penetration (Fig. 1). Distilled water was used as the displacing phase. The volume of kerosene or petroleum displaced by capillary penetration was measured.

1. The Effect of the Permeability of the Sample on Capillary Penetration

Experiments were carried out with sandstone samples, completely saturated with kerosene or petroleum. Figure 2 shows the results obtained. It is clear that the kerosene yield was directly proportional to the permeability of the sample.

2. The Effect of Water Saturation of the Sample on Capillary Penetration

Water saturation of the sample was achieved by capillary evacuation [4]. The sample, saturated with distilled water, was put with a highly dispersed inert powder. Depending on the packing density of the powder, on its humidity, and the time for which it was in contact with the sample, it was possible to achieve any desired water saturation in the sample.

TABLE 1

Surfactant	Solvent	Surfactant concentration, %	Air permeability, darcys	Pore volume, cm^3	Kerosene displaced, cm^3
–	Kerosene	–	1.51	20.4	5.2
–	"	–	0.60	16.45	4.5
Arquad S-50	"	0.05	1.31	20.77	1.1
Arquad T-50	"	0.05	0.48	17.28	1.0
Arquad 2C	"	0.01	1.60	20.34	0.0
Alkamon OS-2	"	0.05	0.84	17.03	2.5
Alkamon OS-2N	"	0.05	0.84	17.03	2.5
Karbozolin O	Water	0.05	1.37	19.91	0.0
Karbozolin S	"	0.5	1.24	19.66	0.0
Katapin A	"	0.05	0.86	–	0.0
Katamin A	"	0.05	0.64	16.67	0.5
Amine C_{II}	Kerosene	0.05	0.61	17.95	0.0
Amine C_I	"	0.05	1.26	–	0.0
Diamine dioleate	"	0.05	2.0	–	0.3
OP-10*	"	0.5	0.63	19.19	2.5
OP-4	"	0.5	0.82	21.91	4.6
Sulfamide OE-10	Water	0.5	0.88	20.40	1.5
Sulfamide OE-5	"	0.5	1.88	–	4.3
GKZh-94	Kerosene	0.5	1.20	21.92	4.6
Mylonaft	Water	0.5	0.58	18.40	4.0
Copper naphthenates	Kerosene	0.5	0.67	18.70	1.9
Copper sulfonates	"	0.5	0.80	17.47	5.6
Substance P	Water	0.5	1.04	18.84	2.1
Alcohols from sperm oil fat	Kerosene	0.5	0.72	19.32	0.5

*To make the sample hydrophobic we used kerosene which had been stirred vigorously with added OP-10 and allowed to stand for a long time. Because of the large amount of sample required for carrying out the tests, it was not possible to carry out all the experiments with samples of the same permeability.

The water saturation was determined, and the sample was then brought into contact with kerosene, of which a definite quantity was absorbed. When absorption ceased, the sample was placed under vacuum for additional saturation by kerosene. Samples with approximately the same permeability were taken so as to be able to determine the relation between the amount of kerosene displaced by capillary penetration and the water saturation.

Figure 3 shows the results obtained. It is clear that Q, the amount of kerosene displaced by capillary penetration, decreased with increasing water saturation of the sample. This is in accordance with literature data [5].

3. The Effect of Making the Sandstone Hydrophobic on Capillary Penetration

In order to approximate to industrial conditions, the investigations on the effects of the water-repelling properties of surfactants were carried out with samples in which the water saturation was established by displacement of water from the sample with kerosene, or with a kerosene solution of surfactant, in the filtration apparatus. The kerosene was filtered through under a pressure drop $\Delta p = 0.5$ atm. The water saturation achieved under these conditions was about 20%.

Various petroleum-soluble and water-soluble surfactants were tested for their effect in making the sandstone hydrophobic. Cationic surfactants were the most effective.

TABLE 2

Surfactant	Saturating liquid	Surfactant concentration, %	Water permeability, darcys	Displacement pressure, cm water/cm
–	Kerosene	–	0.60	8
–	Petroleum	–	0.50	16
Arquad 2C	Kerosene	0.01	0.22	21
The same	"	0.05	0.64	22
" "	Petroleum	0.01	0.32	28
Karbozolin O	Kerosene	0.05	0.73	20
Amine C_{II}	"	0.05	0.68	15
Katapin A	"	0.05	0.11	25
Diamine dioleate	"	0.05	0.30	15

Table 1 shows the results of these laboratory tests.

It is clear from Table 1 that the best properties for rendering sandstone hydrophobic were shown by the cationic surfactants: Arquad 2C, Karbozolin O, Katapin A, Amines C_I and C_{II}, and Diamine dioleate.

Since petroleum is now the most readily available raw material in industry, laboratory experiments were carried out on the ability of petroleum and solutions of surfactants in petroleum to render sandstone hydrophobic.

Tests showed that displacement of petroleum from a sample by capillary penetration was less pronounced than displacement of kerosene.

Using a sandstone from which kerosene was nearly all displaced in the first 10 min, it was found that displacement of petroleum only began after 1.5 to h, and that it was 5 to 7 days before displacement was practically complete.

Experiments with degassed Devonian petroleum showed that the capacity of petroleum to render sandstone hydrophobic showed up after prolonged surface contact (more than 15 days), and capillary penetration did not occur then. However, capillary penetration took place if the sample had been stored in petroleum for only 2 to 3 days.

Although petroleum does not make sandstone completely hydrophobic after a short contact time, still solutions of surfactants in petroleum should be more effective than those in kerosene, owing to blanketing properties of the petroleum itself [6].

Critical Displacement Pressure

Waterproofing of the facial zone of a well and shutting out of the underground water will only be effective if the capillary forces are greater than, or at any rate equal to, the pressure drop in the stratum across the hydrophobic zone.

On the basis of literature data [7], the capillary pressure in pores with a mean diameter of 5 to 25 μ is 0.01 to 0.05 atm.

Thus, it was necessary to create in the stratum a screen of such diameter that the pressure drop in the stratum at the boundary would be less than the capillary forces in the hydrophobic sandstone.

An approximate estimate of the pressure drop at the boundary of the screen was obtained from Dupuis' equation. For a screen radius of 10 m, and a total pressure drop in the stratum of 35 atm, the pressure gradient along 5 cm should be 0.016 cm of water per centimeter.

Laboratory experiments were therefore undertaken to determine the critical pressure for displacement, i.e., the pressure drop at which water penetration began into a sample saturated with hydrocarbon liquid. From the results it was possible to determine the requisite radius of the facial rock zone to be rendered hydrophobic, and to select the most effective surfactant for achieving this.

The hydrocarbon liquids used were kerosene and petroleum.

The samples were first rendered hydrophobic and stored for 18 to 20 h, and filtration of water though the sample was then attempted at the lowest pressure gradient of 2 cm of water per centimeter. The pressure drop was increased in stages and held at each value of 2 h.

The appearance of drops of hydrocarbon liquid on the sample surface indicated the beginning of water penetration, and the pressure drop when this occurred was taken as the displacement pressure.

Displacement pressures were only determined for those products which completely prevented capillary penetration.

Table 2 shows the experimental results.

The best results in this series of tests were when the sandstone was rendered hydrophobic by Katapin A, Arquad 2C, or Karbozolin O. These products are accordingly recommended for field trials.

The sorptive properties of the surfactant used are very important in connection with the technology of rendering the rock of the facial zone hydrophobic. These properties are required in order to calculate the surfactant concentration and the amount of solution required for successful application of the method.

Investigations are being continued in this direction.

Literature Cited

1. A. Michaelis, P. Bergman, and D. Brown, "Prevention of the formation of flooding cones by injection into the stratum of reagents which alter the wetting of the rock," Petroleum Engineer, 29, No. 8, VIII (1957).
2. P. A. Rebinder, "Surface-Active Substances: their importance and application in the petroleum industry," Collection, The Use of Surface-Active Substances in the Petroleum Industry (Gostoptekhizdat, Moscow, 1961).
3. B. G. Loginov and L. G. Malyshev, The Development of Methods for Rupturing a Stratum at a Point and for Increasing the Effectiveness of Acid Treatment of Wells in Reef Formations, UFNII Report on Theme No. 29 (1959).
4. D. A. Antonov and V. M. Berezin, "A new method for determining the residual water content of rock samples," Trudy UFNII, No. II (1957).
5. G. A. Babalyan, I. I. Kravchenko, I. L. Markhasin, and G. V. Rudakov, The Physicochemical Principles of the Use of Surfactants in the Development of Petroliferous Strata (Gostoptekhizdat, 1962).
6. I. I. Kravchenko, E. N. Umrikhina, and G. L. Shapiro, The Development of a Method for Selective Isolation of Underground Water by Chemical Means under the Conditions Prevailing at the Devonian Fields at Tuimazakh and Romashkino (UFNII Publications, 1954).
7. M. M. Kusakov and Sh. K. Gimatudinov, "The capillary displacement of petroleum by water from artificial cores," Trudy MNI, No. 22 (1958).

THE INFLUENCE OF VARIOUS SURFACTANTS ON WETTING AND SURFACE TENSION

N. D. Tairov

A considerable amount of work has been done in recent years in this country and abroad to find surfactants for use in various branches of the petroleum industry [1, 5]. In particular, a search has been made for surfactants to improve the quality of the water injected into petroleum reservoirs. There are now a considerable number of surfactants, differing in their molecular structure and their behavior at hydrocarbon liquid–water, hydrocarbon liquid–mineral, and water–mineral interfaces.

In spite of the fact that surfactants are being used on an ever-increasing scale in the petroleum industry, there are as yet insufficient data on the influence of these substances on the wetting of mineral surfaces by water and petroleum, and on surface tension at different temperatures, although these problems are of considerable importance in the study of the mechanism of petroleum recovery.

The present paper gives the results of a study of the influence of the addition of surfactants to water on the magnitude of the contact wetting angle and surface tension of hydrocarbon liquids at their boundary with water. We have carried out an extensive series of experiments but we given here only the most characteristic curves illustrating these relationships. Each experimental point corresponds to the arithmetic mean of 8-12 measurements.

The hydrocarbon liquids used in the experiments were kerosene and petroleum from the Podkirmakinskii formation of the Leninskii deposit in the Apsheron Peninsula.

Since in Azerbaidzhan flooding and other operations are carried out for the most part using water–sea water from the Caspian Sea and water from the Kura River, these waters and distilled water were used in the experiments.

The contact angle of wetting was measured on quartz, which is the mineral most frequently encountered in petroleum reservoirs.

As is well known [1, 8], surfactants can be classified as ionogenic and nonionogenic. The ionogenic surfactants can in turn be classified as anionogenic and cationic.

Different surfactants are used for different purposes. Thus, for example, for the removal of salt and water from petroleum and for increasing the recovery of petroleum from reservoirs, best results are obtained with nonionogenic surfactants. For opening up an oil-bearing stratum and increasing the pickup of driving wells, best results are given by nonionogenic and anionogenic surfactants. The latter give good results only in the absence of hard stratal waters containing calcium chloride.

In our work on the study of the wetting of quartz surfaces by petroleum and water, we used nonionogenic and anionogenic surfactants.

The nonionogenic surfactants used were $LKFE_{6.1}$ and K_{150}, and the anionogenic surfactants used were Azolyat, Sulfonol, OKPN, and calcium and sodium NChK produced in Azerbaidzhan.

Water	Salt composition of waters, meq/liter								Density, g/cm^3	Salinity, °Be	Alkalinity, meq/liter
	Cl'	SO_4''	HCO_3''	n.k.	Ca..	Mg..	Na.+K.	Σ			
Technical Balaisemskii water intake	122	4	2	—	20	12	96	256	1.004	0.7	2.0
Sea	166	58	2	—	16	56	154	452	1.009	1.3	2.0
Well No. 371, horizon B Kum-Dag	2900	3	1	—	231	181	2492	5808	1.105	14.3	1.0
Well No. 586, horizon Zh Kum-Dag	2525	2	4	—	212	147	2172	5062	1.093	12.3	4.0
Well No. 87, horizon Sh Dagadzhik	520	2.7	16.5	1.8	6	6	529	1082	1.021	2.98	18.3
Well No. 194, Dagadzhik	623	4	18.2	1.0	9.8	14.7	621.7	1292	1.027	3.79	19.2
Well No. 169, group VIII Aligul	243.3	25.1	12.1	2.7	5.7	13.5	264.3	576	1.0092	1.29	14.8

Azolyat and Sulfonol were used to study the influence of temperature on the surface tension of surfactant solutions at the boundary with hydrocarbon liquids.

We studied wetting in water drop – mineral that was made hydrophobic by oil–air and oil drop–mineral–water systems.

The mineral surfaces were made hydrophobic by the method previously described [2, 4, 7]. In the study of the wetting of a hydrophobic mineral by water, the water drop was applied to a section of the mineral in air.

In the measurement of the contact angle of selective wetting, the mineral plates were placed in a vessel containing water, after which the oil drop was applied to the mineral from below.

The contact angles of wetting were measured by the projection method, using Professor M. M. Kusakov's apparatus.

Curves showing the dependence of the contact wetting angle, for quartz made hydrophobic by petroleum, on the concentration of surfactant in Kura water, are given in Fig. 1. It can be seen that increase in the concentration of surfactant in the water leads to an improvement in the wetting of the hydrophobic quartz surface by water. In most cases a sharp change in θ takes place at concentrations of surfactant in water greater than 0.01%. The value of θ changes little at lower surfactant concentrations.

These curves also show clearly the influence of the surfactants on wetting. The dependence of the contact wetting angle, for quartz made hydrophobic by oil, on the concentration of surfactant in sea water is shown in Fig. 2.

The dependence of θ on c is analogous to that of the previous case. Comparison of the curves in Figs. 1 and 2, however, reveals the influence of the quality of the water on the change in the contact wetting angle with change in the surfactant concentration.

In the study of the displacement of petroleum by water and the displacement of water by petroleum, it is important to know the magnitude of the contact angle of selective wetting. The results of these studies are given in Figs. 3 and 4.

The dependence of the contact angle, for the selective wetting of quartz by a drop of petroleum, on the concentration of surfactant in Kura water is shown in Fig. 3.

Although in the case of the water–hydrophobic mineral–air system increase in the concentration of surfactant in water in all cases led to an improvement in the wetting of the mineral surface by water (decrease in θ), in the present case θ may increase or decrease, depending on the nature of the surfactant.

In some cases, increase in the concentration of surfactant in water leads to the production of hydrophobic properties on the mineral surface (the contact angle of selective wetting reaches 90°).

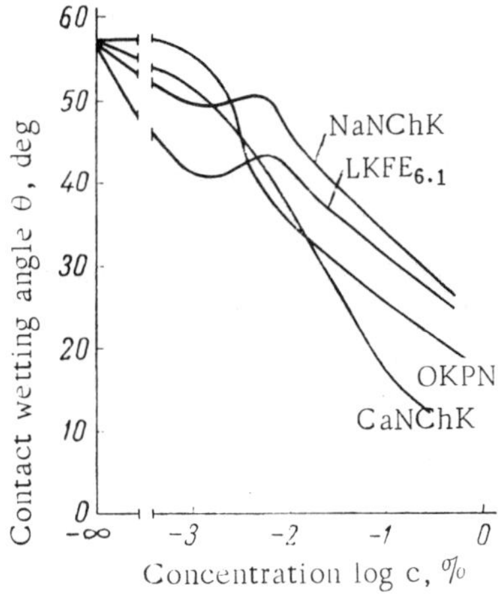

Fig. 1. Dependence of the contact wetting angle, for quartz made hydrophobic by oil, on the concentration of surfactant in Kura water.

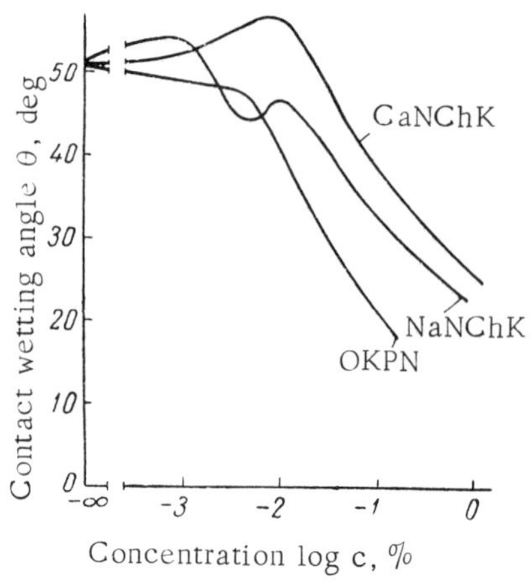

Fig. 2. Dependence of the contact wetting angle, for quartz made hydrophobic by oil, on the concentration of surfactant in sea water.

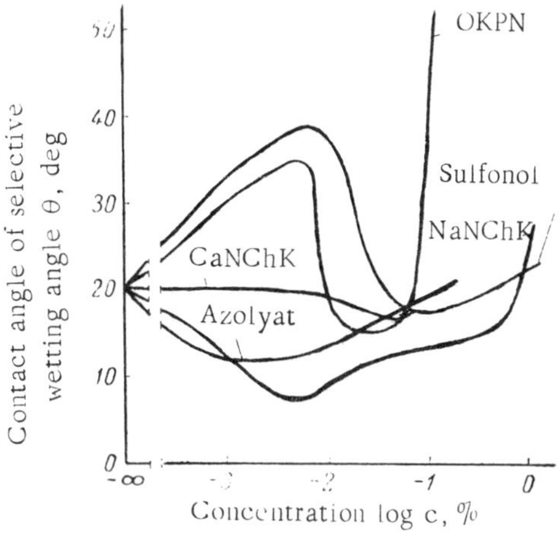

Fig. 3. Dependence of the contact angle, for the wetting of quartz by a drop of petroleum, on the concentration of surfactant in Kura water.

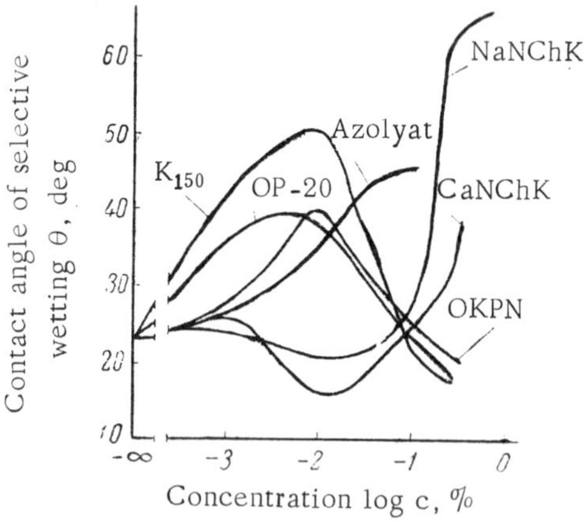

Fig. 4. Dependence of the contact angle, for the wetting of quartz by a drop of petroleum, on the concentration of surfactant in sea water.

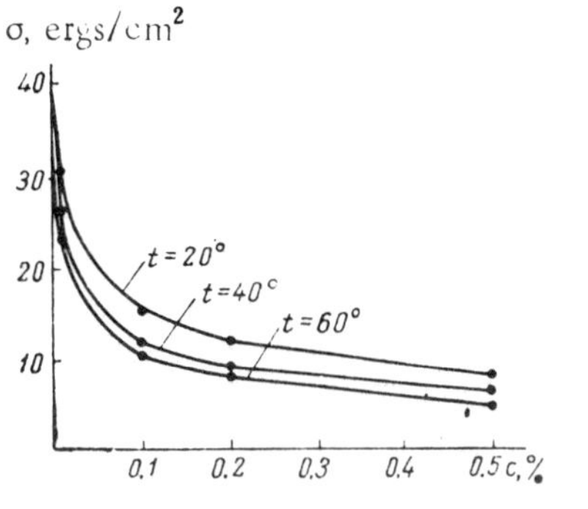

Fig. 5. Isotherms for the surface tension of an aqueous solution of Azolyat (distilled water) at its boundary with kerosene at different temperatures.

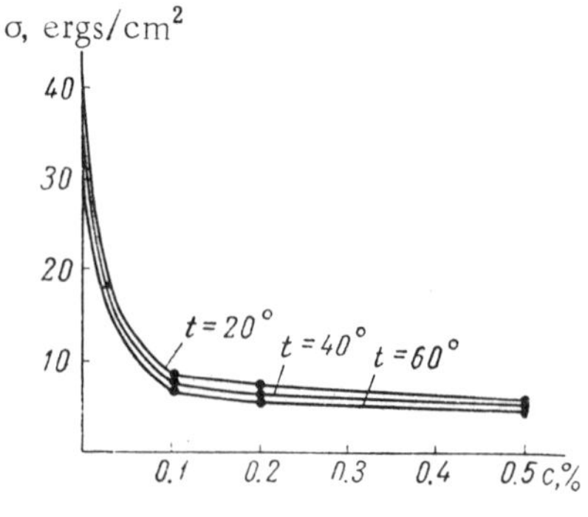

Fig. 6. Isotherms for the surface tension of a solution of Sulfonol in distilled water at its boundary with kerosene at different temperatures.

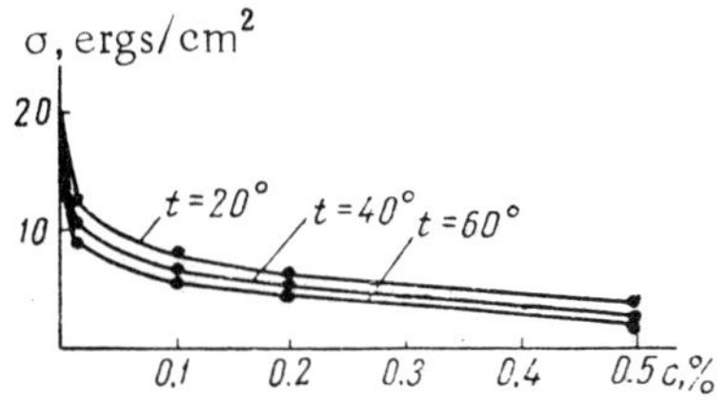

Fig. 7. Isotherms for the surface tension of solutions of Azolyat in sea water at their boundary with petroleum from the PK formation of the Leninskii deposit, at different concentrations.

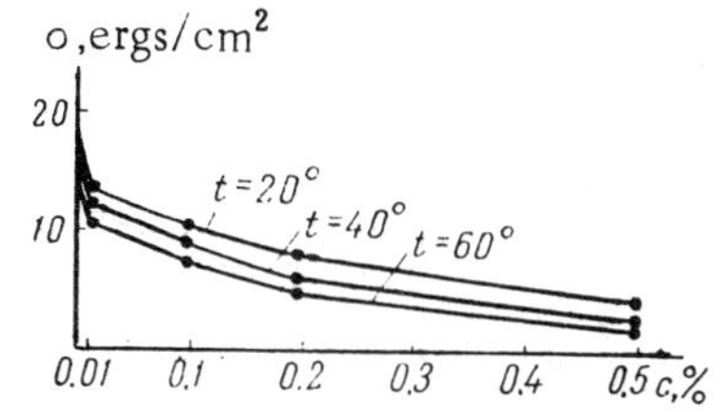

Fig. 8. Isotherms for the surface tension of a solution of Sulfonol in sea water at its boundary with petroleum from the PK formation of the Leninskii deposit, at different concentrations.

The curves (Figs. 3 and 4) show that the quality of the water also has a definite influence on the dependence of θ on c.

Thus for minerals made hydrophobic by petroleum, an improvement in wetting by water in a medium of air is observed for both nonionogenic and ionogenic surfactants. In the case of the wetting of quartz by a drop of petroleum in a medium of water, however, a deterioration is observed when nonionogenic surfactants are added to the water.

When anionogenic surfactants are added to sea water and Kura water, the contact angle of selective wetting almost always increases. At the same time, there is a definite concentration of surfactant in water, above which a sharp increase in θ takes place. This concentration depends on the properties of the water, the petroleum, and the surfactant.

Thus if the use of surfactants is required in processes in which the wetting of a mineral surface by water or a hydrocarbon liquid plays an important part, the concentration of the test surfactant in water must be chosen in such a way that the sign and order of wetting satisfy the required conditions.

In earlier studies [7] it was shown that when petroleum is added to a nonpolar hydrocarbon liquid—toluene—the contact angle at first increases with increase in the petroleum concentration, but later decreases. This complex change in θ with change in c is due to the structure of the adsorption layer formed on the solid quartz surface by the molecules of surfactant in the petroleum, and also to their reorientation.

In the case under consideration, the change in the contact angle of selective wetting with increase in the concentration of surfactant in water is also apparently due chiefly to the structure of the adsorption layer under the drop of petroleum, formed by the surfactant molecules adsorbed on the quartz. The mechanism of this behavior of the contact angle is not yet completely clear but the nature of the curve for the function $\theta = f(c)$ shows that in this case increase in the surfactant concentration apparently leads to the formation of several layers of surfactant molecules orientated in different ways on the quartz surface.

It is known that when surfactants are added to water, the surface tension (σ) at the boundary with hydrocarbon liquids, including petroleum, decreases sharply at comparatively low surfactant concentrations.

Increased pressures and temperatures are found in oil-bearing strata.

The influence of pressure and temperature on the surface tension of hydrocarbon liquids and water has been studied by a number of authors [3, 9]. The influence of temperature and pressure on the surface tension of aqueous surfactant solutions at their boundary with hydrocarbon liquids has not yet been studied, however. The present paper gives some results of a study of the influence of temperature on the surface tension of aqueous solutions of surfactants in distilled water and sea water at their boundary with kerosene and petroleum.

The experiments were carried out at temperatures close to those of the stratal conditions in most of the Apsheron petroleum deposits (20-60°C).

The surface tension was determined on Academician P. A. Rebinder's apparatus.

The results of the studies are given in Figs. 5-8. The isotherms for the surface tension of solutions of Azolyat and Sulfonol in distilled water at their boundary with kerosene at temperatures of 20, 40, and 60°C are given in Figs. 5 and 6.

The surface tension decreases with increase in temperature. The temperature dependence of σ is also influenced by the properties of the surfactant.

This is also observed in experiments with solutions of Azolyat and Sulfonol in sea water, where petroleum was used in place of kerosene (Figs. 7 and 8).

Thus the studies which have been carried out show that with increase in the temperature of the surrounding medium, the surface tension of surfactant solutions at their boundary with hydrocarbon liquids decreases slightly.

The following conclusions can be reached on the basis of the results of the experiments which have been carried out.

1. The properties of hydrocarbon liquids, water, and surfactants have a definite influence on wetting.

2. In selective wetting, the wetting of minerals by water deteriorates when the surfactant concentration exceeds a certain limiting value, which is different for different surfactants.

3. The greatest selective wetting is shown when nonionogenic surfactants are added to water.

4. Increase in temperature leads to a further slight decrease in the surface tension of surfactant solutions at their boundary with hydrocarbon liquids. The decrease in σ with increase in temperature depends on the properties of the hydrocarbon liquids, the surfactants, and the water.

Literature Cited

1. G. A. Babalyan, I. I. Kravchenko, I. L. Markhasin, and G. V. Rudakov, The Physicochemical Principles of the Use of Surface-Active Substances in the Exploitation of Oil-Bearing Strata (Gostoptekhizdat, 1962).
2. M. A. Geiman, A. L. Larin, V. B. Shneerson, and R. A. Fridman, Trudy Instituta Nefti AN SSSR, Vol. 6 (1955).
3. M. M. Kusakov, N. M. Lubman, and A. Yu. Koshevnik, Neft. Khoz., Nos. 2-3 (1954); Trudy Instituta Nefti AN SSSR, Vol. 2 (1952); Trudy MNI im. I. M. Gubkina, No. XIV (1955).
4. M. M. Kusakov and N. D. Tairov, DAN Azerb. SSR, No. 111 (1959).
5. P. A. Rebinder, "Surface-active substances: their importance and applications in the petroleum industry," Collection, The Use of Surface-Active Substances in the Petroleum Industry (Gostoptekhizdat, Moscow, 1961).
6. N. D. Tairov and M. M. Kusakov, Izv. AN Azerb. SSR, No. 4 (1957).
7. N. D. Tairov, ANKh, No. 2 (1956); No. 4 (1957).
8. G. Shtyupel', Synthetic Wetting and Cleaning Agents (Gostoptekhizdat, 1960).
9. E. W. Hough, M. I. Rzaza, and B. A. Wood, Trans. AIME, Vol. 91 (1951), p. 57.

A STUDY OF THE ADSORPTION OF OIL-SOLUBLE SURFACTANTS ON VARIOUS SANDS

R. A. Musaev, G. A. Babalyan, and I. I. Kravchenko

Oil-soluble surfactants may find extensive application in the petroleum industry for improving various technological processes. Thus, for example, oil-soluble surfactants, which decrease considerably the relative permeability of a porous medium for water and improve the conditions for its displacement by hydrocarbon liquids, can be used in the technological processes of oil drilling and recovery, where a decrease in the residual water saturation of the rock is required, to decrease the residual water in the end-face zone of the wells, to decrease the filtration of water from the emulsion solutions into the strata during drilling, etc. [1, 2]. Oil-soluble surfactants can also increase the capillary penetration of oil into water-saturated rock [4].

These surfactants can also be used as additives to emulsion solutions and oil-based solutions in order to increase the rate of drilling [2, 6].

The present paper describes the results of studies of the adsorption of oil-soluble surfactants on sands from the productive strata of the Apsheron Peninsula.

The experiments were carried out with the nonionogenic oil-soluble surfactants OP-4 and Stearox-6. The surfactant OP-4 is the product of the reaction of 4 moles of ethylene oxide and a mixture of mono- and dialkylphenols with alkyl residues containing 8-10 carbon atoms, and Stearox-6 is oxyethylated stearic acid containing 6 ethylene oxide groups, mixed with 5-10% OP-7 or OS-20.

These substances are incapable of undergoing dissociation to form ions. The solvent used was purified kerosene, free from polar substances, with a surface tension at its boundary with water equal to 47.3 ergs/cm^2.

The adsorption of OP-4 and Stearox-6 from kerosene was studied on quartz sand and on sands from surface outcrops from the productive strata of the Apsheron Peninsula—the Kirmakinskii formation (KF) of the Balakhan deposit, the Nadkirmakinskii sand formation (NSF) of the Karadag deposit, and the Apsheron level of the Bibi-Eibat deposit.

TABLE 1

Sand	Fractional composition, %						Density, g/cm^3	Porosity, %	Carbonate content, %
	> 0.5 mm	0.5–0.25 mm	0.25–0.15 mm	0.15–0.1 mm	0.1–0.075 mm	< 0.075			
Quartz sand	—	7.65	73.99	17.80	0.34	0.22	2.62	32.3	—
Sand from the KF formation	—	8.02	29.93	50.59	9.53	1.93	2.65	35.7	4.8
Sand from the NSF formation	—	24.62	37.75	7.30	26.71	3.62	2.68	35.8	1.6
Sand from Apsheron level	5.24	50.73	27.95	11.64	2.62	1.82	2.67	27.7	35.0

TABLE 2

Water	Found by analysis, g/100 g of water	Equivalent		Characteristics of water
		mg/eq	%	
Alkaline water from the PK formation of the Leninskii region	$Na' + K'$	24.17	46.53	First salinity $S_1 = 66.06$ Second salinity $S_2 = 0$
	Ca'' 0.0100 %	0.50	0.96	
	Mg'' 0.0158 %	1.30	2.51	
	Total **cations**			
	Cl' 0.5388 %	15.19	29.24	First alkalinity $A = 27.00$ Second alkalinity $a = 6.94$
	SO_4'' 0.0947 %	1.97	3.79	
	HCO_3' 0.4461 %	7.31	14.08	
	CO_3'' 0.0168 %	0.56	1.08	
	O. K.' 0.0949 %	0.67	1.29	
	HB_4O_7' 0.0417 %	0.27	0.52	
	Total anions +K 51.94			
Hard water from the KF formation in the Leninskii region	$Na' + K'$	171.90	42.20	First salinity $S_1 = 84.40$ Second salinity $S_2 = 14.94$
	Ca'' 0.3314%	16.54	4.05	
	Mg'' 0.1858%	15.28	3.75	
	Total **cations**			
	Cl' 7.1747%	202.33	49.66	First alkalinity $A = 0$ Second alkalinity $a = 0.66$
	SO_4'' 0.0030%	0.06	0.01	
	HCO_3' 0.0563%	0.92	0.23	
	CO_3'' None	—	—	
	O. K. 0.0550%	0.39	0.09	
	HB_4O_7' 0.0030%	0.02	0.01	
	Total anions +K 407.44			

TABLE 3

OP-4 solution				Stearox-6 solution				Notes
Quartz sand	Sand from the KF formation	Sand from the NSF formation	Sand from the Apsheron level	Quartz sand	Sand from the KF formation	Sand from the NSF formation	Sand from the Apsheron level	
0.84	4.05	3.30	4.05	1.05	6.06	5.10	6.60	σ determined at the boundary with distilled water
0.90	4.50	3.75	4.59	1.20	6.66	6.45	6.75	σ determined at the boundary with hard water
0.78	3.90	2.85	3.90	0.9	5.70	3.90	5.79	σ determined at the boundary with alkaline water

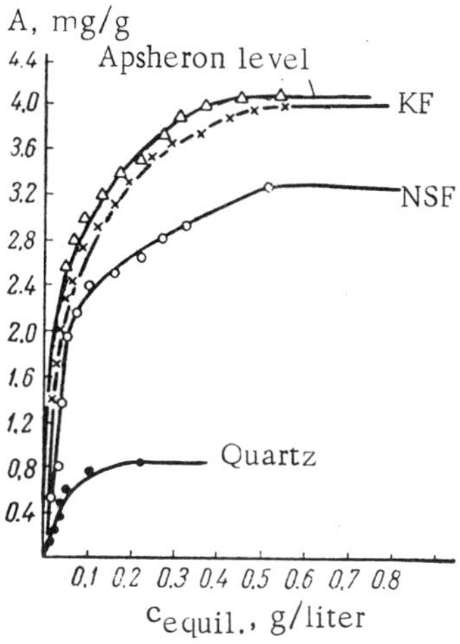

Fig. 1. Isotherms for the adsorption of OP-4 from solutions in kerosene on different sands (σ determined at the boundary with distilled water).

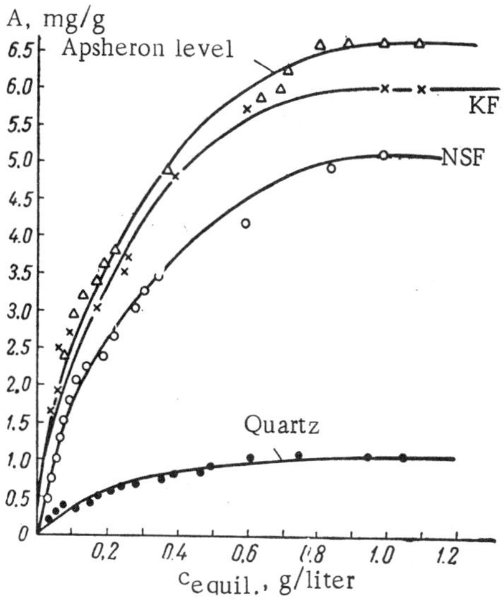

Fig. 2. Isotherms for the adsorption of Stearox-6 from solutions in kerosene on different sands (σ determined at the boundary with distilled water).

Before the experiments the sands were washed with fresh water and extracted with an alcohol-benzene mixture (1:3) until the petroleum, asphaltic, and tarry substances had been completely removed.

After the extraction the sands were washed several times with distilled water and dried to constant weight at a temperature of 100-105°C.

The characteristics of the sands are given in Table 1.

The experiments were carried out using distilled water, hard stratal water from the KF formation, and alkaline water from the PK formation of the above deposits. The physicochemical properties of the waters are given in Table 2.

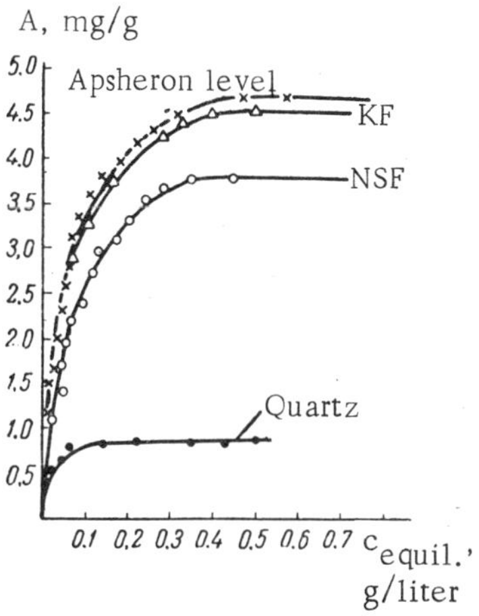

Fig. 3. Isotherms for the adsorption of OP-4 from solutions in kerosene on different sands (σ determined at the boundary with hard water).

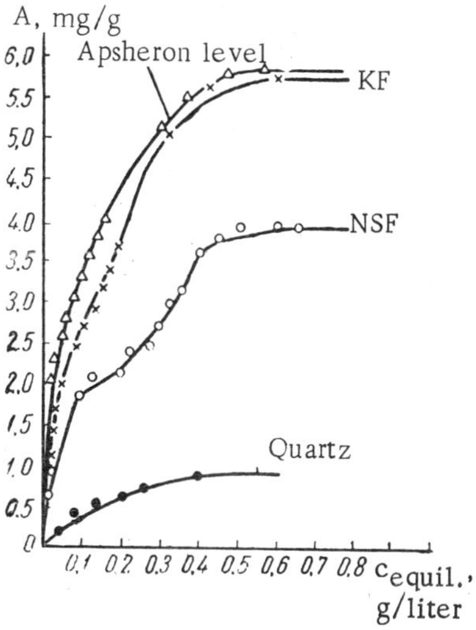

Fig. 4. Isotherms for the adsorption of Stearox-6 from solutions in kerosene on different sands (σ determined at the boundary with alkaline water).

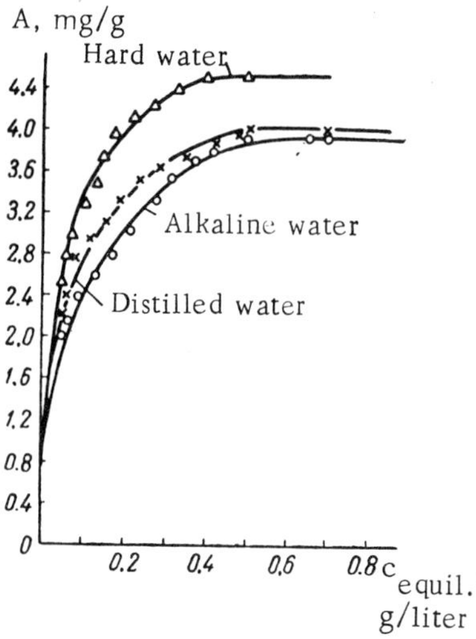

Fig. 5. Isotherms for the adsorption of OP-4 from solutions in kerosene on KF sand (σ determined at the boundary with different waters).

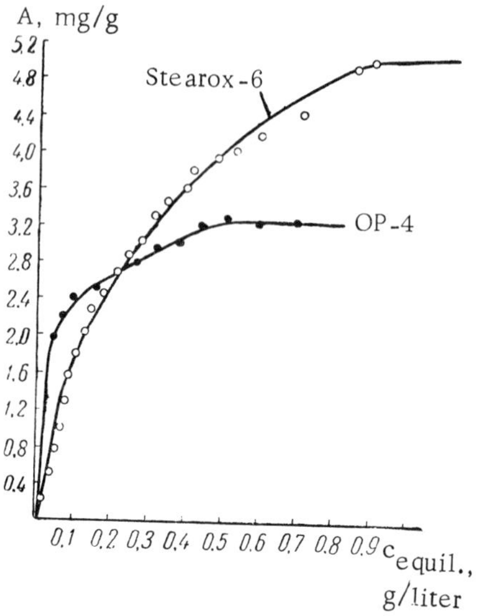

Fig. 6. Isotherms for the adsorption of OP-4 and Stearox-6 from solutions in kerosene on NSF sand(σ determined at the boundary with distilled water.

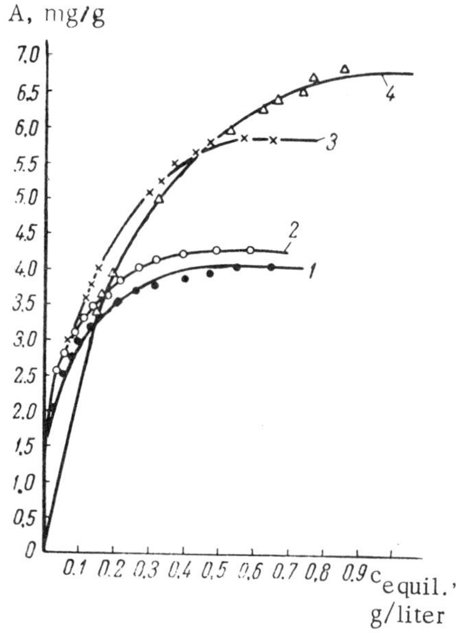

Fig. 7. Isotherms for the adsorption of OP-4 and Stearox-6 from solutions in kerosene on sand from the Apsheron level. 1) OP-4 (σ determined at the boundary with alkaline water); 2) OP-4 (σ determined at the boundary with hard water); 3) Stearox-6 (σ determined at the boundary with alkaline water); 4) Stearox-6 (σ determined at the boundary with hard water).

The adsorption was calculated in the usual way [3]. The surface tension was determined at constant temperature by measuring the volume of the drops produced by means of a micrometric instrument on a UFNII stalagmometer [3].

The surface tension of kerosene solutions of surfactants at different concentrations was determined at their boundary with different waters. An appropriate weight of sand was then introduced into a solution of known concentration and when adsorption equilibrium had been reached (in approximately 2 days) the surface tension was measured again.

The adsorption was calculated from the formula

$$A = \frac{(c_1 - c_{equil.}) V}{p}$$

where A is the quantity of adsorbed surfactant for a given solution concentration in mg/g; c_1 is the concentration of surfactant before adsorption in g/liter; c_{equil} is the equilibrium concentration after adsorption in g/liter; V is the volume of surfactant solution in ml; and p is the weight of adsorbent in g.

The main results of the studies are given in Figs. 1-4, where the magnitude of the adsorption in mg/g is plotted on the y axis and the equilibrium concentration after adsorption in g/liter is plotted on the x axis.

With increase in the concentration of surfactant in kerosene, the adsorption of the surfactant on the sand increases and reaches a limiting value. The limiting values of the adsorption for OP-4 and Stearox-6 are given in Table 3.

The isotherms for the adsorption of kerosene solutions of OP-4 and Stearox-6 on different sands are given in Figs. 1 and 2. The surface tension was determined at the boundary with distilled water.

The lowest adsorption of surfactants in the concentration range studied was observed for quartz sand (OP-4 0.84 mg/g, Stearox-6 1.05 mg/g) and the greatest adsorption was observed on sand from the Apsheron level (OP-4 4.05 mg/g, Stearox-6 6.60 mg/g).

The high adsorption on sands from the oil-bearing strata is due to their high specific surface and to the presence of carbonates.

Earlier studies [5, 6] showed that the surface of carbonate particles is less hydrophilic than a quartz surface. The carbonate particles themselves are porous, however, and their presence leads to a marked increase in the specific surface. Moreover, the possibility of chemical adsorption of surfactants on carbonate particles is not excluded.

Isotherms for the adsorption of OP-4, determined by measuring the surface tension at the boundary with hard water, are given in Fig. 3, and isotherms for the adsorption of Stearox-6 at the boundary with alkaline water are given in Fig. 4. For comparison, Fig. 5 gives isotherms for the adsorption of OP-4 on KF sand, determined by measuring the surface tension at the boundary with different waters. In spite of the fact that the adsorption of the surfactants took place from the same solvent on the same adsorbent, different values were obtained for the limiting adsorption. This is due to the reaction of the surfactants of the hydrocarbon liquids with the salts in the waters and to their different solubilities in these waters. In order to determine the magnitude of the adsorption from data obtained by measuring σ, therefore, it is correct to use distilled water, which eliminates the distortion of the results of the measurement of σ by other waters due to the presence of various salts, alkalies, and organic substances.

The data obtained in the studies show that the adsorption of the surfactants OP-4 and Stearox-6 is different for different sands (Figs. 6 and 7).

Comparison of the curves in Figs. 6 and 7 shows that on all sands the quantities of Stearox-6 adsorbed are greater than the quantities of OP-4 adsorbed.

By using the data obtained for the adsorption of OP-4 and Stearox-6 on different sands from productive strata in the Apsheron Peninsula, it is possible to calculate the necessary quantity of surfactants and their concentration in kerosene for treatment of the end-face zone, in order to improve the different processes for the intensification of petroleum recovery by the method already described [8].

Literature Cited

1. G. A. Babalyan, "The physicochemical principles of the use of surface-active substances in the drilling of wells in order to intensify the recovery of petroleum and increase the petroleum output from strata," Proceedings of the First All-Union Conference on the Use of Surface-Active Substances in the Petroleum Industry (Gostoptekhizdat, 1961).
2. G. A. Babalyan, G. V. Rudakov, N. L. Markhasin, and I. I. Kravchenko, "The use of surface-active substances to increase the recovery of petroleum from strata," Trudy UFNII, No. VII (1961).
3. G. A. Babalyan, I. I. Kravchenko, I. L. Markhasin, and G. V. Rudakov, The Physicochemical Principles of the Use of Surface-Active Substances in the Exploitation of Oil-Bearing Strata (Gostoptekhizdat, 1962).
4. G. A. Babalyan, "The theory of the action of surface-active substances in the establishment of wells," Trudy UFNII, No. VI (1960).
5. G. A. Babalyan, Z. I. Zeinalov, E. F. Korkhova, A. I. Tairov, and M. S. Agalarov, "An example of the flooding of a petroleum deposit by bottom water," Trudy AzNII, No. III (1956).
6. M. M. Kusakov and N. D. Tairov, Dokl. AN Azerb. SSR, Vol. II (1959).
7. P. A. Rebinder, "Surface-active substances: their importance and applications in the petroleum industry," Collection, The Use of Surface-Active Substances in the Petroleum Industry (Gostoptekhizdat, 1961).
8. E. G. Khalilov, I. I. Kravchenko, and G. A. Babalyan, "The problem of the monitoring of surface-active substances in the treatment of the end-face zone" (this volume).

THE ADSORPTION OF ANIONIC SURFACTANTS

I. I. Kravchenko, V. A. Brovenko, and N. A. Rybinsakaya

The amount of surfactant adsorbed on quartz sand can be determined by two methods:

1) from the amount of surfactant adsorbed on the quartz sand.

The (direct) method of determining the amount of surfactant adsorbed on quartz sand is laborious and cannot always be used (irreversible adsorption or chemosorption);

2) the method of determining the amount of surfactant from the differences in the concentrations before and after adsorption; this most widely used method requires an accurate determination of the surfactant concentration in solution before and after solution.

In the majority of cases these concentrations are determined by measuring the surface tension of the surfactant solutions at the boundary with air or a nonpolar liquid. Unfortunately, this method is unsuitable for most anionic surfactants [1].

Multicomponent anionic surfactants actually comprise active components in the form of salts of sulfonic acids and of impurities consisting mainly of resinous substances, nonsulfonated oils, and inorganic sulfates. The resinous substances are formed during the synthesis of anionic surfactants in the alkylation and sulfonation processes, as the reagents used in these processes (H_2SO_4, SO_3, $AlCl_3$, etc.) have a polymerizing effect besides their principal mode of action.

Sulfonate solutions have low surface tensions in contact with air or liquid hydrocarbons [2]. Nonsulfonated oils are virtually insoluble in water, but their solubility is increased in the presence of sulfonates (solubilization). However, both these oils and inorganic sulfates, soluble in water and occurring in small quantities (less than 10%) in anionic surfactants, have practically no effect on the surface tension of sulfonates in contact with air or a nonpolar liquid.

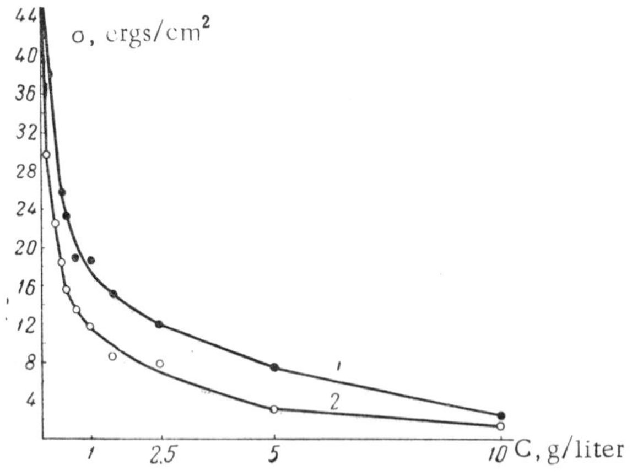

Fig. 1. Surface tension isotherms for solutions of anionic surfactants, and the same solutions after removing resinous substances, in contact with purified kerosene. 1) Anionic surfactant; 2) anionic surfactant after removing resinous substances.

The resinous substances of anionic surfactants are dissolved colloidally in water in the presence of sulfonates, and due to their combining with the basic surfactant (a sulfonate) during dissolution, obviously lead to an increase in the surface tension of aqueous sulfonate solutions in contact with air or liquid hydrocarbons. Thus, the surface tension isotherm on the graph lies considerably above the position of the isotherm which would be obtained in the absence of resinous substances in sulfonate solutions (Fig. 1).

In the adsorption process the resinous substances are adsorbed first on quartz sand, sandstones, and many other solid adsorbents [3] and, by blocking their surface, prevent the adsorption of sulfonates on them.

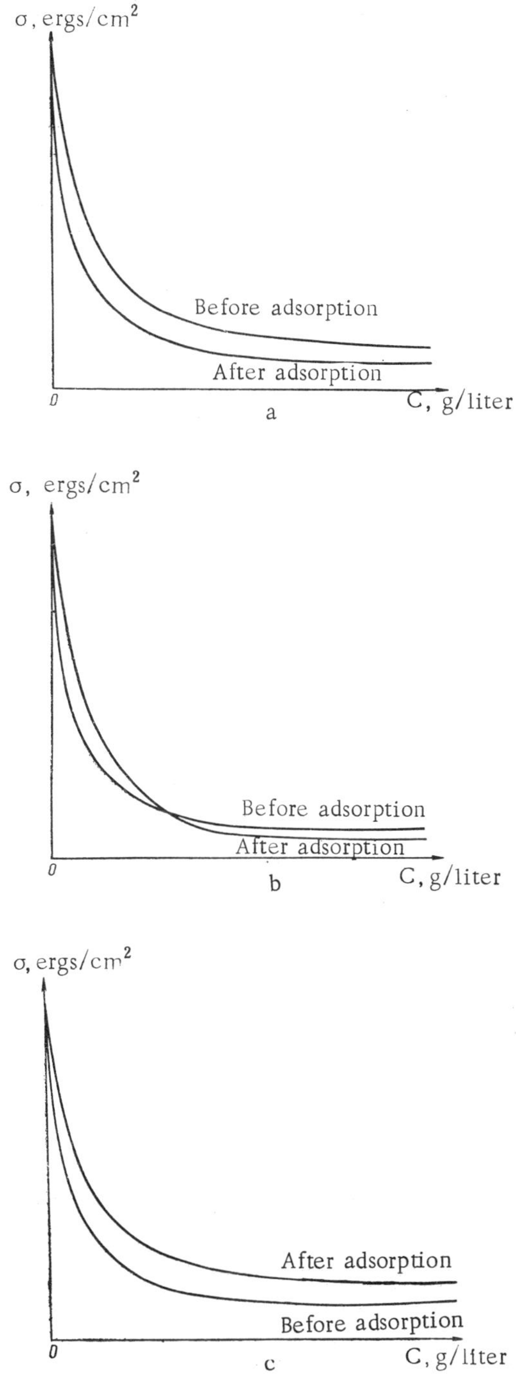

Fig. 2. Possible positions of the surface tension isotherms of solutions of anionic surfactants, obtained before and after adsorption in contact with purified kerosene.

In addition, due to the preferential adsorption of the resinous substances the sulfonates previously combined with the resins are liberated, thus leading to an increase in their concentration in solution and consequently to a lowering of the surface tension of the solution in contact with air or a liquid hydrocarbon.

On the graph the surface tension isotherm obtained after adsorption lies considerably below the isotherm which would have been obtained in the absence of resinous substances in a solution of anionic surfactants.

Thus, a situation can be created in which the surface tension isotherm obtained after adsorption lies below (Fig. 2a) or intersects the isotherm obtained before adsorption (Fig. 2b) but even if this does not happen (the isotherm before adsorption lies below the isotherm obtained after adsorption, Fig. 2c) then the adsorption values obtained will be considerably less than the true values.

Taking account of what has been said above, the concept of "apparent" adsorption has been introduced, i.e., the adsorption determined from the surface tension isotherms recorded before and after adsorption. It is assumed that an anionic surfactant is a one-component system which consists solely of a sulfonate [1].

The method of determining the apparent adsorption consisted of determining the surface tension of aqueous solutions of the anionic surfactant at various concentrations in contact with purified kerosene, the σ of which is not less than 47 ergs/cm^2 in contact with pure water (calibration curve). Then a quartz sand of known fractional composition (Table 1) was added to the surfactant solutions.

After adsorption equilibrium was reached and the quartz sand was centrifuged from the solutions, the surface tension in contact with the kerosene was determined.

From the calibration curve and the curve obtained after adsorption the "apparent" difference in the concentrations of the solutions ($c_1 - c_s$) was calculated and from this the apparent adsorption A.

Thus, the apparent adsorption does not correspond to the quantitative adsorption. However, it does provide the possibility of making a qualitative estimate of the nature and intensity of adsorption of the various components in anionic surfactants, and also of classifying them from the point of view of their content of components slightly soluble in water but readily adsorbed on solid surfaces. It also enables the nature of the adsorption of anionic surfactants from their solutions onto solid adsorbents to be represented. Based on their apparent adsorption isotherms, all anionic surfactants can be divided into three groups (Fig. 3).

TABLE 1

Grain size of fraction, mm	Fractional composition, %
0.5 -0.25 mm	3.0
0.25-0.10 mm	57.0
0.10-0.06 mm	24.5
<0.06 mm	15.5

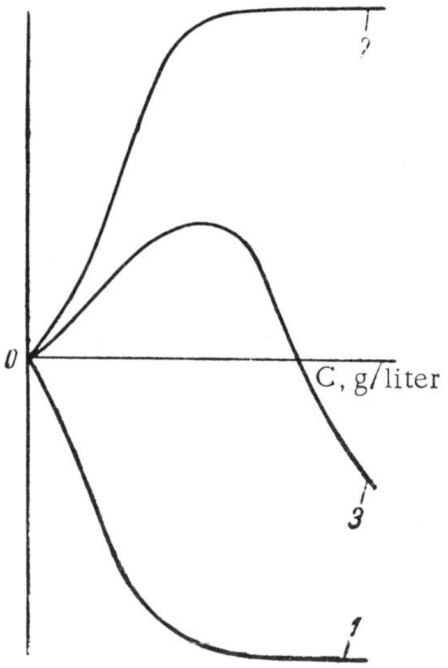

Fig. 3. Apparent adsorption isotherms for anionic surfactants. 1-3) First, second, and third groups.

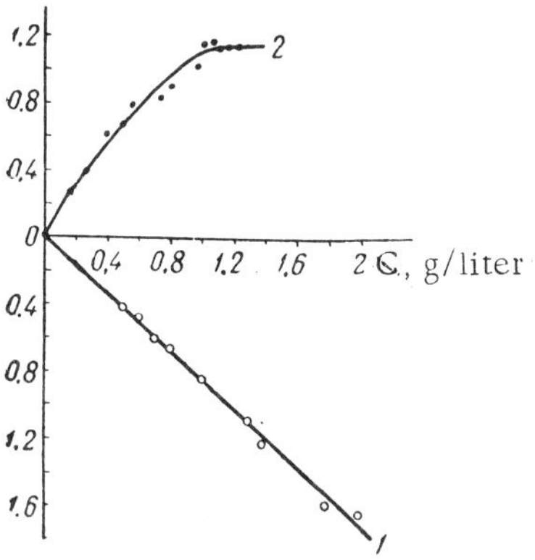

Fig. 4. Adsorption isotherms for samples of Sulfonol TKB, obtained by sulfonating at high (1) and low (2) temperatures.

The apparent adsorption isotherms of the first group of surfactants lie in the region of negative adsorption values and the apparent adsorption reaches limiting values with an increase in concentration. This group of surfactants contains a large amount of resinous substances which are only slightly soluble in water but which may be readily adsorbed on a solid surface.

The isotherms for the second group of surfactants lie in the region of positive values, and the apparent adsorption reaches limiting values with an increase in concentration. The values of this apparent adsorption are close to the true adsorption values, as these surfactants contain comparatively small quantities of resinous substances.

The apparent adsorption isotherms of the third group of surfactants lie in the region of positive values at low concentrations and reach a maximum with an increase in concentration. After reaching the maximum, the apparent adsorption values begin to drop with an increase in concentration and enter the region of negative values. Such surfactants contain a comparatively large quantity of resinous substances which are slightly soluble in water.

Moreover, the apparent adsorption isotherms can serve as the criteria for the accuracy of carrying out the synthesis of anionic surfactants. For example, the apparent adsorption isotherm of Sulfonol TKB, obtained by alkylating toluene with a particular cut from benzene cracking, sulfonating, and neutralizing the sulfonic acid with alkali, belongs to the first or second group depending on the sulfonation temperature (Fig. 4). At a higher sulfonating temperature the polymerizing action of sulfuric acid increases sharply, thus leading to the formation of a large amount of resinous substances. The apparent adsorption isotherms of such surfactants enable us to assign them to the first group. With a low sulfonation temperature a small quantity of resinous substances is formed and the apparent adsorption isotherms enable us to assign them to the second group.

It must be noted that anionic surfactants sulfonated at low temperature are usually considerably more active than the surfactants sulfonated at a higher temperature.

Thus, to determine the true value for the adsorption of anionic surfactants on quartz sand from their aqueous solutions, the surface tension method cannot be used.

Of the other methods for determining the concentrations of anionic surfactants in solutions before and after adsorption, the most promising are the methods based on the reactions of anionic surfactants with certain compounds according to the scheme

TABLE 2

Surfactant content, mg	Optical density					Arithmetic mean	Mean absolute error	Mean relative error, %
	1	2	3	4	5			
0.0200	0.005	0.005	0.007	0.005	0.004	0.0052	0.0008	15.4
0.1500	0.043	0.045	0.046	0.043	0.040	0.0434	0.0017	3.9
0.3000	0.086	0.090	0.097	0.091	0.086	0.0900	0.0033	3.7
0.500	0.130	0.136	0.132	0.130	0.138	0.1132	0.0031	2.7
0.700	0.166	0.175	0.174	0.168	0.168	0.1702	0.0035	2.0
1.000	0.218	0.223	0.221	0.218	0.216	0.2192	0.0023	1.0
1.50	0.285	0.288	0.288	0.284	0.282	0.2854	0.0021	0.8
2.00	0.331	0.351	0.341	0.328	0.330	0.3362	0.0079	2.3

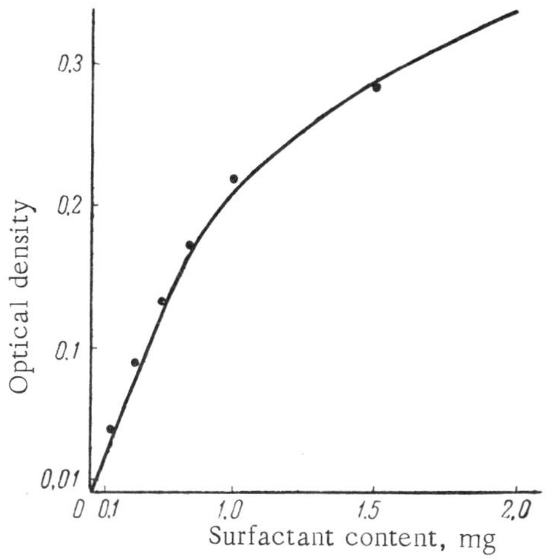

Fig. 5. Variation of the optical density with Sulfonol NP-1 content in aqueous solution.

where R^- is the negatively charged hydrophobic ion of the anionic surfactant; M^+ can be H^+, Na^+, NH_4^+, K^+, etc.; RN^+ is a positively charged hydrophobic ion; X^- is a univalent acid radical Cl^-, Br^-, RSO_3^-, etc.; RN^+X^- can be a dye, a cationic surfactant, a suitable amine salt,or an inorganic metal salt [4].

The most interesting of these reagents (RN^+X^-) are the dyes,as they give colored compounds (R^-RN^+) the solution color intensities which make it possible to determine the concentration of anionic surfactant. Furthermore, the compounds formed between dyes and anionic surfactants are usually less stable than those formed between cationic and anionic surfactants; in the majority of cases this enables the dye to be liberated from its compound with an anionic surfactant by adding a cationic surfactant to the solution. Such a replacement in the dye compound by the cationic surfactant, which involves displacement of the dye, is the basis for a quantitative determination of an anionic surfactant in solution, in which the dye acts as indicator.

The methods based on the reaction given above can be divided into several groups.

The first, most widely used group of methods is based on the extraction of a colored compound of the anionic surfactant with a dye contained in a solvent,and on the determination of the optical density of the colored solution obtained. From the optical density of solutions of known concentrations,a calibration curve is constructed and the adsorption is calculated from this. The dyes used for this group of methods are: methylene blue, methylene green, fuchsine, and rosaniline. An interesting modification of these methods is the adsorption of a complex of an anionic surfactant with methylene blue by quartz sand and the subsequent desorption of the adsorbed compound by 70% ethanol; the intensity of the color of the solution obtained is determined with the aid of the spectrophotometer.

The second group of methods is based on the titration of anionic surfactants with cationic surfactants in the presence of an indicator, e.g., the use of bromphenol blue made alkaline with ammonia where, with an excess of cationic surfactant, the color of the solution changes from purple to a pure blue [6].

The methods based on titration involving the transfer of a dye from one liquid phase to another belong to the third category of methods. These methods make use of the following phenomena [4].

1. The compound of an anionic surfactant with an alkaline dye which is soluble in an organic solvent is decomposed with cationic surfactants and the liberated dye transfers into the aqueous phase.

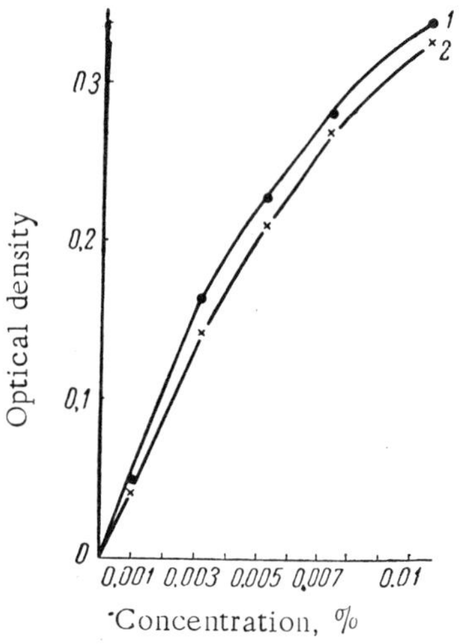

Fig. 6. Variation of optical density with Sulfonol NP-1 concentration in solutions before and after adsorption. 1) Before adsorption; 2) after adsorption.

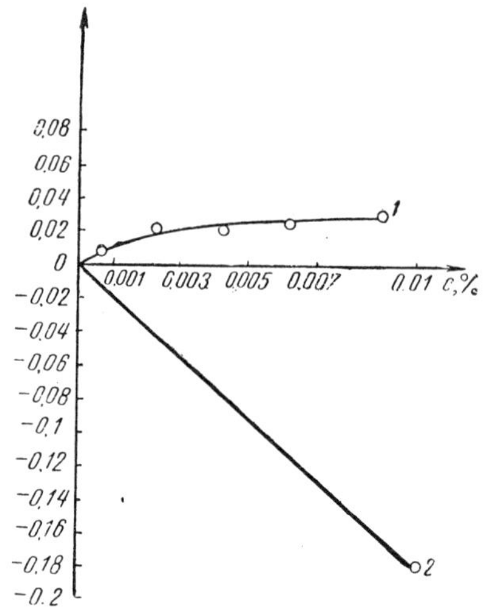

Fig. 7. Isotherms for the adsorption of Sulfonol NP-1 on quartz sand, obtained by determining concentrations colorimetrically (1) and by the surface tension method (2).

2. The compound of a cationic surfactant with an acid dye which is soluble in an organic solvent is decomposed by an anionic surfactant and the dye is released into the aqueous phase.

3. The water-soluble mixture of anionic surfactant and acid dye yields a colored dye compound with an excess of cationic surfactant, which is soluble in the organic solvent.

4. The water-soluble mixture of cationic surfactant and alkali dye yields the dye on titration with an anionic surfactant which forms, in excess, a compound of the anionic surfactant soluble in the organic solvent.

The most widely used solvent is chloroform. Others have been used,however.

For anionic surfactants the dyes used most are: methylene blue, bromphenol blue, methyl yellow (paradimethylaminoazobenzene), dichlorofluorescein, eosin, etc.

It must be noted that in all the methods based on the reaction of a surfactant with dyes, the attainment of quantitative results depends on the choice of solvent, the pH of the aqueous solution; and the insolubility of the dye in the solvent chosen.

In the Surfactant Laboratory at UFNII,the Jones method, improved by Longwell and Maniece [7], has been investigated for its applicability to the determination of the concentration of Sulfonol NP-1 in aqueous solutions. The method is based on the reaction of methylene blue with Sulfonol NP-1(see top of next page).

TABLE 3

Concentration, %		Optical density	
Before adsorption	After adsorption	Before adsorption	After adsorption
0.0010	0.00070	0.049	0.041
0.0030	0.0023	0.163	0.142
0.0050	0.0043	0.228	0,211
0.0070	0,0062	0.282	0.271
0,010	0.0090	0.341	0.330

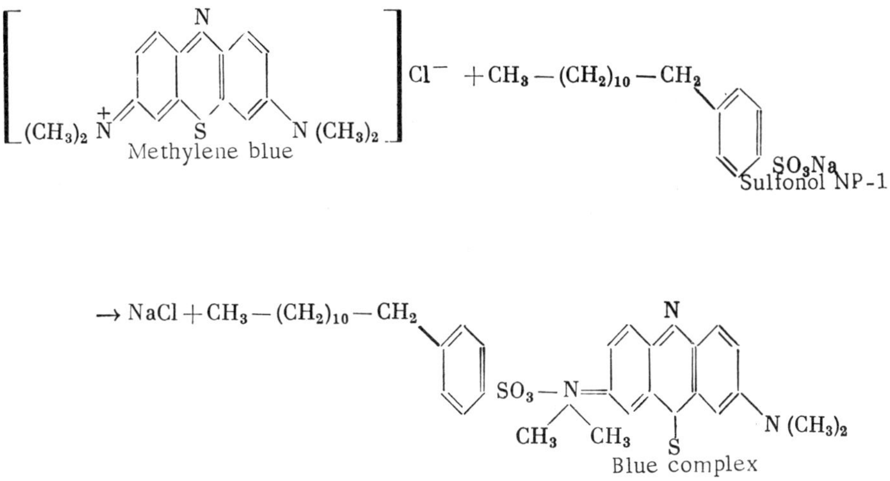

The following reagents were used for analysis: a buffer solution of alkaline disodium phosphate at pH 10, solutions of neutral and acid methylene blue, and chemically pure chloroform.

The method of carrying out the analysis is as follows.

A sample (30 cc) containing a known quantity of Sulfonol NP-1 is transferred by pipette into a 250-ml separating funnel. Then 10 ml of the buffer solution, 5 ml of the neutral methylene blue solution, and 15 ml of chloroform are added. After mixing and separating into two phases, the lower layer (chloroform) is transferred to a second separating funnel containing 110 ml of distilled water and 5 ml of the acid methylene blue solution. The solvent extraction of the aqueous solution in the first separating funnel is continued with 5-ml portions of chloroform until it is colorless (two to three extractions).

After shaking, the clarified chloroform from the second funnel is transferred into a 50-ml measuring flask through a funnel filled with chloroform-moistened cotton wool. Then the funnel is rinsed out with 2 ml of chloroform. The volume of chloroform is made up to 50 ml and the optical density of the solution is determined on a photoelectric colorimeter.

To establish the reproducibility of the analytical results, five determinations of the optical density as a function of concentration were carried out, the results of which are shown in Table 2.

As may be seen from Table 2 the mean error is low, with a 95% confidence [8]. The calibration curve was constructed based on the arithmetic mean of five determinations (Fig. 5).

It must be noted that according to data in the literature [7] this method is applicable only for surfactant contents within the limits 0.02-0.15 mg. For adsorption investigations it was important to find the maximum amount of surfactant in solution at which this method gives reproducible results. It was shown that this quantity is 3.0 mg. For large amounts of surfactants the optical density increases to some extent, but this is hardly noticeable on the curve (it is depressed). In addition, after extraction with chloroform the aqueous solution becomes colorless. Attempts to increase the amount of dye did not lead to positive results, owing to the necessity for a large number of chloroform extractions. The chloroform remains slightly colored.

On examining the calibration curve one can be persuaded that over the range of surfactant contents from 0.15 mg and above it does not obey the Lambert-Beer Law, although over the range 0.02-1.0 mg it is almost linear. Nevertheless, the reproducibility of the results and the satisfactory slope of the curve enable it to be used as a calibration curve, within the permissible limits of error, for the determination of Sulfonol NP-1 concentrations [9].

From what has been said above it follows that the method studied is applicable to the investigation of Sulfonol NP-1 adsorption on quartz sand within the concentration limits from 0.001 to 0.01%, using the following ratio: 30 cc of surfactant solution per 10 g quartz sand.

The curves for the variation of optical density with concentration before and after adsorption are shown in Fig. 6. The figures for the adsorption values as a function of concentration are given in Table 3.

The isotherms for the adsorption of Sulfonol NP-1 on quartz sand after determining the concentrations colorimetrically (curve 1) and by the surface tension method (curve 2) are shown in Fig. 7.

An analysis of these curves confirms the uselessness of using the surface tension method for the quantitative determination of an adsorbed surfactant ("apparent" adsorption), for while curve 2 lies in the region of negative values, curve 1 lies completely in the region of positive values of adsorption, thus showing that Sulfonol NP-1 is adsorbed on quartz sand.

Judging from these preliminary results, the method we have studied can be used within definite limits of concentration for studying the adsorption of alkylarylsulfonate-type anionic surfactants on quartz sand.

Summary

1. Based on literature data [1] and laboratory studies, the reason is given for the unsuitability of the surface tension method for determining the adsorption of anionic surfactants on quartz sand, and the concept of "apparent" adsorption is introduced.

2. The literature is reviewed of the methods based on the formation of colored compounds between anionic surfactants and dyes for determining concentrations.

3. The results are given of an investigation of a method for determining concentrations of anionic surfactants in solutions, based on the reaction between methylene blue and Sulfonol NP-1.

4. The possibility is shown of using this method for Sulfonol NP-1 concentrations 20 times greater than than those indicated in literature sources.

5. The mean relative error of the method is calculated.

6. The data are presented of laboratory investigations on the determination of Sulfonol NP-1 adsorption from its aqueous solutions on to quartz sand, which show the existence of positive adsorption.

7. The possibility is shown of using this method, within definite concentration limits, for studying the adsorption of alkylarylsulfonate-type anionic surfactants from their aqueous solutions on the solid adsorbent.

Literature Cited

1. G. A. Babalyan, I. I. Kravchenko, I. L. Markhasin, and G. B. Rudakov, Principles of the Use of Surface-Active Substances in the Development of Petroleum Beds (Gostoptekhizdat, 1962).
2. P. A. Rebinder, Khimicheskaya nauka i promyshlennost', 4, 5, 554 (1959).
3. J. F. Bates, R. M. Grunner, and S. T. Juster, Oil Weekly, 123, No. 8 (1946).
4. M. J. Rosen and H. A. Goldsmith, Systematic Analysis of Surface-Active Agents (Interscience Publishers, 1960).
5. G. R. Edwards, W. S. Emers, and W. W. Manifield, Analyst, 77 (1952).
6. G. S. Hartley and D. F. Runnices, Proc. Roy. Soc. (London), A168 (1938).
7. J. Longwell and W. D. Maniece, Analyst, 80 (1955).
8. V. N. Alekseev, Quantitative Analysis (Goskhimizdat, 1958).
9. P. S. Kireev and E. V. Zagoryanskaya, Molecular Spectral Analysis (State Press Vysshchaya shkola, Moscow, 1951).

THE INFLUENCE OF ADDED SURFACTANTS ON THE HYDRATION AND SWELLING OF CLAY ROCKS

L. N. Malysheva, I. L. Markhasin, and G. A. Babalyan

Recent years have witnessed an increased interest in the possibility of using surfactant additives to increase the rate of development of oil-bearing strata, to increase petroleum recovery, and to treat the end-face zone.

Surfactant additives, by being adsorbed on the surfaces of separation between the phases, may alter the course of the physicochemical processes accompanying the displacement of one liquid by another from a porous medium in the required direction.

Nonionogenic surfactant additives give best results in the displacement of petroleum by water [4]. Their advantage is that they dissolve completely in mineralized stratal waters without forming a precipitate.

As the surfactant solution penetrates into the depth of the stratum, the concentration of surfactant in the solution will depend to a considerable extent on the magnitude of the adsorption of the surfactant by the rock. Studies [4] have shown that in order to increase petroleum recovery and the rate of development of oil-bearing deposits, it is most advantageous to use the surfactant OP-10 for flooding. This surfactant is weakly adsorbed on sands and sandstones and increases considerably both the rate of displacement and the petroleum recovery.

In order to verify the efficiency of the use of surfactant additives for flooding strata, it is now proposed to carry out industrial experiments in the Arlanskii petroleum deposits, in which an aqueous solution of the nonionogenic surfactant OP-10, at a concentration of 0.05%, will be injected into the stratum [1, 4].

While improving the oil-flushing properties of water, the surfactant also facilitates the removal of petroleum from the surfaces of the clay particles which are almost always present in oil-containing rocks (it also changes its hydrophilic character). The production of hydrophilic properties on the surface of a clay particle will lead to an increase in its swelling (while on the other hand the production of hydrophobic properties on the surface will prevent swelling).

Moreover, the action of the surfactant may lead to a decrease in the adsorption strength of the rock and to dispersal of the rock.

These phenomena should alter the fi'tration properties of the rock to varying degrees, depending on the type and concentration of the surfactant used (and also on the nature of the clay minerals).

In view of the foregoing, we have made a study of the reaction of specimens of stratal clays from the Carboniferous formation of the Arlanskii petroleum deposit with solutions of the nonionogenic surfactants OP-10 and $S_{10}E_{8.5}$ in distilled water. The extent of the interaction was estimated by determining the coefficient of swelling of the test clays in these solutions.

Well No.	Range of removal of specimens, m	Brief lithological description of rock	Absorption capacity, meq	Concentration, meq		Maximum hygro-scopicity, %
				Ca··	Na·	
58	1254-1258	Dark grey argillite, aleuritic with pyritized detritus	20	15.6	–	3.3
81	1337-1345	Dark grey argillite, aleuritic, micaceous with bands of dark grey aleurolite	12.4	9.6	1.9	5.1
188	1244-1246.6	Dark grey argillite, with aleuritic bands, slightly micaceous with inclusions of siderite and pyrite	121	7.2	–	–
325	1288.7-1292.8	Argillite	17	9.6	–	–
343	1262-1264.7	Dark grey argillite, with inclusions of pyrite, strongly aleuritic bands	51.6	9.0	–	–
Talalaevskii			14	14.4	–	–

Experimental Procedure

The specimens of test rock taken from the Arlan Carboniferous formation (with characteristics given in the table) were first extracted thoroughly in a Soxhlet apparatus. They were then washed free from salts, dried, ground in a mortar with a plastic pestle, and passed through a sieve with mesh size 0.1 mm. The Talalaevskii clay was not subjected to this treatment but merely ground in the mortar and sieved. The rock specimens from wells No. 180 and 343 were not washed free of salts.

The swelling properties were estimated by the method of K. F. Zhigach and A. N. Yarov [2].

The instrument used (Fig. 1) consists of a cylinder 1 with perforated bottom 2, piston 3, and lid 4. The bracket 5 and clock-type indicator 6 attached to the lid of the apparatus 7 formed the measuring assembly to which the cylinder was fitted.

In the experiments 2 g of clay powder were sprinkled on the foot of the cylinder (circles of filter paper were placed above and below). When the specimen had been compacted, the piston was introduced into the cylinder and the cylinder was fixed to the measuring assembly. The complete instrument was lowered into a beaker filled with the test liquid. The liquid penetrated into the powder via the holes in the foot of the cylinder. The readings on the indicator recorded the change in the volume of the clay powder.

For each determination of the coefficient of swelling it is necessary to carry out a series of experiments with different degrees of compacting.

The results are used to construct graphs in which the volume of the dry specimen V_0 (in cm^3) is plotted on the x axis, and its volume after swelling V_S (in cm^3) is plotted on the y axis.

The resulting curve is given by the equation

$$V_S = a + \tan \beta \, V_0,$$

where a and tan β are constants for the given experimental conditions and are found from the experimental data on the graph.

The magnitude of the swelling is found from the formula

$$k + 1 = \frac{\gamma a}{g} + \tan \beta,$$

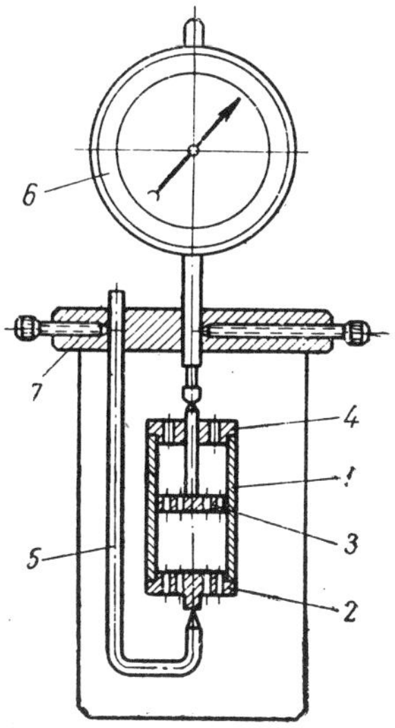

Fig. 1

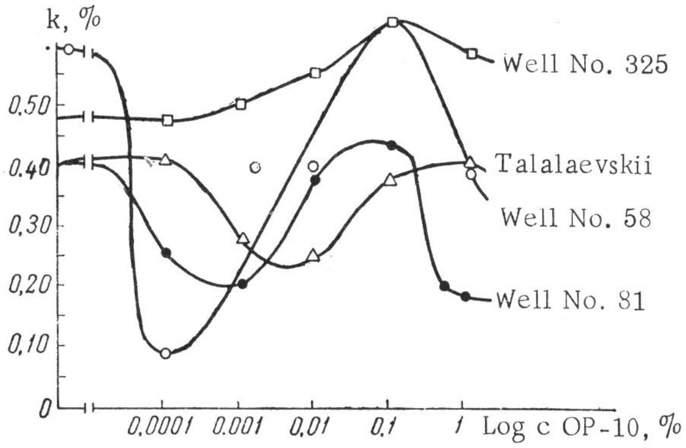

Fig. 2

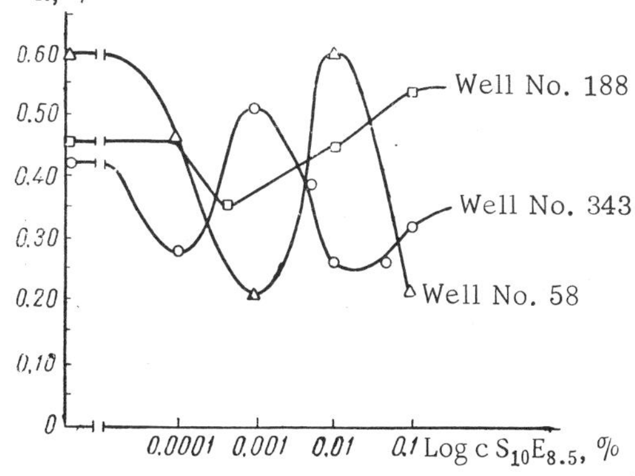

Fig. 3

where g is the weight of specimen, γ is the specific gravity of the clay, a is the coefficient depending on the properties of the clay and tan β, tan β shows the fraction of the volume in the dry specimen preserved in the swollen specimen, and k is a coefficient which characterizes the swelling of the clay and is independent of the initial porosity and quantity of immobilized liquid.

The values obtained for k are plotted on a graph giving the dependence of the magnitude of swelling k on the logarithm of the concentration of surfactant in the distilled water.

Experimental Results

The results obtained are given in Fig. 2 and 3, from which it can be seen that the degree of hydration of the test rocks, as expected, depends on the concentration of surfactant dissolved in the water.

The addition of small quantities of surfactant at first decreases k to a certain value, which is different for different clays and surfactants. With further increase in the surfactant concentration, k increases to its initial value (the value of k in distilled water) or even to a slightly higher value.

Further increase in the surfactant concentration leads to a new decrease in k.

Thus the influence of the concentration of surfactant in water on the coefficient of swelling for the stratal clays is regular in character.

It is difficult to discuss the mechanism of the processes taking place in this case, since there are a number of different points of view regarding the actual nature of the interaction between water and clay.

It is now known [3] that the molecular nature of the surface of clay particles is responsible for the different degrees of hydrophilic character, the strength of the interaction between the particles, and the structural-mechanical properties of the clay–water system.

The swelling of clay is closely related to its degree of hydrophilic character and depends to a marked extent on the nature of the clay.

In addition to the nature of the structure of the clay-particle surface, the polarity of the medium also has a significant influence on the swelling of clays.

When adsorbed on the surface of clay particles, the surfactants are fixed with their polar groups on the clay minerals to form a layer of orientated molecules with their hydrophobic ends directed toward the aqueous phase [5]. The adsorption of surfactants on clay particles thus decreases their hydrophilic character and also decreases the coefficient of swelling.

The adsorption of surfactants also leads to a decrease in the strength of interaction between the particles (the so-called adsorption decrease in the strength of the rock) and dispersal of the rock [4], leading to a marked increase in the specific surface of the clay powder and an increase in the coefficient of swelling.

It may thus be assumed that the existence of these two opposed effects is responsible for the appearance of the maximum on the curve of the function $k = f(c)$.

Further increase in the surfactant concentration may lead to the phenomenon of over-oiling [5], when the layer of adsorbed orientated surfactant molecules responsible for the hydrophobic properties of the surface adsorbs a new layer of surface-active molecules with the opposite orientation, leading to the production of hydrophilic properties on the particle surface. This change in the wettability of the particles should lead to an increase in k.

Summary

1. The addition of surfactants to water may increase or decrease the degree of swelling of clay rocks.

2. The same concentration of surfactant in aqueous solution may produce opposite effects in different rocks.

3. The addition of small quantities of surfactants to water decreases the value of k. Thus the addition of OP-10 to the four specimens studied should not exceed 0.05% if a decrease in the value of k is to be obtained.

Literature Cited

1. G. A. Babalyan, G. P. Ovanesov, A. M. Baikov, and I. L. Markhasin, "The development of new methods for increasing the recovery of petroleum from strata by using surface-active substances," Report on Theme No. 24-61, UFNII Fund (1962).
2. K. F. Zhigach and A. N. Yarov, "The estimation of the swelling properties of clays," Izv. vuz, Neft' i gaz, No. 10 (1959).
3. F. D Ovcharenko, The Hydrophilic Properties of Clays and Clay Minerals (Izd. AN USSR, 1961).
4. Proceedings of the First All-Union Conference and The Use of Surface-Active Substances in the Petroleum Industry (Gostoptekhizdat, Moscow, 1961).
5. P. A. Rebinder, The Physical Chemistry of Flotation Processes (Metallurgizdat, 1933).

NEW REAGENTS FOR DRILLING FLUIDS

R. S. Lerner

Detailed investigations into the chemical treatment of drilling fluids have been carried out over a number of years at VNIIBT. The main trends of these studies have been concerned with the search, synthesis, and testing of new reagents for lowering the viscosity of solutions, stabilizing and protecting them from agressive influences and forconferring a number of new properties on drilling fluids such as higher thermal stability, lubricating action, emulsification, etc. As a result of these studies a number of reagents have been developed, including surfactants, which are already being used or will be used in the near future.

The most pressing problem at present is to create thermally stable clay solutions able to withstand conditions of high face temperatures and pressures. The investigations in this field are associated with the creation of new chemical processing agents possessing increased activity over a wide range of temperature and pressure.

The stabilizing reagents and viscosity reducers usually used cannot ensure a reliable thermal stability of of the drilling fluid.

Consequently, great attention has been paid in world drilling practice to the use of water-soluble polyacrylates.

Commencing in 1959 a large number of different acrylic polymers and copolymers have been prepared and tested at VNIIBT.

An investigation of polymers based on the alkali salts of acrylic and methacrylic acids and their copolymers with acrylamide and methacrylamide in various ratios, as well as hydrolyzed polyacrylamide and polyacrylonitrile, showed that these products had great promise as protective colloids and for increasing thermal stability. The presence of carbon-carbon type bonds in the molecules of these substances confer on them a special stability toward temperature effects. A comparative study of these polymers showed that the most effective ones are the products from the alkaline hydrolysis of highly polymerized polyacrylamide. The studies of other research organizations confirm this.

From economic considerations special attention was paid to the study and introduction of hydrolyzed polyacrylonitrile (Gipan). Its industrial production, based on a flowsheet developed at VNIIBT, has commenced at one of the chemical industry undertakings.

Gipan is a copolymer consisting of 70-80% sodium acrylate with approximately equal amounts of acrylamide and acrylonitrile and is obtained on saponifying polyacrylonitrile with caustic, the degree of hydrolysis increasing from 70 to 90% on increasing the alkali consumption from 0.68 to 2 moles. Gipan ensures the physicochemical stability of a system by maintaining acceptable water losses at temperatures up to 220°C. At lower temperatures this reagent acts as a protective colloid which is more than twice as active as KMTs.

In the presence of salts the efficiency of Gipan is lowered under high temperature conditions. For example, at 150°C stabilization of Gipan is guaranteed for a salinity of 10%, but at 200°C for a salinity of about 3%.

Gipan has been tested in drilling at face temperatures up to 160°C at Stavropol and at 150°C in the western Ukraine. It is being used at present in these regions either alone or in conjunction with other reagents. A test of Gipan in combination with chromates is contemplated in Krasnodar oil.

Other reagents of those developed at VNIIBT for thermally stable solutions are nonionic surfactants, a polyethoxylated phenol (OF) containing 20-30 moles of ethylene oxide, and a monoethoxylated alkylphenol (OP-1).

OF is a typical inhibitor for a drilling fluid at high temperatures in aggressive media. Its main function is to prevent the thickening of high-temperature solutions stabilized with KMTs or Gipan.

OF also appreciably lowers the water loss of solutions stabilized with KMTs. It is essential that formulations containing OF be nonalkaline or only slightly alkaline to decrease peptization of the clay. With an increase in mineralization of the solutions at high temperatures less OF is consumed in dilutions.

OP-1 is an auxiliary surfactant which is used to prevent the foaming of solutions, whereas this does occur with the action of other ethoxylated reagents.

Studies on the modification of starch by conferring on it solubility in water and stability to fermentation occupy a separate place.

As a result of joint studies between VNIIBT and TsNIIKhI (Central Scientific Research Institute for Wood Chemistry), the Academy of Wood Chemistry, and the Institute of Organic Chemistry of the Academy of Sciences, USSR, the following viscosity reducers have been developed, tested, and proposed for treating argillaceous solutions:

1) polyphenol wood chemical (PFLKh), a formaldehyde sulfited product from the condensation with formaldehyde of the total phenols from the thermolysis of wood;

2) phenolic acid pitch which consists of the methoxysulfonated phenolic acids of wood pitch, a waste product from resin processing;

3) oxidized lignin, obtained by treating lignin (a waste product from the wood alcohol industry) with nitric acid, a mixture of nitric acid and sulfuric acid, chlorine, etc.

These viscosity reducers are anionic colloidal electrolytes. Their effect is brought about by the presence of large active anions whose functional groups react with the argillaceous material. As a result, the active parts of the surface of the clay phase are blocked, so that structure formation and an increase in their viscosity are restricted. The main purpose of these viscosity reducers is to thin the solutions which have been thickened by the presence of additional solids from the rock cuttings and weighting materials.

In the investigations it was found that these reagents increase the clay capacity of solutions and make it possible to control viscosity and structure formation by small additions (0.2-0.3%) to the solution. The optimum activity of the viscosity reducers occurs at a clay solution pH of 9-10.

The instability of all viscosity reducers with increased mineralization is common. Therefore, when diluting mineralized solutions it is advantageous to introduce protective colloids at the same time.

The diluting power of viscosity reducers does not deteriorate with elevated temperatures (up to 200°C). Nevertheless, clay solutions treated with viscosity reducers do thicken as a result of heating but can be brought to a workable state by repeated processing of the thinning reagents.

The industrial testing of PFLKh and oxidized lignin under different geological conditions at wells in Turkestan, Azerbaidzhan, and the Ukraine, and at Volgagrad and Stavropol, showed that these reagents improve drilling conditions and enable economies to be made on weighting compounds and alkali.

There was thus achieved both complete continuity and the possibility of the interchange of reagents (UShR, SSB, PFLKh); alternate use of the reagents greatly enhances their efficiency in clay solutions.

At the present time the production of the viscosity reducers PFLKh and oxidized lignin has been put into operation. The production technology of the reagent based on wood-resin pitch is being developed.

Since 1961 VNIIBT have been carrying out studies to give lubricating properties to drilling fluids by introducing appropriate additives into them. The purpose of these additives is to improve the operating conditions of the bit support, to increase its lifetime, and to reduce delays, damage to instruments, and hydraulic resistance during drilling. The increased wear of the bit support bearings and the duration of mechanical drilling associated with it is governed by the fact that they operate in an abrasive medium without lubrication. It is therefore a pressing problem to increase its lubricating properties.

Both nonionic and ionic surfactants, as well as chemically active additives, were investigated as lubricating additives. The most promising in economic characteristics and in efficiency of action are the carboxylic acid type of ionic surfactants, the polyhydroxy acids, esters, and also mixtures of these compounds. Synthetic fatty acids (a $C_{10}-C_{16}$ fraction, a vat residue OZhK), oxidized petroleum, and waste products from the glycerine industry and the fat industry (soap stock) have been tested with promising results under laboratory conditions in water and in a clay-solution composition.

The introduction of these additives into a clay solution does not lead to a deterioration in its main properties.

From laboratory test data on a four-ball friction rolling machine, the introduction of lubricant additives increases the lifetime of the balls (on testing for pitting) by a factor of 8-10 when working with a clay solution and by a factor of 4-6 when working with water. The lubricant additives in the drilling fluid composition improve the sliding friction considerably, which is shown as a decrease in the coefficients of friction and in pitting.

According to the laboratory data the additives mentioned exceed in efficiency the American product "Bitlube," which is used in drilling fluid compositions as a lubricating additive for extremely high-pressure conditions. The amount of additive which ensures the required lubricating effect is 0.5-1.5%.

Among the deficiences of these additives is their sensitivity to mineralization (calcareous, in particular). The methods of protecting drilling fluids from the deleterious effect of increased mineralization will be the subject of further investigation.

The ethoxylated types of nonionic surfactant (OP-10, Stearox, UFE_8, ethoxylated vat residues, OZhK, etc.) do not confer lubricating properties on drilling fluids under rolling conditions, although they do improve sliding conditions to a certain extent. The Sulfonols NP-1 and NP-2 give a similar effect.

All the materials proposed as lubricating additives are being produced indigenously or are inexpensive and readily available industrial wastes.

THE EFFECT OF SURFACTANTS ON THE WASHING OUT OF PETROLEUM PRESENT AS A FILM

F. G. Khatmullin, G. A. Babalyan, and I. I. Kravchenko

One of the promising methods for increasing the petroleum yields from strata is the addition of surfactant to the water injected into a stratum during flooding. Surfactant addition reduces the surface tension, facilitates wetting of the rock surface by water, and thus increases the petroleum yield and the rate of petroleum displacement.

It is well known that the petroleum in a porous medium may be present in the form of drops in pores, lenses encompassing several grains, or films. In the latter case petroleum covers the surface of quartz grains with a thin film [1]. In the displacement process, there is initially a petroleum film adhering to the hydrophobic surface of the trapping rock, behind the water–petroleum boundary. This film is subsequently disrupted by the water, and part of its is converted to the free state, no longer bound to the solid surface. In other words, part of the petroleum is washed away from the surface of the rock particles, and this means that the hydrophobic emulsion existing near the water–petroleum contact is inverted into a direct petroleum-in-water emulsion. Surfactant makes possible a thinning of the film, greater movement along the surface, and rupture. We have investigated the conditions for rupture of the petroleum film and its washing away from various surfaces by surfactant solutions.*

The experiments were carried out with petroleum from the Carboniferous Arlansk field in Bashkiria. This petroleum was selected because the first field trials in the Soviet Union, on the addition of surfactant to water to be injected into a stratum, are planned for this field. Table 1 shows the characteristics of Arlansk petroleum.

These tests were done using the technique of Babalyan and Kravchenko [2].

Plates, wetted with petroleum, were placed in water and in aqueous surfactant solutions of various concentrations. The plates were of glass, marble, and quartz. Calcium and magnesium carbonates and quartz are the main constituents of the rock which forms petroleum traps. Glass plates were used because the hydrophilic properties of their surface are similar to those of a quartz surface. A study of the breakaway of films from marble reflected, to some extent, the conditions for breakaway from carbonate cement and carbonate rocks. It is well known that carbonate rocks consist mainly of calcite and dolomite.

The surfaces of the plates were treated as follows. The glass and quartz plates were left in chromic acid mixture for not less than 5 h, washed with tap water and then with distilled water, and dried with filter paper. The marble plates were washed with benzene, with alcohol, and then with distilled water, and were dried with filter paper.

The petroleum film on the plates was always produced in the same way: the plate was immersed in petroleum; it was then removed and placed in the apparatus (Fig. 1), which was filled with nitrogen; to prevent oxidation of the thin petroleum film, nitrogen, at a low pressure, was passed through the apparatus from a tank. The excess petroleum was allowed to drain off the plate in a nitrogen atmosphere (Fig. 1a), leaving a stable film when drainage had finished (Fig. 1b). The weight of this petroleum film P_{pet} was determined by weighing the original clean plate P_1, the apparatus with the plate wetted with petroleum P_2, and the apparatus without

*V. V. Alekseenko and R. I. Stepanov assisted in this work.

TABLE 1

Liquid	Density, g/cm^3	Viscosity, cP	Surface tension of distilled water at interface with petroleum, $ergs/cm^2$	Contact wetting angle, degrees *
Arlansk petroleum	0.892	30.06	53.0	46.0

* A drop of petroleum on quartz in water.

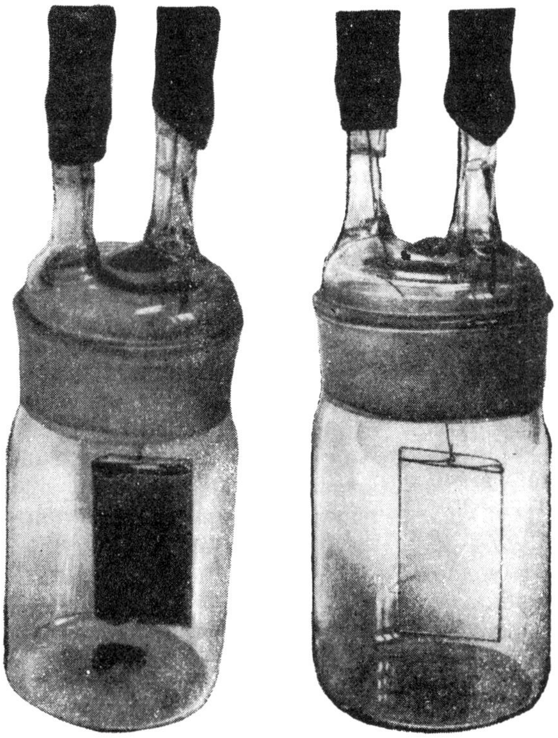

a b

Fig. 1. Apparatus for producing petroleum film.

the plate but with the petroleum which had drained off the plate P_3. Then $P_{pet} = (P_2 - P_3) - P_1$. The plate was then removed from the apparatus and immersed in a solution of surfactant in distilled water of given concentration. In the course of time the petroleum film broke away from the plate, and drops of petroleum floated up to the surface of the solution.

The following method was used at first for determining the amount of petroleum which had been washed off. At the end of a definite time interval, the floating petroleum was removed from the surface of the solution and the plate, with residual adhering petroleum, was lifted out of the water and washed with kerosene. The concentration of petroleum in the kerosene washings was then determined. This was done with a photoelectric colorimeter, using a previously constructed calibration curve relating optical density and petroleum concentration in kerosene (Fig. 2).

The weight of petroleum remaining on the plate was calculated, and the percentage of petroleum washed off the plate could then be obtained. However, this method of measuring the residual petroleum on the plate had serious drawbacks. When the plate was lifted out of the solution, drops of petroleum were pulled off the plate under the action of surface tension at the petroleum–solution or petroleum–water boundary, whereas these drops would have remained on the plate under static conditions. This meant that the results for the amount of petroleum washed off were too high and were not reproducible.

A more satisfactory method was then found for determining the amount of petroleum washed off. The plate with the petroleum film was immersed in surfactant solution, of given concentration, in such a way that the upper edge of the plate was about 2 cm below the surface of the solution. A measured volume (15 ml) of purified kerosene ($\sigma_{w/k} = 47.5$ $ergs/cm^2$) was pipetted onto the top of the solution. The petroleum droplets which broke away from the film floated up and dissolved in the kerosene, forming a petroleum-in-kerosene solution. An appropriate quantity of this solution was then pipetted into the photoelectric colorimeter cell, and its optical density was measured. The previous calibration curve of optical density versus petroleum concentration in kerosene was then used to calculate the weight of petroleum washed off the plate, and this was

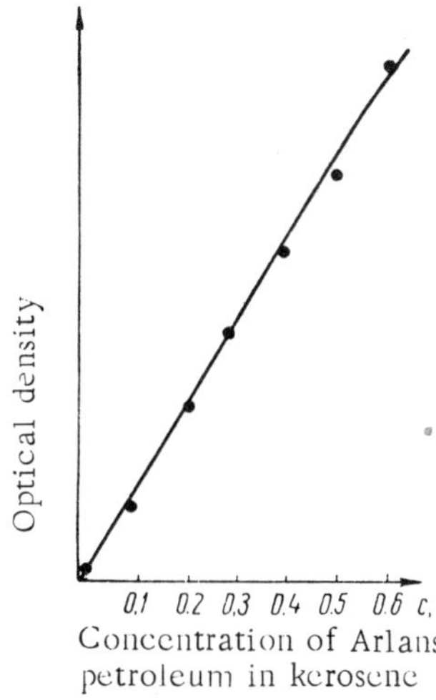

Fig. 2. Relation between optical density and concentration of Arlansk petroleum in purified kerosene.

expressed as a percentage of the petroleum originally on the plate (η = petroleum yield). All subsequent experiments were carried out in this way, using surfactant solutions of concentrations 0.1, 0.075, 0.05, 0.025, and 0.01%, and pure distilled water. The dimensions of the glass plates were 45 x 25 x 1.5 mm, and those of the marble and quartz plates were 45 x 25 x 8 mm.

Figure 3 shows the results of experiments on the breakaway of the petroleum film from glass plates in solutions of the surfactant OP-10 at distilled water. The curves represent the means of three series of tests. Visual observations showed that, in the absence of surfactant, there was no breakup of the film on the surface of a glass plate; there was only a thinning of the film and partial breakup at the edges of the plate.

With surfactant in the solution the film broke up at the surface of the plate. The times for breakup and for the establishment of the equilibrium value of the wetting angle at the petroleum droplets, after breakup, were less the greater the surfactant concentration, within the limits taken. There was a corresponding increase in the amount of petroleum washed off and in the rate of washing off. A narrow strip of petroleum film formed at the narrow edges of the plate; this strip contracted and at the same time broke up into fine droplets. On the wide surfaces of the plate the petroleum film gradually contracted toward the center and, after some time, broke up into droplets which floated up to the solution kerosene interface. These droplets took 2 to 5 min to spread and coalesce with the kerosene. It was noted that, with 0.01 to 0.025% of surfactant in the solution, the drops broke away from the plate about 3 h after the plate with petroleum film was first immersed in the solution. For concentrations of 0.05% and above, the drops broke away after about 2 h.

The following conclusions could be drawn from the results.

1. With either distilled water or surfactant solution, the petroleum drops did not start to break away from the plate immediately, but only after a definite time, which was always 2 h or more in our experiments.

2. The amount of Arlansk petroleum washed off was 8 to 10 times more in the presence of OP-10 than without surfactant. Moreover, the amount washed off by OP-10 solution became practically constant after 15 h, but not until 24 h in the case of distilled water.

3. The breakup of the petroleum film and the breakaway of droplets from the plate began sooner the higher the surfactant concentration. Sometimes separate drops formed on the plate and did not break away in the course of 10 h or more, but in all probability this was due to heterogeneity of the plate surface and the existence of less hydrophilic regions. Accordingly, in order to obtain comparative results, each experiment was repeated three or more times.

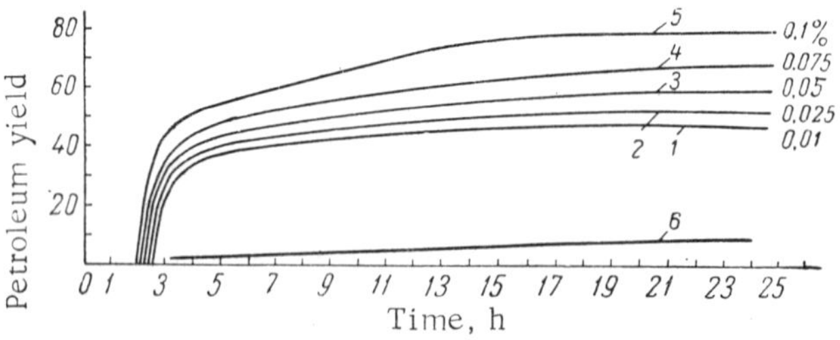

Fig. 3. The relation between "petroleum yield" and the contact time of the glass plate with water or an aqueous solution of OP-10. The concentration of OP-10 in water was: 1) 0.01%; 2) 0.025%; 3) 0.05%; 4) 0.075%; 5) 0.1%; 6) distilled water.

TABLE 2

Surfactant concentration, %	Petroleum yield, % Washing time, h	
	15	24
0.1	22.2	65.3
0.075	21.1	47.1
0.05	20.0	31.3
0.025	21.0	27.6
0.01	2.7	12.2

TABLE 3

Solution	Petroleum yield, % Washing time, h		
	10	15	24
OP-10	8.3	13	52.4
Katapin A	1.3	9.4	9.7
Distilled water	11	16.5	15

4. The greater the OP-10 concentration in the solution the greater was the amount of petroleum washed off the plate.

5. With OP-10 solutions, the main bulk of the petroleum was washed off in the first 2 to 5 h. With distilled water, only a small amount of the petroleum (2-3%) was washed off in this period.

According to literature data, addition of a cationic surfactant sometimes makes it possible to increase the petroleum yield considerably [3]. We therefore examined the effect of a cationic surfactant on the washing off of a petroleum film from the surface of a plate. The experiments were carried out with the cationic surfactant Katapin A and glass and marble plates.

Table 2 gives the results of experiments on the breakaway of the film from glass plates in Katapin A solutions of various concentrations.

Table 2 shows mean values for three series of tests. Visual observation indicated that breakup of the petroleum film was slow in this case. The film broke up and contracted after 3 to 5 h, forming lenses and large drops of petroleum on the plate. The wetting angle and the area of drop adhesion were greater. Analysis of Table 2 and visual observations make it possible to draw the following conclusions.

1. Breakup of the film on a glass plate was considerably slower in a Katapin A solution than in distilled water. Practically no petroleum was washed off by Katapin A solution in 10 h, whereas washing off in distilled water began after 3 h.

2. More petroleum was washed off the greater the concentration of Katapin A solution.

3. Intensive washing off of petroleum from the plate began after 15 h and reached a limit after another 9 to 10 h. About 65% of the petroleum was then washed off, whereas the amount washed off by distilled water in this time did not exceed 10%.

4. Addition of Katapin A to the water retarded breakup of the film and the initial washing off of petroleum from the plate but considerably increased the final amount of petroleum washed off.

Comparison of the results for the breakup and washing off of the petroleum film from glass plates shows that OP-10 was considerably more effective than Katapin A. The amount of petroleum washed off was also greater with OP-10 than wtih Katapin A. This was particularly noticeable at low surfactant concentrations. However, considerably more petroleum was washed off by Katapin A solution than by distilled water alone.

Table 3 gives results for the breakaway of the petroleum film from marble plates in 0.05% OP-10 solution, 0.05% Katapin A solution, and distilled water.

Visual observations showed that a thinner petroleum film was produced on a marble plate in a nitrogen atmosphere than on a glass plate. The film on marble broke up exactly when the plate was immersed in distilled water. Small petroleum droplets, approximately uniform in size, formed on the surface of the plate; these droplets showed a low wetting angle and a low adhesion area. Coarse drops and lenses of petroleum formed after the film broke up when the plate was immersed in 0.05% Katapin A solution; the wetting angle and adhesion area were greater. This was evidently because the Katapin A made the marble surface hydro-

TABLE 4

Surface	Solution	Washing time, h	Petroleum yield, %
Marble	OP-10	5	43.0
"	Katapin A	5	5.5
"	Distilled water	5	14.3
Glass	OP-10	5	51.0
"	Katapin A	5	4.0
"	Distilled water	5	2.8

phobic. In the case of 0.05% OP-10 solution, the petroleum droplets on the plate were small, and their wetting angle and contact area were considerably less than with distilled water. It is clear that OP-10 made the marble surface hydrophilic.

The following conclusions were derived from analysis of Table 3.

1. A 0.05% solution of Katapin A made the surface of a marble plate hydrophobic, and this retarded the washing off of petroleum, as compared with washing off by distilled water.

2. A 0.05% solution of OP-10 promoted the washing off of petroleum from the surface of a marble plate, and the final result of washing was much the same as with a glass plate and OP-10 solution of the same concentration.

3. Only a small amount of petroleum was washed off a marble plate by distilled water, but the amount washed off was rather more than in the case of a glass plate immersed in distilled water.

The main components of the sandstones of petroliferous formations is quartz, so further experiments were carried out on the breakaway of a petroleum film from quartz plates in various surfactant solutions at a range of concentrations. These tests showed that the processes occurring on a quartz plate were essentially the same as those occurring on a glass plate. The final results of washing were approximately the same as on glass.

Under field conditions the disintegration of the film and the washing off of petroleum would occur with the surfactant solution in movement. Some preliminary experiments were therefore carried out on disintegration of the film and washing off of the film, in distilled water and in 0.05% solutions of Katapin A and OP-10, under conditions of gentle movement of the solution.

Five hours after the plates were immersed, the solution was stirred gently with a glass rod so as to create a gentle movement around the plates. Table 4 gives the results.

Visual observations showed that, with both marble and glass surfaces in OP-10 solution, small droplets of petroleum with a low wetting angle formed when the film broke up and were easily washed away by the moving solution.

With both marble and glass surfaces in Katapin A solution, coarse drops and lenses of petroleum formed and were not readily washed away by the moving solution. In distilled water, the film on a marble surface broke up to form fine droplets of petroleum, but the film on a glass surface did not break up, except partially at the edges.

The following conclusions were derived from analysis of Table 4.

1. Under conditions such that the solution was moving, the rate of washing off and the amount of petroleum washed off were considerably increased by the presence of surfactant, particularly by a nonionic surfactant.

2. The use of 0.05% OP-10 solution to wash off the petroleum, under conditions such that the solution was moving, gave a considerable increase in the amount of petroleum washed off, threefold for a marble surface and seventeen-fold for a glass surface, as compared with distilled water.

3. The use of 0.05% Katapin A solution to wash off the petroleum, under conditions such that the solution was moving, impaired the washing off from a marble surface and slight improved the washed off from a glass surface, as compared with distilled water. The reason why there was only a slight increase in the washing off action in this case was that the washing action of the cationic surfactant Katapin A only develops after 10 to 20 h.

Summary

1. Experiments were first made to establish a quantitative method for determining the amount of petroleum washed off a petroleum film on a solid surface by distilled water and surfactant solutions.

2. Addition of the nonionic surfactant OP-10 to the water considerably accelerated the washing off of petroleum from quartz, glass, and marble, and considerably increased the amount of petroleum washed off, as compared with washing with distilled water.

3. Addition of the cationic surfactant Katapin A to the water increased the amount of petroleum washed off glass and quartz, as compared with washing with distilled water, but impaired washing off of petroleum from marble.

4. On the basis of the results obtained it may be concluded that the nonionic surfactant OP-10, if added to the water injected into a stratum, should markedly reduce the amount of petroleum present as a film, increase its rate of displacement, increase the petroleum yield, and speed up the development of the strata of the Arlansk field.

Literature Cited

1. G. A. Babalyan, Problems in the Mechanism of Petroleum Yield (Aznefteizdat, 1956).
2. G. A. Babalyan, I. I. Kravchenko, I. L. Markhasin, and G. V. Rudakov, The Physicochemical Principles of the Use of Surfactants in the Development of Petroliferous Strata (Gostoptekhizdat, 1962).
3. G. N. Dunning, R. G. Joganson, and Liao-Hsiao, "The displacement of petroleum by means of detergents," Petrol. Eng., XXX, 1 (1954).

SOME CHARACTERISTIC FEATURES OF THE INFLUENCE OF DISPERSED SOLID PHASE ON THE EMULSIFCATION AND DEMULSIFICATION OF PETROLEUM PRODUCTS

A. B. Taubman and A. F. Koretskii

The action of surfactants in a variety of processes in petroleum recovery and treatment is shown at different surfaces of separation: a liquid—liquid boundary(e.g., in the case of the phenomena of emulsification or demulsification), or a liquid—solid boundary(as the case of the phenomena of wetting and migration in the rocks of the petroleum-bearing stratum).

It would therefore seem justified to study these processes separately and to select surface-active additives in these two cases independently of one another, e.g., in the case of emulsification, on the basis of a study of the surface activity of emulsifiers, and in the case of the production of hydrophobic properties on rocks, on the basis of selective wetting.

In a number of cases, however, it is found that the molecular interactions at the liquid and solid surfaces of separation are superimposed on one another, so that the separate study of the above phenomena in a model study of the processes becomes unjustified.

Thus, for example, if surfactants are used to produce hydrophobic properties in the rocks of a stratum, this not only changes the conditions of selective wetting of the solid phase by the corresponding aqueous solution and the petroleum, but may also lead to mutual emulsification of the two liquid phases as a result of the fact that the rock particles, after adsorption modification of their surface, acquire the properties of a solid emulsifier, which are often very effective and sometimes show an unexpected influence on the subsequent technological processes of petroleum recovery.

If we carry out the emulsification of "oils" (petroleum, oil, kerosene, etc.) with aqueous suspensions of a highly dispersed solid phase by the usual method involving agitation and estimate the stability of the emulsion from the time required for breakdown τ, it is found that mineral hydrophilic powders (e.g., carbonates, sulfates, aluminosilicates, etc.) readily give emulsions of differenttypes (oil in water O/W, or water in oil W/O) with different degrees of stability.

This possibility of controlling the properties of emulsions over a wide range also arises if the oil phase of the emulsion contains even small quantities of surfactants of the aliphatic or naphthenic acid type, which are capable of undergoing adsorption on the surface of the solid phase with subsequent chemical fixation and the formation of surface soap-like compounds.

These effects involving intensification of the emulsifying power of highly dispersed solids are shown most clearly in those cases where the aqueous phase of the emulsion contains electrolytes, particularly those formed by polyvalent cations (Al, Fe, Cu, and other ions).

These factors may also be encountered under practical conditions of petroleum recovery and treatment, particularly for natural petroleums with high water and salt contents; these usually contain a considerable quantity of natural surfactants and are contaminated by highly dispersed rock particles.

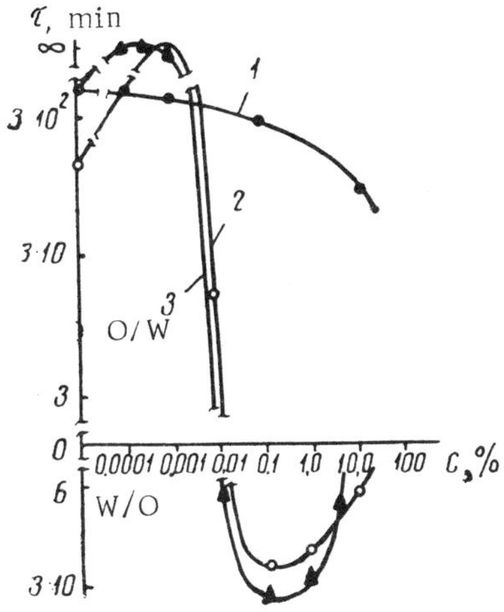

Fig. 1. Emulsifying action of 2% suspension of $BaSO_4$ and $BaCO_3$ for oil containing $C_{17}H_{33}CO_2H$. 1) $BaSO_4$, pH = 6.2; 2) $BaCO_3$, pH = 8.5; 3) $BaSO_4$, pH = 8.5.

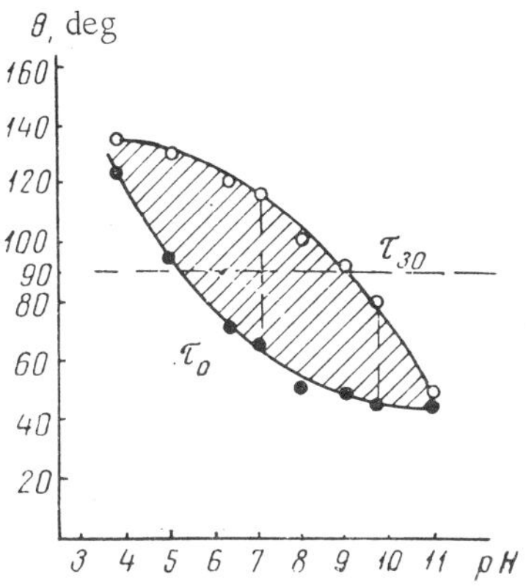

Fig. 2. Influence of pH on the angle and hysteresis of selective wetting of glass by oil containing 0.1% octadecylamine.

These effects can be illustrated by the results of experiments which we have carried out in a study of highly dispersed powders (bentonite clay, calcium carbonate, barium sulfate, aluminum oxide, silicon dioxide, etc.) as solid stabilizers of mineral oil emulsions used as models of emulsions of highly viscous petroleum products.

The mechanism of the emulsifying action of these hydrophilic mineral dispersed solids, which are unable to adhere to a pure liquid paraffin –water surface of separation or to stabilize these emulsions, is as follows. When the solid particles during mixing come in contact with the drops of technical oil containing as contaminants surface-active substances, or, as in our experiments, specially introduced oleic acid, their surface is made partly hydrophobic in isolated regions (e.g., the Ca regions of calcium carbonate) and thus acquires the ability to undergo selective (with intermediate values of the contact angle) wetting by the two phases of the emulsion and hence the ability to stabilize the emulsion.

In this case the type and stability of the emulsion τ are determined by the extent to which the surface is made hydrophobic. As the degree of hydrophobic character increases (with increase in the concentration of oleic acid in the oil), the stability of the resulting emulsions of the O/W type gradually increases to a certain maximum value, after which further increase in the degree of hydrophobic properties leads to a decrease in the stability of the emulsion until reversal of the emulsion takes place with conversion to an emulsion of the W/O type. Curves 2 and 3 in Fig. 1 show that even at an acid concentration $c \approx 10^{-4}$-10^{-3}% the stability of the direct emulsions corresponds to many days (indicated on the ordinate by the symbol ∞), after which it decreases sharply and at $c = 10^{-2}$-10^{-1}% unstable emulsions of the inverse type are formed.

The formation and stabilization of emulsions by solid emulsifying agents depend to a considerable extent on the pH of the aqueous phase, due to the change in the conditions under which their particles are made hydrophobic by the adsorption layers of surfactant.

The influence of the pH of the aqueous phase on the type and stability of emulsions is particularly marked in those cases where the oil phase contains acidic or basic organic soap-like compounds, which at appropriate pH values are capable of dissociating to form ions and are irreversibly sorbed on the surface of the particles. This can be seen from Fig. 1, which gives data characterizing the emulsifying properties of 2% suspensions of barium sulfate and carbonate, which do not form stable emulsions with pure oil (the points on the ordinate show that they exist for not more than a few hours). The influence of the addition of oleic acid to the oil phase is different in the two cases – the emulsifying properties of barium sulfate remain practically unchanged (curve 1), whereas those of the carbonate are altered appreciably (curve 2).

These differences are readily explained if we take account of the part played by the "unavoidable" ions which appear in the suspensions as a result of the solubility of the solid phase and determine the pH of the solution, and also the fact that barium sulfate does not form a compound with the aliphatic acid in the pH range below 7, while barium carbonate, whose suspension has pH = 8.5, reacts readily with the sodium oleate formed at this pH.

It is obvious that under these conditions, i.e., at pH > 7, sodium oleate (unlike free oleic acid) should also react with the surface atoms (ions) of the barium sulfate lattice. In actual fact, if the pH is raised to 8.5 by the addition of NaOH, the emulsifying properties of the sodium oleate suspension become identical to the properties of the barium carbonate suspension (curve 3).

Thus even a slight increase in the pH of the barium sulfate suspension from the acid to the alkaline range makes it possible to obtain emulsions of quite different type and stability with the same emulsifying agent.

The absence of reaction between barium sulfate particles and aliphatic acid at pH < 7 and the resulting absence of emulsifying power in its suspensions are also confirmed by measurements of the contact angles of selective wetting at the surface of the barium sulfate crystals.

At pH < 7 the contact angle of wetting θ for pure oil and for oil containing aliphatic acid lies in the range 30-45°, but with increase in pH to 8-9 the contact angle increases to 100-120°, indicating the production of limiting hydrophobic properties on the surface by the adsorption layers.

Exactly analogous relationships are observed for the emulsifying action of quartz and silicates (in our experiments models consisting of highly dispersed glass powder were used) if the surface of their particles is modified not by anionic or nonionogenic surfactants, which do not undergo chemosorption on these materials, but with surfactants of the cationic type (alkylamines or tetrasubstituted ammonium bases). In this case also the course of the $\tau = f(c)$ curves, e.g., for the glass–octadecylamine system, corresponds exactly to the course of curves 2 and 3 (Fig. 1).

At the same time it was found that in this system it is possible to produce a radical change in the properties of the emulsion by changing the duration of the emulsification process.

On the assumption that this effect was related to the kinetics of the production of hydrophobic properties on the particles by the adsorption layers, which gradually increases with increase in the number of contacts between the particles and the oil drops as the emulsion is agitated, we carried out parallel measurements of the following: 1) the dependence of the contact angles of wetting on glass under selective conditions (at the aqueous medium – "oil" drop boundary) on the duration of contact, concentration of octadecylamine in the oil phase, and pH of the aqueous medium, and 2) the dependence of the stability of the emulsions on the duration of the emulsification process.

It was found (Fig. 2) that as a result of the hysteresis of wetting produced by the retarded adsorption and the "firm" attachment of the adsorption layers of octadecylamine irreversibly attached to the glass surface, the equilibrium value of the contact angles of wetting was reached only after 20-30 min.

A particularly marked influence on wetting and on the hysteresis is shown by change in the pH of the aqueous phase. For example, for oil containing 0.1% octadecylamine at pH = 11, the contact angle of wetting is equal to 45°, whereas in the acid range at pH = 4, θ reaches 120°, corresponding to the limiting production of hydrophobic properties on the glass surface. At these pH values (4 to 11) almost no hysteresis of wetting is observed; the influence of hysteresis is most marked at intermediate pH values. For examples, at pH = 7 in the initial moments of measurement, $\theta = 65°$, but after 30 min $\theta = 115°$, which again characterizes the qualitative change in the nature of the glass surface.

Thus the clearly defined hysteresis of selective wetting of the emulsifier particles is also due to the possibility of formation of emulsions of both types with the same emulsifying agent, depending on the duration and intensity of the emulsification process.

All the above phenomena associated with emulsification by solid highly dispersed particles are even more marked and more characteristic when the aqueous phase of the emulsion contains electrolytes. If at the same time the oil phase contains organic acids (oleic, naphthenic) capable of forming corresponding metallic soaps with di- and trivalent cations, the cation of the solid emulsifying agents is extremely specific and depends on the nature of the cation.

Thus in the presence of Na, Ca, Sr, and Ba salts, inverse emulsions with low stability are formed, whereas in the presence of Al, Fe, Cu, and other salts, at definite pH values for each ion, marked stabilization of

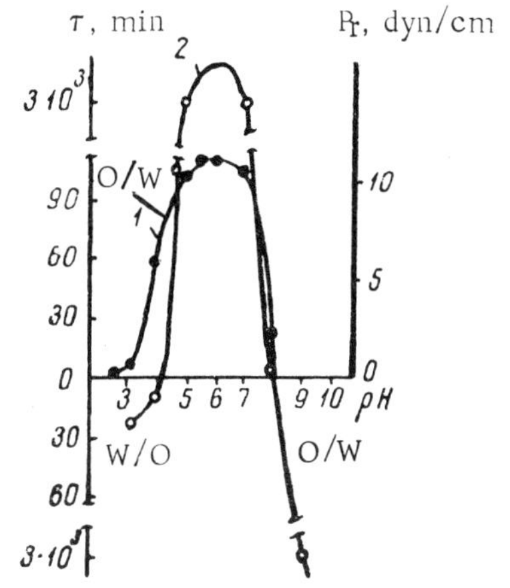

Fig. 3. Influence of pH on the limiting shearing stress P_r and the stability of emulsions τ. 1) Strength of interphase layer at the CCl_4+ 0.1% stearic acid–water+ 0.01 N $AlCl_3$ boundary in the presence of highly dispersed Al_2O_3 particles; 2) stability of the corresponding emulsion.

emulsions of the direct type takes place. These emulsions have a particularly high degree of dispersion and retain their stability over periods of many months.

An important feature is that under these conditions it does not matter which solid phase is used for emulsification. The same effect of marked stabilization is observed for particles on whose surface the formation of chemosorption layers of soaps of aliphatic or naphthenic acids is possible (bentonite clay, calcium and barium carbonates, aluminum oxide, etc.) and also for quartz particles, which do not show this property; a particularly interesting feature is that the effect is shown even by graphite particles, which form only emulsions of the inverse type.

The mechanism forming the basis of this maximum stabilization of emulsions by solid emulsifying agents in the presence of adsorption layers of metallic soaps can be interpreted if we take account of the fact that the narrow range of optimum pH values for emulsification also corresponds to the production, in a solution of the given electrolyte, of nuclei of a new highly dispersed colloidal solid phase consisting of basic salts. For example, in 0.01 N aluminum chloride solution at pH=4.5-7.0, the formation of a basic aluminum salt of complex composition (e.g., $Al_3(OH)_8Cl$, aq) is possible. When the aqueous phase of the emulsion also contains particles of a solid emulsifying agent (including quartz and graphite), their surface, in the above optimum pH range, is covered by a compact layer of a hydrophilic phase film of basic salt particles, which block the surface and give to it new properties characteristic of the phase.

As a result of the subsequent reaction of these modified particles with the aliphatic acid present in the oil, their surface becomes covered with adsorption layers of basic metallic soaps with a characteristic polymeric chain structure, which exhibit high structural-mechanical properties – high viscosity and strength – and bring about not only the adhesion of particles of emulsifying agent to the oil drops but also cohesion of the particles of emulsifying agent to one another with the resultant production of stable coagulation structures on the surface of the drops. Outside the above optimum pH range, the metallic soaps formed have the usual composition and form emulsions of the inverse type in accordance with their oleophilic character.

Figure 3 shows the pH dependence of the stability of the emulsions τ (curve 2) and the limiting shearing stress P_r of the adsorption layers of aluminum soaps stabilizing these emulsions (curve 1). These emulsions are formed with any dispersed solid phase (bentonite, quartz, graphite) and with an oil phase (carbon tetrachloride) containing 0.1% stearic acid, and aqueous phase containing 0.01 N $AlCl_3$. The τ–pH and P_r–pH curves correspond, indicating a direct relationship between the stability of emulsions and the strength of the structural-mechanical barrier responsible for the stabilization of the system.

Soaps with a polymeric structure are also oleophilic but the possible formation of stable emulsions of the direct type by means of these soaps is due to the fact that when the system contains a sufficient quantity of solid phase and electrolyte, a stabilizing action is shown not by the micelles of oleophilic soap but by complex particles formed in the emulsification process and consisting of particles of the initial solid phase covered with a compact hydrophilic layer of a basic salt of the metal, partly modified by an adsorption layer of the basic soap formed on individual regions of the surface at the points of contact between particle and drop.

This complex, as a result of its optimum diphilic (hydrophobic-hydrophilic) structure, can be held at the surface of separation between the phases, with the hydrophilic regions of the surface directed towards the aqueous phase of the emulsion and the hydrophilic sections of the surface directed towards the oil phase of the emulsion. The stability of these emulsions can be broken down by artificially intensifying the oleophilic character of the particles of the protective covering, e.g., by bringing about further contact between its hydrophilic regions and the surface of the oil drops. In actual fact, prolonged intensive mechanical agitation of the emulsion, by breaking down the compact protective coating and successively producing oleophilic properties on new hydrophilic regions of the solid particle, gradully decreases the stability of the emulsion and finally leads to reversal of the emulsion as a result of transfer of the compact hydrophobized particles of emulsifying agent to the oil phase.

These regular features of the emulsification process, which we have established in a study of the physicochemical basis of the emulsion method for removing viscous petroleum products from storage tanks and vessels, should obviously be taken into account in the study of the reactions of surfactant solutions in a petroleum-bearing stratum, since in both cases the aim of the operation is the same – to alter the conditions of selective wetting of the metal or stratal rock respectively by water and petroleum (petroleum products) in such a way that the bond between the "oil" phase and the solid surface is broken.

At the same time it is clear from the foregoing that in a number of cases it is difficult to predict the results of the action of surfactants on the stratum. The presence of electrolytes in the stratal waters and the presence of strongly surface-active substances in the petroleum may lead to a situation where, under the kinetic conditions of contact between the rock particles of the stratum and the two liquid phases, the surface of the solid particles will be modified (its selective wettability will be altered) and the particles will acquire the properties of effective emulsifying agents.

Under these complex conditions, depending on the influence of a large number of factors – the mineralogical composition of the rock, the nature and concentration of surfactant and electrolyte, the pH of the medium, the hysteresis of wetting, the conditions of reaction between the solutions and the rock, etc. – emulsions of either the direct or inverse type (sometimes both types simultaneously) with different degrees of stability up to maximum stability may be formed; this influence may be particularly marked under conditions of turbulent flow with vigorous agitation of the two phases of the emulsions. All this must be taken into account in the study of the various processes involved in the technology of petroleum recovery and treatment.

THE STUDY OF SURFACTANTS AS DEMULSIFYING AGENTS FOR THE PRETREATMENT OF PETROLEUM

A. A. Petrov

Work by P. A. Rebinder [6] has shown that a surfactant acts as a demulsifying reagent firstly if its surface activity is sufficiently high for the polar components adsorbed from the petroleum to be displaced from the petroleum–water boundary of separation and also for hydrophilic properties to be produced on the surface of the solid "armoring" emulsifying agents forming protective layers on the drops of emulsified water, and secondly if the surfactant at the petroleum–water boundary of separation forms adsorption layers with low structural-mechanical properties, i.e., does not act as a strong stabilizer for emulsions of water in petroleum or petroleum in water.

Surfactants exhibiting these properties may belong to different classes; thus, it is of interest to determine the influence of the nature of the polar group on their demulsifying power. For this purpose, studies have been made in Giprovostokneft' of a number of surfactants with different polar groups: 1) alkyl sulfates; 2) alkyl aryl sulfonates; 3) salts of carboxylic acids; 4) tertiary aminophenols; 5) tetrasubstituted ammonium salts; 6) polyethylene glycol ethers; and 7) polyglycerides.

The demulsifying power was tested with petroleums from four deposits in the Kuibyshev region – Zhigulevskii, Zol'nenskii, Mukhanovskii, and Radaevskii. Typical dehydration curves for six surfactants with different

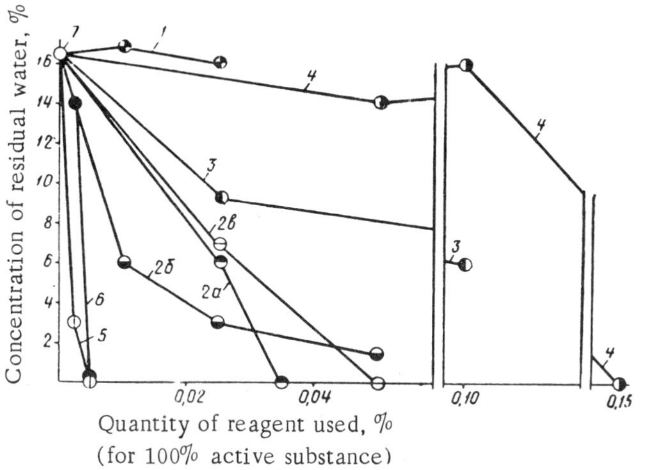

Fig. 1. Petroleum dehydration curves (settling temperature 60°, duration of settling 4 h). 1) Alkyl sulfate, alkyl aryl sulfonate; 2a) Wetting Agent NB; 2b) NChK; 2c) SCa; 3) salt of carboxylic acid; 4) tertiary aminoalkylphenol; 5) polyethylene glycol ether of alkylphenol; 6) polyglyceride of dibasic acid; 7) control test without reagent.

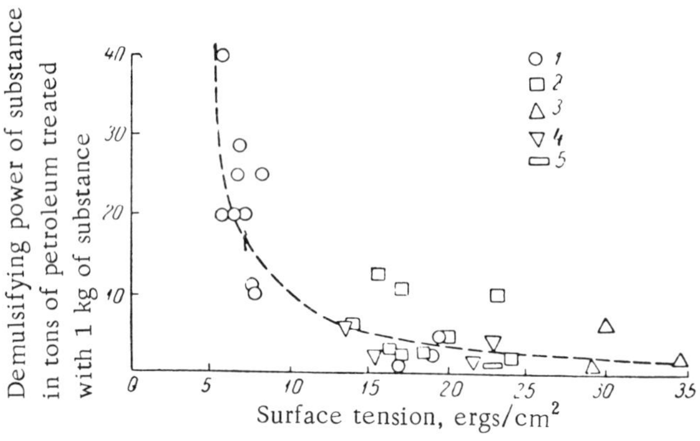

Fig. 2. Dependence of the demulsifying power of surfactants on the surface tension at the petroleum–water boundary of separation. 1) Polyglycol ethers; 2) alkyl aryl sulfonates; 3) alkyl sulfates; 4) tertiary aminophenols, tetrasubstituted ammonium salts; 5) salts of carboxylic acids. Open symbols) Zhigulevskii petroleum; half-shaded symbols) Zol'nenskii petroleum; dark symbols) Mukhanovskii petroleum.

polar groups are given in Fig. 1. The dehydration curves constructed for Zhigulevskii petroleum give the dependence of the concentration of residual water in the petroleum on the quantity of surfactant used. The other petroleums studied gave analogous curves.

In addition to the determination of the demulsifying power of the surfactants, parallel measurements were made of the surface tension of their aqueous solutions, containing 1 g/liter of active substance, at the boundary with the same petroleums.

For comparison of the surface activity and demulsifying power of the substances, Fig. 2 gives the dependence of the demulsifying power of the surfactants with respect to three petroleums – Zhigulevskii, Zol'nenskii, and Mukhanovskii – on the surface tension at the petroleum–water boundary of separation. The demulsifying power of the substances is expressed as the quantity of petroleum in the form of emulsion which can be treated by one part by weight of the substance.

As expected, the substances with the highest surface activity at the petroleum–water boundary of separation are also the most effective demulsifying reagents.

From the positions of the points in Fig. 2 it also follows that the highest demulsifying activity is shown by the nonionic surfactants – the polyglycol ethers formed by alkylphenols and alcohols and the polyglycerides formed by acids. The polyglycol ether of a carboxylic acid – the hexaglycol ether of stearic acid (Stearox 6) – showed a low demulsifying power, possibly due to its low surface activity, which results from the small number of glycol residues in the ether molecule.

The introduction of a sufficient number of glycol residues into the molecule of a surfactant with weak demulsifying properties leads to a considerable increase in its activity. Figure 3 gives dehydration curves for a tertiary aminophenol (Phenamin) and the product of its condensation with 25 molecules of ethylene oxide (Aksamon AF). The same figure also gives dehydration curves for tetrasubstituted glycol residues. The introduction of 25 glycol residues into the molecule of the aminophenol increases the demulsifying power of the substance by a factor of more than 7. Similarly, the additional introduction of eight glycol residues into the molecule of the tetrasubstituted ammonium salt increases the efficiency of the substance by a factor of 4.5.

Systematic studies of the demulsifying power of oxyethylated alkylphenols have been carried out in Giprovostokneft' since 1956 [1, 2]. Laboratory studies with specimens of these reagents (OP-7, OP-10, DB), prepared on the industrial scale or synthesized in VNIIneftekhim from coal and shale phenols, showed that the oxyethylated alkylphenols are highly effective demulsifying reagents for petroleum emulsions.

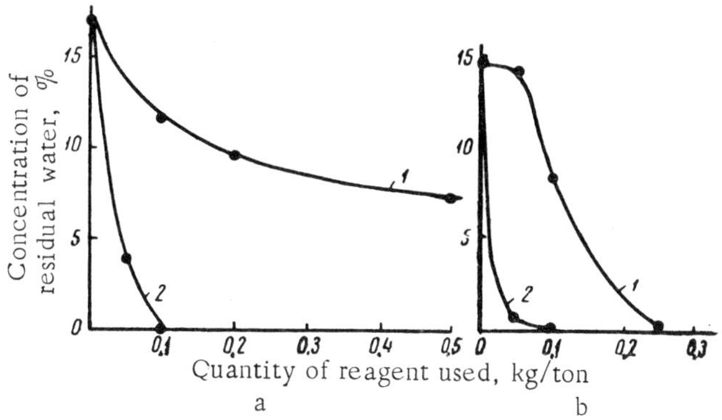

Fig. 3. Influence of the introduction of glycol residues into the molecule of a substance on its demulsifying power. Dehydration curves: a) for Radaevskii petroleum:

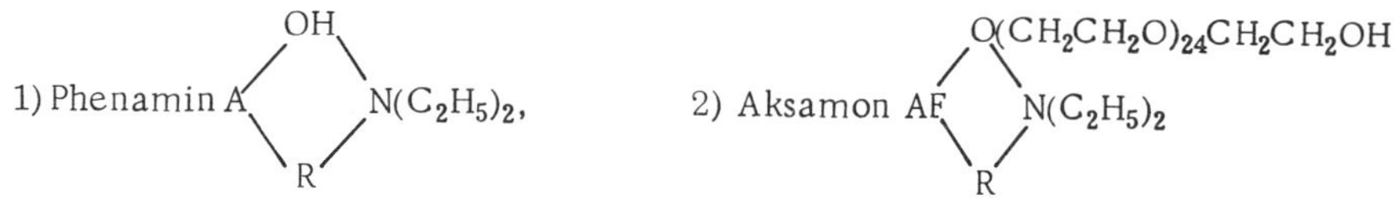

b) for Mukhanovskii petroleum:

1) Alkamon OS-2 [RO $(CH_2CH_2O)_2CH_2N(C_2H_5)_2]C_6H_5SO_3$,
2) Leveling Agent A$[RC_6H_4O(CH_2CH_2O)_{10}CH_2N(CH_3)(C_2H_5)]C_6H_5SO_3$.

The first industrial tests of the demulsifying power of the reagent OP-10, carried out early in 1957 on industrial demulsifying assemblies in the Kuibyshev region, confirmed the high demulsifying activity of this reagent [3-4].

Industrial tests on experimental batches of oxyethylated alkylphenols prepared from coal phenols by the method developed by VNIIneftekhim are being carried out in the petroleum plants of the Kuibyshev Sovnarkhoz. The tests are being carried out with the reagents UFE_8, $KAUFE_{14}$, and OP-7. The first results, obtained on a thermochemical dehydrating assembly in the Zol'nenskii plant, are given in Table 1, which also gives data on the operation of the assembly with the reagent NChK from the Novokuibyshev NPZ.

It follows from the specific quantities used that the demulsifying activities of the reagents UFE_8, $KAUFE_{14}$, and OP-7 are approximately equal to one another and greater than the demulsifying activity of NChK by a factor of 28-32. In spite of the high cost of these reagents – from 2500 rubles/ton for UFE_8 to 3500 rubles/ton for $KAUFE_{14}$, the expenditure on reagents when these demulsifiers are used is decreased considerably.

The advantage of oxyethylated alkylphenols (polyglycol ethers) over other surfactants as demulsifying reagents in the pretreatment of petroleum is obvious. In all work which has been carried out on the demulsifying power of these reagents, however, no attempt was made to study the influence of the structure of oxyethylated alkylphenols on their demulsifying power, although the solution of this problem is extremely important since it should serve as a basis for the industrial production of the most active demulsifying reagents from the class of oxyethylated alkylphenols.

In order to study the magnitude of both the hydrophobic and the hydrophilic sections of the molecules on the demulsifying power of oxyethylated products, studies have been carried out in Giprovostokneft' with a series of polyglycol ethers formed by alkylphenols, with hydrocarbon radicals of different size and with different numbers of glycol residues. In order to eliminate the influence of the structures of the alkyl radical chains, secondary o-butylphenol and secondary nonylphenol were synthesized by the method described by I. P. Tsukervanik [7] from the corresponding normal alcohols and phenol with anhydrous aluminum chloride catalyst. The characteristics of the alkylphenols obtained are given in Table 2.

TABLE 1. Results of Tests on a Thermotechnical Assembly with Reagents UFE_8, $KAUFE_{14}$, and OP-7 and Zhigulevskii Petroleum

Water content of crude petroleum, %	Temperature	Duration of settling, h	Reagent	Quantity of reagent used, kg/ton		Concentration of residual water in commercial petroleum, %	% pure water in drained liquid	Quantity of petroleum in drained water, % of commercial petroleum	Cost of reagent kopeks/ton		Point at which commercial petroleum specimen was removed
				Crude petroleum	Commercial petroleum				Crude petroleum	Commercial petroleum	
38.5	68	13.5	OP-7	0.052	0.09	0.15	100	None	1.63	2.95	1 m from the foot of the reservoir
29.5	67	14.3	OP-7	0.050	0.08	0.35	70.5	1.8	1.56	2.5	From three layers
31.3	69	15	UFE_8	0.050	0.08	0.12	88.5	0.5	1.56	2.5	The same
26.1	69	13.7	UFE_8	0.047	0.07	0.3	76.5	0.9	1.47	2.2	1 m from the foot of the reservoir
27.5	67.6	14	$KAUFE_{14}$	0.058	0.09	0.12	92.5	1.0	2.2	3.45	From three layers
24.9	67.0	14	$KAUFE_{14}$	0.049	0.07	0.3	83.5	0.8	1.87	2.7	1 m from the foot of the reservoir
25	68	14	NChK	1.50	2.3	0.6	82.3	3.0	9.5	14.5	From three layers
27	69	16	NChK	1.35	2.2	0.1	90	1.0	8.5	13.8	The same

Cost of reagents:

OP-7 – 303 rubles
UFE_8 – 290 "
$KAUFE_{14}$ – 355 "
NChK – 58.3 "

} + 8% Additional technical costs

TABLE 2. Characteristics of the Alkylphenols

Alkylphenol	Boiling point		Molecular weight		Elementary composition, %				OH content, %	
	Found	Literature data	Found	Calculated	Found		Calculated		Found	Calculated
					C	H	C	H		
o-Butylphenol	231-237	232-235	148	150	79.23	8.88	79.96	9.39	10.6	11.4
Nonylphenol	165-172	106-109	219	220	83.3	9.55	81.76	10.98	7.97	7.73

TABLE 3. Characteristics of Polyglycol Ethers of Alkylphenols

Alkylphenol	CH_2CH_2O group content			Clarification temperature, deg.	Surface tension at the boundary of separation between petroleum and a solution of the reagent with a concentration of 1 g/liter, $ergs/cm^2$	Arbitrary symbol
	From increase in weight during reaction, wt. %	By analysis, wt. %	In moles per mole of phenol			
Methylphenol	63.5	65.0	4.5	19-20	15.4	$MFOE_{4.5}$
"	78.8	79.4	9.4	*	20.9	$MFOE_{9.4}$
"	89.5	—	24	*	20.8	MFOE
"	93.8	95.5	52	‡	21.1	$MFOE_{52}$
Butylphenol	53.5	52.7	3.7		17.6	$BFOE_{3.7}$
"	77.3	78.5	12.7	89.4-89.6	9.92	$BFOE_{12.7}$
"	85.8	82.6	20	93.8-93.9	10.4	$BFOE_{20}$
"	91.7	91.8	40	‡	14.7	$BFOE_{40}$
Nonylphenol	42.5	47.5	4.2	†	13.1	$NFOE_{4.2}$
"	71.5	75.0	16	89.3-89.5	4.84	$NFOE_{16}$
"	77.5	77.6	19	90-90.2	5.75	$NFOE_{19}$
"	86.3	86.6	32.8	‡	8.17	$NFOE_{32.8}$

*Turbidity increased gradually.

†Substance did not dissolve completely in water.

‡No turbidity observed with heating to 100°.

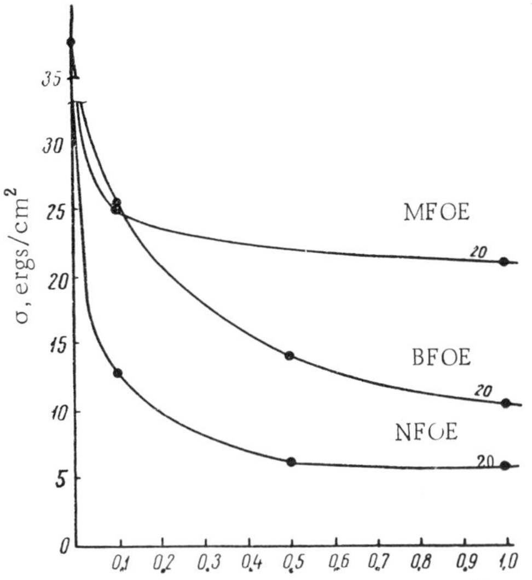

Fig. 4. Surface tension isotherms for alkylphenols condensed with 20 molecules of ethylene oxide (petroleum–water boundary of separation).

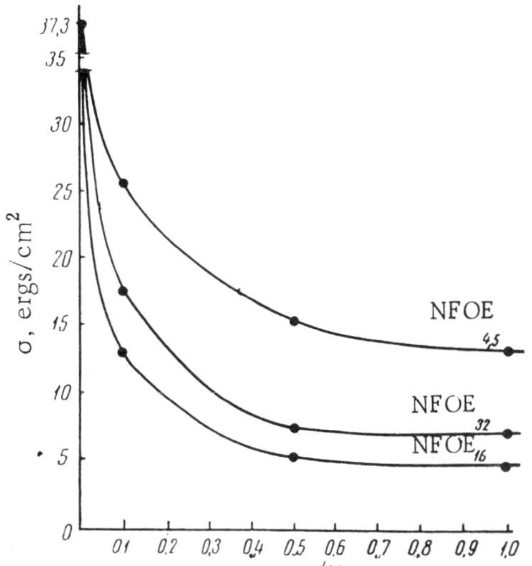

Fig. 5. Surface tension isotherms for oxyethylated nonylphenol (petroleum–water boundary of separation).

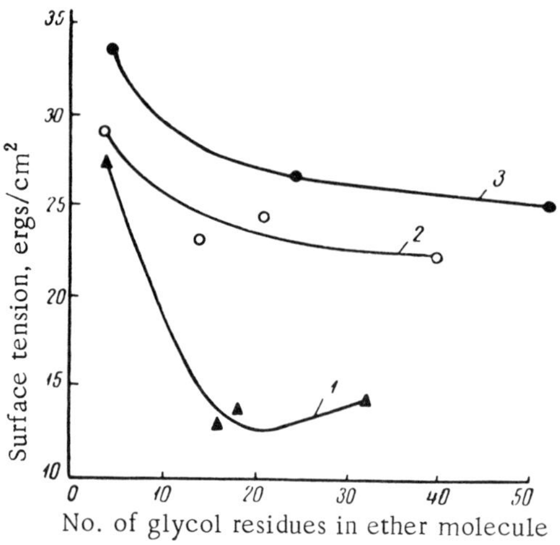

Fig. 6. Dependence of the surface tension of polyglycol ethers of alkylphenols on the number of glycol residues in the molecule at a concentration of $1 \cdot 10^{-4}$ M (petroleum–water boundary of separation). 1) Nonylphenol ether; 2) o-butylphenol ether; 3) o-methylphenol ether.

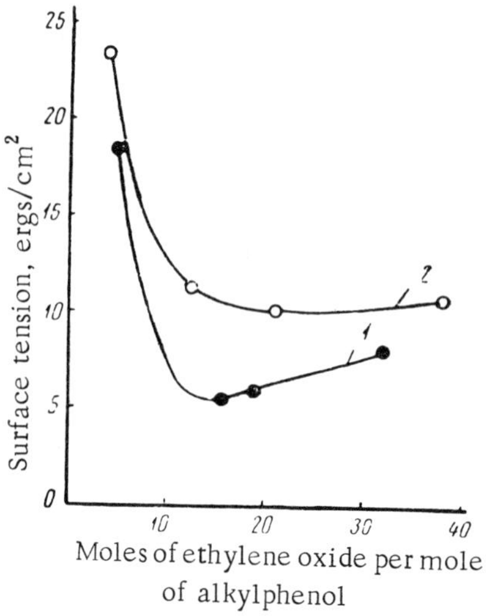

Fig. 7. Dependence of the surface tension of polyglycol ethers of alkylphenols on the number of glycol residues (petroleum–water boundary of separation). 1) Nonylphenol ethers at a concentration of $6 \cdot 10^{-4}$ M; 2) o-butylphenol ethers at a concentration of $1 \cdot 10^{-3}$ M.

The resulting alkylphenols, and also o-cresol, were condensed with different quantity of ethylene oxide in the presence of NaOH as catalyst. A total of 12 polyglycol ethers of alkylphenols were obtained, differing in their hydrocarbon radicals–C_1, C_4, and C_9– and in the number of glycol residues–approximately 4-10-20-40. The number of glycol residues was determined by two methods: from the increase in weight of the reaction mass during the condensation with ethylene oxide, and iodometrically [9] from the quantity of iodine liberated in the reaction between the ether and hydriodic acid. The two methods gave almost identical results, which are given in Table 3.

For the polyglycol ethers of the alkylphenols, we determined the clarification temperatures by the GOST (USSR Standard) method with OP-7 [2], the surface tension isotherms at 20° at a water–Zhigulevskii petroleum surface of separation by the drop weight determination method of Harkins and Brown [1], and the demulsifying

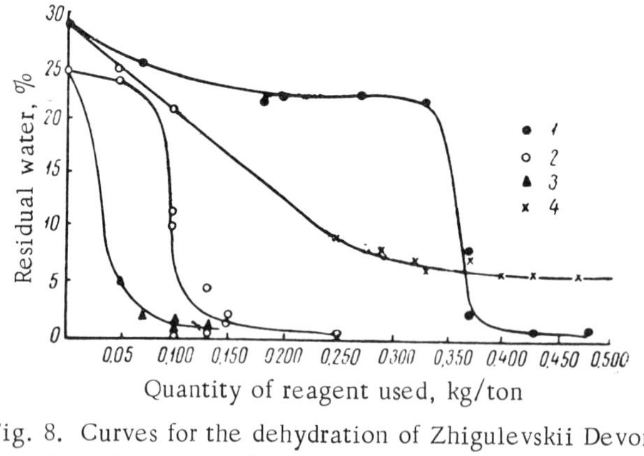

Fig. 8. Curves for the dehydration of Zhigulevskii Devonian petroleum by means of the polyglycol ether of orthomethylphenol $MFOE_n$ (temperature 60°, duration of settling 4 h. 1) $MFOE_{4.5}$; 2) $MFOE_{9.4}$; 3) $MFOE_{24}$; 4) $MFOE_{52}$.

TABLE 4. Demulsifying Power of Polyglycol Ethers of Alkylphenols

Substance	Molecular weight	Specific quantity used, g/ton	Demulsifying activity	Substance	Molecular weight	Specific quantity used, g/ton	Demulsifying activity
o-$MFOE_{4.5}$	306	372	2700	o-$BFOE_{20}$	1030	46	21700
o-$MFOE_{9.4}$	521	126	8000	o-$BFOE_{40}$	1910	258	3700
o-$MFOE_{24}$	1168	83	12000	$NFOE_{4.2}$	403	173	5800
o-$MFOE_{52}$	2393	470	2100	$NFOE_{16}$	922	79	12700
o-$BFOE_{3.7}$	313	372	2700	$NFOE_{19}$	1054	60	16700
o-$BFOE_{12\ 7}$	709	70	14300	$NFOE_{32.8}$	1658	38	26300

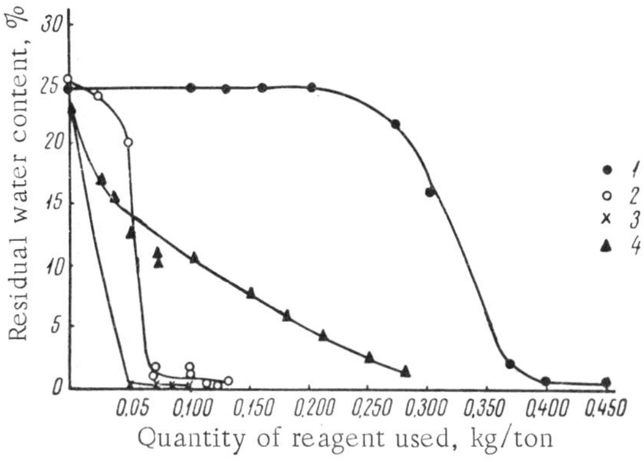

Fig. 9. Curves for the dehydration of Zhigulevskii Devonian petroleum by means of the polyglycol ether of orthobutylphenol o-$BFOE_n$ (temperature 60°, duration of settling 4 h). 1) $BFOE_{3.7}$; 2) $BFOE_{12.7}$; 3) $BFOE_{22}$; 4) $BFOE_{40}$.

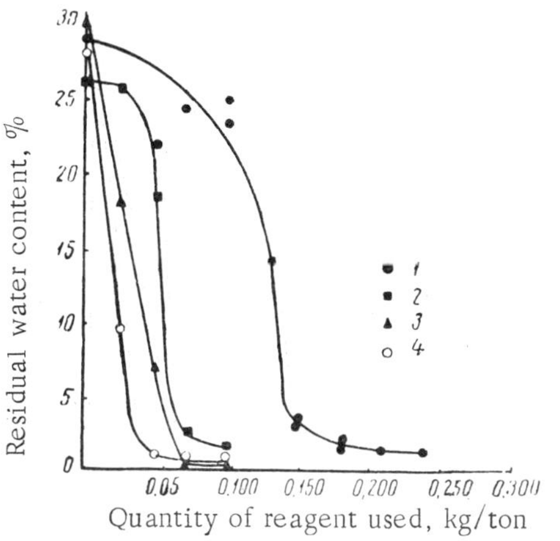

Fig. 10. Curves for the dehydration of Zhigulevskii Devonian petroleum by means of the polyglycol ether of nonylphenol $NFOE_n$ (temperature 60°, duration of settling 4 h). 1) $NFOE_{4.5}$; 2) $NFOE_{16}$; 3) $NFOE_{19}$; 4) $NFOE_{32}$.

power in emulsions formed by petroleum and stratal water from the Devonian horizon of the Zhigulevskii deposit. Table 3 gives the clarification temperatures and the values of the surface tension at the boundary of separation between petroleum and water containing 1 g/liter of the polyglycol ether.

The synthetic polyglycol ethers of the alkylphenols cover an extremely wide range of clarification temperatures. Thus for nonylphenol, ethers were obtained with a clarification temperature lower than the freezing point (for the ether containing 4.5 molecules of ethylene oxide), or higher than the boiling point (for the ether containing 32 ethylene oxide molecules. Behavior different from that of the other polyglycol ethers was shown by the methylphenol ethers for which, at a concentration of 9.7-24.4 glycol residues, a gradual increase in turbidity with increase in temperature was observed and also a gradual production of turbidity with decrease in temperature. The methylphenol ether containing 4.5 glycol residues behaved normally and gave a clarification temperature of 19-20°.

The surface tension σ isotherms for different alkylphenols condensed with the same quantity of etylene oxide – 20 moles per mole of phenol – are given in Fig. 4. As expected, the surface activity of the compound increases with increase in the length of the alkyl radical, i.e., on going from the methylphenol ether to the ethers of butylphenol and nonylphenol (Traube's rule).

The way in which the surface tension is influenced by the number of glycol residues for the same nonpolar section of the ether molecule is shown in Fig. 5, which also gives the surface tension isotherms for nonylphenol ethers containing 4.5, 16, and 32 ethylene oxide molecules. It follows from these data that with increase in the number of glycol residues in the alkylphenol molecule, the surface tension at first

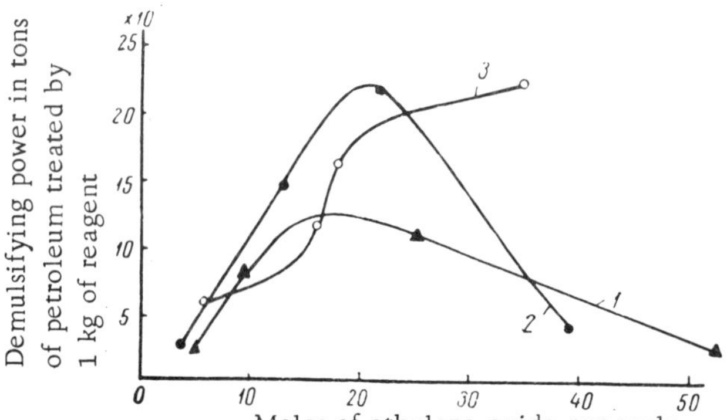

Fig. 11. Curves giving the dependence of the demulsifying power of polyglycol ethers of alkylphenols on the number of glycol residues. 1) o-Methylphenol (o-cresol); 2) o-butylphenol; 3) nonylphenol.

decreases and then begins to increase at identical weight concentrations in the solution. These results agree with data on the surface tension of polyglycol ethers of nonylphenol containing 6, 10, 18, and 30 glycol residues, given by Jelinek and Mayhew [8].

On going to molar concentrations, however, no appreciable increase in surface tension with further increase in the number of glycol residues is observed in the low concentration range. Figure 6 shows the change in the surface tension of solutions of polyglycol ethers of alkylphenols at a concentration of $1 \cdot 10^{-4}$ M with increase in the number of glycol residues. The curves for o-methyl-, o-butyl-, and nonylphenols show that the surface tension at first decreases rapidly but subsequently changes only slightly. An analogous relationship is obtained from the data of the American authors for nonylphenol ethers after conversion of the weight concentrations to molar concentrations. In the range of high molar concentrations, however, (of the order of $1 \cdot 10^{-3}$ M and above) the surface tension of the alkylphenol ethers passes through a minimum with increase in the number of glycol residues (Fig. 7).

The nature of the dependence of the surface tension on the number of glycol residues in the ether molecule is possibly related to the fact that in the low concentration range the dissolved alkylphenol ethers are present in the molecular state and their surface activity increases with increase in the number of glycol residues until the optimum hydrophilic–lipophilic balance in the molecule is reached. Further increase in the number of CH_2CH_2O groups when a sufficient number are present in the molecule, however, produces little change in the balance in the molecule, so that it has little influence on the surface activity (see Fig. 6).

With increase in the molecular concentration of the ether in solution, the surface tension at the boundary with petroleum decreases. When the critical concentration of micelle formation is reached, however, the surface tension remains practically constant. The dimensions of the molecules obviously increase with increase in the number of glycol groups, and micelle formation may take place at lower concentrations of ether in the solution and hence at a higher surface tension value. Since the surface tension of the ether solution for which micelle formation has taken place changes only slightly with further increase in concentration, the surface tension will be lower for ethers containing a smaller number of glycol residues (see Fig. 7), since no micelle formation is observed for these substances at these concentrations.

The demulsifying power of the synthetic polyglycol esters of alkylphenols was studied on emulsions prepared under standard conditions from petroleum and stratal water from the Devonian horizon of the Zhigulevskii deposit.

The emulsions were prepared by mixing petroleum and water by means of a propeller stirrer for 5 min at 4000 rpm. The original water contant of the emulsion was 30%. The prepared emulsions were kept for 24 h before the experiment so that the increase in the strength of the protective layers on the drops of emulsified water as a result of the aging of the emulsion should be the same in all experiments.

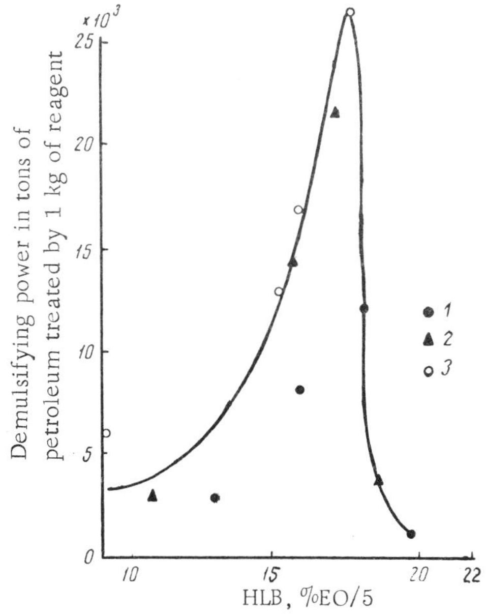

Fig. 12. Dependence of the demulsifying power of polyglycol ethers of alkylphenols on HLB. 1) o-Methylphenol ethers; 2) o-butylphenol ethers; 3) nonylphenol ethers.

The polyglycol ethers of the alkylphenols were added in quantities appropriate for each experiment in the form of a solution (in 5 ml of water) to 150 ml of emulsion. The emulsion specimens were mixed with the reagent and left to settle in a water bath at 60° for 4 h. After settling, the petroleum was separated from the free water formed, and the residual water content of the petroleum was determined.

The results obtained are given in the form of dehydration curves in Figs. 8-10. These curves were used to determine the specific quantities of polyglycol ethers of alkylphenols required to ensure sufficiently complete breakdown of the petroleum emulsion; concentration of residual water in the petroleum was less than 2%. These specific quantities are given in Table 4. It is more convenient to express the demulsifying power of the surfactant in terms of the quantity of petroleum which can be treated by one part by weight of the reagent. Table 4 also gives the demulsifying activity of the polyglycol ethers of alkylphenols.

The dependence of the demulsifying activity of the polyglycol ethers on the composition of their hydrophobic and hydrophilic parts is shown in Fig. 11, which also gives curves showing the dependence of the demulsifying power on the number of glycol residues in the molecule for the ethers of three alkylphenols – o-methyl-, o-butyl-, and nonylphenol. The ethers of all three alkylphenols show a characteristic initial increase in demulsifying power with increase in the number of CH_2CH_2O groups in the hydrophilic part of the molecule. Further increase in the number of groups, however, leads to a flattening of the curve, i.e., to a decrease in the rate of increase in the demulsifying power of the alkylphenol ethers and, in the case of individual alkylphenols – o-methyl- and o-butylphenol – to the production of a maximum, followed by a decrease in activity. With increase in the length of the alkyl radical chain, the maximum on the curve giving the demulsifying power is displaced toward higher values of the glycol residue content. For nonylphenol ethers containing up to 34 glycol residues, the curve giving the demulsifying power shows no maximum. Since the curve flattens out, however, it is to be expected that for ethers with a higher glycol residue content, the demulsifying power will pass through a maximum.

Thus it follows from the data obtained that the demulsifying power of polyglycol ethers of alkylphenols depends on the relationship between the polar and nonpolar parts of the ether molecule, i.e., on its hydrophilic–lipophilic balance (HLB).

Although the qualitative changes in the HLB with change in the polar or nonpolar part of the ether molecule can readily be determined, there is as yet no generally accepted means of expressing the HLB quantitatively. A paper by A. B. Taubman [5] gives a number of methods for expressing the HLB of a surface-active substance.

If we use Griffin's method for expressing the HLB of the oxyethylated products of a given homologous series, in our case the alkylphenols, in the form HLB = %EO/5, it is possible to give a general dependence of the demulsifying power of the polyglycol ethers of different alkylphenols on the relationship between the polar and nonpolar parts of the ether molecule. Figure 12 shows the dependence of the demulsifying power on HLB for the polyglycol ethers of the alkylphenols studied.

The curve shows that the maximum demulsifying power for the ethers corresponds to an HLB value in the range 15-18. An interesting feature is that Griffin gives HLB values of 3-6 for emulsifying agents stabilizing emulsions of water in oil, 7-9 for wetting agents, 8-18 for emulsifying agents stabilizing emulsions of oil in water, and 13-15 for detergents.

The demulsifying power of a polyglycol ether of an alkylphenol, however, can only as a first approximation be defined by the percentage of condensed ethylene oxide present. The form of expressing the HLB given by Griffin makes no allowance for the length, structure, or number of alkyl groups in the benzene nucleus of the alkylphenol, although these should undoubtedly influence the demulsifying power of the polyglycol ether of the alkylphenol.

The absolute value of the demulsifying power of the polyglycol ethers of alkylphenols is fairly high (Table 4). The most effective materials were the ethers of o-butylphenol containing 24 ethylene oxide molecules (demulsifying power 21700, specific quantity used 46 g/ton) and the ethers of nonylphenol containing 19 ethylene oxide molecules (activity 16700, specific quantity used 60 g/ton) or 32 ethylene oxide molecules (activity 26300, specific quantity used 38 g/ton).

Literature Cited

1. A. Weissberg, Physical Methods of Organic Chemistry [Russian translation] (Izd. inostr. lit., Vol. 1, 1950), pp. 178-187.
2. GOST 8433-57.
3. A. A. Petrov and N. P. Borisova, Khimiya i tekhnologiya topliv i masel, No. 2 (1959) p. 66.
4. A. A. Petrov and N. P. Borisova, Tr. in-ta Giprovostokneft', Geologiya i razrabotka neftyanykh mestorozhdenii, No. 2 (1959) p. 413.
5. A. B. Taubman, Khimicheskaya nauka i promyshlennost', 4, 5, 566 (1959).
6. P. A. Rebinder and K. A. Pospelova, Introductory Article in V. Clayton's book: Emulsions (Izd. inostr. lit., Moscow, 1950).
7. I. P. Tsukervanik and Z. N. Nazarova, Zhurnal obshchei khimii, 7, 3-4, 623 (1937).
8. E. F Jelinek and R. L. Mayhew, Textile Res. Journ., 24, 8, 765 (1954).
9. S. Siggia, et al., Analyt. Chem., 30, 1, 115 (1958).

THE FORMS OF COMBINED WATER IN A PETROLEUM RESERVOIR

G. A. Babalyan, V. L. Komarov, and R. B. Morozov

At the present time there is no single point of view regarding the form in which connate water exists [7, 8, 9, 10, 11, 12]. At the same time, this question is of considerable importance for the solution of a number of problems in industrial geophysics and in the technology of petroleum drilling and recovery, since the choice of a given theory – the film theory or the capillary-retention theory – determines the approach to the solution of a given problem.

Because of the well-known difficulties associated with the study of the form in which connate water exists under stratal conditions in a number of cases, it is necessary to study this problem in systems which can be taken as models of the basic element of the pore space in rock – the individual pore channel (capillary). The first work of this kind with glass and quartz capillaries was carried out by M. M. Kusakov [4, 12] and as a result of this work it was concluded that the combined water is found in the capillary-retained form under the conditions of the Devonian deposits in the Western Bashkir Republic. In both glass and quartz capillaries, in contact with Devonian petroleums, rupture of the film of mineralized water took place within a finite, very short time interval.

A stimulus to the carrying out of the present work was provided by the discovery of the vast Arlanskii petroleum deposit in the lower Carboniferous deposits of the northwestern Bashkir Republic. The physicochemical characteristics of the petroleums from this deposit differ significantly from those of the petroleums of the Tuimazinskii deposit which were studied by M. M. Kusakov.

The method of setting up and carrying out the experiments was exactly the same as that used by Boimel'-Mekenitskaya [4]. The only significant change was in the method used to measure the electrical conductivity. Since no quartz capillaries were available, the experiments were carried out with a glass capillary and this is undoubtedly reflected in the degree of completeness of our examination of this problem, although it does not affect its fundamental aspects.

Measuring Circuit

The resistance of the system was determined by an alternating-current bridge system (Fig. 1).

A ZG-10 audio-frequency oscillator was used as alternating current source. The reference resistances were VS constant resistances with nominal resistances of 1, 0.5, and 0.1 meg. The power of the resistances was 2 W.

These resistances were chosen in order to eliminate induction and capacity pickup. The presence of three resistances in the arm of the bridge made it possible to extend the range of measurements.

A 0.1 meg SP resistance was used as compensating resistance. In contrast to M. M. Kusakov's circuit [12], in which a telephone was used as compensation position indicator, we increased the accuracy (sensitivity) of the circuit, in accordance with [6], by replacing the telephone by an EO-7 oscillograph with a "filter-plug" connected to its input to eliminate the influence of currents of industrial frequency on the oscillograph reading.

TABLE 1

Concentration of solution	Sodium chloride				Potassium chloride
	4 N	1 N	0.5 N	0.1 N	0.5 N
Specific electrical resistance at t = 20°C, Ω · m	0.044	0.129	0.238	1.04	0.204
Specific electrical conductivity, $\Omega^{-1} \cdot m^{-1}$	22.7	7.75	4.2	0.96	4.9

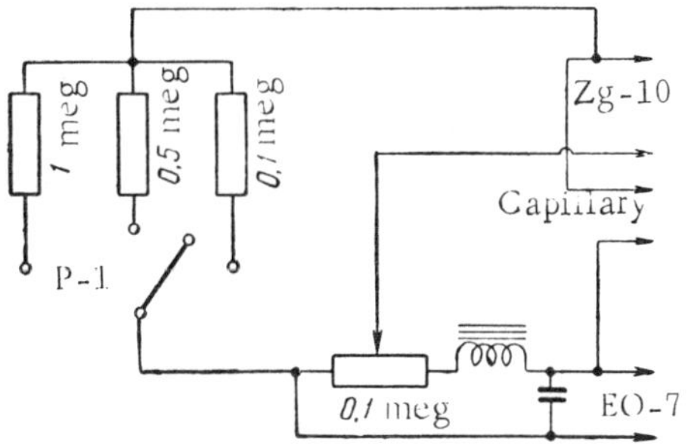

Fig. 1. Basic electrical circuit for measuring the resistance of the aqueous layer.

Materials Studied

The capillaries used for the studies were prepared in the Glass-Blowing Workshop of the UFNII from No. 46 molybdenum glass from the "Svetlana" factory.

Composition of molybdenum glass No. 46 [5] (wt. %)

SiO_2	68.25	Na_2O	6.00
B_2O_3	16.5	ZnO	5.00
Al_2O_3	4.00	Ar_2O	0.25

The length of the capillary 1 was equal to 150-170 mm. The capillary diameter was 1.35 mm. The constancy of the diameter was checked from the length of a mercury column.

The length of a drop of hydrocarbon liquid or air bubble, equal to the length of the thin layer, was measured by means of a microscope.

Preparation of the Capillaries for the Experiments

The capillaries in which the measurements were carried out were washed in a hot concentrated solution of potassium dichromate in sulfuric acid (concentration ~30%), then with distilled water, and finally with twice-distilled water. Washing was continued until no drops were formed on the walls of the capillary.

Electrolytes and Liquids Used in the Studies

Aqueous Salt Solutions

The electrolytes used consisted of aqueous solutions of sodium chloride (analytically pure grade) and potassium chloride (analytically pure grade). The solutions were prepared with twice-distilled water. We prepared solutions with concentrations of 0.1, 0.5, 1.0, and 4 N in the case of sodium chloride and 0.5 N in the case of potassium chloride.

The specific electrical resistances and electrical conductivities of these electrolytes are given in Table 1.

Hydrocarbon Liquids

The hydrocarbon liquids used in the studies were heptane and stratal petroleums from the Arlanskii, Tuimazinskii, and Shkapovskii deposits in the Bashkir Republic. The petroleum specimens were taken from the trap. The characteristics of the hydrocarbon liquids are given in Table 2.

Experimental Results

In the development of our procedure and in our preliminary experiments, we used systems which are readily comparable with the systems studied by M. M. Kusakov [12], as follows:

TABLE 2. Characteristics of the Hydrocarbon Liquids (UFNII Data)

Hydrocarbon liquid	Density, g/cm³	Viscosity at 20°, cP	Concentration, %					Surface tension, ergs/cm²
			Sulfur	Hydrogen sulfide	Asphaltenes	Tar	Paraffins	
Heptane	0.684	–	–	–	–	–	–	47.3
Arlanskii deposit petroleum	0.897	58.1	3.94	None	4.0	15.1	2.96	25
Tuimazinskii deposit petroleum D II	0.853	11.8	1.48	"	3.72	8.95	5.41	17
Shkapovskii deposit petroleum . .	0.819	6.6	0.67	"	0.92	4.4	5.0	19

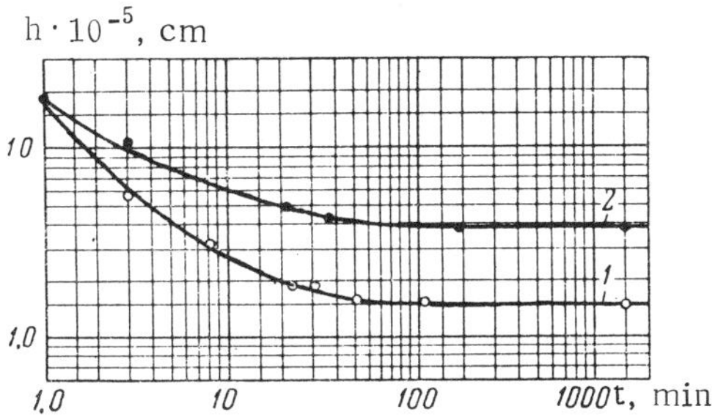

Fig. 2. Kinetics of the change in the average thickness of a thin layer of aqueous 0.5 N KCl solution at its boundary with air (2) and heptane (1) in glass capillaries.

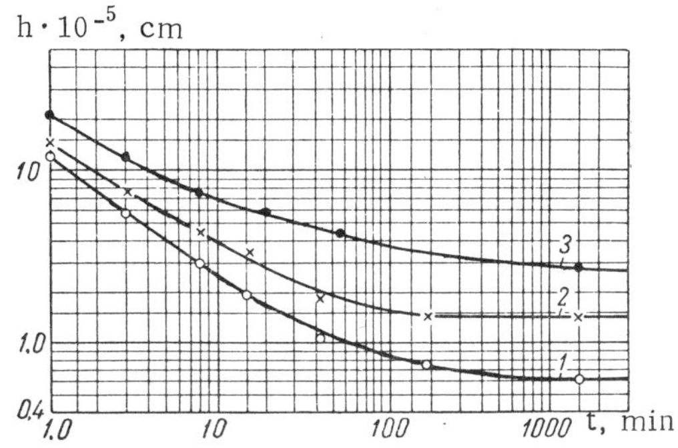

Fig. 3. Kinetics of the change in the average thickness of a thin layer of aqueous sodium chloride solutions of different concentrations, formed between heptane and glass. 1) 4 N NaCl solution; 2) 0.05 N NaCl solution; 3) 0.1 N NaCl solution.

1) the air – 0.5 N potassium chloride solution – glass system;

2) the heptane – 0.5 N potassium chloride solution – glass system;

3) the Tuimazinskii petroleum – 4 N sodium chloride solution – glass system.

In addition to the above, the following systems were studied:

1) heptane–sodium chloride solutions of different concentration–glass systems;

2) the Arlanskii deposit petroleum – 4 N sodium chloride solution – glass system;

3) the Shkapovskii deposit petroleum – 4 N sodium chloride solution – glass system.

The study of thin layers of aqueous electrolytes in the above systems and the comparison of our results with M. M. Kusakov's results [12] are described below.

The Air – 0.5 N Sodium Chloride Solution – Glass System

Studies of the thin films of aqueous solution formed between an air bubble and glass showed that these layers reach equilibrium values, i.e., their thickness becomes constant with time. Figure 2 (curve 2) shows the kinetics of the change in the thickness of thin layers formed by aqueous KCl solution (concentration 0.5 N) at its boundary with air. An interesting feature is that the curve is analogous to the corresponding curve obtained by M. M. Kusakov, the only difference being that the magnitude of the equilibrium layer observed by M. M. Kusakov [12] is slightly smaller. This discrepancy may be attributed to the difference in the temperatures at which the studies were carried out.

The Heptane – 0.5 N Aqueous Potassium Chloride Solution – Glass System

Figure 2 (curve 1) shows the kinetics of the change in the thickness of thin layers of aqueous electrolyte solutions (potassium chloride) formed between heptane and glass.

It can be seen from the graph that the thickness of the thin layer at first decreases but is subsequently stabilized (100 min after the start of the measurements) and reaches an equilibrium value.

The Heptane – Aqueous Sodium Chloride Solution – Glass System

In addition to the studies with aqueous potassium chloride solution, studies were carried out with heptane and aqueous sodium chloride solutions of different concentrations (0.1 N, 0.5 N, and 4 N). Figure 3 shows the kinetics of the change in the average thickness of the thin layer formed between heptane and glass for solutions of different concentrations. In this case, as can be seen from Fig. 3, the layer at first becomes thinner, after which an equilibrium thickness is established. The concentration of the solution has a significant influence on the magnitude of the equilibrium layer: the lower the concentration of the solution, the greater the thickness of the equilibrium layer.

Point measurements were made for a period of 1500 min but the magnitude of the equilibrium layer remained constant even after several days. No rupture of the thin layer was observed.

The Petroleum – 4 N Aqueous Sodium Chloride Solution – Glass System

In the study of the thickness of a thin layer of wetting liquid between petroleum and a solid phase, we carried out experiments to measure the thickness of a thin layer of 4 N aqueous sodium chloride solution between glass and stratal petroleum from different Bashkir deposits (Arlanskii, Tuimazinskii, and Shkapovskii). The kinetics of the change in the thickness of the thin layer for these systems is shown in Fig. 4. It can be seen from the graphs (Fig. 4) that a change in the thickness of the thin layer is observed only in the initial period, after which the thickness of the thin layer reaches equilibrium values. The magnitude of the equilibrium layer is different for different petroleums.

The low values for the aqueous layer of Tuimazinskii petroleum are apparently due to the more intensive interaction between the active components of this petroleum and aqueous sodium chloride solution, and to their partial dissolution with the formation of a thin layer of aqueous surfactant solution under the drop. The surfactant is adsorbed on the solid surface and makes it hydrophobic, as a result of which there is a sharp decrease in the thickness of the aqueous layer. The slightly higher activity of Tuimazinskii petroleum compared with Shkapovskii and Arlanskii petroleums is also indicated by surface tension data. The values for Tuimazinskii petroleum at its boundary with stratal water are slightly lower than those for Shkapovskii petroleum, while the latter are slightly lower than those for Arlanskii petroleum [15].

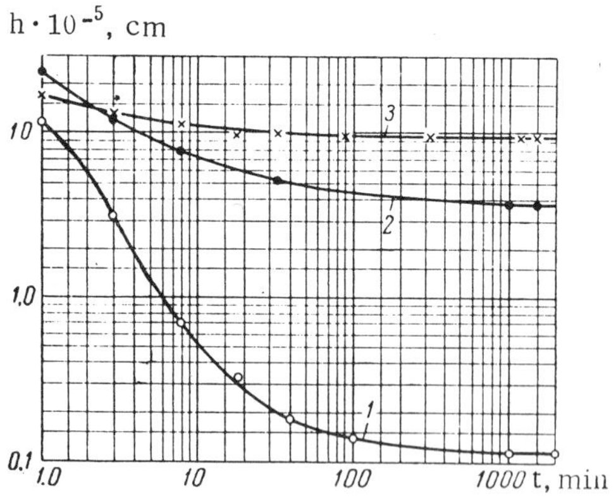

Fig. 4. Kinetics of the change in the average thickness of a thin layer of aqueous 4 N NaCl solution formed between petroleum and glass. 1) Tuimazinskii petroleum; 2) Arlanskii petroleum; 3) Shkapovskii petroleum.

Thus the measurements which we have made have shown that in a hydrophilic glass capillary, no rupture of an aqueous layer under an air bubble, drops of heptane, or drops of different petroleums takes place.

The equilibrium thickness of the aqueous layer depends on the quality of the petroleum and the degree of mineralization of the water. Other conditions being equal, it is greater for air bubbles and smaller for heptane, and decreases with increase in the degree of mineralization of the water.

With increase in the degree of mineralization of the water (although the thickness of the hydrate layer decreases as a result of compaction of the electrical diffusion double layer) there is a decrease in the possibility of rupture of the layer.

Since the equilibrium thickness of the aqueous film determines the magnitude of the pressure drop and the rate of displacement of water by petroleum, the following conclusion may be reached: the production of a given degree of water saturation in a porous medium requires a greater pressure drop when the water is displaced by air than when the water is displaced by heptane [2].

With increase in the degree of mineralization of the water the necessary pressure drop will decrease. Its magnitude will be smaller for the displacement of water by Tuimazinskii petroleum than for its displacement. At high pressure drops the quantity of residual water may be the same in all cases, since considerable compression of the residual layer of water under the influence of the pressure drop is possible, although this was not taken into consideration in our experiments.

It is difficult to reach any conclusions regarding the possible state of connate water in the stratum on the basis of the experiments which we have carried out, since the conditions of oil migration produced not only drop migration but also migration of continuous masses of petroleum, i.e., displacement of petroleum by water [3]. The nature of the behavior of aqueous films under these conditions in a porous medium is not yet clear. Rupture of the films is obviously most probable at the points of the granules [2]. The probability of adsorption of the active components of the petroleum is greatest at these same points, which will thus correspond to the zone in which direct adhesion of petroleum to the solid surface takes place.

In order to obtain a more objective estimate of the possible state of connate water, future studies should be carried out in quartz capillaries, since the majority of oil-bearing strata are composed for the most part of quartz particles. It should be noted, however, that considerable changes in the state of connate water may be produced by carbonate, carbonaceous, ferriferous, and other particles which are less hydrophilic than quartz particles.

In the general case it is necessary to assume the presence of both film water and capillary-retained water in the strata.

In those cases where hydrophilic particles and clay predominate in the stratum, the connate water is present for the most part in the form of a film.

On the basis of earlier work [1, 2, 3, 13, 14] it may be concluded that with decrease in the quantity of hydrophilic particles, e.g., with increase in the carbonate content of strata, the connate water is found for the most part in the capillary-retained state, since in the migration process the possibility of direct contact between the petroleum and the solid particles increases. There is a corresponding increase in their hydrophobic properties and in the probability of phase reversal in the pore space, i.e., the conversion of the water on the surface of the particles from the dispersion medium to the disperse phase.

The opposite picture should be observed in the displacement of petroleum by water (the case most frequently observed in geophysical studies). With increase in the carbonate content of rocks and particles which are difficult to wet by water and on which chemical fixation of the active components of the petroleum is possible, the residual petroleum will be found chiefly in the film state. With increase in the concentration of particles which are readily wet by water (quartz, clay), however, the residual petroleum will be found chiefly in the capillary-retained state.

Since with increase in the activity of the petroleums the wettability of rocks by these petroleums increases, the possibility of rupture of the aqueous layer increases. Thus with increase in the activity of the petroleums, the surface area of the solid particles occupied by film water decreases.

In future studies, for the complete solution of the problem by the method proposed by M. M. Kusakov, it would be advantageous to examine the state of the aqueous layer during the movement of a continuous mass of petroleum in a capillary, the influence of roughness on the walls of the capillary, and the presence of solid particles of different shape and hydrophilic character adhering to the wall of the capillary.

Literature Cited

1. Sh. N. Aliev and G. A. Babalyan, The Mobility of Oil Pool Outlines (Aznefteizdat, 1956).
2. G. A. Babalyan, Problems of the Mechanisms of Petroleum Recovery (Aznefteizdat, 1956).
3. G. A. Babalyan, I. I. Kravchenko, I. L. Markhasin, and G. V. Rudakov, The Physicochemical Principles of the Use of Surface-Active Substances in the Exploitation of Oil-Bearing Strata (Gostoptekhizdat, 1962).
4. L. I. Boimel'-Mekenitskaya, "A study of the film state of the combined water in petroleum reservoirs," Thesis (1957).
5. S. F. Vsevolodskii, Glassblowing (Izd. AN SSSR, 1952).
6. J. N. Hummel, Principles of Geoelectrical Prospecting [Russian translation] (Aznefteizdat, 1936).
7. B. V. Deryagin, ZhFKh, 2, 29 (1932).
8. B. V. Deryagin and E. Obukhov, Kolloidnyi zhurnal, 1, 385 (1935).
9. B. V. Deryagin and M. M. Kusakov, Izvestiya AN SSSR, Otdelenie matematiki i estestvennykh nauk, 5, 741 (1936); 5, 1119 (1937).
10. S. L. Zaks, Neft. khoz., 4 (1947).
11. S. L. Zaks, Izv. AN SSSR, ONT, 7, 787 (1947).
12. M. M. Kusakov and L. I. Mekenitskaya, "The thickness of thin layers of combined water," Trudy MNI, 16 (1956).
13. P. A. Rebinder and K. A. Pospelova, "Modern ideas on the stability, formation, and breakdown of emulsions and methods for their study," Introductory article to Clayton's book: The Theory of Emulsions and Their Technical Treatment [Russian translation] (IL, 1950).
14. A. B. Taubman and A. F. Koretskii, "Some characteristic features of the influence of dispersed solid phase on the emulsification and demulsification of petroleum products," Proceedings of the First All-Union Conference on the Use of Surface-Active Substances in the Petroleum Industry (Gostoptekhizdat, 1961).
15. R. I. Cherkasov, U. Shch. Kurbanaev, and G. A. Babalyan, "The stability of the structure of adsorption layers of Bashkir deposit petroleums at an oil-water interface" (this volume).

DEMULSIFYING POWER OF A SURFACTANT IN RELATION TO MOLECULAR STRUCTURE

D. N. Levchenko and N. M. Nikolaeva

Separation of Ethoxylated Fatty Acids By a Chromatographic Method

When a mixture of high-molecular-weight synthetic fatty acids is ethoxylated, a complex mixture of surface-active compounds is obtained.

In the present work, an attempt has been made to recover individual fractions of the compounds from a mixture of ethoxylated high-molecular-weight fatty acids and to determine their ability to demulsify hydrophobic petroleum emulsions.

To separate the mixtures into fractions, use was made of a chromatographic method on silica gel, based on the principle that differences in molecular structure of substances affect their adsorption affinity.

If a difference exists in the sorption interaction between the mixture components and the sorbent, then, when the mixture moves through the sorbent bed, it is separated into components; this is also the principle of the chromatographic method. For the desorption of groups of components or fractions, one must choose appropriate solvents as desorbents. The sorbent used in the present work was Type ASK silica gel of the Voskresensk Chemical Combine (after drying 5 h at 150°C).

The column was charged with silica gel in amount ten times the weight of the sample material. Before introducing the sample of surface-active substance (surfactant) into the column, the silica gel was wetted with isopentane. A 100-g sample of the surfactant was dissolved in benzene, and the solution was poured into the column containing the silica gel. The column was 2500 mm long and 20 mm in diameter. After complete adsorption of the solution by the silica gel, desorption of the individual components of the surfactant was effected by the following solvents: isopentane (or heptane), benzene, ethanol, acetone, and water. This order of solvents for desorption of the individual components of the nonionic surfactants was established by preliminary experiments. The first solvent removed hydrophobic substances from the silica gel, and the subsequent solvents removed fractions with increasing hydrophilicity. The end of desorption of a given fraction of the surfactant was determined by the refraction of the solution and by the residue on a watch glass after evaporation of the solution.

After driving off the solvent and drying, the resulting fractions of the substance were investigated and tested to determine their demulsifying ability.

Data are shown in Table 1 from the chromatographic separation of OZhK demulsifier, and in Table 2 from the chromatographic separation of a substance obtained by ethoxylation of oleic acid.

As is evident from Tables 1 and 2, a very large fraction of material was obtained on desorption with ethanol; this material evidently included esters of the fatty acids with long ethoxy chains. These substances have the highest demulsifying power.

The demulsifying power of the materials was determined under laboratory conditions by washing the crude petroleum with 10% distilled water, mixing in a shaker under identical conditions, and then allowing to stand in a water-bath thermostat at identical temperature; consumption of demulsifier was identical in all cases.

TABLE 1. Chromatographic Separation of OZhK Demulsifier

Fractions recovered by desorbents	Plant samples of OZhK demulsifer								Laboratory samples of OZhK demulsifier					
	Yield, %	Content of ethoxy groups, %	Demulsifying power, %	Surface tension of aqueous solution, ergs/cm²		Elemental composition, %			Yield, %	Content of ethoxy groups, %	Demulsifying power, %	Elemental composition, %		
				0.03%	0.005%	H	C	O				H	C	O
Original OZhK demulsifier	100	65.0	85	50.2	53.5	10.02	60.43	29.55	100	65.7	87	–	–	–
Fractions obtained on desorption:														
By isopentane	4.1	20.0	20	–	–	11.39	74.96	13.65	6.5	31.0	–	11.0	73.02	15.98
By benzene	18.0	38.1	21	48.5	66.6	11.32	70.7	17.98	19.0	35.0	47.0	11.1	69.64	19.26
By ethanol	65.5	74.15	86	49.1	53.2	9.9	58.0	32.1	60.7	68.8	94.0	9.92	59.52	30.56
By acetone	1.9	79.1	66	66.0	71.6	–	–	–	2.3	75.5	50.0	8.31	46.87	44.82
By water	5.2	89.4	3.4	70.1	72.3	5.16	27.62	67.22	10.0	–		–	–	–
Losses	5.3								2.5					

TABLE 2. Chromatographic Separation of Ethoxylated Oleic Acid

Fractions recovered by desorbents	Ethoxylated oleic acid					
	Yield fraction, %	Ethoxy groups, %	Surface tension of aqueous solution, ergs/cm²		Demulsifying power, %, with demulsifier consumption:	
			0.03%	0.005%	50 g/metric ton	100 g/metric ton
Starting material	100	81.4	45.2	48.3	40.0	75
Fractions obtained on desorption with:						
Isopentane	4.5	17.8	–	–	–	–
Benzene	4.8	50.6	–	–	–	–
Ethanol	62.0	78.2	45.47	52.54	64.0	91.0
Acetone	11.8	96.9	51.61	71.39	38.2	54.0
Water	7.2	97.0	72.6	73.2	4.6	2.5
Losses and residue on silica gel	9.5					

The quantity of water recovered after four hours of settling in the thermostat, expressed in percent of the quantity of water added, is arbitrarily termed the demulsifying power of the substance.

A small amount of material was recovered from the acetone solution; this material was close to the alcohol-recovered material in content of ethoxy groups but had relatively low surface activity and demulsifying power. In the case of the OZhK, all of the asphaltic-resinous substances fell into the acetone fraction; these were soluble in benzene only with difficulty.

The material going into the benzene solution was more hydrophobic than that in the alcohol; the benzene-desorbed material had short ethoxy chains (ethoxy number half that of the alcohol-recovered material) and lower demulsifying power. In the case of the ethoxylated oleic acid, the fraction recovered from benzene was very small. This may be explained on the basis that for each acid there is a definite rate of ethoxylation under given conditions. Therefore, a mixture of acids will give more substances differing in the number of ethoxy groups in the chain; an individual acid will give substances with chains containing an identical number of ethoxy groups.

The isopentane or heptane dissolves hydrocarbons and hydrophobic ethoxylated substances with a low content of ethoxy groups. The water desorbs polyoxyethylene glycols with high content of ethoxy groups, formed mainly on account of the water present in small amount in the acids.

These substances do not possess surface activity or demulsifying power and represent ballast [inert material] that consumes a large amount of the ethylene oxide during the synthesis of OZhK demulsifier.

Therefore, before ethoxylation, it is necessary to dry thoroughly the initial fatty acids, distilling off the residual water under vacuum.

By means of the chromatographic method, it is possible to recover from nonionic surfactants narrow fractions of substances having practically identical structure, and to investigate these fractions.

It may be considered that the chromatographic method can be used to examine the characteristics of composition of surfactants obtained by ethoxylation.

Demulsifying Abilities of Ethoxylated Stearic and Dihydroxystearic Acids

To establish the relation between demulsifying power of a substance and its structure and functional groups, an investigation was made of the polyoxyethylene glycol esters of stearic and hydroxystearic acids.

The materials selected for investigation were chemically pure stearic acid and 9,10-dihydroxystearic acid. The latter was synthesized by a method described in an article of D. Swern, J. T. Scanlan, and G. B. Dikel [1]. The synthesis was based on the interaction of oleic acid with formic acid and hydrogen peroxide according to the following reaction:

$$CH_3(CH_2)_7CH{=}CH(CH_2)_7COOH + H_2O_2 + HCOOH \rightarrow CH_3(CH_2)_7\underset{OH}{CH}{-}\underset{OCHO}{CH}(CH_2)_7COOH$$

$$CH_3(CH_2)_7\underset{OH}{CH}{-}\underset{OCHO}{CH}(CH_2)_7COOH + NaOH \rightarrow CH_3(CH_2)_7\underset{OH}{CH}{-}\underset{OH}{CH}{-}CH(CH_2)_7COONa + HCOONa$$

$$CH_3(CH_2)_7\underset{OH}{CH}{-}\underset{OH}{CH}(CH_2)_7COONa + HCl \rightarrow CH_3(CH_2)_7\underset{OH}{CH}{-}\underset{OH}{CH}(CH_2)_7COOH + NaCl$$

TABLE 3

Demulsifiers	No. of ethoxy groups	% ethoxy groups	HLB	Molecular weight (calc.)	Demulsifying power for emulsion of Romashkino crude, %, with demulsifier consumption		Demulsifying power for emulsion of Bavly crude, %, with demulsifier consumption		Surface tension of 0.05% solution vs. air, ergs/cm²
					50 g/metric ton	100 g/metric ton	50 g/metric ton	100 g/metric ton	
Polyethoxylated ester of stearic acid	11	64.0	12.7	769	27	51	40.5	25.3	57.3
	22	78.5	15.7	1254	25	18	43.0	24.6	57.5
	31	82.2	16.4	1644	51	50	52.7	43.8	49.6
	42	84.6	16.9	2134	37	52	41.0	43.4	45.1
	50	88.4	17.6	2484	32	40	56.4	41.5	47.4
Polyethoxylated ester of 9, 10-dihydroxystearic acid	10	57.5	11.5	758	44	40	56.2	51.0	44.5
	22	75.0	15.0	1410	48	72	57.0	54.0	40.5
	31	86.0	17.2	1679	34	44	61.0	52.4	46.5
	42	89.5	17.9	2186	49	52	50.0	40.6	45.2
	52	91.0	18.1	2599	64	51	63.5	58.5	43.9
Polyethoxylated esters of stearin	12	28.2	5.65	798	7.5	39	40	64	38.8
	25	66	14.0	1370	32	48	36.5	70	37.6
	31	76.0	15.1	1632	39	48	38.0	59.3	43.9
	40	81.0	16.2	2030	59	61	42.4	47.0	49.8
	49	90.0	18.0	2420	47	58	38.2	55.5	46.4
OZhK	22	67.2	13.5	1420	85	96	96	98	50.1
OP-10	10	63	12.5	646	50	56	35.6	45.6	43.1

TABLE 4

Demulsifiers	% ethoxy groups	HLB	Demsulsifying power, %, with demulsifier consumption	
			50 g/metric ton	100 g/metric ton
Polyethylene glycol ester of 9,10-dihydroxy-stearic acid with 10 molecules of ethylene oxide	52.02	10.4	56.2	51.0
Same, after saponification	26.58	5.3	64.0	51.5
Polyethylene glycol ester of 9,10-dihydroxy-stearic acid with 31 molecules of ethylene oxide	84.64	16.9	61.0	52.4
Same, after saponification	48.1	9.6	60.0	47.4

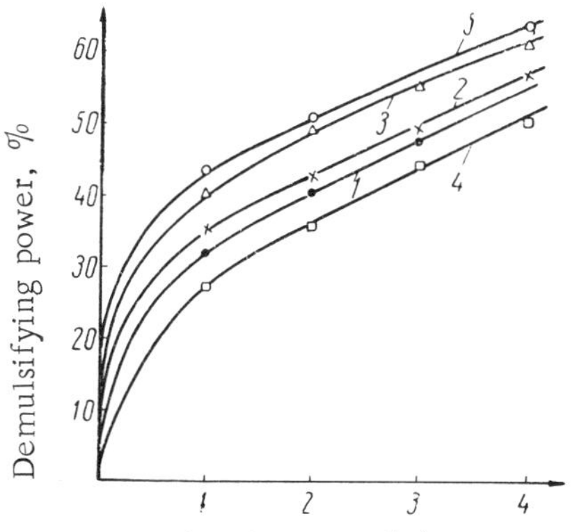

Demulsifying power as a function of settling time for esters of 9,10-dihydroxystearic acid containing different numbers of ethylene oxide groups. For esters of 9,10-dihydroxystearic acids: 1) with 10 groups; 2) with 22 groups; 3) with 31 groups; 4) with 42 groups; 5) with 50 groups.

The melting point of the acid obtained was 92°C.

The polyethylene glycol esters of stearic and hydroxystearic acids were obtained by condensation with ethylene oxide. Alkali was used as the catalyst; the temperature of ethoxylation was 180°C.

With the esters that were obtained, demulsifying power was determined on emulsions of Romashkino and Bavly crudes. The results are shown in Table 3.

From these data it follows that the esters of 9,10-dihydroxystearic acid have better demulsifying ability than the esters of stearic acid; i.e., with the appearance of branching in the fatty acid molecule, its demulsifying ability is improved. The demulsifying power of the esters of both stearic acid and of 9,10-dihydroxystearic acid are increased with increase in the number of ethoxy groups; however, on condensation with 40 molecules of ethylene oxide, the demulsifying power decreases slightly and then again begins to increase [with further addition of ethylene oxide]. The efficiency of esters condensed with 31 molecules of ethylene oxide is slightly inferior to that of those condensed with 50 molecules (see the figure).

There are indications in the literature [2] that when a hydroxystearic acid is ethoxylated the ethylene oxide adds only at the carboxyl group, and the hydroxyl group does not participate in the reaction.

To investigate the possibility of ethoxylation not only of the carboxyl groups, but also the hydroxyl groups contained in hydroxy acids, we carried out the saponification of the esters obtained from 9,10-dihydroxystearic acid, using alcoholic caustic solution. The polyoxyethylene esters of stearic acid, as true esters, are completely saponified by alcoholic caustic solution, whereas ethers are not saponified by caustic.

Saponification was performed on the esters of 9,10-dihydroxystearic acid condensed with 10 and 31 molecules of ethylene oxide. The saponification was effected with 3 N alcoholic KOH solution, boiling 10 h on a water bath. Then the reaction mixture was treated with 3 N HCl, and the liberated acid was separated, purified, and dried. The products that were recovered differed in appearance from the original hydroxy acid. They were brownish in color and did not crystallize. The esters condensed with 31 molecules of ethylene oxide were readily soluble in water; the esters with 10 molecules of ethylene oxide were less soluble. On the polyethoxylated esters of 9,10-dihydroxystearic acid and the products obtained after saponification, the percentage of groups (method of S. Siggia) and the demulsifying power were determined. The results are shown in Table 4.

From Table 4 it is evident that, in the case of dihydroxystearic acid, the condensation with ethylene oxide proceeds at both the carboxyl and the hydroxyl groups; here about 50% of the ethylene oxide adds at the carboxyl group and about the same amount at the hydroxyl groups.

It is interesting to note that a twofold reduction in the number of ethoxy groups (by saponification with caustic) does not change the demulsifying power of the product.

Hence, a surfactant obtained by ethoxylation of only the hydroxyl radicals of dihydroxystearic acid, with a free carboxyl radical, has the same demulsifying power as substances containing twice as much ethylene oxide with both hydroxyl and carboxyl radicals ethoxylated to give compounds of the following type:

$$\begin{array}{l} CH_3(CH_2)_7\,\underset{\displaystyle H(C_2H_4O)_aO}{\underset{|}{CH}}-\underset{\displaystyle O(C_2H_4O)_bH}{\underset{|}{CH}}\,(CH_2)_7COO(C_2H_4O)_nH \end{array}$$

Literature Cited

1. D. Swern, J. T. Scanlan, and G. B. Dikel, Synthesis of Organic Waxes [Russian translation] (IL, 1961).
2. J. Am. Oil Chemists' Soc., 31, 1 (1954).

INVESTIGATION OF POLYGLYCOL ETHERS OF ALKYLPHENOLS AS DEMULSIFIERS FOR PETROLEUM EMULSIONS

A. A. Petrov and Yu. S. Smirnov

The search for new demulsifying reagents can be subdivided into two basic directions. The first of these consists of tests under laboratory and field conditions of known surface-active substances (surfactants) obtained in industry or in laboratories especially for demulsification and also for other purposes, from raw material that is available in the present national economy or that represents mostly a waste from any production process.

However, a substance thus obtained will hardly prove to be the most effective demulsifier of a given class of compounds, and raw material from wastes cannot provide constancy of the demulsifying reagent.

The second direction consists of a detailed study of the effect of molecular structure and demulsifying power of a surfactant belonging to a given class of compounds. In this case, a general correlation is established, relating the structure of the hydrophilic and hydrophobic portions of the molecule to its demulsifying activity.

Such a correlation can reveal the most effective group of substances belonging to a given class of compounds, and, knowing their structure, the raw material can be selected and the production technology developed. This approach is a longer process, but it ensures the production of a surfactant with high demulsifying power and defines the required quality of raw material and the technology for producing this surfactant as the desired end-product.

At the present time, there have been proposed a number of more active demulsifying reagents of the polyglycol ether type (1 – 4), which can be applied successfully in the petroleum industry.

However, these compounds are still insufficiently effective as demulsifying reagents; therefore, further work is necessary in searching for reagents with higher activity, as can be accomplished by systematic searches.

At the same time, still another question can be partly resolved.

At present, there has been very little study of the mechanism of chemical breaking of petroleum emulsions, and there is no method for determining the nature of structure of the protective layers on the drops of emulsified water. However, if it is established that effective breaking of a petroleum emulsion can be performed by a reagent with well-known structure and properties, then the predominant action of the reagent as a surfactant can be used to judge the structure of the protective layers in the mechanism of their destruction.

In the present work, polyglycol ethers of alkylphenols were investigated systematically as to the effect of surfactant molecular structure on demulsifying power. The choice of this class of compounds was related in the first place to the fact that among these compounds one may expect representatives with high demulsifying power; in the second place, in these compounds, the size of the hydrophobic part can be changed easily by changing the alkyl radical, and the size of the hydrophilic part by changing the length of the polyglycol chain. As a result, it is possible to obtain compounds of a single class having a wide range of surface properties. Studies on the relation between demulsifying power of the reagent and its structure have also been conducted previously.

Characteristics of

Name of alkylphenol	Boiling point, °C		Elemental composition, %			
	Found	Literature data	Found		Calculated	
			C	H	C	H
p-sec-Butylphenol. . . .	239-240	230-240	79.40	9.38	79.95	9.39
p-tert Butylphenol. . . .	236-237.5	235-240	79.61-80.02	9.35-9.41	79.95	9.39
di-sec-Butylphenol . . .	271-274	–	81.63	10.17	81.50	10.74
p-sec-Nonylphenol . . .	150-152*	–	81.29	11.12	81.76	10.98
p-sec-Dodecylphenol. .	165-168*	–	81.93	11.43	82.44	11.44
p-sec-Octadecylphenol.	210-216*	–				

*At pressure of 3-4 mm Hg abs.

Labbe [15] studied the demulsifying properties of surfactants containing different numbers of phenol nucleic in the hydrophobic part of the molecule and polyglycol chains of different length in the hydrophilic part. He demonstrated that there is a relation between demulsifying power of the product and its chemical composition.

D. N. Levchenko [6] used for the hydrophobic portion of the surfactant molecule a mixture of wastes from the production of DBPC antioxidant [2,6-di-tert-butyl-4-methylphenol], consisting of the monocresol fraction and the still bottoms. From these he obtained by ethoxylation a whole series of products differing from each other in the length of the polyglycol chain. He showed that in the desalting of Romashkino crude the most effective product proved to be that containing 25-30 glycol groups per mole of phenol.

Interesting studies were conducted in the laboratory of the Chernikovsk NPZ [Petroleum Refinery] by Ya. I. Nel'kenbaum [8]. A series of demulsifying reagents was synthesized there. The most effective proved to be a reagent obtained by ethoxylation of the condensation product of a fatty acid monoglyceride with phthalic anhydride. This material contained two polyglycol chains located at a considerable distance from each other. Nel'kenbaum proposed that the presence of two or more hydrophilic groups in a molecule, located at some distance from each other, permits the reagent to occupy at the phase boundary a considerably greater section of the surface, and that this apparently is responsible for its greater effectiveness.

A. G. Gobzhila and F. S. Popesku [1], on samples of polyglycol ethers of alkyl- and dialkylphenols, showed that products containing a shorter alkyl chain (C_8) are more effective reagents than those with a longer alkyl chain (C_{12}). According to these same data, derivatives of alkylcresols were more active than derivatives of alkylphenols; this was explained by the more highly branched structure of the alkylcresols.

All of the enumerated investigators, except Labbe, made their ethoxylated products not from individual compounds but from mixtures for which the composition of the starting materials was not always known accurately. As a result, the ethoxylation gave a complex mixture of surfactants including components differing from each other in size and structure of the hydrophobic and hydrophilic portions of the molecule. Therefore, to speak of the effect of structure of these compounds on their demulsifying power can be only very arbitrary.

In Giprovostokneft, over a number of years, studies are being carried out on the effect of structure of ethoxylated products on their demulsifying power. To carry out this work in the institute, a series of polyglycol ethers of alkylphenols was synthesized; these differed from each other in the size and structure of the hydrophobic and hydrophilic portions of the molecule. Alkylphenols were used as the hydrophobic part of the molecule. The alkylphenols were obtained by the method of V. I. Isagulyants [4, 5], which involves condensation of normal primary alcohols with a synthetic phenol in the presence of ion exchange resin KU-2 as a catalyst.

The following alkylphenols were synthesized: p-sec-butylphenol, p-tert-butylphenol, p-sec-nonylphenol, p-sec-dodecylphenol, and p-sec-octadecylphenol.

Synthesized Alkylphenols

Mol. wt.		Content of hydroxy, %		Refractive index		Density		Molar refraction	
Found	Calc.	Found	Calc.	Found	Calc.	Found	Calc.	Calc. per Lorentz-Lorenz	Calc. per Eisenlohr
149	150	11.8	11.32	1.5189	1.5182	0.9746	0.9860	46.37	46.30
149-151	150	11.51	11.32	—	—	—	—	—	—
207	206.3	8.21	8.24	1.5140	—	0.9605	—	64.9	64.75
221	220	7.62	7.72	1.4970	—	0.9273	—	69.7	69.40
262	263	6.21	6.50	1.4979	—	0.9208	—	82.24	83.25
346-349	246.6	4.8	4.9						

The characteristics of the alkylphenols obtained are given in the table. The following were determined on the synthesized alkylphenols: boiling point, elemental composition, molecular weight by cryoscopy in benzene, content of hydroxyl (per Tserevetinov), refractive index, and density.

The molar refraction was calculated by the Lorentz-Lorenz formula. The values obtained agreed with the values calculated from the atomic refractions given in the tables of Eisenlohr [3].

The physicochemical characteristics of the products obtained are in satisfactory agreement with literature or calculated data for these products.

The polyglycol ethers of the alkylphenols were obtained by two methods. The ethers with more than six glycol groups were synthesized by condensation of the alkylphenols with ethylene oxide in the presence of caustic. The ethers with less than six glycol groups were obtained according to Hibbert [16] by condensation of the β-chloroethyl ether of the alkylphenol with the monosodium salt of a suitable polyethylene glycol. The β-chloroethyl ether of the alkylphenol was obtained through the β-chloroethyl ester of p-toluenesulfonic acid according to Clemo and Perkin [10].

As is well known, the polyglycol ethers of alkylphenols obtained by condensing an alkylphenol with ethylene oxide are not individual compounds; they represent mixtures of substances with different polyglycol chain lengths.

Flory [11], starting from a Poisson distribution and assuming that the condensation reaction of ethylene oxide proceeds by successive addition of monomer molecules to the polymer that is being formed, gave a formula for calculating the distribution of the weight and molar fractions of polyglycol ethers in a mixture according to the number of added moles of ethylene oxide.

Kelly and Greenwald [14] carried out a chromatographic separation of polyglycol ethers of octylphenol with average content of ethylene oxide 9.7 moles per mole of alkylphenol; they showed that the weight distribution of the polymers in the mixture agreed satisfactorily with the values calculated by the Flory formula.

In the present work, ethers were synthesized with average number of glycol groups from 2 to 100.

Using the Flory formula to calculate the weight content of the individual components in the mixture obtained by condensation of ethylene oxide with an alkylphenol, it can be shown that the principal amount of the mixture consists of polyglycol ethers differing from each other by a small number of glycol groups.

Thus, for example, for the ether of nonylphenol with average number of glycol groups equal to 9, about 70 wt. % of the total material in the mixture consists of compounds with chain length from 8 to 13 glycol groups; for the ether of nonylphenol with average content of glycol groups equal to 29, this same percentage (70%) of the total material consists of compounds having from 26 to 35 glycol groups.

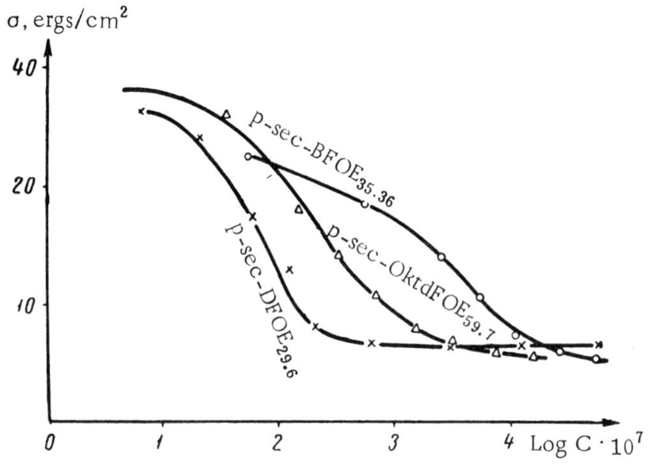

Fig. 1. Isotherms of surface tension for solutions of polyglycol ethers of alkylphenols at the interface with Zhigulevsk crude at 20°C.

For a characterization of the synthesized polyglycol ethers of alkylphenols, the following were determined: percentage content of ethylene oxide (glycol groups), cloud point, and surface tension. Based on the amount of ethylene oxide condensed, the molecular weight and the hydrophilic-lipophilic balance (according to Griffin) were calculated.

The percentage content of the glycol groups in the products obtained was determined by the iodometric method of Siggia [17] and according to the increase in weight of the reaction mass. Both methods gave the same results.

The cloud point of the synthesized polyglycol ethers of alkylphenols was determined in accordance with GOST 8435-57 for the auxiliary substances OP-10 and OP-7. The cloud point of a solution of a polyglycol ether is one of the indices of the hydrophilic-lipophilic balance of a substance. However, the cloud point determined in accordance with GOST 8435-57 cannot by any means characterize all substances. Compounds with marked hydrophilic properties have a cloud point that often proves to be higher than the boiling point of a solution of this substance. Compounds with marked hydrophobic properties, being insoluble in water, eliminate the possibility of determining the cloud point.

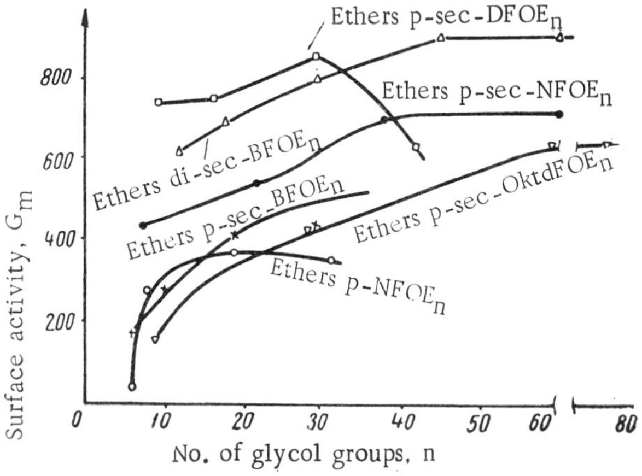

Fig. 2. Surface activity G (ergs · liters/cm^2 · mg) of polyglycol ethers of alkylphenols as a function of the number of glycol groups in the ester molecule, at the phase boundary with Zhigulevsk crude.

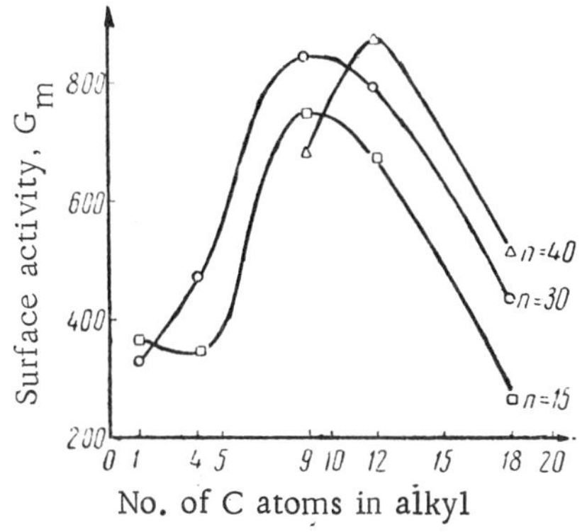

Fig. 3. Surface activity G(ergs · liters/cm^2· mg) of polyglycol ethers of alkylphenols as a function of number of carbon atoms in the alkyl, at the phase boundary with Zhigulevsk crude, with 40, 30, and 15 glycol groups in the molecule.

The surface tension of the solutions was determined at the interface with Zhigulevsk Devonian crude by the suspended-drop method of Harkins and Brown. The measurements were made at 20°C, with a drop formation time of 120 sec.

The surface tension isotherms for the polyglycol ethers of alkylphenols on semilogarithmic coordinates σ vs log C have a characteristic S-shaped form. Some of these curves are shown in Fig. 1. Such a form of the isotherm is related to the concentrations of micelle formation, which possibly is caused by the inhomogeneous composition of the ether.

For characterizing the ability of a surfactant to adsorb at a phase boundary, Academician P. A. Rebinder [9] has proposed the quantity $G = -d\sigma/dc$ and has termed it the "surface activity."

The value of surface activity of polyglycol ethers of alkylphenols in the concentration range from 0 to 0.2 mg/liter is shown in Fig. 2 as a function of the number of glycol groups in the ether molecule.

The range of concentrations within which the surface activity values were determined was chosen below the critical micelle concentration. This was based on the following considerations: with the formation of micelles in solutions of nonionic surfactants, as the total concentration is increased, the concentration of free molecules first increases slowly and then remains practically constant. The surface tension also changes correspondingly with this change in concentration of free molecules. As is evident from the curves of Fig. 1, the surface tension isotherms first flatten out and then go parallel to the concentration axis. Since micelle formation for different polyglycol ethers of alkylphenols sets in at different concentrations, an arbitrary selection of the concentration range in which the surface activity G_m is determined may result, for some ethers, in the onset of micelle formation, which will lead to unduly low values of G_m.

As is evident from Fig. 2, the value of surface activity (and hence also the capability of adsorption at the petroleum-water interface) for polyglycol ethers of alkylphenols, as the number of glycol groups is increased, shows at first an increase, then a leveling out and even a slight tendency to decrease. Such a nature of the change in surface activity is evidently caused by the increase in hydrophilicity of the polyglycol ether as the number of glycol groups in the ether is increased. On reaching its optimum value that is characteristic for each hydrophobic radical, the surface activity of the ether becomes maximum. With further increase in hydrophilicity, the activity changes only slightly, with a tendency to lower values appearing here.

If one examines the change in surface activity of the polyglycol ethers of alkylphenols as a function of the alkylating radical with constant number of glycol groups in the molecule, the curves in Fig. 3 show that there is an optimum value of radical size that corresponds to maximum surface activity. It is interesting to note that, with increasing number of glycol groups, the maximum in activity is shifted in the direction of ethers with larger size of alkylating radical.

The existence of an optimum size of the hydrophobic portion of the molecule for obtaining ethers with maximum surface activity can be explained on the basis that, with a significant increase in size of the hydrophobic portion, the required hydrophilicity can be attained only by the introduction into the molecule of an extremely large number of glycol groups. The increase in size of the resulting molecule is connected with a decrease in its structural assymmetry and hence with a reduction in adsorption ability and surface activity.

Thus, for polyglycol ethers of alkylphenols, optimum sizes exist both for the alkyl radical (9-12 carbon atoms) and for the length of the polyglycol chain (30-40 CH_2-CH_2O groups), at which their adsorption capability at the water–petroleum interface reaches a maximum value.

It is interesting to note that surface sctivity is reduced by the introduction of two alkyls instead of one, the total number of carbon atoms being the same. This is confirmed by the position of the surface activity curves in Fig. 2 for the polyglycol ethers of dibutylphenol and nonylphenol. The curve for the ethers of dibutylphenol, containing eight carbon atoms in two alkylating radicals, passes below the curve for ethers of nonylphenol, containing one alkyl radical with nine carbon atoms.

Thus, there have been synthesized a series of polyglycol ethers of alkylphenols, the physicochemical properties of which vary over wide limits.

This wide variation evidently makes it possible to select from this series of compounds the most active demulsifier for the given petroleum emulsion and, on the basis of a study of the properties of this demulsifier, to evaluate approximately the structure of the protective layers stabilizing the emulsion in question.

The investigation of the demulsifying power of the synthesized polyglycol ethers of alkylphenols was carried out on emulsions prepared under standard conditions from petroleum and formation water of the Devonian horizon of the Zhigulevsk oil field. The water content of the emulsion was 20%. The prepared emulsions were held for one day before the experiment so that the strengthening of the protective layers on the drops of emulsified water as a result of emulsion "aging" would be identical in all experiments. The emulsified water existed to the extent of 86% in the form of globules of radius less than 4 μ (analysis according to Figurovskii).

The polyglycol alkylphenol ethers in suitable quantities governed by the dosage adopted in the given experiment, in the form of a solution in 5 ml of water, were added to 150 g of the emulsion.

Mixing of the reagent with the emulsion was carried out in a mechanical mixer of the "cement mixer" type for ten minutes. After mixing, the emulsion samples with the reagent were placed in a water bath heated to 60°C and held four hours at this temperature. After standing, the petroleum was separated off from the freed water, and the petroleum was analyzed for residual water content in Dean and Stark apparatus. The identity of the "stability" of the emulsion was checked by means of OP-10 reagent.

The obtained results were expressed in the form of dehydration curves on coordinates of percent residual water vs reagent consumption in kg/metric ton of crude petroleum. On the basis of the dehydration curves, a critical specific consumption was determined for each polyglycol alkylphenol ether; this is defined as the specific consumption required to give a sufficiently complete breaking of the petroleum emulsion (residual water content below 2%).

The demulsifying activity of surfactants is more conveniently expressed as the quantity of crude petroleum that may be treated with one weight unit of reagent.

The effect of the hydrophobic and hydrophilic portions of polyglycol alkylphenol ethers on their demulsifying power is shown in Fig. 4, giving curves of demulsifying activity of the synthesized alkylphenol ethers as a function of the number of glycol groups in the molecule. For the polyglycol alkylphenol ethers there is a characteristic initial increase in demulsifying power in parallel with an increase in the number of glycol groups in the hydrophilic part of the molecule; then the curve passes through a maximum and, with further increase in the number of glycol groups, the demulsifying power decreases. As the size of the alkyl chain of the alkylphenol increases, the maximum demulsifying power is shifted in the direction of increasing number of glycol groups.

For a comparison of the demulsifying power of the synthesized reagents with that of commercially produced reagents, straight lines are plotted on this same graph showing the demulsifying power of reagents OP-10, NChK, OZhK, and foreign [non-Soviet] reagents Separol 221 and Disolvan 4411.

From a comparison of these data, it is evident that the demulsifying power of the synthesized reagents is comparable with the effectiveness of the foreign [non-Soviet] reagents Separol 221 and Disolvan 4411, and some of the synthesized reagents even surpass these materials. The reagents synthesized in the laboratory exceed significantly in effectiveness the reagents OP-10, OZhK, and NChK.

When using the polyglycol alkylphenol ethers in breaking emulsions of Devonian crude from the Zhigulevsk oil field, the maximum demulsifying power is obtained with those ethers in which the ratio of number of glycol groups to number of carbon atoms of the alkyl varies from 3 to 5.

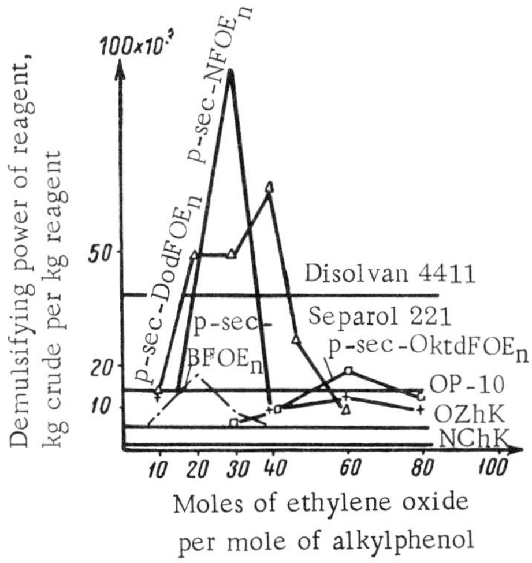

Fig. 4. Demulsifying power of polyglycol alkylphenol ethers as a function of the number of glycol groups.

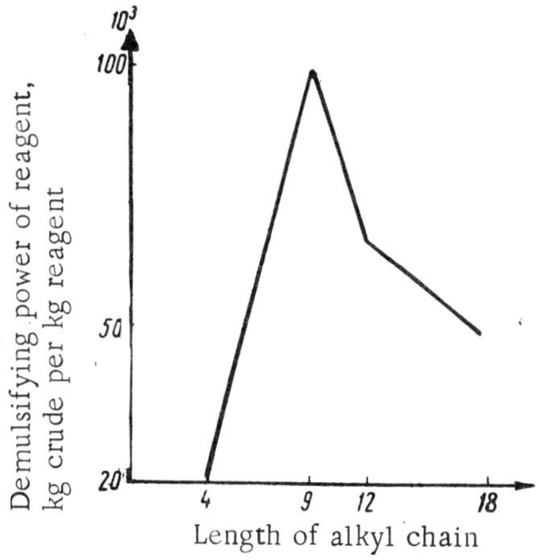

Fig. 5. Demulsifying power of polyglycol ethers of alkylphenols as a function of length of alkyl radical.

This ratio approaches the lower limit as the hydrocarbon radical of the hydrophobic portion of the molecule becomes larger; and, conversely, this ratio approaches the upper limit as the number of carbon atoms is decreased.

Thus, for butylphenol ethers, this ratio is equal to 5 and is independent of the structure of the alkyl chain of the hydrophobic radical. Both for the ethers of p-tert-butylphenol and of p-sec-butylphenol, the most effective compound proved to be the condensation product with 20 moles of ethylene oxide. For the compounds of higher molecular weight of the hydrophobic portion, the most effective proved to be those compounds in which the ratio of number of glycol groups of the ether to the number of carbon atoms of the alkyl chain approached 3.

The maximum demulsifying power in the series of polyglycol alkylphenol ethers was shown by the laboratory-synthesized polyglycol ether of p-sec-nonylphenol containing 29 glycol groups.

To clarify the effect of alkyl chain length, the curve in Fig. 5 shows the demulsifying power as a function of alkyl chain length for those polyglycol ethers in which the ratio of number of glycol groups to the number of alkyl carbon atoms varies from 3 to 5. From the graph it is evident that the demulsifying power of the polyglycol alkylphenol ethers increases with increasing size of the alkyl chain, reaches a maximum, and then decreases.

Thus, it may be said that within the class of polyglycol alkylphenol ethers the maximum demulsifying power is observed for compounds with hydrocarbon radicals and polyglycol chains in a ratio that lies within certain rather narrow limits.

The identical nature of the variation in demulsifying power and in surface activity of the polyglycol alkylphenol ethers in relation to Zhigulevsk crude is evidently explained on the basis that the protective layers in the emulsion of this crude are formed primarily as a result of adsorption of the natural surface-active emulsifiers of the crude. Therefore, an emulsion of Zhigulevsk crude is broken on account of adsorptive displacement of such emulsifiers from the petroleum-water interface by the more highly surface-active polyglycol ethers of alkylphenols. The intensity of this process, as follows from its nature, will depend to a considerable extent on the surface activity of the demulsifying reagent.

From the data obtained, it follows that the demulsifying power of polyglycol alkylphenol ethers depends on the ratio between the polar and nonpolar parts of the ether molecule, i.e., on its hydrophilic-lipophilic balance (HLB).

The qualitative change in the HLB with a change in the polar or nonpolar portion of the ether molecule can be determined readily; however, there is not yet any generally accepted expression for quantitative char-

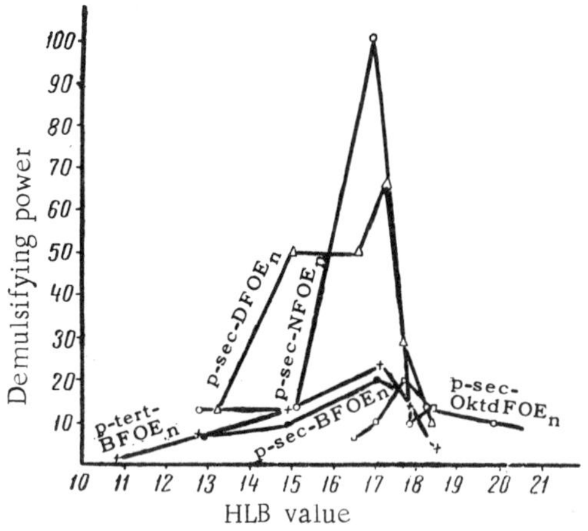

Fig. 6. Demulsifying power of polyglycol ethers of alkylphenols as a function of HLB.

acterization of the HLB. At the present time, the most suitable method for quantitative characterization of the HLB is the method proposed by Griffin [12].

Griffin proposed empirical formulas [13] for calculating the HLB value for certain classes of compounds. In particular, for ethoxylated alcohols and phenols, this formula has the form:

$$\text{HLB} = \frac{\%\ \text{ethoxy}}{5}$$

It is interesting to note that Griffin, on the basis of a large number of experiments, established certain limits of HLB value for different substances having the same type of action. Thus, emulsifiers for water-in-oil type emulsions have HLB 4-6, wetting agents 7-9, detergents 13-15, and emulsifiers of the oil-in-water type 8-18.

The relation between demulsifying power of polyglycol alkylphenol ethers and the HLB value calculated from the above formula is shown in Fig. 6. These curves demonstrate that ethers with HLB values in the range 16-18 have the maximum demulsifying power.

However, although the maximum demulsifying power of the polyglycol ethers of a single alkylphenol occurs at an HLB value of 16-18, the value of the maximum itself may be very different for different alkylphenols. From the practical point of view, a given series of ethers, even with "maximum" demulsifying power, may not be of any interest as demulsifiers.

An example of this situation occurs with the polyglycol ethers of p-methylphenol and p-butylphenol, with respective demulsifying powers of 10,000 and 20,000 with HLB values of 18.5 and 17.0. By comparison, the maximum demulsifying powers of nonylphenol and dodecylphenol ethers with the same HLB values are respectively 100,000 and 66,700, i.e., 5 to 7 times higher.

Thus, the HLB value of Griffin, even though it is within the required interval (16-18), still cannot serve as an index of high efficiency in a demulsifying reagent. Such an HLB value is evidently only a necessary condition but still insufficient to ensure that a substance will possess high demulsifying power and will be suitable for practical application.

Literature Cited

1. A. G. Gobzhila and F. S. Popesku, Khim. i tekhnol. topliv i masel, 12, 21 (1961).
2. N. D. Gudaskina, D. M. Rudakovskii, and E. L. Yarzhemskaya, ZhPKh, 30, 1 (1957).
3. B. V. Ioffe, Refractometric Methods in Chemistry (Goskhimizdat, 1960).
4. V. I. Isagulyants and V. N. Tishkova, Chemical Processing of Petroleum Hydrocarbon, Collection (Izd. AN SSSR, 1956).
5. V. I. Isagulyants and V. N. Tishkova, "Problems in petroleum and gas," Trudy MINKh i GP, No. 24 (1959).
6. D. N. Levchenko, A. D. Khudyakova, et al., Khim. i tekhnol. topliv i masel, 4, 24 (1960).
7. A. A. Petrov and N. P. Borisova, Test results on reagent OP-10 for pretreatment of crude oil before refining, Trudy Giprovostoknefti, No. 2 (Gostoptekhizdat, 1959).
8. Ya. G. Sorkin, Ya. I. Nel'kenbaum, and F. A. Mamina, Khim. i tekhnol. topliv i masel, 2, 28 (1961).
9. P. A. Rebinder, Surface phenomena, adsorption, and properties of adsorption layers, Chapter IX in book of V. Naumov: Colloid Chemistry (Goskhimizdat, Leningrad, 1932).
10. G. R. Clemo and W. H. Perkin, J. Chem. Soc., 122, 176 (1922).
11. P. J. Flory, J. Am. Chem. Soc., 62, 6 (1958).
12. J. Griffin, J. Soc. Cosmetic Chemists, 1, 5, 311-326 (1949).
13. J. Griffin, J. Soc. Cosmetic Chemists, 5, 249 (1954).

14. J. Kelly and H. Greenwald, J. Phys. Chem., 62, 9, 1096-1098 (1958).
15. C. Labbe, Demulsification of Crude Petroleums, Vol. 5 (1956), 584-598.
16. S. Z. Perry and H. Hibbert, Canadian J. Research, 14, Sec. B., 3 (1936).
17. Siggia, et al., Anal. Chem., 30, 115 (1958).

THE STRUCTURAL STRENGTH OF ADSORPTION LAYERS OF CRUDE PETROLEUMS OF THE BASHKIR OIL FIELDS

R. I. Cherkasov, U. Sh. Kurbanaev, and G. A. Babalyan

Introduction

As is well known, the majority of petroleum emulsions possess high stability. The stability is determined mainly by the structural-mechanical properties of the adsorption layers [1-5]. In the studies of P. A. Rebinder andco-workers [2], an investigation was made of the structural-mechanical properties of the adsorption layers of various disperse and colloidal systems in the field of unbroken structures. P. A. Rebinder posed the problem of establishing the smallest number of invariant quantities that would characterize adequately the deformation and relaxation properties of these systems. For this purpose, primary attention was turned to experimentally determined families of curves of deformations of pure shear ε vs. time τ, giving the increase in shear with respect to time under the action of a constant stress P.

As a result, by means of the simplest equations derived from an analysis of simple mechanical models, it proved possible to characterize the systems under investigation by the smallest number of invariant constants and also to predict semiquantitatively the nature of the deformation and relaxation processes in such systems, having extended the phenomena of elastic relaxation to typical coagulation-colloid systems, and, finally, in the simplest case, to describe quantitatively the development of processes of deformation and relaxation.

Here it is necessary to turn attention to the fact that the choice of the field of unbroken structures was occasioned by the basic direction of the work of P. A. Rebinder on structure formation, work on investigating the stability of disperse and colloidal systems.

The strength properties of the structures of these systems, phenomena developing at the approach to breakdown, fall outside the field of application of the given method, i.e., outside the field of unbroken structures [1, 2].

However, it must be noted that the field of structure breaking is inherently related to the problem of breaking colloidal and disperse systems and hence with the breaking of petroleum emulsions.

Developing the ideas of P. A. Rebinder on the strength properties of colloidal structures, A. A. Trapeznikov [5] investigated the structural-mechanical properties of the adsorption layers also in the field of structure breaking; he established a marked dependence between the ultimate strength P_r and the rate of application of the load for adsorption layers at the petroleum-water boundary.

The ultimate strength of the adsorption layer is one of the important parameters determining the stability of emulsions.

Experimental Methods

The investigations were carried out in a modified SNS-2 apparatus used for determining the static shear stress of clay suspensions. The apparatus (Fig. 1) is a variant of the Shvedov apparatus, in which the loads that develop are determined by the elastic properties of the filament suspension of the disc. The disc 2 transmits the force arising on turning the container 3, through the adsorption layer at the petroleum–water interface, to the elastic filament 1. The disc is made of stainless steel. The lateral surfaces are polished. The mirror 4 is

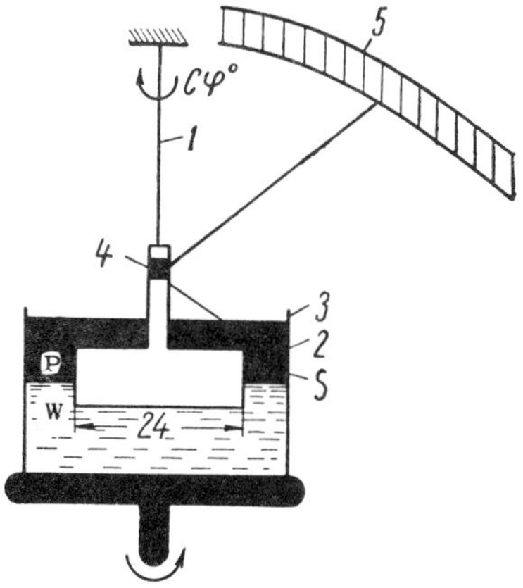

Fig. 1. Basic scheme of apparatus for measuring the ultimate strength of an adsorption layer. 1) Elastic filament; 2) disc; 3) container; 4) mirror; 5) scale.

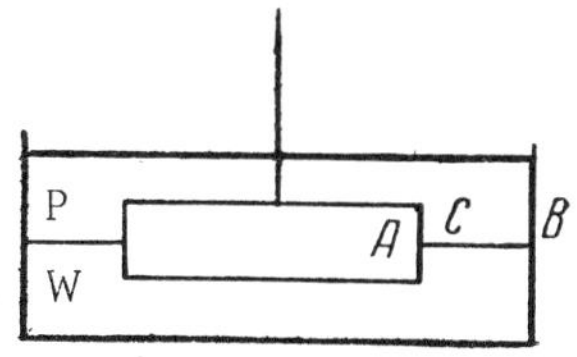

Fig. 2. Adsorption regions.

mounted on the disc for registering the turning angle; the mirror reflects a beam from the light source S to the centimeter-graduated scale 5. The elastic filament 1 is made of nichrome wire of 0.1 or 0.2 mm diameter and 22 cm long. The entire apparatus is thermostated in an air chamber and mounted on a heavy table to eliminate the effect of external vibrations. The experiments were conducted at a constant temperature of 20°C.

The container 3 is rotated until breakdown of the adsorption layer occurs. At the moment of breaking, the movement of the disc is stopped, and it begins to move in the reverse direction.

The limiting value of the angle of rotation $\varphi°$ will determine the ultimate strength of the adsorption layer, P_r.

Let us use M to designate the turning moment applied to the disc 2, α to designate the rotation angle of the light beam reflected from the mirror on rotation of the disc through the angle $\varphi°$, and C to designate a constant of the elastic filament 1. Then the counteracting force $C\varphi°$ arising in the elastic filament 1 will be determined by the moment M, and the stress arising at the perimeter of the disc 2 for the system disc–water–petroleum will be determined by the value M/r referred to the perimeter $2\pi r$.

In other words,

$$M\text{[dynes/cm]} = C\varphi°; \quad P_r = \frac{C\varphi°}{r}\,\frac{1}{2\pi r} = \frac{C\varphi°}{2\pi r^2}\text{[dynes/cm]}\,.$$

It is also evident that $\varphi = \alpha/r$; hence,

$$P_r = \frac{C\alpha}{4\pi r^2}\,.$$

The angle α can be expressed in terms of an arc segment γ, knowing the distance R from the mirror 4 to the scale 5:

$$\alpha = \frac{180\gamma}{\pi R}\,,$$

where γ and R are expressed in cm.

Then

$$P_r = \frac{C}{4\pi r^2}\,\frac{180\gamma}{\pi R} = \frac{45C}{\pi^2 r^2 R}\,\gamma.$$

Taking $K = (45C)/(\pi^2 r^2 R)$ as a new constant, the value of which is determined primarily by the filament constant C, we find that the value of P_r is a linear function of the value of γ.

Thus,

$$P_r = K\gamma.$$

The filament constant was determined by well-known methods used in similar cases [8].

Let I_k be the moment of inertia of a plane ring applied to the disc. Let us designate the radii of this ring (external and internal) by r_2 and r_1, respectively.

Also, let I_0 and I be respectively the moments of inertia of the disc 2 before and after application to the disc of the ring with mass M_k, and T_0 and T the corresponding periods of torsional oscillation.

TABLE 1

Material of disc	Arlanskii crude vs distilled water		
	Dia. of glass container, cm	P_r, dyn/cm	Note
Stainless steel	6.5	0.218	Average of three experiments
Same	9.1	0.210	
Stainless steel coated with methacrylate plastic	9.1	0.211	

TABLE 2

Material of disc	Chekmagush crude vs distilled water		
	Material of container (diameter constant)	P_r, dyn/cm	Note
Stainless steel	Methacrylate plastic	0.412	Average of three experiments
Stainless steel coated with methacrylate plastic	Same	0.450	

TABLE 3

C, $g \cdot cm^2/sec^2$	K	P_r, dyn/cm	Note
59	0.097	0.475	Average of three experiments
165.8	0.272	0.476	
799	1.31	0.480	

Then

$$T_0 = 2\pi \sqrt{\frac{I_0}{C}}; \quad T = 2\pi \sqrt{\frac{I}{C}},$$

whence

$$C = \frac{4\pi^2 I_K}{T^2 - T_0^2},$$

where

$$I_K = I - I_0 = \frac{M_K (r_2^2 - r_1^2)}{2},$$

or

$$C = \frac{2\pi^2 M_K}{T^2 - T_0^2} (r_2^2 - r_1^2).$$

The adsorption layer consists of: region A, a three-phase wetting perimeter (disc—water—petroleum); region B, a three-phase wetting perimeter (container—water—petroleum); region C, a two-phase boundary (Fig. 2).

Each of these regions is characterized by a certain strength of the adsorption layer. Let us designate the limits of strength of these regions as Pr_A, Pr_B, and Pr_C, respectively.

On the basis of the experiments that were conducted, it was confirmed that in the system of solid body—water—hydrocarbon liquid the absorption layer is broken not in the region of the three-phase boundary, but in the region of the water—petroleum boundary, and hence that the results of determining P_r indicate the strength limit of this layer. It was established that the value of P_r is not affected by the diameter of the external container or by the material of the disc and the container.

Tables 1 and 2 present the results of these experiments.

The formation time of the adsorption layer was 2 hours.

The value of P_r is likewise unaffected substantially by the adsorption layers on the surface of the disc and the walls of the container [8].

The value of P_r is not affected substantially by changing the filament constant within the limits of 59 to 799 $g \cdot cm^2/sec^2$ (Table 3). Crude from the Arlanskii oil field and distilled water were used in these experiments. The time of formation of the layer was six hours.

The value of P_r, as is well known, depends on the rate of application of load. In view of this, and in order to make it possible to compare the results of the investigations, the angular rate of rotation of the container was held constant at 0.61 deg/sec.

TABLE 4

Oil field	Horizon formation	Density at 20°C	Content, %			Viscosity, cs	Salt content, mg/liter	Well No. from which petroleum was taken
			Water	Silica-gel resins	Asphaltenes			
Shkapovskii	D-IV	0.8266	0.05	6.45	1.22	6.94	5.18	164, 526
"	D-I	0.8616	0.02	8.50	4.60	14.1	1.2	525
Tuimazinskii	D-I	0.8535	0.043	9.75	2.56	11.2	30.6	840, 862, 1237, 1093, 863
"	D-II	0.8527	0.129	11.80	3.00	11.9	102.7	100
Arlanskii	Carboniferous formation	0.8956	0.17	13.81	6.30	58.0	88.7	236, 233, 226, 7 17, 32
Tuimazinskii (Alexandrov area)	Same	0.8860	0.07	13.00	5.02	29.9	44.8	790, 1117, 1368, 800
Chekmagush	"	0.8886	0.03	16.3	2.40	32.9	36.4	101
Yuzhno-Vvedenov	Sakmaro-Artinsk deposits	0.8440	0.03	9.6	0.64	6.1	11.0	351, 352, 56, 83
Staro-Kazankovskii	Same	0.8680	0.63	11.1	1.65	13.2	90.2	132, 128

TABLE 5

Oil field	Horizon formation	Density at 20°C	pH	O_2, mg/liter	Well No. from which water was taken
Shkapovskii	D-IV	1.1884	4.80	48.0	185, 188
"	D-I	1.1906	5.75	60.0	60
Tuimazinskii	D-I	1.1825	5.60	88.0	1008
"	D-II	1.1885	5.90	52.0	1294
Arlanskii	Carboniferous	1.1807	6.05	48.0	63
Tuimazinskii (Alexandrov area)	Same	1.1686	6.48	104.0	1068, 1069, 1007
Yuzhno-Vvedenov	Carboniferous	1.1644	6.05	184.0	54
Staro-Kazankovskii	Same	1.0749	5.46	64.0	583

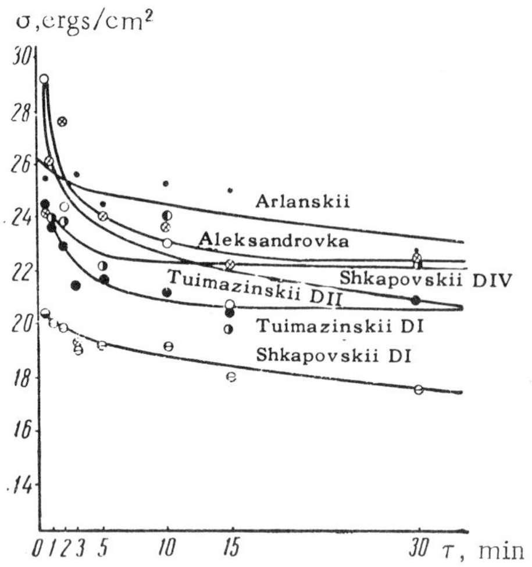

Fig. 3. Variation in surface tension with time (petroleum–distilled water) at 20°C.

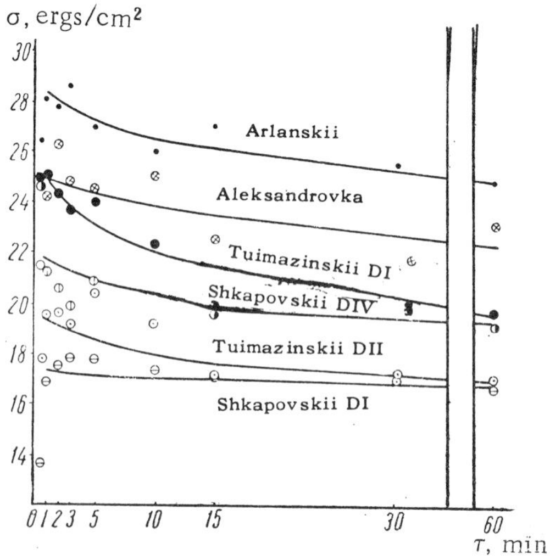

Fig. 4. Variation in surface tension with time (petroleum–formation water) at 20°C.

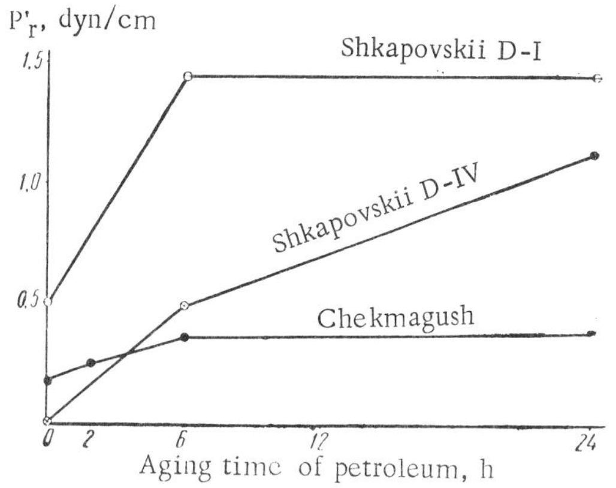

Fig. 5. Variation of static shear stress with time (for system petroleum–solid body).

Characteristics of Petroleums and Waters

The investigations were conducted with crude petroleums and formation waters from various Bashkir oil fields and also with a nonpolar hydrocarbon liquid (refined kerosene) containing various concentrations of Arlanskii crude.

The characteristics of the petroleums and waters are shown in Tables 4 and 5.

All of the crudes shown in Table 4 were taken from wells that were practically water-free. For the majority, the water content did not exceed 0.05%. The crudes from the Carboniferous formation of the Arlanskii oil field are characterized by the highest content of asphaltenes and resins. These substances are somewhat lower in amount in the crudes of the Aleksandrov area of the Tuimazinskii oil field.

The crude from the Chekmagush oil field is distinguished by a considerably lower content of asphaltenes with a large quantity of resins. In the Devonian crudes, the quantity of asphaltenes and resins is considerably lower than in the crudes of the Carboniferous formation. The content of these substances in crudes from Horizons D-I and D-II of the Tuimazinskii oil field is higher than in the crudes from Horizons D-I and D-IV of the Shkapovskii oil field.

These data on the content of asphaltenes and resins do not permit the drawing of conclusions as to the strength of the adsorption layer, since the latter is determined not only by the content but also by the chemical composition of the resins and asphaltenes, and also by the presence of other components in the crude.

All of the waters used in the experiments were hard, highly mineralized, with pH values from 4 to 6.8.

Measurement of Surface Tension

Data are shown in Table 6 for measurements of surface tension of distilled water at the interface with crude petroleums. The measurements were carried out in the apparatus of M. M. Kusakov [6] by the sessile drop method using photography.

TABLE 6

Oil field	Horizon formation	Surface tension, ergs/cm^2, at various times (min)									Well No. from which petroleum was taken
		0.5	1	2	3	5	10	15	30	60	
Arlanskii	Carboniferous formation	25.5	26.1	23.9	25.6	24.4	25.3	25.1	22.7	22.0	7, 32, 233, 236, 226, 17, 39
Tuimazinskii (Alexandrov area . . .	Same	24.2	26.1	27.7	19.3	23.9	23.6	22.2	22.5	22.5	79, 1117, 1068, 1069
Shkapovskii	D-IV	24.4	23.6	24.1	21.5	22.2	24.0	19.8	22.2	21.8	526, 164
"	D-I	20.4	20.0	19.9	19.0	19.2	19.1	18.0	17.5	16.9	525
Tuimazinskii	D-I	24.4	23.8	23.0	21.5	21.6	21.2	20.5	20.9	20.0	1093, 862, 1237
"	D-II	29.2	25.6	23.7	24.3	24.0	23.0	20.7	21.5	19.3	100

TABLE 7

Oil field	Horizon fromation	Surface tension, ergs/cm^2, at various times (min)									Well No. from which petroleum was taken
		0.5	1	2	3	5	10	15	30	60	
Arlanskii	Carboniferous formation	26.4	28.1	27.8	28.6	27.0	26.0	27.1	25.5	24.9	7, 32, 233, 236, 226, 17, 39
Tuimazinskii (Alexandrov area)	Same	21.5	24.3	26.3	24.9	24.5	25.1	22.5	21.7	23.1	790, 1117, 1068, 1069
Shkapovskii	D-IV	24.8	21.3	20.6	19.9	20.9	20.5	19.7	19.9	19.1	526, 164
"	D-I	13.7	16.9	17.5	17.8	17.8	17.3	17.1	17.1	16.7	525
Tuimazinskii	D-I	24.9	25.1	24.3	23.7	24.0	22.4	19.9	19.9	19.7	1093, 862, 1237
"	D-II	17.8	19.5	19.6	19.1	20.4	19.2	17.0	17.0	17.0	100

TABLE 8

Oil field	Horizon formation	Solution aging time, h					
		0.05	2	6	12	18	24
Shkapovskii	D-I	0.49	1.45	1.45		1.45	
"	D-IV	0.00	0.49	0.49		1.14	
Chegmagush	Carboniferous formation	0.170	0.243	0.36	0.39	0.39	

TABLE 9

Concentration of crude in kerosene, %	0.000	0.015	0.031	0.062	0.125	0.250	0.500	1.0	5.0	10.0	100.0
P_r, dyn/cm	0.00	0.00	0.00	0.00	0.05	0.05	0.07	0.02	0.05	0.05	0.07

TABLE 10

Concentration of crude in kerosene, %	0.000	0.015	0.031	0.062	0.125	0.25	0.50	1.0	5.0	10.0	100.0
P_r, dyn/cm	0.00	0.05	0.3	5.2	7.0	8.6	9.4	9.6	10.3	10.8	9.2

TABLE 11

Concentration of crude in kerosene, %	0.000	0.015	0.031	0.062	0.125	0.25	0.5	1.0	5.0	10.0	100
P_r, dyn/cm	0.00	7.1	–	9.0	9.2	9.1	9.4	–	–	–	–

TABLE 12

Time, h	0.05	1	2	4	6	12	24	48	72
P_r, dyn/cm	0.00	0.22	0.78	2.3	5.7	6.6	7.8	8.8	9.4

TABLE 13

Water	P_r, dyn/cm	
	For freshly prepared mixture	For mixture after 24 h
Distilled	0.370	–
Distilled + 0.5% NaCl	3.35	1.29

Note: Data are averages of several determinations.

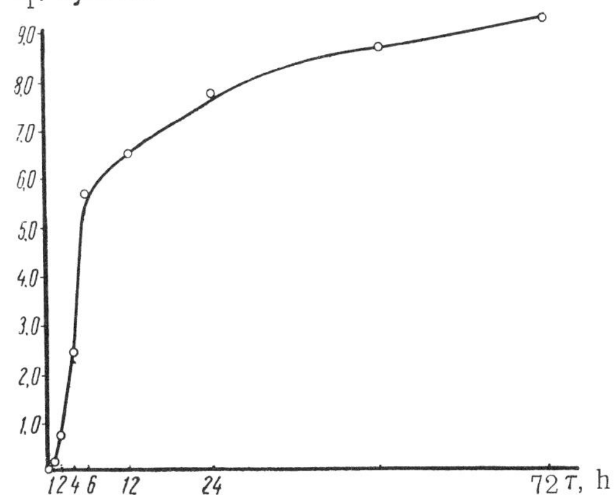

Fig. 6. Variation in ultimate strength with formation time of adsorption layer (0.5% solution of Arlanskii crude in kerosene with distilled water.

In all cases, a lowering in surface tension with time was observed, with subsequent stabilization.

Surface tension isotherms (Fig. 3) were plotted from the data of the table.

The petroleums of the Carboniferous formation, in spite of the high content of asphaltenes and resins, have a lower activity (judging by the surface tension value) than the Devonian petroleums of the Tuimazinskii and Shkapovskii oil fields.

Data are shown in Table 7 for measurements of surface tension of formation waters at the interface with various crudes. As in the case of distilled water, here also there is observed a timewise decrease in the surface tension value and subsequent stabilization (Fig. 4).

For all crudes, the final values of surface tension were higher with formation water than with distilled water.

From the surface tension data, it can be concluded that the surface activity of Devonian crudes is higher than that of the crudes of the Carboniferous formation.

Measurement of Ultimate Strength of Adsorption Layer, P_r

Time Dependence of Static Shear Stress, P_r'

The petroleums shown in Table 4 contain one quantity or another of asphaltic-resinous substances. This indicates that the petroleums are colloidal solutions and hence have a definite value of static shear stress P_r', which is low in magnitude. The value of P_r' is evidently dependent, other conditions being equal, on the time of aging of the solution.

Values of P_r' are shown in Table 8 for various aging times of the petroleums. These data are used to plot in Fig. 5 curves relating P_r' to aging time (for a comparison, the forces are referred to the disc diameter and expressed in dyn/cm).

These data show that the magnitude of P_r' for the petroleum crudes is small and does not exceed 1.45 dyn/cm. With the passage of time, P_r' increases and is stabilized. The stabilization time of the light Shkapovskii crude is greater than that of heavy crude.

P_r for Constant Formation Time of Adsorption Layer for the System Solid Body–Water–Nonpolar Hydrocarbon Liquid, as a Function of Concentration of Arlanskii Crude in the Nonpolar Hydrocarbon

Distilled water was used in these experiments. The value of P_r was determined immediately after formation of the adsorption layer (Table 9) and after 24 hours (Table 10).

In all subsequent experiments, unless specifically stated otherwise, the materials of the disc and container were stainless steel and glass, respectively.

TABLE 14

Well No.	Oil field	Horizon	Formation time of adsorption layer, h							Water
			0.05	1	2	4	6	12	24	
116	Arlanskii	Carboniferous	0.05	0.63	—	2.70	4.20	6.00	8.70	Distilled
116	"	Same	0.121	0.268	—	—	1.43	—	4.06	Distilled + 0.5% NaCl
116	"	"	0.19	0.243	—	—	0.49	—	2.52	Formation water, Well No. 63
116	"	"	0.07	0.70	—	—	1.80	—	7.05	Distilled + HCl (pH = 2)
116	"	"	0.05	0.292	—	—	0.36	—	2.05	Distilled + NaOH (pH = 12)
116	"	"	0.07	—	—	—	0.14	—	0.316	Distilled + 0.05% $KAUFE_{14}$ (nonionic surfactant)
101	Moncharovsk	"	0.05	0.07	—	—	0.49	—	3.18	Distilled
101	"	"								Distilled + NaCl
101	"	"	0.08	0.10	—	—	0.12	—	0.27	Formation
101	"	"	0.05	—	—	—	1.24	—	4.86	Distilled + HCl (pH = 2)
101	"	"	0.02	—	—	—	0.146	—	0.195	Distilled + NaOH (pH = 12)
101	"	"	0.05	0.07	—	—	0.07	—	0.07	Distilled + 0.05% $KAUFE_{14}$
109	"	D-I	0.17	0.32	—	—	0.73	—	4.93	Distilled
109	"	"	0.10	0.19	—	—	1.26	—	3.36	Formation
109	"	"	0.17	0.32	—	—	1.02	—	2.09	Distilled + HCl (pH = 2)
109	"	"	0.10	0.15	—	—	0.29	—	0.46	Distilled + NaOH (pH = 12)
109	"	"	0.15	0.19	—	—	0.27	—	0.29	Distilled + 0.5% $KAUFE_{14}$

TABLE 15

Surfactant concentration, %	Formation time of layer, h						
	0.05	1	2	4	6	12	24
0.0000	0.08	0.10	—	—	0.12	—	2.68
0.0005 NChK	0.05	—	—	0.07	—	—	0.22
0.0005 DS	0.05	0.09	—	—	0.15	—	0.36
0.0005 $KAUFE_{14}$	0.05	—	0.07	0.07	—	—	0.07

TABLE 16

Oil field	Horizon formation	Formation time of layer, h							Note
		0	1	2	4	6	12	24	
Arlanskii	Carboniferous	0.07	0.12	0.12	0.20	0.22	0.51	1.8	—
Chekmagush	Same	0.06	2.00	4.20	13.50	18.60	23.1	25.9	—
Cheraul	"	0.07	0.68	1.36	0.66	1.63	4.00	10.3	—
Shkapovskii (Well No. 216)	D-I	—	5.0	6.6	8.1	8.3	8.6	13.4	Average of three experiments
Shkapovskii (Well No. 138) etc.	D-IV	—	7.6	7.7	9.6	11.5	12.1	16.3	Same

With a short formation time of the adsorption layer (Table 9), the ultimate strength of the adsorption layer became appreciable at a concentration of 0.125% and subsequently, in spite of increase in concentration, did not undergo any significant change.

With a substantial formation time (Table 10), the ultimate strength of the adsorption layer became appreciable at a concentration of 0.015%. Subsequently, it increased with increasing concentration up to 0.5% and then became essentially constant. This indicates that saturation of the adsorption layer sets in at a concentration of 0.5%.

To establish the effect of the nature of the solid surface on the rate of saturation of the adsorption layer, experiments were performed with a disc and container made of methacrylate plastic. With prolonged formation of the layer (24 h), in contrast to the preceding case, here there was observed a sharp increase in P_r at low concentrations (Table 11), which indicates a faster rate of saturation of the adsorption layer. This is probably explained by the better wetting of the methacrylate plastic by the crude petroleum, in comparison with its wetting of steel. On reaching a concentration of 0.062%, the value of P_r became constant.

Relation Between P_r and Formation Time of Adsorption Layer for the System Solid Body—Water—Kerosene Solution of Crude

Table 12 shows the variation in P_r with formation of the adsorption layer for a 0.5% solution of Arlanskii crude in kerosene.

These data are used to plot the curve in Fig. 6 showing the relation between P_r and the formation time of the adsorption layer. Stabilization of the value of P_r set in after a prolonged period.

Effect of Water Quality on P_r

Studies were made to determine the effect of water mineralization on the magnitude of P_r. The hydrocarbon phase was kerosene with the addition of 0.5% Arlanskii crude. Table salt was added to distilled water to give a different degree of mineralization. The value of P_r was measured with a freshly prepared mixture of kerosene and crude and after 24 h.

The time of formation of the adsorption layer was 2 h both in the case of the distilled water and of the mineralized water.

The results of the studies are shown in Table 13.

With the freshly prepared mixture, the value of P_r was higher for the NaCl solution than for the distilled water. For the freshly prepared mixture, the value of P_r was higher than after aging. Similar results were obtained with a mixture of kerosene with 1% crude from the Carboniferous formation of the Khazinsk area of the Arlanskii oil field. The crude was taken from the tank farm. For the freshly prepared mixture, $P_r = 25.0$ dyn/cm; for the mixture after 24 h, $P_r = 5.0$ dyn/cm. In this case, distilled water was used. On aging of the colloidal solution (mixture), the strength of the adsorption layer decreases, which is evidently explained by coagulation of the micelles [7].

Relation Between P_r and Formation Time of Adsorption Layer with the System Solid Body—Water—Petroleum

The results of P_r measurements are shown in Table 14. For all of the crudes, an increase in the formation time of the adsorption layer gives an increase in the magnitude of P_r, regardless of the nature of the water.

The highest P_r value is observed with distilled water. Increasing the mineralization of the water by adding NaCl gives some decrease in P_r for the crudes of the Carboniferous formation. The P_r value is reduced still more with the highly mineralized formation water for the crudes of the Carboniferous formation. The addition of hydrochloric acid to the distilled water decreases P_r for the Arlanskii crude, but increases P_r for the crude of the Moncharovsk oil field.

The addition of alkali to the distilled water gives a marked reduction in P_r.

The addition of the surfactant $KAUFE_{14}$ leads to a substantial decrease in P_r. The timewise increase in P_r is insignificant; after 24 h, the value of P_r is many times smaller than in the other cases.

All of the trends noted also apply to the high-viscosity Devonian crude of the Chekmagush oil field.

Table 15 shows the variation in P_r for crudes of the Carboniferous formation of the Moncharovsk oil field upon the addition of various surfactants to the formation water.

Ammoniacal NChK containing 10% sulfonic acid salts was used in the experiments.

With the nonionic surfactant $KAUFE_{14}$, no timewise increase in P_r was observed, whereas with the other surfactants P_r increased with time.

Table 16 shows the variation in P_r for crudes from various oil fields with formation waters of these same fields. The P_r values vary over wide limits.

Summary

The ultimate strength of the adsorption layer P_r is one of the highly important characteristics of emulsions. A study of this property can bring some clarity to the complex mechanism of formation and breaking of petroleum emulsions and aid to the search for cost-reducing measures in the pretreatment of crude by the selection of suitable surfactants.

1. The magnitude of P_r depends on the nature of the crudes and waters, the time of formation of the adsorption layer, etc.

2. For the system solid body—water—nonpolar hydrocarbon liquid, the value of P_r is essentially zero.

When a small quantity of crude petroleum is added to the nonpolar hydrocarbon liquid, P_r becomes substantial. On further increasing the concentration of the crude to 100%, P_r does not undergo any significant change.

As the aging time of the kerosene—crude solution is increased, P_r decreases. This is related to decreased adsorption in connection with coagulation of the micelles and also a possible decrease in concentration of asphaltenes in the bulk of the solution owing to partial adsorption on the container walls while the mixture is stored in it. In the system solid body—water—petroleum, an increase in the mineralization of the water gives a decrease in P_r in the majority of the cases. Addition of alkali to the water, and in certain cases the addition of HCl, reduces the P_r value.

3. Surfactants reduce P_r markedly. The greatest reduction is observed on the addition of $KAUFE_{14}$.

In conclusion, we consider it our duty to express gratitude to Laboratory Director M. Z. Mavlyutova for kindly releasing the data obtained in the work reported in [9-13], which was carried out under her direction.

Literature Cited

1. P. A. Rebinder and K. A. Pospelova, "Modern concepts of the stability, formation, and breaking of emulsions and methods of investigation," Introductory article to the book by V. Clayton: Emulsions—Their Theory and Application Techniques (IL, 1950).

2. P. A. Rebinder, New methods of physicochemical investigation of surface phenomena, Trudy Instituta fizicheskoi khimii, No. 1, Izd. AN SSSR (1950).
3. P. A. Rebinder and N. A. Serb-Serbina, Physicochemical principles of the demulsification of petroleum emulsions, Referaty nauchno-issledova-tel'skikh rabot za 1944 gg., Otdelenie khimicheskikh nauk. (Izd. AN SSSR), p. 133.
4. M. M. Kusakov and A. A. Petrov, DAN SSSR, 116, 4 (1957).
5. A. A. Trapeznikov, New methods of physicochemical investigation of surface phenomena, Trudy Instituta fizicheskoi khimii, No. 1 (Izd. AN SSSR, Moscow, 1950).
6. M. M. Kusakov, N. M. Lubman, and A. Yu. Ksheevnik, Determination of the Surface Tension of Hydrocarbon Liquids and Petroleum Crudes by the Drop Measurement Method (Izd. AN SSSR, 1952).
7. G. R. Kryut, Colloid Science [Russian translation], Vol. 1 (IL, 1955).
8. A. A. Petrov, Fondy instituta Giprovostokneft (1952).
9. M. Z. Mavlyutova, et al., Development of a rational scheme for pretreatment of Bashkir crudes, Fondy UFNII (1958).
10. M. Z. Mavlyutova, et al., Dehydrating and desalting Bashkir crudes in the oil fields, Fondy UFNII (1959).
11. M. Z. Mavlyutova, et al., Study of the physicochemical properties of emulsions and combating their formation in oil fields, Fondy UFNII (1960).
12. M. Z. Mavlyutova, et al., Improvement of methods for dehydrating and desalting Devonian crudes, Fondy UFNII (1961).
13. M. Z. Mavlyutova, et al., Study of the physicochemical properties of emulsions of Bashkir crudes, Fondy UFNII (1962).

RESULTS OF COMMERCIAL TESTS OF DEMULSIFIERS $KAUFE_{14}$, UFE_8, AND OP-10 IN THE DESALTING UNITS OF THE OIL FIELDS OF THE NPU [OIL FIELD ADMINISTRATION] TUIMAZANEFT'

M. Z. Mavlyutova

In 1956, after we had investigated a considerable number of samples of surfactants to serve as possible demulsifiers for Devonian crude, we proposed for commercial test the reagents UFE_8, $KAUFE_{14}$, and OP-10 [1].

As was noted previously [2], tests of the reagent $KAUFE_{14}$ in the ELOU VCh [high frequency electrode-salting unit] of Field No. 4 of NPU Tuimazaneft' in June 1959 did not yield favorable results.

There was great interest in testing an experimental batch of these reagents in a commercial-frequency electrodehydrating unit. In 1960 at NPU Tuimazaneft' in the Naryshev ELOU [electrodesalting unit], operating according to the engineering plan "thermochemical dewatering with desalting in the electrodehydrators of the NZP system," commercial tests were performed on the reagents UFE_8 and $KAUFE_{14}$ and mixtures of these materials with ammoniacal NChK; in 1961, tests were run in this same unit on the reagent OP-10.

The experimental batches of UFE_8 and $KAUFE_{14}$, before the industrial test, were investigated in detail under laboratory conditions in order to select the optimum dosage for the Naryshev emulsion and to determine the optimum concentration of reagents in the working solution, the intensity of mixing, the temperature of heating the crude, and the settling time; this was done because the limited quantity of reagents made it impossible to select these parameters in the commercial unit.

The best results were obtained with solutions of $KAUFE_{14}$ at 1% concentration and UFE_8 at 2.5% concentration.

For the Naryshev crude, taken from the discharge line of the raw crude pump, the required supply of $KAUFE_{14}$ was 80-120 g/metric ton, and for UFE_8 350-500 g/metric ton, depending on the water content of the crude. The demulsification results were better at 70°C than at 60°C.

The settling time and the intensity of mixing were practically the same as for NChK.

The tests on the reagent were conducted in one of the units, consisting of six settlers of 80 m^3 capacity and 12 commercial-frequency electrodehydrators, heat exchangers, steam heaters of the "tube-in-tube" type, etc.

In normal operation using NChK, the unit processes up to 260 m^3/h (5000 metric tons/day); the raw crude is treated in two stages (thermochemical dewatering, and desalting in one step in the electrodehydrators); in this operation, marketable export-quality petroleum is produced with a water content below 0.3% and a salt content of 34-45 mg/liter.

In view of the limited quantity of the UFE_8 and $KAUFE_{14}$ reagents, the capacity of the unit was cut in half. All of the heat exchangers were used in the operation, but only three settlers and six dehydrators. The other six dehydrators were used for the second stage in desalting the crude.

The raw crude is fed from the tanks by centrifugal pumps at a pressure up to 16 kg/cm^2 to the desalting unit.

TABLE 1

Run No.	Length of run, h	Type of demulsifier	Dosage of emulsifier, g/metric ton of raw crude	Feed rate, m^3/h of raw crude	Temperature of crude after preheating, °C	Consumption of fresh water ahead of electrodehydrator, m^3/h: First stage	Consumption of fresh water ahead of electrodehydrator, m^3/h: Second stage	Pressure drop in mixing valve ahead of first-stage dehydrator, kg/cm^2	Quality of treated crude: Thermochemically dewatered: Water, %	Quality of treated crude: Thermochemically dewatered: Salts, mg/liter	Quality of treated crude: Desalted in electrodehydrators: First stage: Water, %	Quality of treated crude: Desalted in electrodehydrators: First stage: Salts, mg/liter	Quality of treated crude: Desalted in electrodehydrators: Second stage: Water, %	Quality of treated crude: Desalted in electrodehydrators: Second stage: Salts, mg/liter	Water content of raw crude, Wt. %: Minimum	Water content of raw crude, Wt. %: Maximum
1	26	Ammoniacal NChK, UNPZ [Ufa refinery]	3.3*	230	70	13	–	0.5	0.03	200	0.2	28	–	–	8	44
2	23	Same	4.0*	147	80	5	2.5	1.6	0.02	155	Not analyzed		0.13	20	5	13
3	5	Mixture of NChK . . UFE_8	2.45* 110	147	76	5	3	1.4	0.21	550	0.15	46	0.14	16	7	14
4	26	5% UFE_8	210-220	140	71	7	3	1.4	0.16	670	0.23	138	0.08	48	6.4	16
5	6	5% UFE_8	360	132	70	6.5	3	1.2	0.03	600	0.26	137	0.05	34	9.6	20
6	17	2.5% UFE_8	340-380	130	71.5	7	3	1.2	0.11	545	0.35	94	0.12	38	12	26
7	14	1.1% $KAUFE_{14}$	85	130	76	7	3	1.1	2.3	7000	0.34	450	0.32	127	9.6	27.2
8	25	1-1.5% $KAUFE_{14}$ . . .	116-128	120	82	7	3	1.1	0.61	1525	0.45	315	0.20	83	3	28
9	32	1% $KAUFE_{14}$	118	126	87	9	3	1.1	0.22	820	0.45	130	0.17	43	8	12
10	66	Same	118	126	85	8	3	0.9	0.95	3200	0.5	275	0.25	82	1.6	36
11	30	Mixture of NChK . . 2% $KAUFE_{14}$	2.66* 49	140	75-90	8	3	0.2-1.0	0.16	371	0.33	69	0.11	27.4	8	18
12	24	Same	2.2*	233	75	12	–	1.0	0.05	270	0.18	37	–	–	7	35

*Given in kg/metric ton.

TABLE 2

Crude	Content	
	Water, %	Salts, mg/liter
Dewatered	0.21	550
Desalted:		
First stage	0.15	46
Second stage	0.14	16

TABLE 3

Crude	5% solution		2.5% solution	
	Water, %	Salts, mg/liter	Water, %	Salts, mg/liter
Dewatered	0.16-0.03	670-600	0.11	545
Desalted:				
First stage	0.23-0.3	138-137	0.35	94
Second stage	0.05-0.08	34-48	0.12	38

TABLE 4

Date sample drawn	Time drawn	Dewatered		Desalted			
				First stage		Second stage	
		Water, %	Salts, mg/liter	Water, %	Salts, mg/liter	Water, %	Salts, mg/liter
July 28	19	0.3	790	0.6	190	0.24	68
	21	0.06	520	0.4	190	0.06	45
	23	0.12	620	–	160	0.18	34
July 29	1	2.4	3,100	0.27	160	0.03	34
	21	0.27	970	0.21	120	0.15	39
	23	0.24	855	0.24	94	0.15	36
July 30	1	1.6	6,555	0.21	270	0.12	34
	15	0.15	855	0.4	110	0.5	45
	17	0.3	990	0.5	113	0.15	34
	19	0.3	990	0.3	125	0.12	45
	21	0.24	798	0.13	112	0.09	50
	23	3.0	12,430	0.3	280	0.15	102
July 31	9	0.24	900	0.5	170	0.09	34
	11	0.24	680	0.27	130	0.3	45
	13	2.1	5,000	0.5	148	0.4	45
August 1	5	0.4	820	0.3	102	0.2	56
	7	0.3	990	0.3	120	0.27	45
	9	0.5	1,113	0.38	240	0.05	48
	21	0.27	900	1.2	340	0.18	50
	23	0.18	740	0.8	121	0.21	43
August 2	1	1.6	3,620	0.45	165	0.18	45

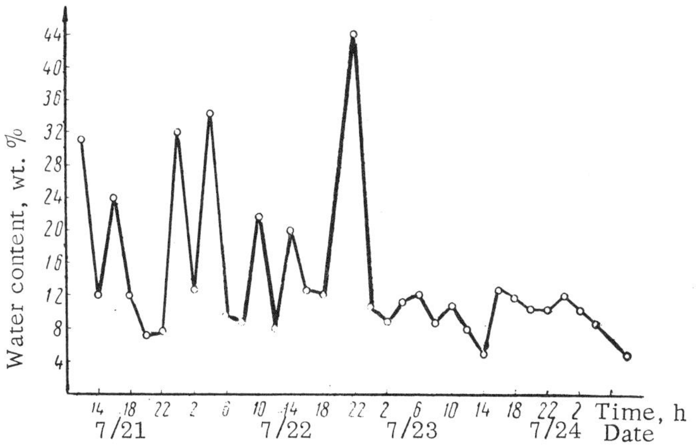

Fig. 1. Variation in water content of raw crude admitted to the Naryshev desalting unit in the period of testing UFE_8 and $KAUFE_{14}$.

The demulsifier is introduced before the inlet to the heat exchangers. The mixture of crude with demulsifier passes through a mixing valve into the heat exchanger, where it is heated by exchange with the desalted crude leaving the unit for the marketable crude tank. From the heat exchanger the mixture passes into the steam heaters and thence, with the temperature near 70°C, it is distributed among the horizontal settlers, operating at a pressure up to 6 kg/cm^2.

Here the crude is separated from the formation water. Dewatered crude from the settlers goes to the dehydrators for desalting. Ahead of the dehydrators, cold fresh water is supplied to the crude from an industrial water main. The intensity of contacting the fresh water with the crude is regulated by settling the pressure drop across a mixing valve.

The mixture of dewatered crude and fresh water passes into the first series of six dehydrators, where the first stage of desalting is accomplished.

At the exit from the first-stage dehydrators, a second portion of fresh water is supplied to the crude, and the mixture passes to the remaining six dehydrators. The desalted crude from the second-stage dehydrators passes to a surge tank and thence is transferred by pump through the heat exchangers and then, at a temperature of 35-40°C, to the marketable crude tank.

During the test of the reagents UFE_8 and $KAUFE_{14}$, the voltage across the electrodes in the six first-stage dehydrators was 16 kV, and 33 kV in the six second-stage dehydrators. In normal operation of the unit on NChK, the voltage is the minimum of 16 kV in all dehydrators, and desalting is accomplished in a single stage.

Analysis of the crude dewatered and desalted petroleum was carried out on line samples every 2 hours.

Establishing the temperature conditions and regulating the control and measuring instruments, in connection with the transition to lower capacity and two-stage operation, was carried out while operating the unit on NChK.

The reduction in capacity of the unit down to 147 m^3/h (with changing the settling time in the settlers and dehydrators) made it possible to include valves regulating the raw crude feed rate, the intensity of contacting the crude with reagent and fresh water, the temperature of the crude after the steam heater, and the system pressure, and also to close all bypass valves, etc.

Passing all of the crude through the heat exchangers led to a reduction in the temperature of the marketable crude going to the tanks, down to 30-35°C, and an increase in the heating temperature of the raw crude to 80°C and higher.

The quality [salt content] of the dewatered crude became more constant (120-180 mg/liter), and the salt content in the desalted crude after the second stage (daily average) amounted to 20 mg/liter, with a water content of 0.13% (see Table 1, Run 2).

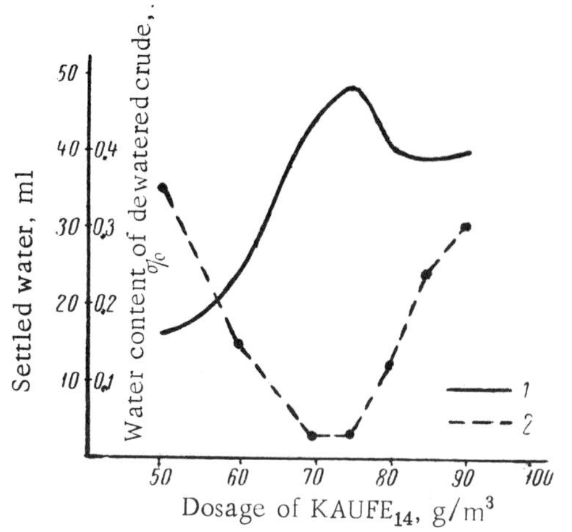

Fig. 2. 1) Volume of settled water, ml; 2) water content of dewatered crude, wt. %.

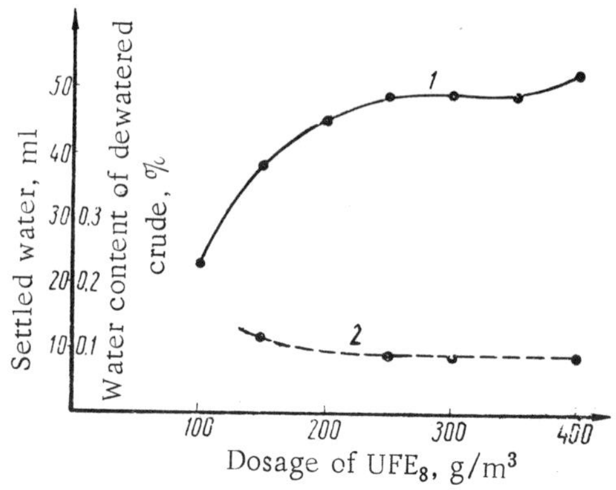

Fig. 3. 1) Volume of settled water, ml; 2) water content in dewatered crude, wt. %.

Subsequently, after reaching the given range of parameters (pressure in settlers 5.5-6 kg/cm^2, drop across mixing valve ahead of the first-stage dehydrators 1.2-1.4 kg/m^2, pressure in first stage dehydrators 4 kg/m^2, second stage 3 kg/m^2, throughout about 130 m^3/h, temperature of heating the crude 70-80°C), the NChK being fed to the crude was replaced by a mixture of NChK and UFE$_8$ solution, and then a 5% solution of UFE$_8$ (Table 1, Runs 3, 4, and 5).

The use of the 5% UFE$_8$ solution was dictated by the desire to accelerate washing of the system free from deposits formed during the period of unit operation on NChK.

During operation of the unit on the mixture of NChK and 5% UFE$_8$ solution, in 1:1 ratio, the consumption of NChK was 2.45 kg/metric ton, and the consumption of UFE$_8$ was 110 g/metric ton.

The quality of the crude at different stages in the treatment is shown in Table 2.

The quality of the pretreated crude during the test period when using 5 and 2.5% solutions of UFE$_8$ (see Table 1, Runs 4, 5, and 6) is shown in Table 3.

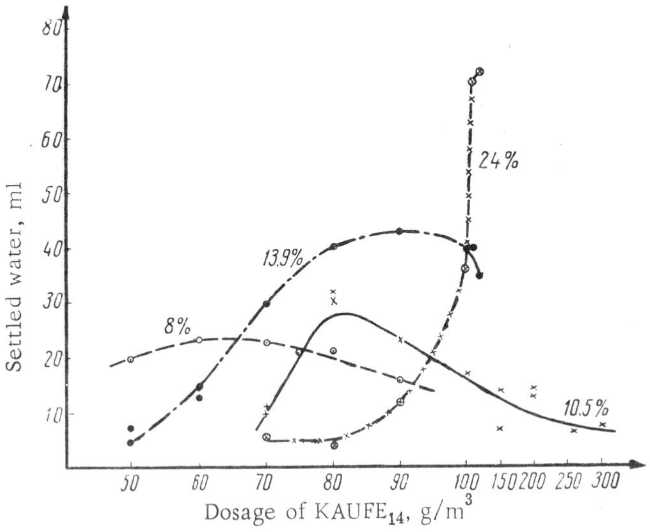

Fig. 4

TABLE 5. Analysis of Drainage Water from Settlers (Dewatering Stage)

Date of taking sample	Time of sampling (h-min)	Demulsifer	pH	Dissolved iron, mg/liter (Fe^{++})
July 26	14-30	UFE_8–2.5%	2.85	136.4
August 1	24-15	$KAUFE_{14}$–1%	2.85	129.4
August 2	18-10	Mixture (1 : 1) of NChK and 2% $KAUFE_{14}$	3.60	68.5
August 2	1-30	Ammoniacal NChK (Vessel II) . .	5.95	Absent

The crude pretreated with UFE_8 was turned over as marketable export grade with water content of 0.37% and salts 40 mg/liter.

The dosage of UFE_8 (100% product) per metric ton of raw crude in the beginning of the run was 220 g and then was increased to 380 g.

When the transition was made to $KAUFE_{14}$ after the UFE_8, the dewatering process was upset. The water content of the crude after the settlers rose to 3.6% and the salts to 9900 mg/liter. Dirty water began to be drained from the settlers.

In proportion to increasing the reagent dosage from 85 to 115 g/metric ton, the dewatering process began to improve.

After 29 hours from the start of the experiment with $KAUFE_{14}$, the water content in the dewatered crude became 0.15% and the salts 790 mg/liter. There was a gradual improvement in the quality of the desalted crude after the first- and second-stage dehydrators.

With more or less stabilized conditions, the stream analysis showed crude quality as indicated in Table 4.

The experiments using $KAUFE_{14}$ lasted 137 hours. The throughput of the unit varied from 110 to 140 m^3/h. The water content of the raw crude varied from 1.6 to 36%. Reagent dosages from 85 to 128 g/metric ton were tested.

The average indices of operation of the unit are shown in Table 1 (Runs 7, 8, and 9).

The marketable crude obtained during this period had the following quality (average sample from tanks):

Tank No.	Water	Salt, mg/liter
5	0.22	82
7	0.20	66
4	0.16	59

During the course of the test on $KAUFE_{14}$, the crude treating process would at first be in good adjustment and then would become worse. The worsening occurred as a rule immediately after cutting in a new tank of raw crude. These tanks did not have any special pretreatment before cutting them into the raw crude line of the desalting units; the free water that was accumulated in the period of admitting crude from the oil field to these tanks was not drained off (there was no suitable arrangement for such draining).

On July 23 and 24, when the raw crude tank was cut in after preliminary drainage of water below the emulsion and when the new tank was cut in at the moment when the other tank (from which the crude was getting into the unit) still had a certain level of relatively low-water crude, this eliminated the sharp jump in water content of the raw crude and also reduced considerably the absolute quantity of formation water admitted to the unit (Fig. 1).

Making up a mixture from the bottom layer of the new tank and the upper layer of the previous tank ensured a more or less uniform water content of the raw crude admitted to the desalting unit. This in turn made it possible to maintain constant conditions and to obtain more steady quality indices of the marketable crude.

Sharp jumps in the water content of the raw crude disrupt the heat and material balances of the demulsification and desalting processes. On supplying raw crude to the unit with high water content, the temperature of the crude after the heat exchangers and steam heaters decreases markedly. The increased drainage of water from the settlers at this moment leads to a sharp reduction in quality of the dewatered crude passing to the electrodehydrators, the heads of which are inflexibly adjusted for the case of operation at high throughput. The pressure and temperature in the system drop.

At the moment of cutting in a new raw crude tank with a high fill level, along with increasing the water content of the crude, there is also an increase in the pressure at the pump discharge, which leads to a reduction in demulsifier dosage.

In final account, all of this disrupts the operating conditions and leads to inconstancy in quality of the treated crude.

When operating the unit on NChK, the treated crude quality is relatively constant regardless of large variations in the water content of the raw crude; this is not observed when nonionic demulsifiers are used, especially $KAUFE_{14}$. This reagent exerts a demulsifying action only when added in a certain optimum quantity. It was established by laboratory experiments that if less than the required quantity is added, the demulsification does not go to completion, and if more than the required quantity is added, this in itself favors the formation of a very fine emulsion.

Figures 2 and 3 show the results of laboratory experiments on demulsifying one and the same sample of raw crude with water content of 14%, from the NPU Tuimazaneft' Oil Field No. 1, with solutions of $KAUFE_{14}$ and UFE_8. The optimum usage of $KAUFE_{14}$ in the given case is 75 g/m^3 and UFE_8 is 250 g/m^3.

For the UFE_8, increasing the dosage from 250 to 400 g/m^3 had practically no effect on the volume of settled water or on the quantity of water remaining in the dewatered crude. However, for the $KAUFE_{14}$, changing the dosage even by 5 g/m^3 (from 75 to 80 m^3) gave a sharp change in quantity of settled water and especially in the water remaining in the crude.

The optimum dosgae of $KAUFE_{14}$ in turn depends on the concentration of the emulsion. Curves are given in Fig. 4 showing the volume of settled water separated from emulsion with different water contents as a function of $KAUFE_{14}$ reagent dosage. The samples of emulsion were taken from the discharge line of the raw crude pump of the Naryshev ELOU. From these curves, it is evident that the optimum dosage for various water contents are: 8% water – 60 g/m^3; 10.5% water – 80 g/m^3; 13.9% water – 90 g/m^3; 24% water – more than 120 g/m^3.

If the dosage is more or less than the optimum, the separation of water from the emulsion is incomplete, and a large amount of water remains in the crude in the form of emulsion.

A similar phenomenon apparently occurred in the unit. Therefore, strict observance of the optimum dosage established experimentally for a specific case (concentration and degree of dispersity of the emulsion) is the basic determining factor in the process of breaking petroleum emulsions when $KAUFE_{14}$ is used.

Based on the results of previous studies [3], experimental runs were made on treating crude with a mixture of the test reagents UFE_8 and $KAUFE_{14}$ with ammoniacal NChK.

In the case of the mixture of NChK and UFE_8, the run was short (5 h), and it was made in order to prepare the unit for operation on pure UFE_8 reagent (Table 1, Run 3).

However, in the case of the mixture of NChK and $KAUFE_{14}$, the unit was operated for 54 hours – first at low throughput with two-stage desalting (Table 1, Run 11), and then at increased throughput with the dehydrators operating in a single stage (Run 12).

The use of NChK diluted with 2% solution of $KAUFE_{14}$ in 1 : 1 ratio permitted a twofold reduction in the consumption of NChK and a threefold reduction in the consumption of $KAUFE_{14}$; it also made possible an improvement in quality of the dewatered crude and hence a transition to single-stage desalting of the crude in the electrodehydrators. The optimum consumption for the Naryshev crude with the mixture was NChK 2 kg/metric ton and $KAUFE_{14}$ 40 g/metric ton, instead of 3.5-4.0 kg/metric ton of NChK and 120 g/metric ton of $KAUFE_{14}$ when these reagents are used individually.

When the unit was operated on $KAUFE_{14}$ and UFE_8, the drainage water from the settlers and dehydrators was clear, and, upon draining them with the emulsion, the crude settled rapidly.

It should be noted that when these reagents are used the drainage water from the dewatering stage has a lower pH than when using ammoniacal NChK and contains dissolved ferrous iron (Table 5).

Summary

1. The nonionic surfactants UFE_8 and $KAUFE_{14}$ are active demulsifiers for emulsions of Devonian crudes. The activity of $KAUFE_{14}$ is three times that of the UFE_8.

2. When treating crude from NPU Tuimazaneft' Oil Field No. 1 in the Naryshev commerical-frequency ELOU in combination with thermochemical dewatering, the optimum consumption of these reagents is as follows (average values):

$KAUFE_{14}$	110-120 g/metric ton
UFE_8	300-360 g/metric ton

In order to obtain in the Naryshev unit crudes with salt content below 50 mg/liter when using $KAUFE_{14}$ and UFE_8 reagents, it is necessary, after dewatering the raw crude, to use two-stage treatment in commercial-frequency dehydrators; at the same time, sharp variations in the unit throughput or in the water content of the raw crude cannot be allowed.

3. The UFE_8 and especially the $KAUFE_{14}$, provide the required degree of desalting only when the optimum dosage is strictly observed; this dosage depends markedly on the water content of the raw crude and the throughput of the unit.

4. The most acceptable concentrations of reagents in the working solution are:

$KAUFE_{14}$	1%
UFE_8	2.5%

5. The use of nonionic surfactants of the type of UFE_8 and $KAUFE_{14}$ in mixture with ammoniacal NChK, in addition to reducing the consumption of both reagents, ensures thorough desalting when treating the dewatered crude in electrodehydrators in a single stage; this is impossible to perform when using $KAUFE_{14}$ or UFE_8 in the pure form.

In August 1961, an industrial test on the reagent OP-10 was performed in the Naryshev ELOU.

By preliminary experiments in the dewatering unit of NPU Tuimazaneft' Oil Field No. 2, carried out in June 1961, and also by laboratory studies, it was established that when OP-10 is used, just as is the case with $KAUFE_{14}$, more complete dewatering of the crude is attained only when the reagent is fed in such a quantity as is required for breaking the given emulsion.

In order to provide more or less constant quality of the emulsion passing to the unit during the OP-10 test period, the raw crude from all four sections of Oil Field No. 1 was gathered in a common collector and fed through an intermediate tank "with a rush" to the unit. Taking into account the technical conditions of the heat exchange apparatus and settlers for the test on OP-10, the second building of the Naryshev ELOU was set aside for use. The components in operation were all of the heat exchangers and steam heaters, six settlers (total volume 480 m^3), and 12 electrodehydrators.

The process flow plan for treating crude during the test was the same as when the unit was operating on NChK (dewatering under pressure by the thermochemical method, and desalting in electrodehydrators in a single stage).

The OP-10 was introduced into the crude ahead of the heat exchangers in the form of a 2% aqueous solution; the consumption varied from 130 to 180 g/metric ton of raw crude. The intensity of mixing varied with the pressure drop across the mixing valve, which varied from 0.8 to 4.5 kg/cm^2.

The temperature of the crude at the exit of the steam heater varied from 60 to 65°C.

The raw crude throughput of the unit averaged 235 metric tons per hour.

The water content of the raw crude, in spite of the measures indicated above, varied from 8 to 26%, with an average content of 16.6%.

The consumption of cold fresh water supplied to the dewatered crude ahead of the electrodehydrators amounted to 10-12 m^3/h (5-6% of the crude).

The unit operated on OP-10 reagent for 48 hours. During this time, 11,300 metric tons of raw crude were treated. Marketable crude in the amount of 9,404 metric tons was obtained, with water content 0.27-0.37% and salt content of 300-400 mg/liter.

During the OP-10 test period in the Naryshev unit, where the crude in the first stage of treatment was subjected to thermochemical dewatering, the content of residual water in the crude after the settlers varied from 1 to 3.4%, and the chloride content from 3,800 to 12,000 mg/liter. In spite of this, the regime of the electrodehydrators was not disrupted; such stability was not the case with even slight deterioration of the crude dewatering process when operating the unit on NChK.

The quality of the crude after first-stage treatment, i.e., after thermochemical dewatering of the crude with separation of the water in the settlers, operating under pressure, was considerably worse in the experiments with OP-10 than when using NChK, operating conditions being the same.

Hence, with a view to obtaining desalted crude of the required condition when using OP-10 as the demulsifier, oil-field,high-water-content crude after thermochemical dewatering must be subjected to two-stage treatment in commercial-frequency dehydrators, which is not taken into account in existing desalting units in the oil fields.

The consumption of OP-10 was 70-180 g/metric ton for the Tuimazinskii Devonian crude, depending on the emulsion composition.

The same as $KAUFE_{14}$, the OP-10 reagent requires selection and observance of accurate dosage, which is practically impossible to accomplish under oil-field conditions.

Thus, OP-10 reagent, the same as $KAUFE_{14}$, in spite of its high surface activity, has little acceptability as a demulsifier and hence cannot be recommended for introduction into existing oil-field crude desalting units.

Literature Cited

1. M. Z. Mavlyutova, Dewatering and desalting of Bashkir crudes in the oil fields, Fondy UFNII (1957).
2. M. Z. Mavlyutova, Investigation of surface-active substances for demulsifying and desalting Bashkir crudes, Trudy Pervogo Vsesoyuznogo soveshchaniya po primeneniyu PAV v neftyanoi promyshlennosti (Gostoptekhizdat, 1961).
3. M. Z. Mavlyutova, Development of a rational scheme for treating Bashkir crudes, Fondy UFNII (1958).

SYNTHESIS OF NONIONIC SURFACTANTS – DERIVATIVES OF FATTY ACIDS – AS DEMULSIFIERS OF PETROLEUM EMULSIONS

D. N. Levchenko, A. D. Khudyakova, N. V. Bergshtein,
V. I. Khokhlov, A. L. Kalitaeva, and E. A. Shklyaruk

Chemical reagents for demulsifying crude petroleum have already been applied for several decades. A method of chemical demulsification of crude was patented in the USA in 1914 by Barnickel [1], and in Russia by L. F. Berkgan [2].

From that time, in the USA and in other countries, many different demulsifying reagents have been synthesized, but little has been published on their composition, if one does not include the patent literature. The firms producing demulsifiers during this time have changed their composition many times. Judging by the data of a review article by W. Franke [3], interest in Germany in the production of demulsifiers appeared only in 1936, after the use of sulfonated castor oil became well known in the industries of Rumania. In Germany, along with castor oil, oleic acid was also used over a number of years for synthesizing effective anionic demulsifiers. At the end of the war, a new direction appeared in the synthesis of demulsifiers – ethoxylation.

Ethoxylated castor oil was the first nonionic demulsifier in Germany. Ethoxylation was also carried out on alcohols, phenols, and naphthols. The most effective demulsifier was obtained in this case from alkylphenols. Another raw material used for synthesizing nonionic demulsifiers was rosin, the principal constituent of which is abietic acid.

W. Franke noted that for a number of emulsions, the ethoxylated rosin is still of value at present. Plants of the firm Farbenindustrie in Ludwigschafen, before the war, began to produce demulsifiers under the name of "Dismulgan." After the Second World War, production of demulsifiers under the name of "Separol" was begun in the plants of the firm "BACF" in the FRG [Federal Republic of Germany]. Other firms in various European countries also began to produce demulsifiers. During this same time, certain demulsifiers of several USA firms were imported into Europe.

Several years after the Second World War, along with ethylene oxide, propylene oxide began to be used for the synthesis of nonionic demulsifiers. Particularly effective demulsifiers were obtained by synthesizing compounds containing mixed chains of propylene oxide and ethylene oxide.

TABLE 1. Characteristics of Synthetic Fatty Acid Fractions

Fraction of acids	Yield, %	Acid No.	Ester No.	Unsaponifiable, %
I C_5—C_6	53.1	455	None	None
II C_7—C_9	10.5	398	0.6	0.4
III C_{10}—C_{16}	36.7	258	2.2	1.6
IV C_{17}—C_{20}	30.5	196	4.2	4.6
V above C_{20}	19.2	107	30.5	13.8

Table 2. Efficiency of OZhK Demulsifier as a Function of Ethoxy Groups (on Romashkino Crude)

Content of ethoxy groups in OZhK		HLB	Mol. wt.	Demulsifier added, %	Water separated, %	
No. of groups	%				After 2 h settling	After 4 h settling
10.5	50.7	10.1	912	0,005	65,7	66.6
15.4	60.0	12.0	1128	0.005	76,2	88.5
22.2	68.7	13.7	1435	0.005	95,0	98.0
26	71.8	14.3	1594	0.005	96,0	98.0
29.7	74.3	14.8	1757	0.005	94.5	99.2
35.6	77.7	15.5	2016	0.005	83.0	92.0
41.2	80.1	16.0	2263	0,005	77.0	81.0
50.3	83,0	16.6	2660	0.005	81.0	86.8
60	85.3	17.0	3090	0.005	79.0	86.8
100.3	90.6	18.1	4863	0.005	74.3	82.0

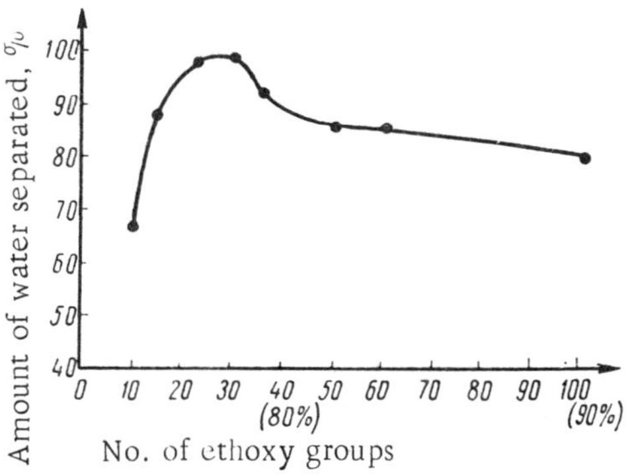

Fig. 1

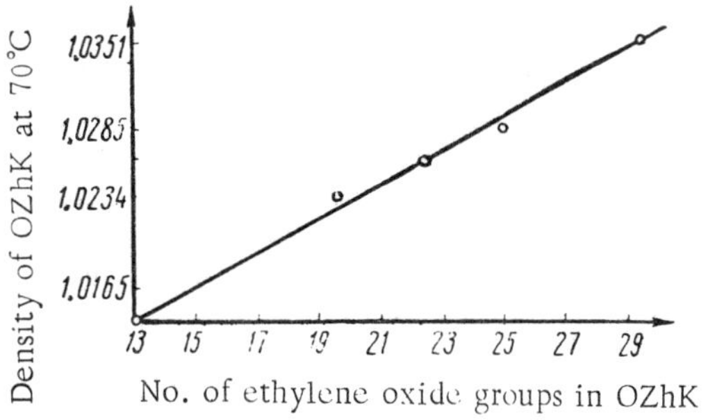

Fig. 2

It has been established by extensive experimentation that surfactants consist of a hydrophobic part and a hydrophilic part, the hydrophobic part of the molecule containing not less than 8 to 12 carbon atoms. In synthesizing nonionic surfactants, the hydrophobic part of the molecule may be fatty acids, partial esters of fatty acids, alcohols, alkylphenols, or other compounds containing a reactive hydrogen. For this purpose, use is often made of specially synthesized substances with a definite content of carbon atoms, which are then ethoxylated. For example, phenol may be alkylated (synthesis of OP-10), or polyoxypropylene glycol (synthesis of Pluronics) or polybutylene glycol may be obtained, or ethylene oxide and propylene oxide may be condensed with amines, etc.

For the synthesis of nonionic surfactants as demulsifiers for petroleum emulsions, the hydrophobic portion of the molecule has been taken from materials already existing in our industry – synthetic high-molecular-weight acids, alcohols, and other substances not requiring special outlay for their production.

For the synthesis of surfactants, the hydrophilic part of the molecule was made by condensation with ethylene oxide, adding on the necessary ethoxy chain. Depending on the composition of the substance taken for ethoxylation, it was experimentally established how much ethylene oxide was required to attain hydrophilicity, surface activity, and high demulsifying power of the synthesized surfactant.

In the plants of the Soviet Union, large quantities of synthetic fatty acids are produced by the oxidation of paraffin.

The acids are separated by fractional distillation into the following fractions: C_5-C_6, C_7-C_9, $C_{10}-C_{16}$, $C_{17}-C_{20}$, and above all C_{20} (still bottoms) (see Table 1).

We tested all of the indicated fatty acid fractions, except the C_5-C_6 fraction, as raw material for producing nonionic demulsifiers.

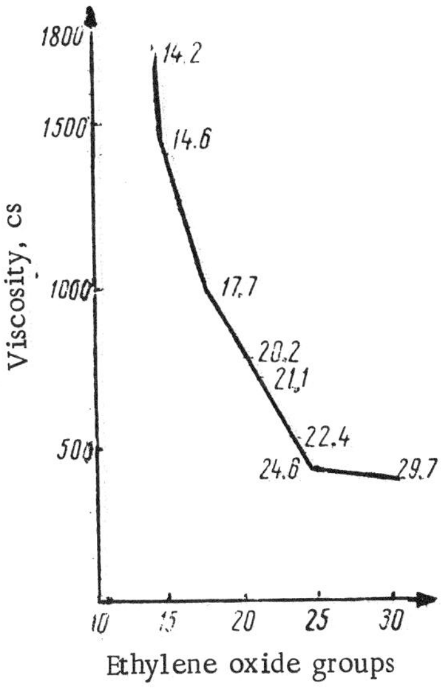

Fig. 3

The best demulsifying properties with respect to petroleum emulsions were shown by the surfactants obtained from the high-boiling fatty acid fractions, above C_{20}.

Up to the present time, the synthetic fatty acids above C_{20} remain in the still bottoms from distillation of the acids and do not find regular application in industry. Therefore, our primary attention was directed toward the synthesis of a nonionic demulsifier from the synthetic fatty acids above C_{20}. The yield of these acids amounts to 19% of the acids distilled or 14.5% based on the paraffin taken for oxidation.

The synthesis of nonionic demulsifiers from the other fractions of the synthetic fatty acids was carried out only for comparison, since there is a shortage of these fractions, and they are used in soap production to replace the fatty acids derived from edible fats.

The fatty acids were ethoxylated in the presence of caustic as a catalyst (1.5-2%) at a temperature of 180-200°C.

For studying the effect of ethoxy chain length on the demulsifying power of the substance, samples were made with content of 7 to 100 molecules of ethylene oxide per average molecule of the fatty acids taken for ethoxylation.

The samples of the synthesized surfactants were tested under laboratory conditions for demulsifying power in the process of thermochemical desalting of Romashkino crude in accordance with the following methods.

A weighed sample of the crude (200 g) in a separatory funnel was heated in a water bath to 70°C, and 10% distilled water was added, containing the given quantity of demulsifier dissolved in the water.

The crude and water with the demulsifier were shaken for 10 minutes on a laboratory (250 strokes per minute), after which the funnel was placed in a water thermostat at 70°C and allowed to stand.

The separated water was measured every hour for 4 hours. In case the water settled completely, the treated crude was analyzed for content of water and chloride salts.

From the results of the test (Table 2, Fig. 1), it is evident that, for producing an effective demulsifier from the C_{20}+ fraction of the fatty acids, the addition of 10-15 molecules of ethylene oxide is inadequate. The surfactant samples containing 19-30 ethylene oxide groups (65-75%) proved to be the most effective demulsifiers. Adding more than 75% ethylene oxide to the fatty acids did not improve the demulsifying properties of the substance, and even a slight reduction in efficiency was observed.

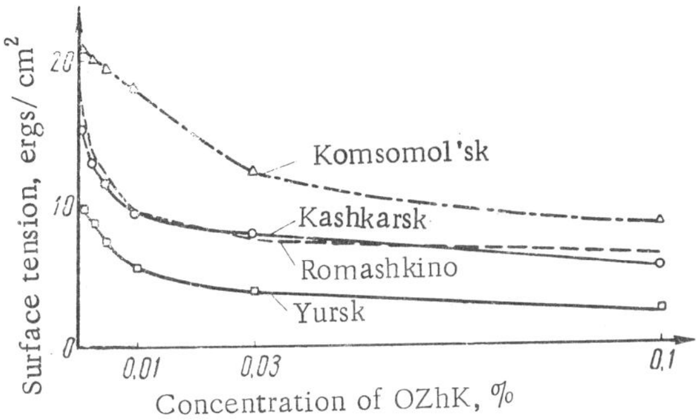

Fig. 4

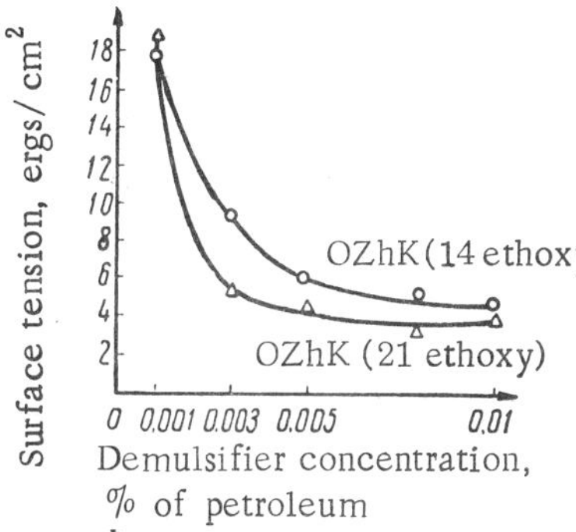

Fig. 5

TABLE 3. Analysis of Composition of Effective Demulsifiers

Demulsifier	Ethoxy groups, %*	Content, %			Note
		C	H	O by difference	
OZhK	67.0	60.43	10.02	29.35	Sulfur and nitrogen content in all samples varied from 0.00 to 0.06%
OZhK after removing hydrocarbon with naphtha	75.0	57.28	9.74	32.98	–
German (FRG) 4411, 1961 sample	68.5	58.17	10.00	31.83	–
German (FRG) 4422, 1961 sample	66.8	59.54	10.27	30.19	–
German (FRG) 4433, 1961 sample	71.0	58.00	10.07	31.93	–
German (FRG) 4400, 1961 sample	71.0	58.48	10.2	31.32	–
Separol	68.5	55.08	9.40	35.52	–
Foumskol (English)	74.1	57.39	9.90	32.73	–
Proksanol 146	–	56.54	9.82	33.64	Sample obtained from laboratory of NIOPIK (K. G. Mizuch, Director)

*Ethoxy number of imported demulsifiers calculated arbitrarily, assuming that there were no propoxy or other groups present.

The greatest reductions in surface tension of the aqueous surfactant solutions at an interface with the crude were also observed for substances containing 65-75% ethoxy groups.

The properties of nonionic surfactants depend on the chemical nature of the hydrophobic portion and on the ratio of hydrophobic and hydrophilic portions (HLB).

The HLB values of the substances having the greatest demulsifying power were between 13 and 15 (calculated according to Griffin).

There is no information in the literature on the optimum value of the HLB for effective nonionic demulsifiers. W. Franke [3] noted that the following applies to nonionic surfactants having high demulsifying power: for 35-55 carbon atoms of the hydrophobic portion of the molecule, there are 30-40 molecules of ethylene oxide. It is true that the author stipulates that this statistic does not show that all nonionic surfactants with such a structure are effective demulsifiers; it merely serves to eliminate substances that are not applicable for demulsification.

We consider that there cannot be a single universal ratio of hydrophilic and hydrophobic portions for all compounds; this depends to a considerable extent on the chemical nature of the compounds. Therefore, only for compounds of very similar nature can a definite HLB be established.

The HLB index can have one value for ethoxylated alkylphenols and another for ethoxylated fatty acids.

There are indications in the literature [5] as to the detergent properties of ethoxylated fatty acids; it was established that the optimum detergent properties of a substance are reached at a content of approximately two-thirds of an ethoxy group per carbon atom of the fatty acid.

The quality and properties of OZhK demulsifier are related directly to the content of ethylene oxide groups in the molecule. The following physical properties were determined for a series of samples of ethoxylated fatty acids (C_{20}+) with different contents of ethylene oxide groups: solubility in water, number of ethylene oxide groups, freezing point, density, viscosity, surface tension, and demulsifying power.

All of the samples of ethoxylated fatty acids (C_{20+}) containing more than 10 ethylene oxide groups per molecule of acid were soluble in water, forming slightly turbid solutions. When less than 10 groups were added, the samples dissolved only on heating, forming very turbid solutions, from which part of the material precipitated on standing.

The number of ethylene oxide groups was determined on the basis of the increase in weight of the fatty acids taken for the synthesis and was checked by the method of Sidzhin [11] with good agreement of results. The solidification point was determined by a rapid and quite convenient method based on congealing on the bulb of a thermometer. It was noted that the OZhK samples with fewer ethylene oxide groups had higher solidification points than the samples with more ethylene oxide groups. For example, the ethoxylated acids containing 14 ethylene oxide groups per molecule of acid solidified at 48°C, and the compounds with 20-25 groups had solidification points of 40-43°C.

The density of the ethoxylated fatty acids increased sharply as the number of ethylene oxide groups was increased (Fig. 2), showing an almost linear relationship.

The viscosity of the OZhK demulsifier samples was very high with low numbers of ethylene oxide groups and decreased sharply as the number of groups was increased up to 25. As more ethylene oxide groups were added, there was no further significant change in viscosity.

For the demulsifier samples with 20-23 ethylene oxide groups, the viscosities were in the range of 500 to 760 centistokes (Fig. 3).

The results that were obtained point out a definite and regular relationship among solidification point, viscosity, density, number of ethylene oxide groups, and demulsifying power of the OZhK samples.

The surface tension was determined at the interface between an aqueous solution of the demulsifier and crude petroleum. The surface tension of the OZhK demulsifier with 21 ethylene oxide groups was lower than the surface tension of the sample with 14 groups (Figs. 4 and 5). This is in agreement with experimental data on the demulsifying power of the OZhK materials.

The work on studying the physical properties of the OZhK samples made it possible to establish a number of correlations among these properties, the number of ethylene oxide groups, and demulsifying power.

For comparison with the OZhK, we investigated the most effective samples of imported nonionic demulsifiers; analyses of these materials are shown in Table 3.

As is evident from Table 3, the imported demulsifiers are close to the OZhK demulsifier in elemental composition and percentage of ethoxy groups and hence also in HLB value; the German demulsifiers contain 66.6-71% ethoxy groups (HLB 13.3-14.2), and the English Foumskol 75% (HLB 15).

The relative efficiency of these demulsifiers and OZhK was determined in the laboratory according to the methods described above and in a pilot electrodesalting unit on desalting Romashkino crude.

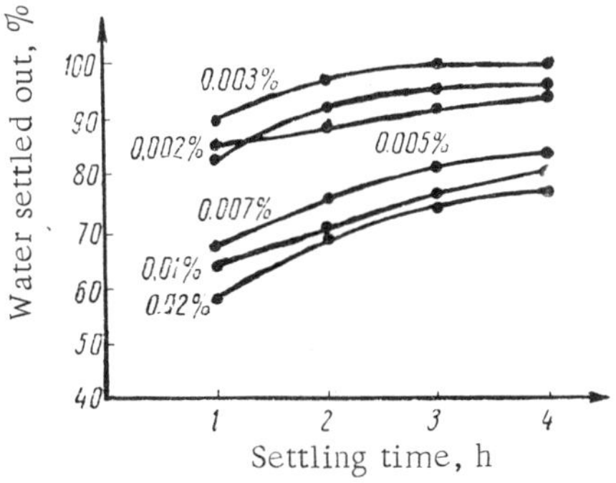

Fig. 6

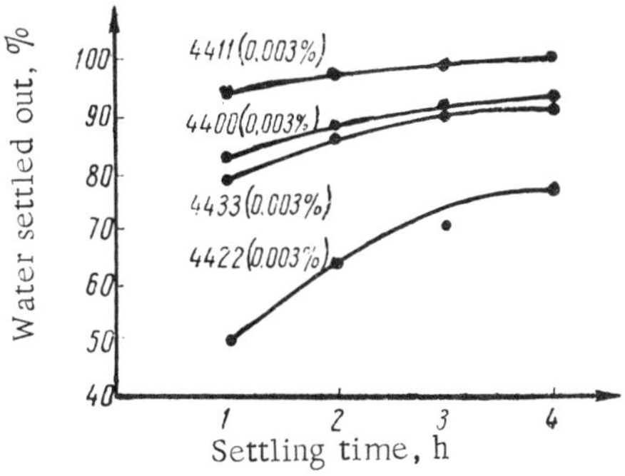

Fig. 7

TABLE 4. Desalting of Sernovodsk and Buguruslan Crudes in Pilot-Plant ELOU of VNII NP

Demulsifier	Demulsifier consumption, g/metric ton	Treated Sernovodsk crude		Treated Buguruslan crude	
		Water, %	Chloride salts, mg/liter	Water, %	Chloride salts, mg/liter
OZhK	20	0.2	2	0.1	11
German (FRG) 4400	20	0.1	2	0.1	5
German (FRG) 4411	20	0.3	2	0.1	2
German (FRG) 4422	20	Traces	40	0.1	8
Dutch, firm "Servo"	20	0.2	28	—	—

The Romashkino crude used for the demulsifier tests contained 2-2.5% water and about 2500 mg/liter of chloride salts. The density of the crude $d_4^{20} = 0.868$.

All of the nonionic demulsifiers that were tested showed an optimum consumption, above or below which the demulsifier efficiency decreased.

In desalting this particular sample of crude, the optimum consumption for OZhK demulsifier was 0.003-0.005% of the crude (30-50 g/metric ton of crude); for the German 4411, the optimum consumption was 0.003% (Fig. 6). The demulsifiers 4400, 4422, and 4433 were less effective on the Romashkino crude (Fig. 7).

For the other samples of mixed Tatar crudes, the optimum demulsifier consumption was somewhat higher (OZhK 0.005-0.007% and 4411 0.003-0.005%).

When OZhK was extracted with naphtha to remove admixtures of hydrocarbons and compounds with few ethylene oxide groups, the composition was changed markedly (Table 3) and the efficiency was increased.

In the thermochemical desalting of Romashkino crude under the indicated conditions (4-hour settling at 70°C), standard-quality crude (containing not more than 40 mg/liter of salts) was not obtained with any of the demulsifiers. The thermochemical desalting was carried out only to determine the relative efficiency of the different demulsifiers.

In the ELOU pilot-plant desalting of different crudes (Romashkino, Sernovodsk, Buguruslan, Anastasievka, and ten samples of Emba crude), the consumption of both the imported demulsifiers (4400, 4411, and others) and the OZhK amounted to 20-30 g/metric ton of crude (Table 4). In desalting heavy Chernushka crude, the consumption of these demulsifiers was somewhat higher (30-40 g/metric ton of crude).

In desalting the Sernovodsk and Buguruslan crudes in the pilot-plant ELOU, the most effective demulsifier was OZhK. With a consumption of 20 g/metric ton of OZhK, complete desalting of the crude was attained, the water separated out in the electrodehydrators without the formation of intermediate emulsion, and the current was low in comparison with the other demulsifiers. Good results were also obtained with the German demulsifiers, but with the same demulsifier consumption (20 g/metric ton), the desalting process in the electrodehydrator proceeded more intensely; the current was increased.

When desalting Sernovodsk and Buguruslan crudes with OP-10, the demulsifier consumption was higher (40-50 g/metric ton), the process was unsteady, the water separated poorly, and emulsion accumulated, resulting in short-circuiting of the electrodes.

When desalting Chernushka crude ($d_4^{20} = 0.886$, chloride salts 550-800 mg/liter, water 0.2-1.5%) in the pilot-plant ELOU in a single stage with the nonionic demulsifiers OZhK and the above-indicated imported materials, with a demulsifier consumption of 30-40 g/metric ton, the treated crude contained 0.2% water and 20-40 mg/liter of chloride salts. With this same consumption of OP-10 (30-40 g/metric ton), the current increased and poor settling of the water was observed. In the normal course of the desalting process, the consumption of OP-10 was required to be not less than 50 g/metric ton.

Desalting of Chernushka crude using OZhK demulsifier has not yet been carried out in the plant unit. However, judging by the results obtained in the pilot unit, it would be expected that desalting with OZhK should be no worse than with the imported demulsifiers.

For desalting Emba crudes, such as Karsaksk, Sagiz, Baichunas, Dossor, Iskine, Dzhaksymaisk, and Permotriassk, the consumption of OZhK amounted to 20 g/metric ton. After treating the indicated crudes in a single electrical stage, the treated crudes contained 1-33 mg/liter of chloride salts. The consumption of OP-10 required to obtain the same results was higher (40 g/metric ton).

Similar results were obtained in desalting Anastasievka crude.

High-viscosity Emba crudes—Kashkarsk and Komsomol'sk—were desalted with more difficulty (in two stages with demulsifier consumption 20-30 g/metric ton).

In 1961, pilot-plant samples of OZhK demulsifiers were prepared (with 65-67% ethylene oxide content).

These samples of OZhK were tested in desalting various crudes in the Moscow and Kuibyshev refineries and in a Tatar ASSR oil field(Bavlyneft'). The OZhK proved to be highly effective in all tests.

In the Moscow refinery, desalting was carried out on mixed Tatar crudes (density of crude 0.866, salt content 1900-2200 mg/liter, water 0.4-0.8%) over a two-week period, operating the ELOU in two stages with a throughput of 350 m^3/h and with OZhK demulsifier consumption of 20 g/metric ton of crude. During the OZhK test, the salt content of the desalted crude averaged 26 mg/liter and the water content 0.1%.

The commerical test of OZhK in the Kuibyshev refinery ELOU on the desalting of heavy sulfurous crudes—Sernovodsk and Buguruslan—also demonstrated its high efficiency.

The Buguruslan crude had density 0.876, water content 1.4%, chloride salts 3200 mg/liter; the Sernovodsk crude had density 0.894, traces of water, and salts 470 mg/liter.

During the OZhK demulsifier test, the Kuibyshev refinery ELOU was operated as single-stage. With an OZhK consumption of 40-50 g/metric ton, the treated Sernovodsk crude contained 0.1% water and 175 mg/liter of salts; the Buguruslan crude contained 0.1% water and 230 mg/liter of salts. Two-stage operation of the ELOU with OZhK demulsifier could undoubtedly attain almost complete desalting of the indicated heavy sulfurous crudes. This is confirmed by the results cited previously from the test on desalting these crudes in the pilot unit (Table 4). It should be noted that nonionic demulsifiers synthesized on the basis of alkylphenols (OP-10, KAUFE, UFE), are unsuitable for desalting of such crudes as Buguruslan and Sernovodsk. Unfavorable results were obtained in tests conducted previously on these demulsifiers in the Kuibyshev refinery ELOU.

The Bavly crude coming in for demulsification and desalting in the oil field ELOU contains 20-30% water and 50,000-100,000 mg/liter of chloride salts. Naturally, more demulsifier is required for pretreating such crude than in the desalting of crude (containing not more than 2% water) in refineries.

The unit available in the Bavlyneft' oil field for conducting the OZhK test consisted of two stages — thermochemical and electrical. During the test, the electrodehydrators were converted from parallel to two-stage operation, increasing the load on the electrodehydrators by a factor of 2. Under these conditions and with an OZhK demulsifier consumption of 50-80 g/metric ton of crude, the Bavly field crude was desalted to a content of 47 mg/liter of chloride salts and 0.2-0.3% water. With normal loading of the electrodehydrators, the desalting results will of course be still better.

Summary

1. An effective nonionic demulsifier, OZhK (ethoxylated fatty acid), has been synthesized. The starting material for synthesis of this demulsifier was synthetic fatty acids with more than 20 carbon atoms per molecule, which are contained in the still bottoms from the fractionation of acids obtained by paraffin oxidation.

The still residues, which up to now have not found regular application, are obtained in high yield and represent a cheap and plentiful raw material and hence are extremely promising for the production of demulsifiers.

2. The OZhK demulsifier was tested in a pilot-plant ELOU and then in refineries and oil fields for desalting various crudes differing in quality and properties.

The tests demonstrated the high efficiency of OZhK and its suitability for pretreatment of both light crudes and the difficultly dewatered and desalted heavy, viscous, and high-sulfur crudes.

3. For pretreatment of crude in an ELOU, the OZhK demulsifier is as good as the best imported demulsifier.

Literature Cited

1. W. S. Barnickel, Crude oil treating process. U.S. Pat. 1,093,098 (April 14, 1914); U.S. Pat. 1,223,659 (1917), and others.
2. L. F. Berkgan, Russ. Pats. 26,671 and 26,676. Method of Berkgan – demulsification by means of naphthenic acids.
3. Walter Franke, "Demulsifiers for crude oil emulsions," Erdöl und Kohle, No. 1, 18 (1960).
4. A. I. Kudryashov, Khimicheskaya nauka i promyshlennost, 4, 5 (1959).
5. A. T. Ballun, et al., J. Am. Oil Chemists Soc., 31, 1, 20 (1954).
6. D. N. Levchenko, A. D. Khudyakova, and A. L. Kalitaeva, "Production of an effective nonionic demulsifier," Novosti neftyanoi i gazovoi tekhniki (neftepererabotka i neftekhimiya), No. 1, 8 (1961).
7. D. N. Levchenko, et al., "Synthesis of demulsifiers from monoesters of xylitan and their testing," Novosti neftyanoi i gazovoi tekhniki (neftepererabotka i neftekhimiya), No. 4 (1961).
8. D. N. Levchenko, A. D. Khudyakova, et al., "Nonionic surfactants – demulsufiers for crude oil emulsions," Khim. i tekhnol. topliv i masel, No. 4 (1960).
9. S. A. Feigin, L. I. Buchina, and D. N. Levchenko, "Prospects for the use of new demulsifiers," Khim. i tekhnol. topliv i masel, No. 12 (1961).
10. P. Becher, "Emulsions," New York–London (1957).
11. Chem. Age India, No. 1, 47-60 (1961).
12. "Nonionic surface-active substances" Ekspress-informatsiya "Promyshlennyi organicheskii sintez," No. 28, abstract No. 248 (1961).
13. "Nonionic detergents," Ind. Eng. Chem., 46, 1930-1934 (1954).

PROSPECTS FOR THE USE OF NEW DEMULSIFIERS

Ya. G. Sorkin and Ya. I. Nel'kenbaum

Serious attempts to convert to the use of new highly effective demulsifiers were started in 1959.

In the Gor'kii district, the auxiliary substance OP-10 has been used as a demulsifier. In Giprovostokneft', VNII NP, and VNIIneftekhim, a large number of different surfactants have been synthesized on the basis of other ethoxylated alkylphenols (VNII NP-58, $KAUFE_{14}$). The shortage of phenol has given rise to a number of proposals for its replacement. Thus, proposals have been made for the use of coal tars, wastes from the production of Ionol [2,6-dibutyl-4-methylphenol], and other products rich in alkylphenols [3]. Systematic work in this direction has been carried out by Giprovostokneft'.

Commercial tests were carried out on some of the synthesized products, in refineries and oil fields. The consumption of these materials proved to be considerably lower than that of NChK. This may also be said of other tests on alkylphenol derivatives. Recently, in Giprovostokneft', still more active demulsifiers have been synthesized on the basis of alkylphenols.

Apart from alkylphenols, alcohols may serve as another potential base for demulsifier production [4].

An outstanding raw material base for producing demulsifiers is the synthetic fatty acids obtained by air oxidation of paraffin. Apart from fatty acids, alcohols, or alkylphenols, other reagents are necessary in demulsifier production. In particular, there is a requirement for ethylene oxide, which is in great shortage. On the basis of fatty acids, it has been possible to obtain samples of demulsifiers with a lower consumption of ethylene oxide (based on the crude treated) than is the case when producing demulsifiers on the basis of alkylphenols.

In view of the diversity of petroleum emulsions, it is difficult to hope that there will be found a single universal demulsifier, equally economical and effective for all types of crude. It is evident that production of

Scheme 1

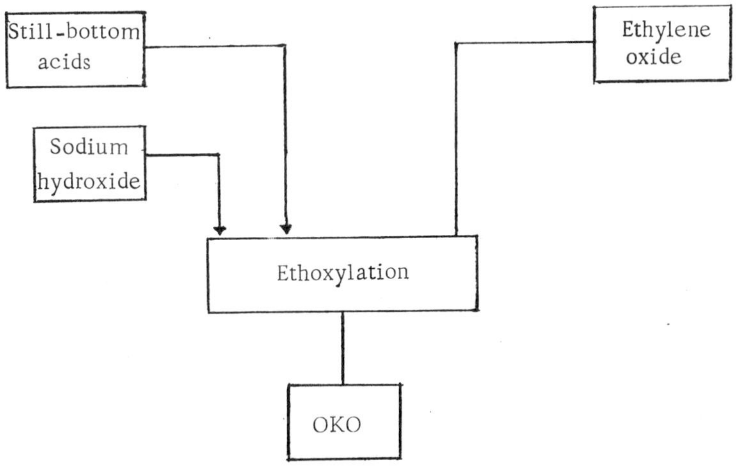

Fig. 1. Scheme of synthesis of OKO demulsifier.

Scheme 2

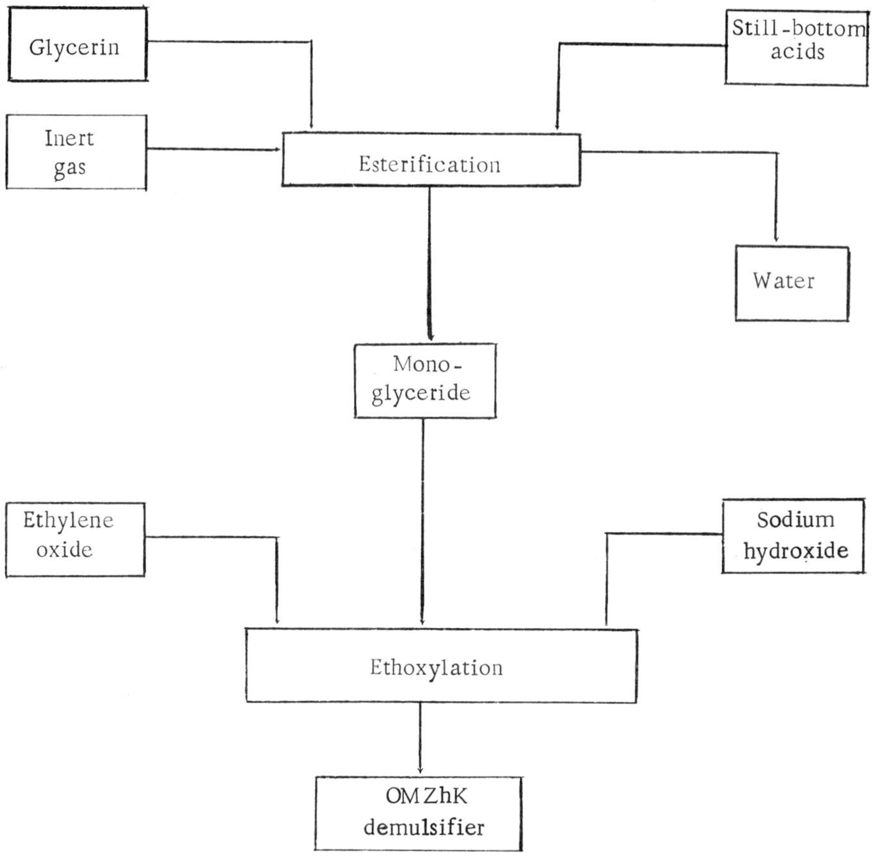

Fig. 2. Scheme of synthesis of OMZhK demulsifier.

a certain number of demulsifiers will be necessary, preferably with identical technology and on the basis of an available inventory of starting materials.

The still bottoms of synthetic fatty acids are typical raw material [8] for the production of demulsifiers. On the basis of this material, at least several types of demulsifiers can be obtained, differing in their properties and, as shown by tests, suitable for pretreatment of all crudes currently being refined in Bashkiriya. Naturally, the products obtained may also find other applications.

These studies were presented for the first time at the First Conference on Application of Surfactants in the Petroleum Industry, in Ufa in February 1960 [7].

VNII NP also carried out work on producing demulsifiers on the basis of the fatty acids obtained by oxidation of petroleum paraffin. The ethoxylated monoester of xylitan (KS-59) and fatty acids went through a commercial test in a petroleum refinery. However, its cost [9] proved to be too high. The material cheaper than the others proved to be demulsifier OZhK, obtained similarly to the demulsifier OKO [5] by ethoxylation of fatty acid still bottoms (above C_{20}).

At UNPZ, three schemes of demulsifier synthesis have been developed on the basis of still-bottom fatty acids (Figs. 1, 2, and 3).

1. Direct ethoxylation of the still bottoms from the synthetic fatty acids.

2. Ethoxylation of the monoglycerides of these acids.

3. Ethoxylation of the di-(mono) glycerides of these acids with orthophthalic acid.

Scheme 3

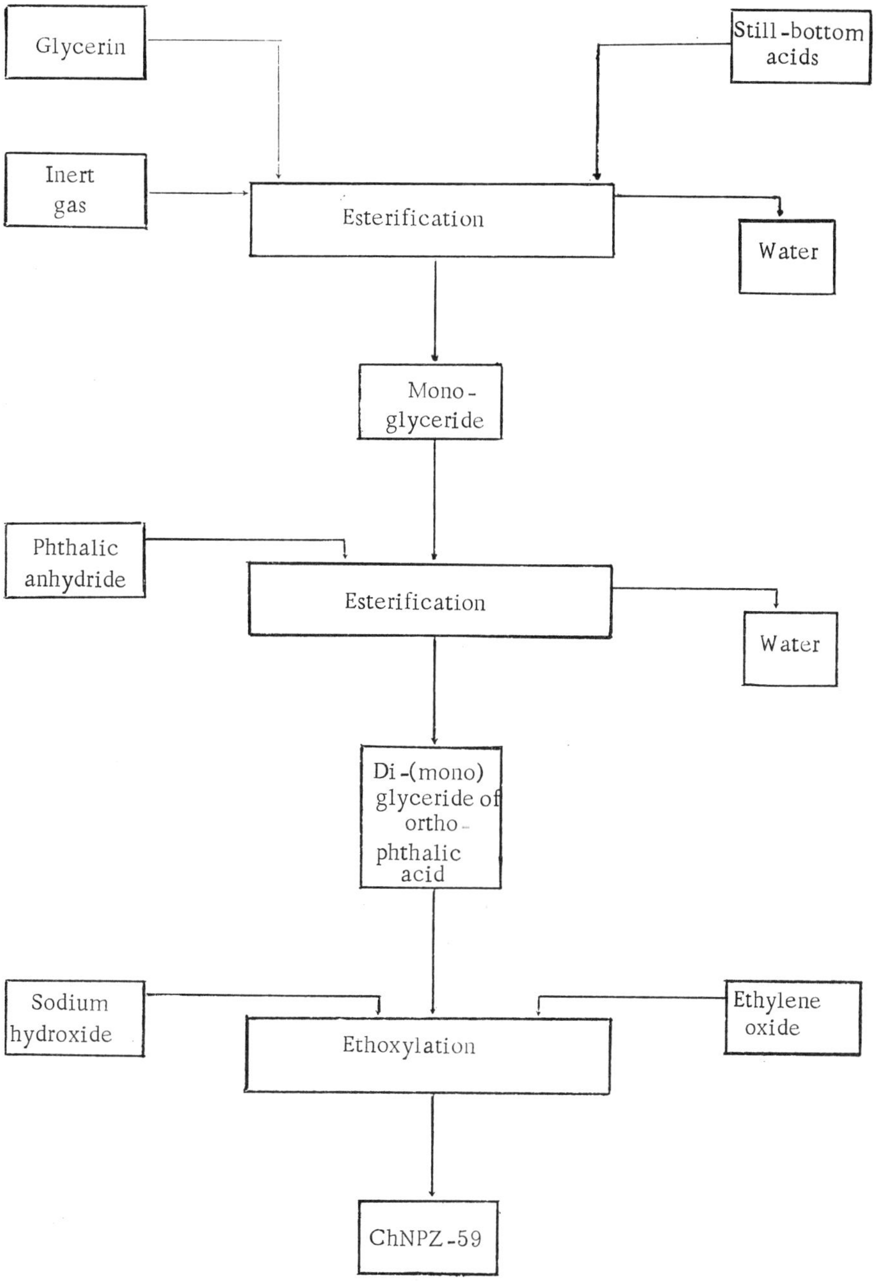

Fig. 3. Scheme of synthesis of ChNPZ-59 demulsifier.

The reagents for all of these syntheses will be produced by Bashkir industry.

The higher fatty acids contained in the still bottoms react with ethylene oxide in the presence of a small quantity of sodium hydroxide (actually present as the fatty acid salt) according to the following equation:

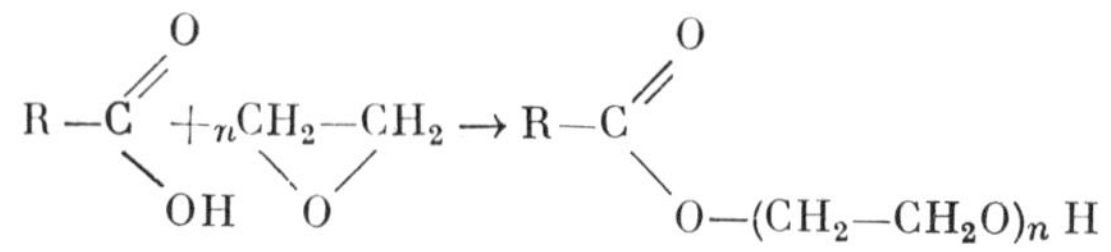

As is evident from the scheme, the formation of the monoester of the polyglycol proceeds as an irreversible addition reaction, not accompanied by the evolution of any by-products. As a result of the synthesis, a mixture of products is formed with different degrees of ethoxylation, i.e., with different numbers of molecules of ethylene oxide added.

The reaction product includes a certain amount of sodium salts of the fatty acids as well as esters present in the original still-bottom acids.

In producing demulsifiers according to the second scheme, it is necessary to obtain first a monoglyceride and in the following stage to subject this monoglyceride to ethoxylation:

1. $CH_2OH{-}CHOH{-}CH_2OH + RCOOH \rightarrow CH_2OH{-}CHOH{-}CH_2{-}COOR + H_2O$

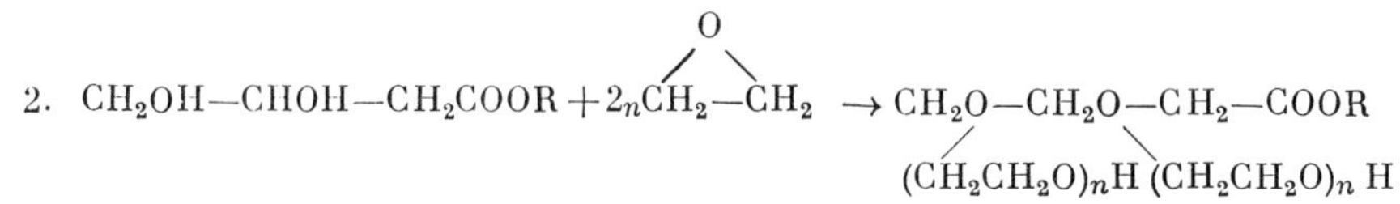

The production of demulsifiers according to the third scheme is carried out in three stages. The first stage is the interaction of the carboxylic acids with glycerin to obtain a monoglyceride as described above. In the following stage, the monoglyceride is condensed with phthalic anhydride (the anhydride of orthophthalic acid). The resulting diester is ethoxylated in the third stage:

2. $2CH_2OH{-}CHOH{-}CH_2{-}COOR + C_6H_4(CO)_2O \rightarrow C_6H_4(COO{-}CH_2{-}CHOH{-}CH_2{-}COOR)_2$

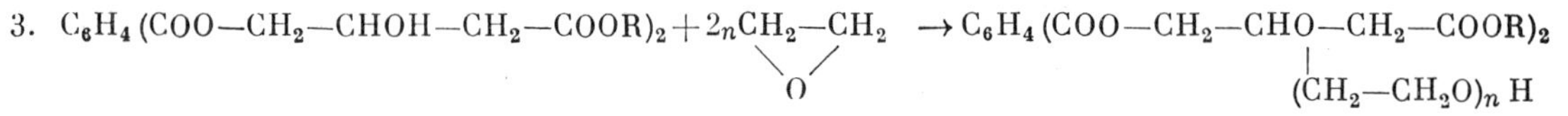

In these syntheses, the glycerin may be replaced by other polybasic alcohols (e.g., metriol), and the o-phthalic acid by other dibasic acids (e.g., adipic acid).

The process of monoglyceride production as well as the condensation of the monoglyceride with phthalic anhydride proceed without catalyst on heating the reaction mixture in a stream of inert gas to 150-180°C. In the ethoxylation, sodium hydroxide or other alkali is used as catalyst.

The products obtained represent a mixture of ethoxylated products of differing degree of ethoxylation.

The products that we synthesized were tested for demulsifying power under standard laboratory conditions (temperature 50°C, settling time 4 hours) and were compared with the well-known demulsifiers NChK, OP-10, and $KAUFE_{14}$. Curves giving the relation between demulsifier consumption and degree of desalting for the demulsifiers investigated (and the corresponding crudes) illustrate graphically the high efficiency of the demulsifiers OKO and ChNPZ-59 in comparison with the other materials.

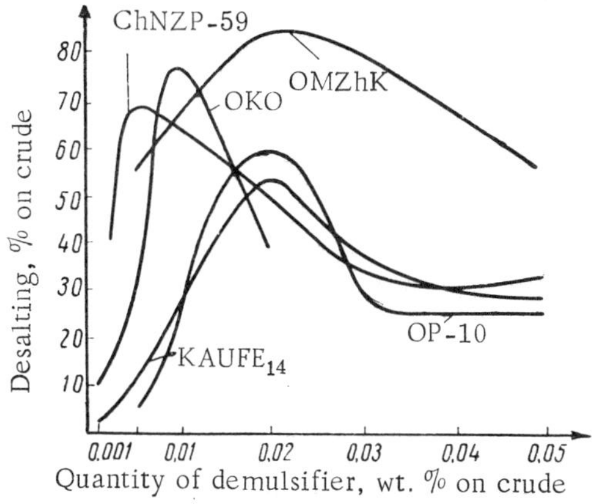

Fig. 4. Degree of desalting as a function of demulsifier consumption.

These tests show that a high degree of desalting can be obtained with low consumptions of the OKO and ChNPZ-59 demulsifiers (Fig. 4). Comparing the OMZhK demulsifier with OP-10 and $KAUFE_{14}$ at the same demulsifier consumption, the OMZhK gave the better results.

Of the series of products that we synthesized, demulsifier OKO is the most simple to prepare.

Here a few words are needed on the concept of "demulsifier efficiency."

Other conditions being equal, demulsifier efficiency is greater the higher the degree of dewatering or desalting and the lower the demulsifier consumption.

This may be expressed by the formula

$$E = D/C,$$

where E is efficiency, D is the dewatering or desalting in %, and C is the consumption.

The curve of degree of desalting (%) vs. demulsifier consumption (%) usually has a maximum in some section.

The ratio of the degree of desalting to the consumption at the maximum point in this curve is arbitrarily termed the "efficiency." At this point, the emulsion under the test conditions shows maximum instability with minimum consumption of demulsifier.

The action of the above-listed reagents was tested with the same crude from the Romashkino field. It was also of interest to investigate the action of one selected reagent (OKO-100) in other crudes refined in Bashkiriya– Vvedenov, Arlanskii, and Shkapovskii, (Figs. 5-7). The demulsifier proved to be effective only in the tests on emulsions of Arlanskii and Vvedenov crudes. For the Arlanskii crude, the optimum dosage was 0.007-0.01%; for the Vvedenov crude, the demulsifying action was shown over a wider range (0.005-0.03%).

With Shkapovskii crude, the demulsifier proved to have little effect; even with a consumption of 0.03%, the desalting amounted to only 54% (Fig. 5). It is interesting to note that the demulsifier OP-10, which has absolutely no effect on Arlanskii crude, desalts Shkapovskii (D-I) crude far better than does the OKO-100.

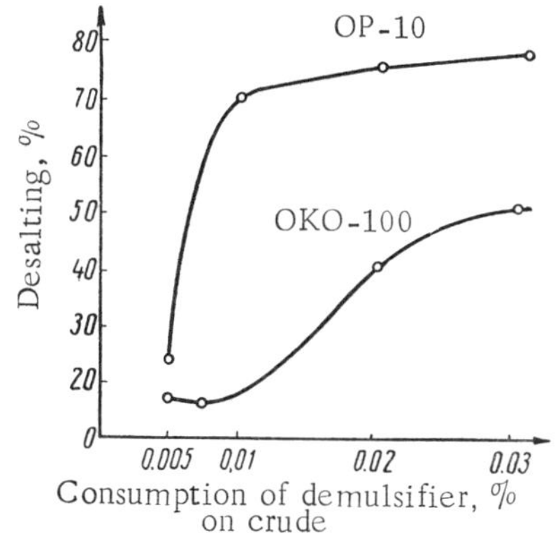

Fig. 5. Degree of desalting of Shkapovskii crude as a function of demulsifier consumption for OKO-100.

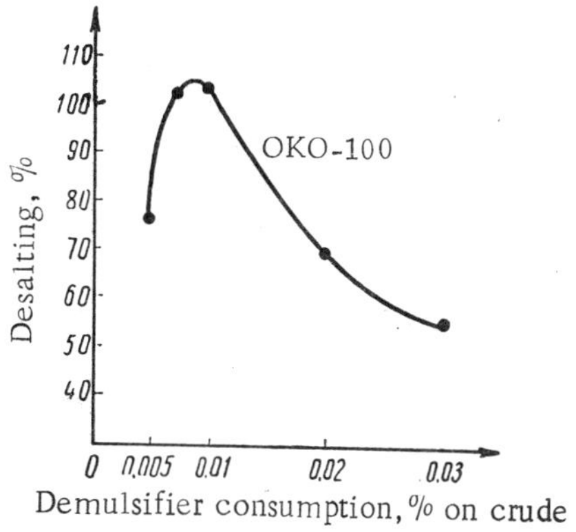

Fig. 6. Degree of desalting of Arlanskii crude as a function of demulsifier consumption for OKO-100.

TABLE 1

Name	Ethylene oxide content, wt. %	Laboratory tests		Commercial tests	Ratio A/B
		Optimum demulsifier consumption, g/metric ton	A = consumption of ethylene oxide, g/metric ton	B = consumption of ethylene oxide, g/metric ton	
$KAUFE_{14}$	72	200	144	–	
OP-10	69	200	138	35	3.95
OMZhK	66	200	132	–	
OKO-100	61	100	61	16	3.8
ChNPZ-59	40	50	20	5?	4?
4422	51*	50	25	6	4.2
4411	51*	30	16	4	4

*Corresponding to the data of VNII NP instructions on the use of these demulsifiers, taking the solvent into account.

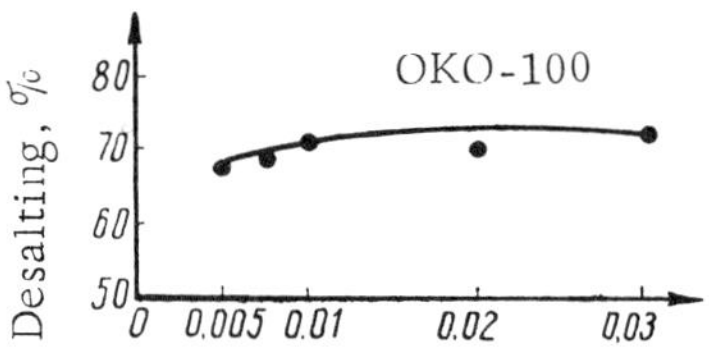

Fig. 7. Degree of desalting of Vvedenov crude as a function of demulsifier consumption for OKO-100.

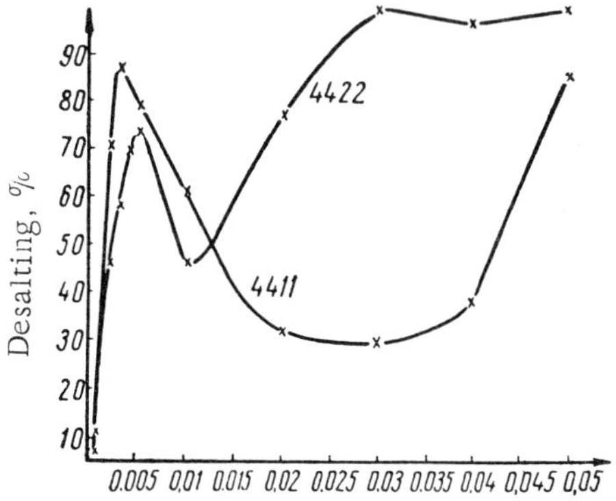

Fig. 8. Degree of desalting of Romashkino crude as a function of demulsifier consumption for 4411 and 4422.

For comparison, results are shown from tests on reagents 4422 and 4411 with Romashkino crude (Fig. 8). From the unusual nature of the curves, it may be concluded that these demulsifiers are suitable for dewatering and desalting crudes from the Romashkino oil field in the region of 0.003% concentration for the 4411 and 0.005% for the 4422.

The tests on these demulsifiers indicate that their action is specific.

The hydrophilic groups in the OKO and OMZhK demulsifier molecules are concentrated at one end of the molecule, as is observed in the case of ethoxylated alkylphenols. The molecules of such a surfactant are arranged at the phase boundary in the form of a "palisade." It is well known [1] that with the presence of several hydrophilic groups located at some distance from each other the molecules at the phase boundary are arranged in a "lying" position. Hence, they occupy a considerably greater section of the surface. The consumption of a "lying" demulsifier required to occupy all or the same part of the entire surface will be smaller, and the efficiency higher.

Corresponding to this working hypothesis, the demulsifier ChNPZ-59 was synthesized, and by its use it was possible to achieve a very low consumption of ethylene oxide for pretreating the crude.

The consumption of ethylene oxide for treating 1 metric ton of Romashkino crude with nonionic demulsifiers is shown in Table 1.

TABLE 2

Name of reagent	Cost per metric ton	Consumption, g/metric ton	Demulsifier cost per metric ton of crude, rubles
OZhK	192*	30	0.57
ChNPZ-59	207	20†	0.40

*According to the data of [9].
†Consumption is overestimated.

Judging by the laboratory tests, at the optimum dosage, the ethylene oxide consumption for treating 1 metric ton of Romashkino crude is 144 and 138 g with $KAUFE_{14}$ and OP-10, respectively; when using OMZhK, the ethoxylated monoglyceride, the consumption is 132 g. For the ethoxylated still-bottom fatty acids (OKO-100), this consumption is 61 g, for ChNPZ-59 only 20 g, and for demulsifiers 4422 and 4411, 25 and 16 g, respectively.

The laboratory tests were conducted under very primitive conditions – settling in separatory funnels for 4 hours at 50°C. Under commercial conditions, when using surfactants as auxiliary substances for electrode-salting, the demulsifier consumption is of course lower, thanks to the use of improved techniques. In commercial tests at UNPZ on the demulsifiers OKO-100, 4422, and 4411, the consumption of ethylene oxide in grams per metric ton of crude proved to be one-fourth of that found under the standard conditions that were adopted for the laboratory tests. This also proved to be the case on recalculating the known indices of OP-10 consumption in the Novo-Gor'kii refinery. It can be assumed with some confidence that the use of ChNPZ-59 under commercial conditions will also result in a consumption that is one-fourth of that obtained under laboratory conditions, i.e., 5 g of ethylene oxide per metric ton of crude.

The article of S. A. Feigin, L. I. Buchina, and D. N. Levchenko [9] examines the question of the prospects for the use of new demulsifiers.

In this article, no heed is given to the work on synthesis of demulsifiers at Giprovostokneft' and UNPZ. The authors limited themselves to a review of the prospects for use of surfactant samples synthesized at VNII NP and also at VNIIneftekhim, and to a comparison of the cost of these surfactants and the capital expenditures in their production. At the end of their article, the authors came to the unfounded conclusion that of all the nonionic demulsifiers "the most efficient both as to cost and to capital expenditures" proved to be the demulsifier obtained by ethoxylation of still bottoms. Proceeding from this, calculations were made on the savings by complete replacement of the currently used demulsifiers by this reagent.

Such a formulation of the question is basically incorrect.

The demulsifiers synthesized previously at VNII NP (VNII NP-58, KS-59) were also intended for exclusive replacement of all existing demulsifiers. In the article cited above, the authors abandoned these in favor of a more recent demulsifier. However, as indicated above, OZhK is not the most efficient of "all" existing demulsifiers. For Shkapovskii crude, it is even inferior to OP-10 demulsifier. The technical data on OZhK can be improved, although it does not attain maximum efficiency. With a higher efficiency, even a more expensive product may prove economical. Calculations show that the costs in pretreating crude using the more efficient demulsifier ChNPZ-59 are lower than the costs when using ethoxylated still bottoms.

A comparison of the crude pretreatment costs using demulsifiers is given in Table 2.

The practice of foreign [non-Soviet] refineries using nonionic demulsifiers shows that, owing to the diverse nature of crudes, a certain assortment of demulsifiers is necessary. It is difficult to expect that a single demulsifier would successfully satisfy all demands in the pretreatment of crude. Therefore, it is necessary to approach more cautiously the prospects for the use of ethoxylated still-bottom acids.

It is also impossible to agree with the calculations of [9] on the amount of ethylene oxide required for the preparation of nonionic demulsifiers. In the first place, the quantity of ethylene oxide is overstated in the batch formula for the ethoxylated still-bottom acids. Actually, 60-61% ethylene oxide is sufficient in the finished product, and not 67% as stated in the above-indicated article. In the second place, even if this latter overstated figure is assumed, then the requirement of ethylene oxide for 12,000 metric tons of demulsifier will be 8000 tons and not the 10,000-15,000 tons stated in the article. It may be considered that this quantity (8000 tons) is also high, since there are already data indicating that demulsifiers can be used in which the specific consumption of ethylene oxide per metric ton of pretreated crude will be lower than for the ethoxylated still bottoms by a factor of at least 2 (ChNPZ-59).

The question of the required quantity of ethylene oxide is one of the decisive points in any discussion of the prospects for production of demulsifiers in the immediate future, since the introduction of effective demulsifiers into production is retarded by the absence of sufficient quantity of ethylene oxide. An unwarranted overstatement of the required quantity by no means favors a solution of the problem.

Fatty acids, as follows from what has been discussed, can be a typical raw material for the production of a certain assortment of demulsifiers. It has been shown that there is a possibility of producing three types of demulsifiers: OKO, OMZhK, and ChNPZ-59. This does not exhaust all the possibilities of this raw material. With the same technology, one can obtain demulsifiers of different quality by changing the composition of the acids used. Other possible variations involve the extent of esterification and ethoxylation and the quality of the polybasic alcohols and dicarboxylic acids that are used, obtaining surfactants that are diverse in composition of the hydrophobic part and thus also differ in their properties. Regarding the hydrophilic part of the substances synthesized, the studies have showed that for each case there are one or two optimum amounts of ethylene oxide added for which the surfactant shows the best demulsifying power. The hydrophilic part may also be influenced, using propylene oxide. Finally, there is also the possibility of producing a mixture of demulsifiers.

Practice in commercial operation on crude pretreatment by means of nonionic demulsifiers indicates that an assortment is advisable and also will indicate whether fatty acids alone are adequate to solve all problems in crude pretreatment or whether other raw materials are also necessary.

Literature Cited

1. N. K. Adam, Physics and Chemistry of Surfaces [Russian translation] (Gostoptekhizdat, 1947).
2. V. S. Edrenkin, P. A. Chirimanov, I. I. Spasskii, Z. A. Bernadyuk, G. V. Konyaev, and M. S. Matveev, "High-efficiency demulsifier – the basis for improvement in waste water quality," Novosti neftyanoi tekhniki, No. 6 (1960).
3. D. N. Levchenko, A. D. Khudyakova, A. A. Kalitaeva, E. A. Shklyarchuk, V. I. Khokhlov, and A. S. Chugreeva, "Nonionic surfactants – demulsifiers for crude oil emulsions," Khim. i tekhnol. topliv i masel, No. 4 (1960).
4. D. N. Levchenko, I. V. Bergshtein, D. V. Ivanyukov, and L. E. Zlotnikov, "Efficient use of nonionic demulsifiers in an electrodesalting unit with spherical electrodes," Novosti neftyanoi i gazovoi tekhniki, No. 5 (1961).
5. D. N. Levchenko, A. D. Khudyakova, and A. L. Kalitaeva, "Production of an effective nonionic demulsifier," Novosti neftyanoi i gazovoi tekhniki, No. 1 (1961).
6. A. A. Petrov and N. P. Borisova, "Investigation of demulsifying power of ethoxylated alkylphenols," Khim. i tekhnol. topliv i masel, No. 2 (1959).
7. Proceedings of the First All-Union Conference on the Use of Surface-Active Substances in the Petroleum Industry and Their Production (Gostoptekhizdat, 1961).
8. Ya. G. Sorkin, Ya. I. Nel'kenbaum, and F. A. Mamina, "Still bottoms of fatty acids – raw material for the production of nonionic demulsifiers," Khim. i tekhnol. topliv i masel, No. 2 (1961).
9. S. A. Feigin, L. I. Buchina, and D. N. Levchenko, "Prospects for the use of new demulsifiers," Khim. i tekhnol. topliv i masel, No. 12 (1961).

SEARCH FOR NONIONIC DEMULSIFIERS WITH INCREASED ACTIVITY FOR DEVONIAN CRUDES

M. Z. Mavlyutova

Commercial tests under oil-field conditions with Devonian crudes using the demulsifiers $KAUFE_{14}$ and UFE_8, synthesized at VNIIneftekhim, showed that they provide breaking of the emulsion at the following consumptions: $KAUFE_{14}$ 70-120, UFE_8 300-360, and OP-10 70-180 g/metric ton of raw crude. However, the quality of the treated crude depends on strict observerance of the optimum dosage, which for the $KAUFE_{14}$ and OP-10 depends to a considerable extent on the water content of the crude and also on the dispersity of the emulsion [1].

We have carried out laboratory experiments on determining the optimum dosage of demulsifier and settling time, and also on establishing the dependence of the demulsification process on the intensity of mixing the reagent with the emulsion.

Samples of crude for our experiments were taken from the discharge line of the raw crude pump in the Naryshev ELOU [electrodesalting unit].

In order to ascertain the dependence of water settling rate on reagent dosage and to find the optimum dosage, experiments were carried out under the following conditions: OP-10 was used in the form of a 2% aqueous solution; constant (60°C); time of contacting the emulsion with the reagent was 3 minutes under conditions of intensive mixing with a mechanical stirrer or 5 minutes by hand; the raw crude was taken in amount of 200 g. Experiments were set up in settlers to determine the "excise tars" [asphaltic tar].

After mixing the emulsion with the demulsifier, the settlers were placed in the water bath, and the volume of water settled out was noted every hour.

After 2 hours of settling the settlers were removed from the bath and held 1 hour at room temperature, after which the upper layer of dewatered crude from each settler was sampled (150 ml) for analysis for residual water content.

TABLE 1

Dosage of OP-10		Volume of water settled out after 2 hours, ml	Water content of dewatered crude, wt. %
% of raw crude	g/metric ton		
0.003	30	0	Not determined
0.004	40	0	Same
0.005	50	0	"
0.006	60	11.5	0.9
0.007	70	13.0	0.27
0.008	80	12.0	1.09
0.009	90	8.2	2.7

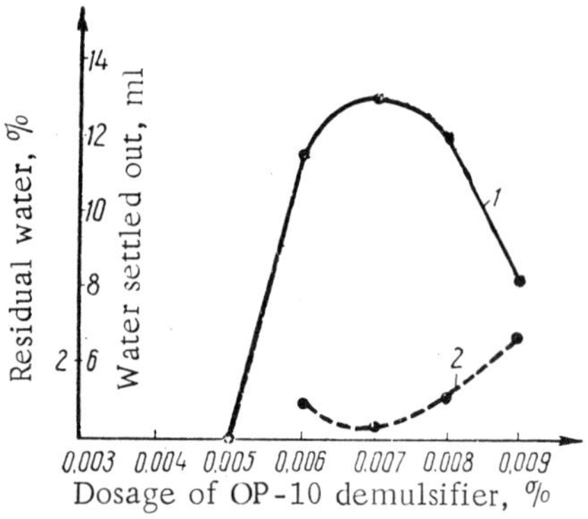

Fig. 1. Volume of water settled out and quality of dewatered crude as a function of OP-10 reagent dosage. 1) Volume of water settled out; 2) residual water content of dewatered crude.

Data are shown in Table 1 and Fig. 1 from the experiment on choosing the optimum dosage of demulsifier OP-10 for a crude oil emulsion with water content 8%.

From these data it is evident that, for the Naryshev crude containing 8% water, the optimum dosage of OP-10 is 70 g/metric ton of raw crude. With dosages below or above this optimum, the demulsification process becomes worse. On increasing the reagent consumption, a sharp increase in the residual water content of the dewatered crude is characteristic.

Figure 2 shows the kinetics of water settling from an emulsion of Naryshev crude containing 12.3% water, and Fig. 3 shows the settling of water in 1 and 2 hours and the water content of the dewatered crude as a function of the dosage of demulsifier OP-10.

Increasing the dosage of OP-10 from 125 to 170 g/metric ton accelerates the water settling. A further increase in dosage up to 185 g/metric ton gives a marked deterioration of the demulsification process, decreasing the quantity of water settled out and increasing the water content of the dewatered crude.

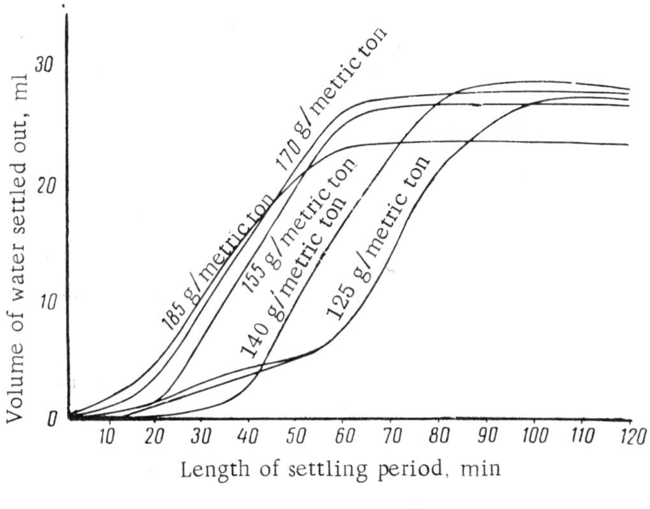

Fig. 2. Rate of water settling as a function of dosage of OP-10.

Dosage of OP-10, g/metric ton	Residual water content of crude, %
125	0.18
140	0.35
155	1.2
170	0.6
185	2.0

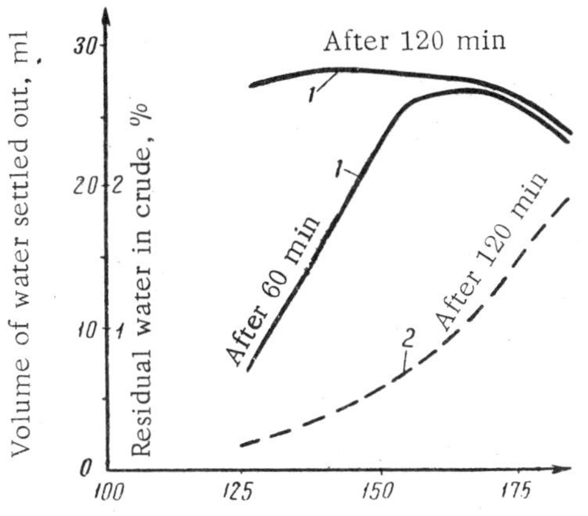

Fig. 3. Volume of water settled out in 1 and 2 hours and quality of dewatered crude, in relation to dosage of OP-10. 1) Volume of water settled out; 2) residual water content of dewatered crude.

In order to obtain satisfactory results with OP-10, it must be introduced into the petroleum emulsion in strictly determined quantities. Also, a prolonged settling period is required, which implies a sharp reduction in the throughput of existing units and provision for a raw crude feed with constant content of emulsified water.

Unfortunately, even under conditions of one and the same oil field, these requirements are practically impossible to accomplish, since the emulsions produced in various sections of the oil field are not identical either in water content or in extent of dispersal of the emulsion.

Besides the indicated shortcomings, OP-10 has a lower demulsifying power in comparison with well-known nonionic demulsifiers, for example, reagent 4411 from the FRG [Federal Republic of Germany].

Data are shown in Fig. 4 on the demulsification of Naryshev crude containing 12.3% water, under the action of three types of nonionic surfactants (OP-10, $KAUFE_{14}$, and reagent 4411).

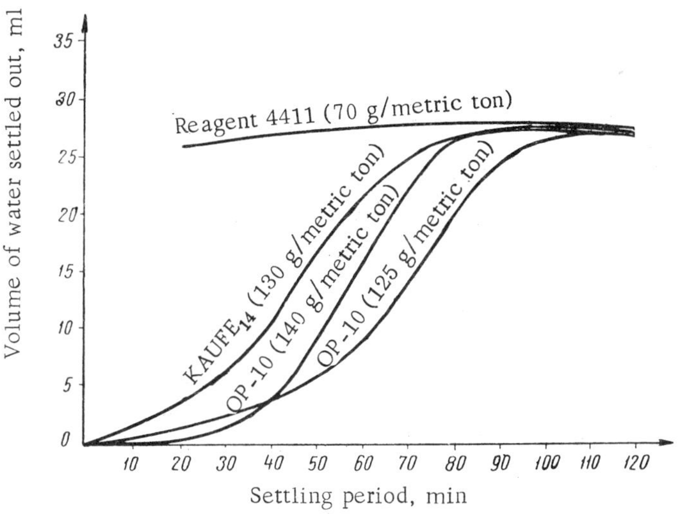

Fig. 4. Rate of settling of water separated from emulsion of Devonian crude by the action of three types of demulsifiers.

Reagent	Reagent consumption, g/metric ton	Residual water content in dewatered crude after 2 h settling, %
$KAUFE_{14}$	130	0.1
OP-10	125	0.18
OP-10	140	0.35
4411	70	0.03

TABLE 2

Dosage, wt. %	OK-13		OK-34		OK-49.6		OK-55	
	Volume of water settled out, ml	Water content of dewatered crude, %	Volume of water settled out, ml	Water content of dewatered crude, %	Volume of water settled out, ml	Water content of dewatered crude, %	Volume of water settled out, ml	Water content of dewatered crude, %
0.006	1.5	Water not determined	50.0	0.06	25.0	3.9	30.0	4.4
0.007	3.3	"	52.0	0.24	38.0	1.35	30.0	1.65
0.008	4.5	"	52.0	0.60	50.0	0.06	40.0	1.17

Note. Water content of original crude 13.1%. Settling time 2 h.

TABLE 3

Dosage, wt. %	DK-15		DK-41		DK-52	
	Volume of water settled out, ml	Water content of dewatered crude, %	Volume of water settled out, ml	Water content of dewatered crude, %	Volume of water settled out, ml	Water content of dewatered crude, %
0.001	0	–	1.0	Water not determined	0	9.04
0.002	0	–	2.0	"	1.1	7.85
0.003	0	–	4.6	"	3.7	3.93
0.004	0	–	7.5	0.06	14.0	4.36
0.005	0	–	40	0.24	19.0	2.75
0.006	0	–	45	0.06	32	0.78
0.007	0	–	50	0.06	40	0.17
0.008	16	0.06	50	0.06	50	0.06
0.009	16	0.12	50	0.06	–	–
0.01	17	0.23	50	0.06	–	–

Note. Water content of original crude 13%.

TABLE 4

Dosage, wt. %	OF-10		OF-34		OF-54	
	Volume of water settled out, ml	Water content of dewatered crude, %	Volume of water settled out, ml	Water content of dewatered crude, %	Volume of water settled out, ml	Water content of dewatered crude, %
0.001	None	Water not determined	0.6	–	0.4	–
0.002	Same	Same	11.0	–	5.25	–
0.005	"	"	42	0.06	33	0.06
0.006	"	"	43	0.13	41	Traces
0.007	"	"	43	0.24	42	0.06
0.008	"	"	42	0.24	43	Traces
0.009	"	"	42	0.35	–	–

Note. Water content of original crude 12.7%.

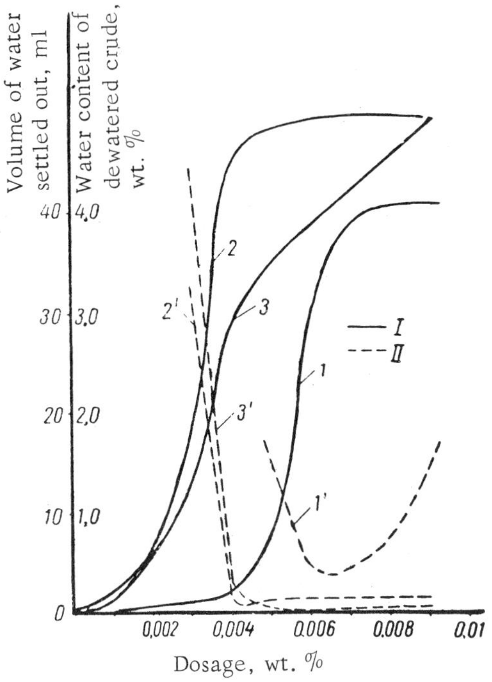

Fig. 5. Volume of water settled out and quality of dewatered crude in relation to dosage of dodecylphenols condensed with various amounts of ethylene oxide. I) Volume of water settled out; II) water content of dewatered crude; 1, 1') DF-15; 2, 2') DF-32; 3, 3') DF-51.

It can be seen that OP-10 has a lower demulsifying power than $KAUFE_{14}$. The specific consumption of these demulsifiers is considerably higher than for the reagent 4411. This latter reagent, at a consumption of 70 g/metric ton, after 20 minutes had separated off the principal mass of the water and made it possible to obtain, after 2 hours' settling, crude with residual water content of 0.02%.

In connection with this situation, we continued studies on selecting nonionic demulsifiers of high activity. The following surfactants were investigated for demulsifying properties.

1. Condensation products of iso-octylcresol (OK) with 13, 34, 50, and 55 molecules of ethylene oxide; iso-octylphenol (OF) with 10, 34, and 54 molecules of ethylene oxide; dodecylcresol (DK) with 15, 41, and 52 molecules of ethylene oxide; and dodecylphenol (DF) with 15, 32, and 51 molecules of ethylene oxide. These were synthesized by O. Voiku in the Kympinsk Institute of Drilling Production of Oil and Gas of the Rumanian People's Republic.

2. Derivatives of the commerical product OP-10—OP-20, OP-45, and a mixture of the two latter reagents in 1:1 ratio [3].

3. Condensation products of coal tar phenols alkylated with 130-140°C and 140-170°C fractions of thermally cracked naphtha, with various numbers of ethylene molecules (from 6.6 to 49 molecules per molecule of alkylphenol. These products were designated $K_{130-140}AUFE_n$ and $K_{140-170}AUFE_n$ [3].

4. Products from the ethoxylation of alcohols and fatty acids, synthesized by VNII NP and others.

Studies of the demulsifying properties were performed on Devonian crude of the Tuimazy oil field, taken from the discharge line of the raw crude pump of the Naryshev ELOU.

The results of the experiments on dewatering crude under the action of ethoxylated alkylcresols OK and DK are shown in Tables 2 and 3, results on the alkylphenol OF in Table 4, and on DF in Fig. 5 [2].

All of the reagents were added in the form of 1% aqueous solutions to the crude being demulsified. The dosages given in the tables are calculated on the basis of 100% active material and are expressed in weight percentages on the raw crude.

The reagents, except the OK series, were tested at dosages from 0.001 to 0.01%. Reagent OK-13 (iso-octylcresol condensed with 13 molecules of ethylene oxide) at dosages below 0.015% did not show any demulsifying action. The OK-34 at a dosage of 0.006% and the OK-49.6 at a dosage of 0.008% provided complete demulsification of the crude, but the OK-55 at these dosages left a considerable amount of water in the crude.

The demulsifying action of dodecylcresol with 15 molecules of ethylene oxide began to appear only at dosages above 0.008%, but even with increasing the dosage to 0.01%, it did not give complete breaking of the emulsion.

The optimum dosage of DK-41 was 0.006-0.007%, and for DK-52 it was 0.008%.

Of the test surfactants obtained from alkylcresols and ethylene oxide, the greatest demulsifying activity was shown by OK-34 and DK-41.

TABLE 5

Dosage, wt. %	OF-34		DF-32		OK-34		DK-41		4411	
	Volume of water settled out, ml	Water content of crude, %	Volume of water settled out, ml	Water content of crude, %	Volume of water settled out, ml	Water content of crude, %	Volume of water settled out, ml	Water content of crude, %	Volume of water settled out, ml	Water content of crude, %
0.004	27	6.5	14.5	–	–	–	–	–	–	–
0.005	43.5	0.03	21.5	–	14	–	9	–	–	–
0.006	43.5	0.03	43.5	0.03	40.5	Traces	41.5	0.06	–	–
0.007	44	0.03	45	0.03	43.5	0.03	43.5	0.06	43	0.18
0.008	–	–	45	0.03	44	0.03	44	0.18	–	–

Note. Crude sample 200 g; water content (Dean and Stark) 20.4%; mixed with propeller stirrer for 5 min; bath temperature 60°C; settling time 2 h.

A similar relation was detected in studying the condensation products of alkylphenols with various amounts of ethylene oxide.

Iso-octylphenol condensed with 10 molecules of ethylene oxide began to show demulsifying action at dosages above 0.01%; for the OF-34 at dosages above 0.005%, and for the OF-54 at dosages of 0.006%, the crude was obtained with insignificant content of residual water.

Dodecylphenol with 15 molecules of ethylene oxide (DF-15) was more active than DK-15 and at dosages of 0.007-0.009% separated a considerable part of the water from the emulsion, but as occurred with OP-10 and $KAUFE_{14}$, increasing the dosage led to an increase in the residual water content of the crude.

The highest demulsifying power was shown by DF-32. Its optimum consumption was 0.004%.

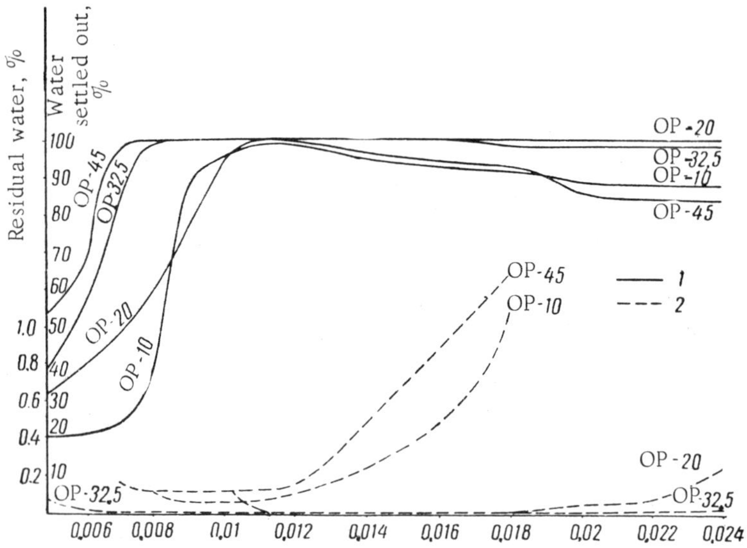

Fig. 6. Settling of free water and quality of dewatered crude as a function of dosage of various OP reagents. 1) Water settled out, in % of volume of water recovered under the action of 0.007% reagent 4411; 2) residual water content of dewatered crude.

TABLE 6

Dosage, Wt. %	OP-10		OP-20		OP-45		Mixture of OP-45 and OP-20 (1:1)	
	Water settled out, %	Water content of dewatered crude, %	Water settled out, %	Water content of dewatered crude, %	Water settled out, %	Water content of dewatered crude, %	Water settled out, %	Water content of dewatered crude, %
0.005	21.1	–	32.2	8.6	53.5	4.6	39.3	0.075
0.006	18.1	–	39.6	7.4	65.5	3.0	55.8	Traces
0.007	23.5	–	48.4	6.0	100	0.16	82.6	"
0.008	38.4	0.12	59.0	4.3	100	0.12	100	"
0.009	89.4	0.06	77.0	2.0	100	0.12	100	"
0.01	95.3	0.06	93.5	0.12	100	0.12	100	"
0.011	99.4	0.06	100	Traces	100	0.12	100	"
0.012	99.3	0.12	100	"	100	0.12	100	"
0.014	95.4	0.30	100	"	97.0	0.48	100	"
0.016	93.2	0.36	100	"	94.7	0.9	100	"
0.018	93.0	1.1	100	"	94.0	1.3	99.0	"
0.02	89.0	1.7	100	0.06	86.0	1.9	98.5	"
0.022	91.5	1.2	100	0.06	85.2	2.4	99.4	"
0.024	88.0	2.2	100	0.24	84.6	2.8	99.0	0.03
0.007	100.0	0.12	100	0.12	100	0.3	100	0.12

Volume of water settled out is expressed in percentage of volume of water recovered from given emulsion by the action of reagent 4411 at a dosage of 0.007%.

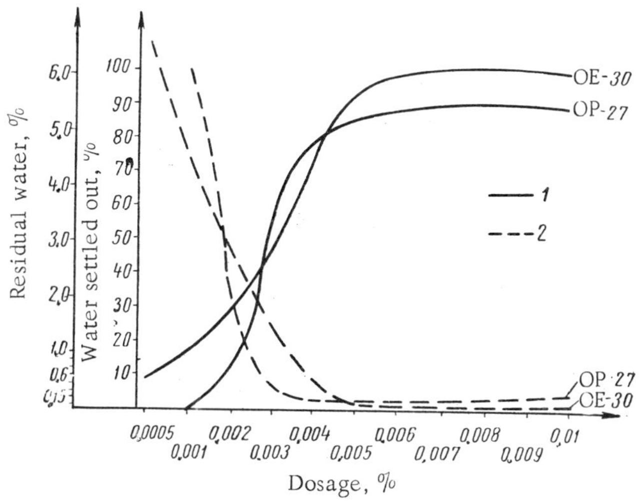

Fig. 7. Nature of action of ethoxylated and propoxylated alcohols on emulsion of Tuimazy Devonian crude (water content 10.8%). 1) Water settled out, % of volume of water recovered by the action of 0.007% reagent 4411; 2) water content of dewatered crude (wt. %).

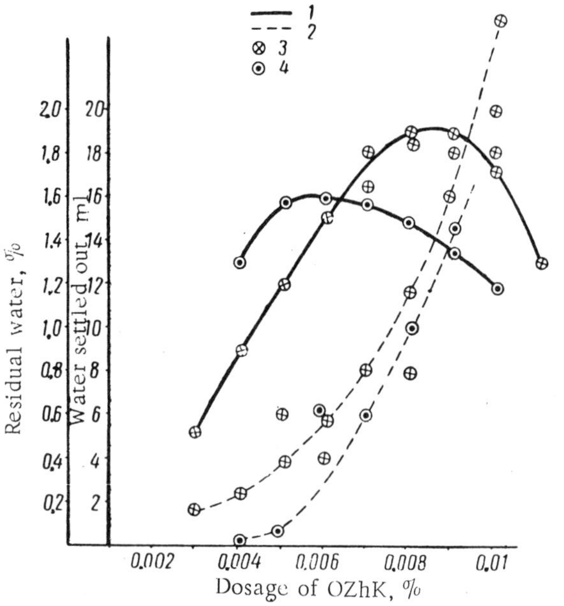

Fig. 8. Volume of water settled out and quality of dewatered crude in relation to dosage of OZhK. 1) Water settled out, ml; 2) residual water, %; 3) for crude with 10% water content; 4) for crude with 8% water content.

A further lenghening of the hydrophilic chain of dodecylphenol to 51 molecules of ethylene oxide somewhat weakens the demulsifying properties of the reagent. Dodecylphenol condensed with 32 molecules of ethylene oxide shows the greatest demulsifying activity, and, most important, increasing the dosage does not result in worsening the demulsification process.

Results are shown in Table 5 for an experiment set up for comparative characterization of the demulsifying properties of reagents OF-34, DF-32, OK-34, and DK-41, and also reagent 4411. All of them proved to be active demulsifiers for Devonian crude. The specific consumptions of these materials for a crude containing 20.4% emulsified water were withim the limits of 0.005-0.007%.

The iso-octyl and dodecylphenols condensed with 32-34 molecules of ethylene oxide show a relatively higher activity than the iso-octyl- and dodecylcresols condensed with the same or even larger numbers of ethylene oxide molecules.

It is well known that the commercial detergent OP-10, which is being produced by domestic industry, is the condensation product of an alkylphenol with alkyl group containing 8-10 carbon atoms with 10 molecules of ethylene oxide.

It was of interest to verify the established relationships on reagents that are derivatives of the commercial product OP-10.

We had at our disposal samples of OP-20 and OP-45. From the results of our investigations of demulsifying power of reagents obtained from Rumania, it was of great interest to investigate a mixture of OP-20 and OP-45 in 1:1 ratio.

The demulsifying properties of the reagents OP-10, OP-20, OP-45, and the indicated mixture were characterized by the volume of water settled out after 2 hours at 60°C, expressed as a percentage of the volume of water recovered from the same crude on adding 0.007% reagent 4411, and also by the content of residual water in the dewatered crude. These data are shown in Table 6 and Fig. 6. As is evident from the settling of free water, OP-45 shows the highest activity, followed by the mixture of OP-20 and OP-45.

The optimum dosages of reagents for the Naryshev crude (with water content 14.8-15.1%) are: OP-45 – 0.007%, mixture of OP-45 and OP-20 – 0.008%, OP-20 and OP-10 – 0.011%.

In quality of the dewatered crude, the best indices were obtained for the mixture of OP-45 and OP-20. These reagents, over a wide range of dosages (0.006-0.024% for the mixture, and 0.011-0.02% for OP-20), left only traces of water in the demulsified crude.

The specific consumption of the OP-20 is considerably higher than for the mixture.

In contrast to these reagents, the OP-45 and OP-10 leave a significant amount of water in the crude, and this increases with increasing dosage.

Thus, of the investigated reagents, the best demulsifying properties are shown by the mixture of OP-45 and OP-20, which at relatively low dosage (0.007-0.008%) provides complete separation of the water and almost anhydrous crude after 2 hours of settling at 60°C.

All of the indicated surfactants were synthesized from technical-grade, relatively pure products.

In order to provide the oil fields with highly effective and at the same time low-cost demulsifiers, there is great interest in investigating the demulsifying power of products from the ethoxylation of coal-tar phenols, unalkylated and alkylated with suitable fractions of thermally cracked naphtha (surfactants of the type of UFE_8 and $KAUFE_{14}$), with different hydrophilic groups. For this purpose, at VNIIneftekhim (D. M. Rudkovskii and N. D. Gadaskina) there were synthesized for UFNII in 1960 seven samples of UFE possessing compositions of from 13.2 to 43.7 molecules of ethylene oxide per molecule of phenol, 11 samples of $K_{130-140}AUFE$ (coal-tar phenols alkylated with 130-140°C fraction of thermally cracked naphtha) with a content of 10.2 to 49.6 molecules of ethylene oxide per molecule of alkylated coal tar phenol, and several samples of $K_{140-170}AUFE$ with content of ethylene oxide from 6.6 to 24 molecules per molecule of alkylphenol.

The investigation of the demulsifying properties of these surfactants also established a direct relationship between the demulsifying power of these materials and the length of the hydrophilic chain. Increasing the length from 8-13 up to 23 molecules of ethylene oxide for reagents of the UFE type reduces the specific consumption from 0.04-0.05 to 0.02-0.025%, i.e., a twofold reduction.

In spite of this, the activity of the reagents of the UFE type is lower than for the KAUFE reagents.

The products from the coal-tar phenols alkylated with the heavier fraction of the cracked naphtha were the more active and, with the same number of moles of ethylene oxide per mole of the hydrophobic part, were not at all inferior in demulsifying power to the ethoxylated synthetic alkylphenols.

The amount of ethylene oxide molecules for the KAUFE reagents depends upon the length of the alkyl group; the optimum number of molecules is 20-30.

It is also interesting to note that the ethoxylated secondary alcohols C_{12-18} of the production of the Shebekino Combine with 30 molecules of ethylene oxide show high demulsifying power for Devonian crudes (specific consumption 0.004-0.005%) and with increasing dosage do not harm the demulsification process (Fig. 7) [3].

Investigations showed that nonionic surfactants obtained by ethoxylation of fatty acids C_{20} and higher (still bottoms), including OZhK synthesized by VNII NP and OKO-100 synthesized by the research group of the Chernikovsk refinery, are similar to $KAUFE_{14}$ and OP-10 in that they provide complete demulsification only at a definite optimum dosage. With dosages lower or higher than this optimum for the given crude, the process of emulsion breaking deteriorates (Fig. 8) [2].

The source of such behavior of the reagents OP-10, $KAUFE_{14}$, the test batches of OZhK and OKO-100, and also all samples of octyl- and dodecylcresols, octyl- and dodecylphenols with short hydrophilic polyoxyethylene chains, is their predominantly hydrophobic character, in consequence of which they are partially dissolved in the crude.

This evidently explains the increased specific consumption of these reagents and their unfavorable action on the demulsification process with consumptions exceeding the optimum. Here, as hydrophobic emulsifiers, they themselves favor the formation of a very highly disperse and hence rather stable emulsion of the water and crude.

As a result of studying the demulsifying power of a series of nonionic surfactants obtained on the basis of alkylated phenols and ethylene oxide, it has been established that for Devonian crudes the most effective are dodecylphenol condensed with 32 molecules of ethylene oxide (DF-32), a mixture of OP-20 and OP-45 (OP-32.5), and $KAUFE_{30}$. This last-named reagent, obtained from coal-tar phenols alkylated with a suitable fraction of thermally cracked naphtha, evidently will be a more readily available and lower-cost demulsifier for use in oil fields.

The higher content of ethylene oxide in the mixture of OP-20 and OP-45 and in the $KAUFE_{30}$ demulsifier does not lead to increasing the specific consumption of ethylene oxide in crude pretreatment, but on the contrary, owing to the increased activity of the indicated demulsifiers, reduces its consumption almost twofold. At the same time, there is eliminated the extreme sensitivity to dosage that is characteristic of the reagents

OP-10 and $KAUFE_{14}$; this facilitates their practical application, especially in existing units, where the processes of dewatering and desalting are carried out in settlers under pressure, without the application of an electrical field.

Literature Cited

1. M. Z. Mavlyutova, "Results of commercial tests of reagents $KAUFE_{14}$, UFE_8, and OP-10 as demulsifiers in the desalting units of NPU Tuimazaneft' " (see this volume).
2. M. Z. Mavlyutova, "Selection and testing of new demulsifiers," Fondy UFNII (1961).
3. M. Z. Mavlyutova and I. I. Chanysheva, "Selection and testing of various demulsifiers," Fondy UFNII (1962).

USE OF NONIONIC SURFACTANTS AS DEMULSIFIERS FOR DEWATERING AND DESALTING CRUDE PETROLEUM

D. N. Levchenko

The laboratory of VNII NP has been occupied for several years with the synthesis and the technology of production of surfactants to act as demulsifiers, with special attention being given to dewatering and desalting of various crudes.

On the basis of commercial tests conducted on demulsifiers and studies in the pilot desalting unit of VNII NP, it may be considered that nonionic surfactants are the principal type of effective demulsifiers for desalting of crude.

This is fully substantiated by the three years of work in the operation of the ELOU of the Novo-Gor'kii refinery (NGNPZ); after a test conducted jointly with VNII NP in 1959, the nonionic material OP-10 has been used as demulsifier. The Novo-Gor'kii NPZ can be an example for other refineries both in desalting of crude and in purification of waste water. The consumption of OP-10 demulsifier at the NGNPZ does not exceed 40-50 g/metric ton of crude; the chloride content of the desalted crude amounts to about 40 mg/liter. The good desalting of the crude using OP-10 and the stable results on water and salt content has made it possible to feed to the combination AVT [atomospheric-vacuum pipestill] from the ELOU.

Such results can be obtained in all of our refineries, using nonionic demulsifiers in an ELOU.

This is in agreement with foreign practice: imported samples of demulsifiers of various firms (Federal Republic of Germany, English, Dutch, Italian, etc.)that have been successfully obtained and tested in the laboratory have been nonionic in the surfactants in the majority of cases.

The nonionic surfactants that we have employed as demulsifiers, such as NChK, SNS, etc., in many respects are inferior to the nonionic materials. The amount of NChK required to demulsify crude oil emulsions is 2-5 kg/metric ton of crude.

The high consumption of NChK to a considerable degree determines its shortcomings. At such relatively high concentrations, the sulfonaphthenates act not only as demulsifiers of the hydrophobic crude oil emulsion, but also as emulsifiers for hydrophilic emulsion of the "oil-in-water" type, which leads to major contamination of the waste water with crude and also to the formation of difficultly broken entrained emulsions of the multiple type. In addition, the ionic demulsifiers, which are salt-like substances, enter into exchange reactions with the calcium, magnesium, and iron salts contained in the formation water and give highly dispersed precipitates that favor the formation of stable emulsions.

In the USA and other capitalistic countries, notwithstanding the production of many sulfonaphthenates of high quality, they are used for entirely different purposes (as additives for various oils, greases, etc.), whereas it is chiefly the nonionic surfactants that are used for the demulsification of crude oil.

Attempts have been made to evaluate nonionic surfactants for various purposes in the drilling and production of petroleum. However, as inhibitors of acid corrosion, they will hardly be better than the well-known inhibiting materials such as nitrogen bases, amines, etc.

Speaking of the effectiveness of demulsifiers, it is unfortunate that we cannot name any one universal demulsifier of definite composition that is suitable for the pretreatment of all crudes.

It is known, for example, that many demulsifiers that are suitable for breaking emulsions of Devonian crudes give poor results with Carboniferous crudes. Therefore, for evaluating the efficiency of new substances, it is necessary to test each proposed demulsifier on several typical crudes and then to draw conclusions as to its suitability.

From the experience of foreign firms, it is known that they have many brands of demulsifiers, and, before recommending a demulsifier for a specific crude, the firm will conduct a screening test in a pilot unit with this particular crude.

For example, the firm Anorgana Munchen of the Federated Republic of Germany proposes four brands of nonionic demulsifiers, Nos. 4400, 4411, 4422, and 4433, from which, on their recommendation, one can select a suitable efficient demulsifier for any crude and for different temperature conditions of treatment.

Therefore, when purchasing any one demulsifier of these four, it is impossible to guarantee its suitability for demulsifying all types of crudes; it is necessary either to conduct preliminary screening tests on the principal types of crudes or to have all four brands.

A selective demulsifying action on various crudes is also shown by our domestic nonionic demulsfiers synthesized on the basis of alkylphenols, fatty acids, or alcohols.

For example, demulsifier OP-10, which is being used successfully in the ELOU of NGNPZ for the desalting of Romashkino crude, proved to be unsuitable in a similar ELOU of the Kuibyshev NPZ for desalting heavy Sernovodsk and Buguruslan crudes, while the nonionic demulsifier OZhK (ethoxylated high-molecular-weight synthetic fatty acids) gave favorable results with the same crudes.

In the Soviet Union, there are too many petroleum deposits, and the crudes that are produced differ markedly among themselves in quality. The crudes that are difficult to demulsify are, for example, such crudes as Sernovodsk, Buguruslan, Arlanskii, Chernushka, Kotur-Tepe, Ozeksuatsk, etc.

Therefore, in choosing a demulsifier for each crude, tests of the proposed surfactants should be conducted under laboratory and refinery conditions. In the Soviet Union, it is also necessary to develop a set of efficient demulsifiers, differing in composition, from which by test one could choose and recommend the most suitable for a crude of any deposit. Thus far there is no such set, and no one of the demulsifiers can pretend to the role of a universal demulsifier. Nonionic demulsifiers have obtained general acceptance in recent years as effective demulsifiers of hydrophobic crude oil emulsions, but our industry does not produce them in adequate quantities. Even the preparation of experimental batches thus far has involved great difficulty.

Within the shortest time, it is necessary to organize the production of nonionic demulsifiers in our petroleum refineries and petrochemical plants in adquate quantities and to supply these materials to the oil fields and refineries. Recently, a decision was made on equipping units in a petrochemical combine, now producing OP-10, for the production of demulsifiers of the OZhK type.

As a result of carrying out this decision, the petroleum industry will obtain several thousand metric tons per year of nonionic demulsifier that will to a considerable degree facilitate the pretreatment of crude in oil fields and refineries.

In addition, it is necessary now to provide the construction of units for the production of nonionic demulsifiers in new petrochemical plants, on the basis of satisfying within the immediate future the demand of the petroleum industry for such material. If we will provide all electrodesalting units (commercial frequency) with nonionic demulsifier, the desalting of the separator crude entering the ELOU will be completely guaranteed; i.e., the content of chlorides in the desalted crude will be below 30-40 mg/liter with a consumption of 20-50 m/metric ton of nonionic demulsifier. The demulsifier consumption varies as a function of crude and demulsifier quality.

The operating flow plan of the ELOU using nonionic demulsifiers was developed by tests at NGNPZ, jointly with VNII NP; this flow plan has been used successfully for several years.

Tests were conducted with this flow plan on OP-10 and on the VNII NP-synthesized demulsifiers KS-59, OZhK, and OS in the Moscow refinery, on demulsifier OZhK in the ELOU of the Kuibyshev refinery, desalting heavy Sernovodsk and Buguruslan crudes, and in the ELOU of the Bavly oil field.

In this flow plan, the desalting of separator crude is carried out in electrodehydrators in two stages without the use of preliminary thermochemical settlers.

The nonionic demulsifier and wash water are fed to the inlet of the centrifugal pump ahead of the first stage of the electrodehydrator; wash water only is fed in ahead of the second stage.

The total quantity of wash water amounts to 5-10%, depending on the quality of the crude and its chloride content.

Such a desalting flow plan ensures that the demulsifier, the electric field, and temperature will act simultaneously on the emulsion in the electrodehydrators; this flow plan is applicable to any nonionic demulsifier. Depending on the efficiency of the demulsifier and the stability of the emulsion, the demulsifier consumption will vary from 20 to 50 g/metric ton. The introduction of caustic is eliminated completely.

At present, the most difficult and urgent problem is oil-field demulsification and dewatering of emulsions with high water content. In the oil fields, we have only thermochemical units for demulsifying the crude, often very primitive, which cannot ensure the required quality of even separator crude with 2% water.

For the first time in the Soviet Union, VNII NP in 1958 began the synthesis of nonionic surfactants for the production of highly effective demulsifiers of crude oil emulsions.

Before this time, nonionic surfactants were synthesized as detergents, and the synthesis technology was directed toward producing substances with high detergency. However, it is well known that not every high-quality detergent surfactant is an effective demulsifier for crude oil emulsions. The best detergency of surfactants is usually obtained at concentrations higher than the critical concentration of micelle formation (CMC), whereas the best demulsifying properties are shown for surfactants at concentrations below the CMC.

The ratio of the hydrophilic portion of the molecule to the hydrophobic (HLB) for nonionic surfactant demulsifiers is always higher than for detergents and is approximately 13-18 (calculated according to Griffin).

All this is evidence that for nonionic surfactant demulsifiers it is necessary to develop their own production technology and to select more suitable starting materials. When entering into the synthesis of nonionic demulsifiers, we proceeded primarily from those types of raw material suitable for this purpose that are available in our industry or will be available in the immediate future.

As is well known, the starting materials for ethoxylation and the production of nonionic surfactant demulsifiers may be alkylphenols, high molecular weight synthetic alcohols, high molecular weight synthetic acids and their partial esters, amines, thioalcohols, and other substances having a reactive hydrogen and containing not less than 8-12 carbon atoms per molecule.

To produce such compounds from phenol, cresols, and other substances, use is made of the process of alkylation with unsaturated hydrocarbons. In our industry there are available in adequate quantities substances that contain the required number of carbon atoms and a reactive hydrogen and do not require additional expense in their alkylation.

These substances include a mixture of alkylcresols and alkylxylenols obtained in production of the inhibitor DBPC (Ionol), and synthetic high-molecular-weight fatty acids and alcohols. There is especially great interest for this purpose in high-molecular-weight synthetic fatty acids and alcohols.

In connection with the building and initial operation of plants in several Soviet combines and refineries for the oxidation of paraffin, there is available a considerable quantity of high-molecular-weight fatty acids and alcohols.

Fractions of the synthetic fatty acids and alcohols containing more than 20 carbon atoms are not finding regular application and up to now are not used economically.

Synthetic fatty acids containing more than 20 carbon atoms remain in the still bottoms after fractionation of the acids. We consider that at present these substances are the most available and promising for the production of nonionic surfactant demulsifiers for crude oil emulsions.

By ethoxylation and propoxylation of these substances, several demulsifiers have been synthesized that are effective in breaking crude oil emulsions, such as OZhK, OPS, and KS-59.

Comparative tests on these materials in the laboratory and in commercial electrodesalting units have shown the highest efficiency for the demulsifier OZhK, obtained by the ethoxylation of fatty acids above C_{20} (still bottoms) and containing 65-70% ethoxy groups. Preliminary economic calculations made at VNII NP for several surfactant demulsifiers have shown the advantage of the OZhK over other nonionic substances. The operating expense in demulsifying one metric ton of crude with OZhK demulsifier is five times less than when using NChK and almost two times less than when using OP-10.

The capital investment in production of OZhK demulsifier is more than 8.7 times less than in the production of NChK and two times less than in the production of OP-10 and other nonionic demulsifiers for which the alkylation process is used ($KAUFE_{14}$, and others). All this is evidence that the petroleum industry must in the shortest time organize the commercial production of OZhK demulsifier in place of NChK.

The synthesis of nonionic surfactants and their testing as demulsifiers of crude oil emulsions were also carried out with various fractions of high-molecular-weight synthetic acids, hydroxy acids, and alcohols, and also with certain individual acids and alcohols. By means of ethoxylation and propoxylation, substances were obtained with varying content of alkoxy groups. Comparative tests on these materials were conducted in the laboratory on demulsification of crudes. Here also the best results in crude demulsification were obtained with the OZhK demulsifier, containing 67-70% ethoxy groups.

The material taken for ethoxylation was still bottoms from the fractionation of synthetic fatty acids, with fatty acid content not below 86%. In 1961, in a plant unit of a facility producing OP-10, a pilot plant batch of OZhK demulsifier was prepared; this was tested in the ELOU of the Moscow refinery for desalting a mixture of Tatar crudes, in the ELOU of the Kuibyshev refinery for desalting heavy crudes (Sernovodsk and Buguruslan), and in the Bavly oil field for the dewatering and desalting of Bavly field crude, containing 30% or more formation water.

Favorable results were obtained in all tests; the consumption of OZhK demulsifier in desalting crude in the commercial-frequency electrodehydrators did not exceed 20-40 g/metric ton of crude.

Good results were also obtained at VNII NP and TsNII Morflot for the dewatering of residual fuel oil using OZhK demulsifier.

In order to provide desalting of crude in the oil fields, it is necessary to accelerate the construction of high-throughput commercial-frequency electrodesalting units and the supplying of effective demulsifiers for these units. The existing thermochemical units in oil fields are inadequate, have low throughput, and cannot provide well-desalted crude in the quantities necessary for our industry.

It is impossible to agree with those who consider that dewatering and desalting of crudes can be carried out in single thermochemical units and who repudiate the use of electrodehydrators. This incorrect direction is a step backward; without the use of electrodehydrators, we cannot ensure high-quality pretreatment of this enormous quantity of crude that is produced in our country at the present time. From the practical experience of our own and foreign refineries and from numerous investigations, it is known that the technically most advanced and economically most favorable method is electrodemulsification in combination with the action of effective demulsifiers and preheating of the emulsion. Electrodesalting units are present in all of the most advanced foreign plants recently started up (in the USA, Federal German Republic, etc.); they are combined with AVT (atmospheric-vacuum pipestill) units, as is evident from articles and the photographs shown in foreign technical journals. Practically complete desalting is attained in these units, down to a chloride salt content of 5-10 mg/liter.

When the AVT and ELOU are combined, additional consumption of heat in desalting is completely eliminated, since the preheated crude passes directly from the ELOU to the AVT.

I consider as premature and incorrect the attempt of our design organizations to design refineries without an ELOU, relying on complete desalting of the crude in the oil fields.

THE POSSIBILITY OF USING SURFACTANTS FOR DOWN-THE-HOLE DEMULSIFICATION IN THE EASTERN REGIONS

S. A. Lebedev, S. G. Yakhin, and G. A. Babalyan

Introduction

Down-the-hole demulsification has found widespread usage in the Baku oil fields [1, 2].

The first detailed studies of this question were made in 1947 by G. A. Babalyan, V. Ya. Mosumyan, and R. G. Gamarnik. These studies showed that with down-the-hole demulsification the production rate of water-cut wells is increased and the specific consumption of working agent is decreased in air-lifting the liquid. In connection with the future application of highly effective nonionic surfactants which might serve the function of demulsifiers [3, 4, 5], it is necessary to establish the possibility of using them for down-the-hole demulsification in the oil deposits of the eastern regions with the various methods of oil-well operation (flowing-well, pump-compressor, and gas-lift). For these regions, down-the-hole demulsification is now acquiring special importance, since the use of this process eliminates the need for constructing expensive dewatering units and opens up the possibility of wide use of forced recovery of fluid from water-cut wells, as one of the methods of increasing the production rate and the rate of development of oil deposits.

The investigations cited in the following are a part of the Institute's complex plan of work for a solution of the demulsification problem as a whole. The basic aim of this investigation was studying the mechanism of emulsion formation and breaking of emulsion in lift pipes and its further dispersion on passage through the flow bean in the head of a flowing well.

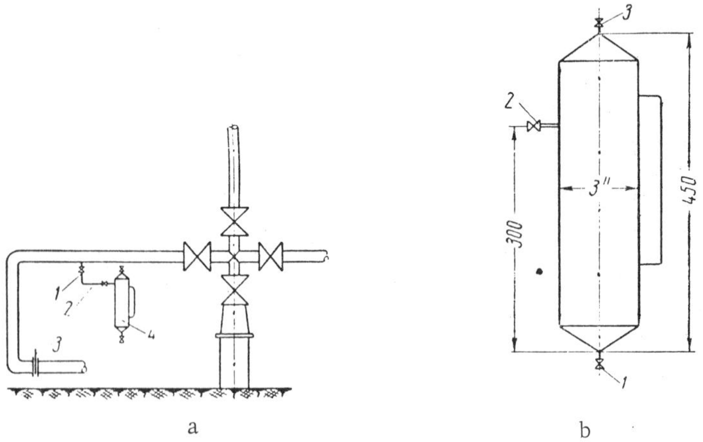

Fig. 1. a) Scheme for connecting well-head sampler to well head. 1) Sampling cock; 2) copper tube, 6 mm diameter; 3) flow bean; 4) well-head sample. b) Well-head sample. 1) $\frac{1}{2}$ in. valve; 2) $\frac{1}{2}$ in. cock; 3) $\frac{1}{8}$ in. valve.

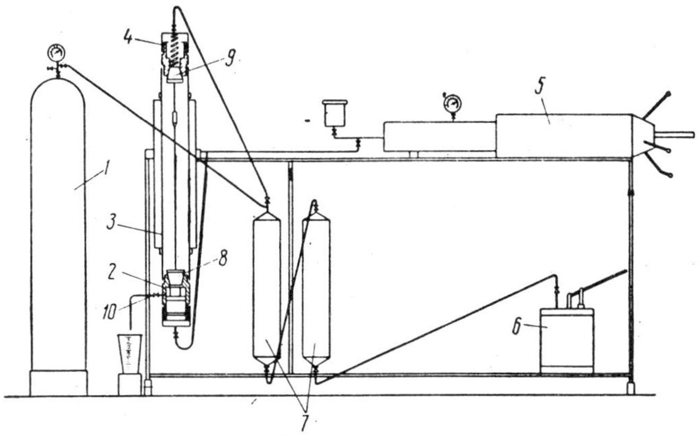

Fig. 2. Unit for transferring depth samples of emulsions to a glass vessel. 1) Nitrogen cylinder; 2, 4) packings; 3) sampler; 5) press; 6) hand-operated pump; 7) surge chambers; 8) lower valve of sampler; 9) upper valve of sampler; 10) drain cock.

Methods of Investigation

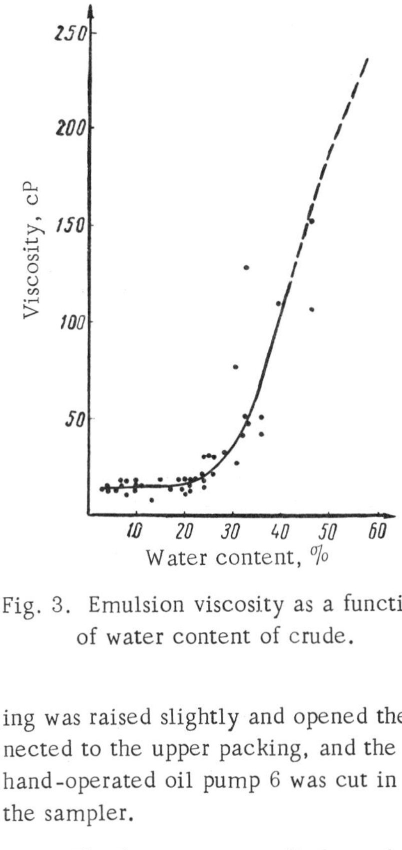

Fig. 3. Emulsion viscosity as a function of water content of crude.

The oil-field studies were conducted in Devonian oil wells of the Tuimazy deposit. In order to establish the change in properties of the emulsion along its route in the lift pipes of flowing wells and wells equipped with bottom-hole electric pumps, samples were taken from various depths and at the well head before and after the flow bean. In the latter case, use was made of a well-head sampler (Fig. 1b).

Before connecting the sampler to the cock of the manifold, the sampler was filled with formation water. After connecting, the water was displaced with emulsion. The emulsion was admitted to the sampler from the manifold at a low rate, to avoid changing the properties of the emulsion. After filling with emulsion, the sampler was disconnected from the manifold, the free water was discharged slowly, and the pressure was reduced to atmospheric by discharging gas through the cock 3. The emulsion remaining in the sampler was subjected to investigation. The samples of emulsion from the lift pipes were taken by a depth sampler. The free water was discharged without reducing the pressure in the sampler (Fig. 2). For transferring the sample to a glass vessel, the packings 2 and 4 were connected to the sampler (Fig. 2). Under the pressure of the hydraulic press 5, the piston of the lower packing was raised slightly and opened the lower valve of the sampler, 8. Then the nitrogen cylinder was connected to the upper packing, and the upper valve of the sampler 9 was opened by the nitrogen pressure. The hand-operated oil pump 6 was cut in if the cylinder pressure was not sufficient to open the upper valve of the sampler.

The free water was discharged to a glass vessel through the drain cock 10. After discharging the water, the nitrogen cylinder was disconnected from the system, and the pressure in the sample was reduced to atmospheric by slow discharge of gas through the surge chambers 7.

TABLE 1. Results of Investigation of Emulsion at Wells

Well No.	Operating method	Production rate, metric tons/day	Pressure on buffer, kg/cm²	Diameter of flow bean, mm	Location of sample withdrawal	Water content, %	Settling, %: Without heating (20°C)	Settling, %: With heating (60°C)	Emulsion viscosity at 20°C, cP	Surface of globules, μ²	Radius of globules, μ	Note
125	Flowing	70	13,5	9	Before flow bean	4—5	0	0	10	—	—	—
125	»	70	13,5	9	After flow bean	4—5	0	0	13—14	—	—	—
129	»	100	15	9	After flow bean	33	0	0	—	—	—	—
336	»	73	10	9	Before flow bean	8	0	0	14—29	89—180	2.7—3.8	—
336	»	73	10	9	After flow bean	8	0	0	11—26	10—28	0.9—1.5	
336	»	73	10	9	Depth	8	0	0	11—14	132—324	3.2—5.1	The difference in water content before and after the flow bean is explained by different time of sampling and pulsations of the well.
377	»	70	6	8	Before flow bean	65—69	27—64	50	100—168	—	—	
377	»	70	6	8	After flow bean	31—50	0	2	105—129	—	—	
498	»	43	8	5	Before flow bean	6—7,5	0	0	13	—	—	
498	»	43	8	5	After flow bean	6—7,5	0	0	13	—	—	
1126a	»	51	20	9	Before flow bean	60—75	45—74	70	—	—	—	—
1220	»	101	—	7	After flow bean	33	0	0	—	—	—	—
1249	»	164	32	6	Before flow bean	22—26	11.2—20	2—20	19—31	—	—	—
1249	»	164	32	6	After flow bean	22—26	0	0	14—22	—	—	—
1249	»	164	32	6	Depth	—	13,6	13,6	—	—	—	—
1269	»	52	4—5	6	Before flow bean	67	25	28	—	606	7.0	—
1269	»	52	4—5	—	After flow bean	67	0—17.8	6—18	234	114	3.1	—
1033	Submerged electric pump	29	0,4—2		—	32—57	0	0	—	—	—	—
1089	»	75	—		—	30—40	0	0	21—22	212—285	4.1—4.8	—
1293	»	85	0—4		—	38—44	0	0	112—155	—	—	—
1293	»	85	0—4		—	64	0	0	—	417	5.8	—
283	»	105	0.2—2		—	32—36	0	0	43—52	152—181	3.5—3.8	—

Results of Investigation

Investigation of Well-Head Samples of Flowing Wells

Data are shown in Table 1 from studies of emulsion samples at the head of flowing wells before and after the flow bean. The data on each well were determined on the basis of several samples. Therefore, Table 1 shows the range of variation in water content, emulsion viscosity, and its dispersity.

As is evident, in the samples taken ahead of the flow bean in the flowing wells 377, 1126a, and 1269, with considerable content of water in the emulsion, water separation was observed on standing, even without heating.

In the similar samples from wells 125, 336, and 498, with water content below 10%, no settling of water was observed, with or without heating.

The samples of emulsions taken after the flow bean, with water content amounting to less than 30%, did not settle out on heating. Cases in which water did settle out were observed with water content of 60% or more. Thus, samples with water content below 10%, taken ahead of the flow bean, have a high stability toward breakdown. Such samples that contain a high percentage of water (60-70%) are less stable. The flow bean gives a marked increase in the dispersity of the emulsion and its stability.

The data on measurements of water globule size in the emulsion before and after the flow bean in wells 336 and 1269 show that the total surface of the globules in unit volume of the emulsion is increased six- to eightfold after the flow bean.

It is interesting to note that the viscosity of the emulsion on passage through the flow bean does not undergo any substantial change, which apparently is explained by the retention of relatively large distances between the globules.

As is well known [6], structural viscosity begins to show itself strongly at disperse phase contents of 73% or more. Such emulsions were not present in our investigations.

We made an attempt to establish a relation between emulsion viscosity and water content, based on the data on the well-head and depth samples (Fig. 3). The viscosity was determined at 20°C in a Hoeppler apparatus. A sharp increase in viscosity was noted at 25-30% water content.

Investigation of Well-Head Samples in Wells Equipped with Bottom-Hole Electric Pumps

The results of these investigations are shown in Table 1. All of the wells had a relatively high water cut (above 30%), but nevertheless the emulsions had high stability against breakdown. No settling out of water was obtained either with or without heating. The dispersity of the emulsion of wells equipped with bottom-hole electric pumps, with water content up to 30-40%, was little different from the dispersity of emulsions taken before the flow bean of flowing wells, with water content below 8%.

Investigation of Depth Samples of Flowing Wells and Wells Equipped with Bottom-Hole Electric Pumps

The depth samples were taken in wells 336, 729, and 1033.

In flowing well 336, the samples were taken from 11 different depths in the 63-mm lift string in the interval from 10 to 1740 meters from the well head. The shoe of the lift string was set 5 meters below the filter [screened pipe] of the well. There was no free water in any of the samples. They did not separate on standing or on heating. Thus, emulsion of high stability to breakdown is observed directly at the screened pipe of the well, which gives a basis for speaking of formation of this emulsion in the formation itself, since at the low velocities of discharge from the screen pipe openings, in the absence of gas evolution ($p_{bot} > p_{sat}$) and in the absence of lifting of the fluid in the lift string, within the limits of the screened pipe and the shoe of the string a stable emulsion cannot be formed.

Such emulsions, as shown by laboratory experiments, often arise in a porous medium during displacement of water by oil and oil by water [7]. On movement of the emulsion in the pipes to the well head, its dispersity increases. The sizes of the globules decrease, and the total surface of the globules per unit volume of emulsion increases.

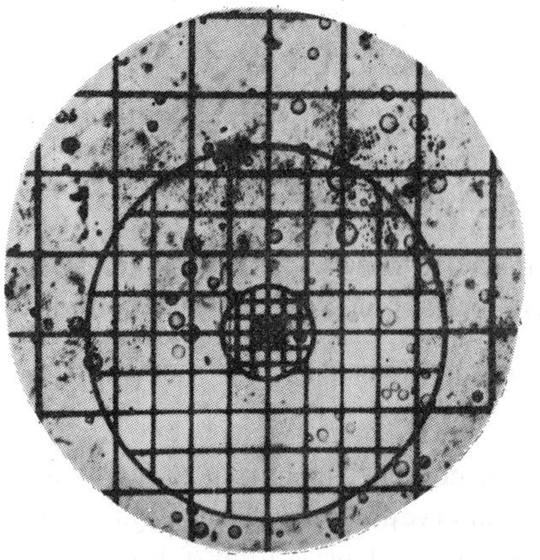

Fig. 4. Microphotograph of emulsion from depth of 450 meters.

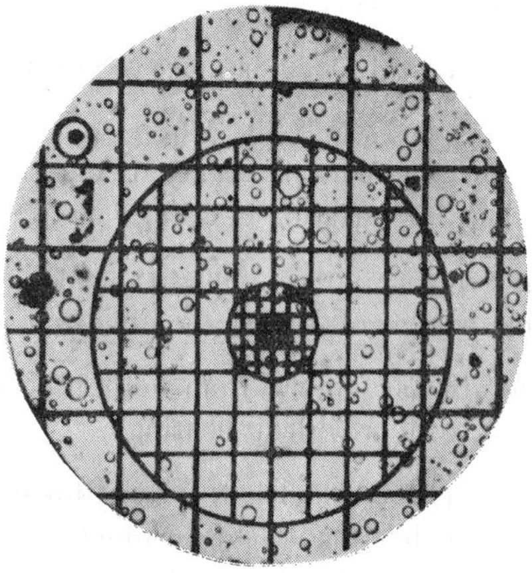

Fig. 5. Microphotograph of emulsion at well head before flow bean.

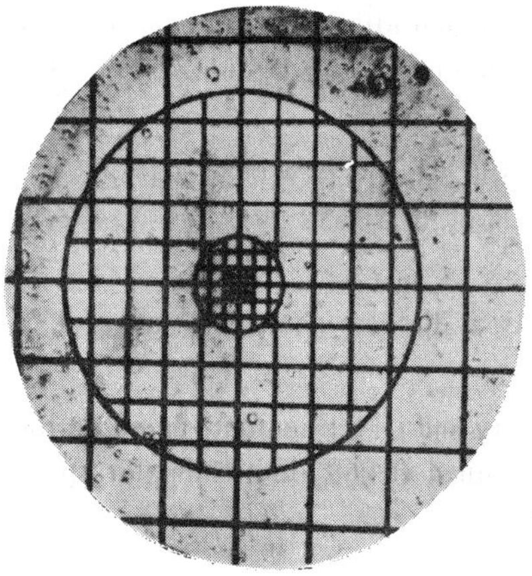

Fig. 6. Microphotograph of emulsion after flow bean.

Figures 4, 5, and 6 are microphotographs of emulsions taken from a depth of 450 meters and at the well head before and after the flow bean (200 magnification). As is evident, the increase in the degree of dispersity of the emulsion is not so great as during its passage through the flow bean. Thus, it can be concluded that the flow bean is the principal source of the sharp increase in emulsion dispersity and hence stability.

From flowing well 729, with a production rate of 52 metric tons/day of crude oil and water and a pressure at the buffer of 15 kg/cm^2, a sample was drawn from a depth of 1169 meters. All of the water separated completely from the emulsion. Complete settling of water took place, with a content of 45%. In the sample taken from a depth of 50 meters from the well head in this same well 729, there was no longer any free water at the moment of transferring the sample to the glass vessel; however, after 3 hours of settling without heating, water was recovered to the extent of 10%, and after 20 hours to the extent of 28% of the total volume of the fluid. This points out the increase in emulsion stability during its movement toward the well head.

In well 1033, equipped with a bottom-hole electric pump, the depth samples were taken above the pump, at depths of 1240 and 10 meters from the well head. In this well the quantity of water in the fluid reached 30%. On settling for 24 hours without heating, no water was recovered from these depth samples.

Thus, the results of the investigation show that the depth of formation of emulsion in flowing wells and the emulsion dispersity depend on the water content in the well production and the intensity of mixing of the emulsion by gas in the lift string and during its passage through the flow bean. The flow bean has an especially great effect on increasing emulsion stability. Stable emulsion in wells equipped with bottom-hole electric pumps is formed mainly in the pump itself.

Investigation of the Change in Emulsion Character with Down-the-Hole Demulsification

In order to ascertain the effectiveness of down-the-hole demulsification, demulsifier was pumped into the annular space and into the discharge at the well head.

The demulsifier was pumped by a piston metering pump of UFNII design with a capacity up to 1.5 liters/min. The metering pump drew NChK from a 160-liter tank into which NChK entered from a 14 m^3 metering tank (Fig. 7). From the pump, the NChK solution passed through a 1 in. pipe into the annular space of the well under

Fig. 7. Unit for pumping surfactant into well in down-the-hole demulsification.

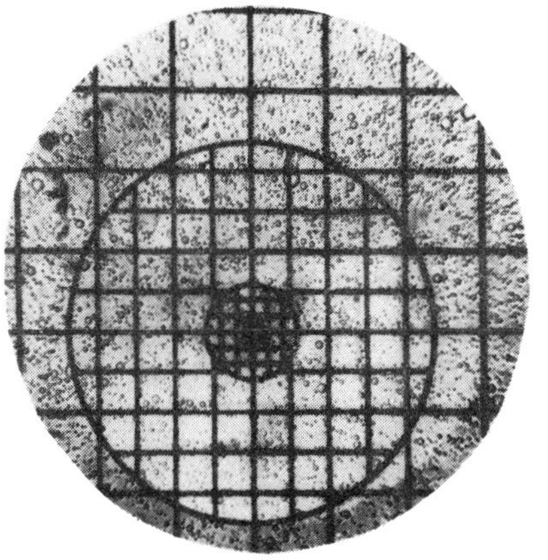

Fig. 8. Microphotograph of emulsion after flow bean, after pumping NChK to the well head (200 magnification).

a pressure up to 15 kg/cm^2. The material used in the experiments was Salavat ammoniacal NChK with a content of 9.6% sulfonate salts in the solution. The NChK was pumped into well 336, which had an oil production rate of 73 metric tons/day, water cut 8% (diameter of flow bean 10 mm, buffer pressure 10 kg/cm^2). The NChK was fed to the well at a rate of 6-7 liters/m^3 of crude. With such a feed rate, the NChK appeared at the well head after 44 hours. Samples taken ahead of the flow bean of the well after feeding the NChK indicated complete layer separation of the emulsion without heating, which did not occur in the absence of NChK feed (see Table 1). The samples taken after the flow bean indicated high stability of the emulsion. It did not separate into layers even with heating.

It was noted that the NChK during movement from the bottom hole to the well head shows a considerable loss in demulsifying properties. A comparison of microphotographs of emulsion after the flow bean, before and after feeding NChK, indicates the lack of any substantial changes in dispersity (Figs. 8 and 6).

The emulsion without the NChK differed little from the emulsion treated with NChK. A considerable amount of NChK was held up in the annular space. After stopping the feed to the well, NChK continued to appear at the well head for an additional 7 days. Subsequently, the observations for the appearance of NChK were discontinued, even though it was continuing to appear.

Subsequently, studies were made to establish the effect of NChK on demulsification when it was fed to the well head before the flow bean. The NChK was pumped in the same as when feeding it to the annular space. The NChK was introduced ahead of the flow bean through a coupling or gage cock specially welded into the barrel of the manifold. In the wells equipped with bottom-hole electric pumps, the NChK was fed through the gage tube of the lubricator.

The operations with pumping to the well head were carried out on four wells (336, 1269, 283, and 1089), of which the first two were flowing wells and the last two were equipped with bottom-hole electric pumps.

Well 1269 had a water cut of 70-80%, and wells 283 and 1089 were 30-40%.

TABLE 2. Data from Study of Emulsion when Feeding NChK to the Discharge at the Well Head

Well No.	Method of operation	Production rate, metric tons/day	NChK feed, liters/m³ water / liters/m³ emulsion	Emulsion water content, %	Free water, %	Water settling, %: Without heating	Water settling, %: With heating	Emulsion sampling location	Viscosity, cP	Surface of water globules, μ^2	Radius of globule, μ
336	Flowing	73	79/6.3	9.6	0	0	3	After flow bean	—	188	3.9
336	»	73	79/6.3	5.6	0	0	6,3	From gage tank	—	102	2.8
336	»	73	79/6.3	8.7	4.1	0	2	Same	—	88,8	2.7
336	»	73	79/6,3	7.8	0	0	4	After flow bean	—	80	2.5
336	»	73	34.4/2.75	10	0	0	5	Same	—	188	3.9
283	Bottom-hole electric pump	103	—	20	0	0	0	At well head	19.1	181	3.8
283	»	103	—	36	0	0	0	From gage head	52.2	88	2.7
283	»	103	—	32	0	0	0	At well head	42.8	173	3.7
283	»	103	—	33	0	0	0	From gage tank	39.8	164	2.6
283	»	103	—	28	0	0	0	At well head	33.8	152	3.5
283	»	151	16.5/4.3	26							
283	»	151	96/4.3	43.6	43,6	0	0	At well head	—	—	—
283	»	170	16/5.9	31	0	32	36	From gage tank	18.3	—	—
283	»	170	15.1/5.9	39	38.6	0	0	At well head	—	—	—
283	»	118.8	6.75/2.2	32	0	10—24	27	From gage head	—	—	—
283	»	118.8	5.2/2.2	41.3	33.7	0—4	8	At well head	—	84	2.6
1269	Flowing	52	—	67.1	17.8	—	—	After flow bean	234	—	—
1269	»	52	12.1/10	82.3	81,3	—	—	Same	—	—	—
1269	»	52	18,0/10	55,6	38.7	—	—	From gage tank	—	—	—
1269	»	52	15.7/10	63.9	55	—	—	After pump	—	—	—
1269	»	52	8,6/7	81.6	80,6	—	—	After flow bean	—	—	—
1269	»	52	7.9/7	88.2	85,7	—	—	After pump	—	—	—
1089	Bottom-hole electric pump	83	—	24	0	0	0	At well head	21	—	—
1089	»	83	—	26	0	0	0	Same	22	—	—
1089	»	88.3	—	20	0	0	0	"	12	229	4.3
1089	»	120	—	22	0	0	0	"	14	83	2.6
1089	»	120	11.5/4.5	31	31	2—7	—	"			
1089	»	120	8/4.5	50	38,8	—	—	From gage tank	11.8	—	—

The results of the studies are shown in Table 2. With pumping of 6.3 liters of 10% Salavat NChK per m^3 of emulsion into the discharge (at the well head before the flow bean) of well 336 with emulsion water content of 8%, the emulsion sample taken after the flow bean showed separation of up to 50% of its water without heating, and up to 100% with heating.

Before feeding NChK to well 1269, the emulsion samples after the flow bean showed practically no separation of water on standing without heating. On standing with heating, 3-15% water was recovered (based on the total quantity of fluid). When pumping in 7 liters of 10% NChK per m^3 of emulsion, practically all of the water separated out on standing without heating (no tests were made with lower dosages of NChK).

In wells 283 and 1089, prior to feeding NChK, no water separated out on standing with heating. On feeding 4 liters of 13% Salavat NChK per m^3 of emulsion, complete settling of water without heating was observed. Hence, it is concluded that with identical feeds of NChK to the discharge at the well head and in the annular space, the demulsification is more effective in the first case.

The cause of this may be the following: as indicated, the NChK being pumped into the annular space, as it moves along in the lift, loses to some degree its demulsifying properties, being adsorbed continuously at the oil–water phase boundary that is formed in the process of lifting. The NChK itself is not homogeneous as to its composition of surfactant. The most active components of the NChK are adsorbed first at the phase boundary. The continuous formation of phase boundary (owing to dispersion of water) and the disappearance of this surface (owing to coalescence) lead to an unusual separation and dispersion of the constituent components of the NChK in the water–oil stream and a resulting decrease in its activity as it moves along in the lift to the well head. Some portion of the NChK will in general interact with the chemically active components of the water and be salted out.

On passage through the flow bean, the emulsion with spent NChK, which has weak demulsifying properties, is again strongly emulsified, and the surface of the oil–water phase boundary increases markedly. The force of interaction of the water globules therefore decreases, and the emulsion stability increases. When NChK is introduced into the discharge at the well head before the flow bean, it possesses strong demulsifying properties both at the flow bean itself and after it during flow from the flow bean; it counteracts the formation of emulsion, decreasing the strength of the adsorbed films. It must be said that these films at the exit from the flow bean are very weak, owing to the instantaneous formation of a large surface of oil–water, oil–gas, and water–gas. With such a rapid formation of new surface, adsorption of the active components does not have time to take place completely at these phase boundaries. The very thickness of the adsorbed layer of NChK and its concentration are low, owing to the large surface of the phase boundaries.

Under these conditions, the intensity of coalescence is determined by the rates of adsorption of the active components of the crude and the NChK, which are determined (other conditions being equal) by their concentrations and the degree of activity. From this viewpoint, nonionic surfactants, which are homogeneous in composition and completely dissolved in the formation water, will have a somewhat different action in down-the-hole demulsification. The demulsifying power of such surfactants will change relatively little on movement of the water–oil mixture to the well head, even though nonuniformity of surfactant concentration in the stream will be observed in this case, also. In addition, since such surfactants have a considerably higher rate of adsorption and a greater activity, they will demulsify the oil more intensively after the flow bean.

Summary

1. The emulsification of oil in wells depends on a number of factors. In flowing wells, it is determined by the intensity of mixing of oil and water by gas in the lift strings and the rate of passage in the flow bean and the rate of flow from it. The effect of the flow bean on emulsification is great. The dispersity of the emulsion (total surface of the globules) increases several times on passage through the flow bean. In wells equipped with bottom-hole electric pumps, emulsification is related mainly to pump operation.

2. When feeding NChK into the annular space, the demulsifying power of the NChK becomes poorer as the water–oil mixture moves toward the well head; therefore, NChK does not play any substantial role in demulsification on passage of the emulsion through the flow bean.

When the emulsion passes through the flow bean, either with or without feeding NChK into the annular space, the emulsion acquires high stability to breakdown. Therefore, it is more advisable to feed NChK to the well discharge at the well head before the flow bean.

In wells equipped with bottom-hole electric pumps, it is possibly more advisable to feed NChK to the pump intake to increase its output.

3. To resolve the problem of down-the-hole demulsification in oil deposits of the eastern regions, it is necessary to carry out wide-scale commercial testing on nonionic surfactants, which are completely soluble in the formation waters and have many times higher activity in comparison with the various NChK materials.

The widespread introduction of these surfactants for down-the-hole demulsification will make it possible to prolong the flowing period of wells and their productivity and will to a considerable extent solve the problem of forced recovery, which, as is well known, is not widely practiced at present because of the absence of enough dewatering units.

Literature Cited

1. I. M. Asadov, Compressor Method of Oil Production and Means of Its Improvement (Aznefteizdat, 1955).
2. R. G. Gamarnik, Dewatering and Demulsification of Petroleum in Oil Fields (Aznefteizdat, 1951).
3. P. A. Rebinder, "Surface-active substances: their importance and applications in the petroleum industry." Collection, Use of Surface-Active Substances in the Petroleum Industry (Gostoptekhizdat, Moscow, 1961).
4. A. A. Petrov, "The study of surface-active substances as demulsifying agents for the pretreatment of petroleum," Collection, Use of Surface-Active Substances in the Petroleum Industry (Gostoptekhizdat, Moscow, 1961).
5. M. Z. Mavlyutova, et al., Fondy UFNII (1961, 1962).
6. P. A. Rebinder and K. A. Pospelova, Introductory article to book by V. Clayton: Emulsions (IL, 1950).
7. Sh. N. Aliev and G. A. Babalyan, The Mobility of the Oil-Content Contour (Aznefteizdat, 1956).

PRODUCTION OF SURFACTANTS BY OXIDATIVE CHLORINATION OF CONCENTRATES OF ORGANIC SULFUR COMPOUNDS

V. I. Dronov and R. S. Enikeev

Having studied the oxidative chlorination of certain organic sulfur compounds present in petroleums and petroleum products, it was established that the oxidative chlorination of aliphatic and cyclic sulfides gives the corresponding alkane- and chloroalkanesulfochlorides with yields of 61-88% of theoretical.

Alkanesulfochlorides containing 12-18 carbon atoms per molecule are starting materials for the synthesis of useful classes of compounds: surfactants, solvents, and plasticizers [1]. Since the organic sulfur compounds present in sulfurous petroleums and petroleum products are represented mainly by aliphatic and cyclic sulfides [2-5], this opens up one possible route for their utilization. Unfortunately, up to the present no effective method has been developed for recovering organic sulfur compounds from petroleum products in pure form, which greatly impedes any studies on determining their application possibilities.

To ascertain the possibility of producing sulfochlorides from the sulfides contained in petroleum distillates, a study was made of the oxidative chlorination of a concentrate of organic sulfur compounds recovered from diesel distillate of mixed Bashkir crudes by means of chromatography on brand ShSM silica gel. The oxidative chlorination was conducted by chlorine-saturating 100 g of the concentrating dispersed in 100 ml of water. The content of sulfochlorides in the products of the oxidative chlorination was estimated by determining the content of hydrolyzable chlorine [1].

As a result of studying the oxidative chlorination of the concentrate, it was shown that varying the temperature in the range 0-40°C and varying the rate of passing in chlorine from 17 to 30 liters/h had no great effect on the yield of sulfochlorides. The best yield of sulfochlorides occurred in the temperature range 0-10°C and with more rapid passage of chlorine. The chlorine consumption had the greatest effect on the yield of sulfochlorides. With increasing chlorine consumption, the yield of sulfochlorides increased, reaching a maximum at a sixfold chlorine consumption based on the sulfide sulfur in the unpurified concentrate and at a threefold chlorine consumption based on the purified concentrate.

Further increase in chlorine consumption did not increase the yield of sulfochlorides, but for complete saturation with chlorine, an 11- to 13-fold quantity was required. This can be explained only on the basis that the chlorine is consumed not only in the oxidative chlorination of the sulfides, but also in some other reactions. Such a reaction may be the chlorination of the hydrocarbons present in the concentrate. It was established by analysis that the content of total chlorine in the products from the oxidative chlorination of the concentrate was 5 to 8 times the content of hydrolyzable chlorine (Table 1), whereas, in accordance with the reaction equation for the oxidative chlorination of sulfides, the total chlorine content should be only 2 times that of hydrolyzable chlorine.

By means of specially set up experiments on the oxidative chlorination of decalin, o-xylene, 1-methylnaphthalene, and biphenyl, it was shown that, under the conditions at which the oxidative chlorination of the concentrate was conducted the chlorination of hydrocarbons does take place; this is in agreement with the work of Gustavson [6], who has established that water is a catalyst in the hydrocarbon chlorination reaction.

TABLE 1. Chlorine Content of Products of Oxidative Chlorination of Concentrate, as a Function of Chlorine Consumption

Content of sulfide sulfur* in concentrate, %	Chlorine consumption in oxidative chlorination of concentrate, % of theoretical	Chlorine content of oxidative chlorination products, %	
		Hydrolyzable chlorine	Total chlorine
4.5	200	2.4	13.2
4.5	400	3.7	25.3
4.5	600	4.5	31.5
4.5	800	4.5	33.6
4.5	1000	4.5	35.2

*Sulfide sulfur was determined by potentiometric titration.

TABLE 2. Surface-Active Properties of Sodium Sulfonates and Detergent Formulations Prepared on the Basis of the Sulfonates

Name of sample	Concentration of solution, %	pH of solution	Surface tension vs air, ergs/cm^2	Foaming power at moment of foam formation; after 30 sec, 3 min, and 6 min	Wetting power, cos θ	Emulsifying power
Sodium sulfonates	0.5 0.25 0.125	3.6 3.8 4.0	37.18 41.40 44.44	75—25—25—25 75—15—15—15 25—10—10—10	0.6156 0.5464 0.5000	Emulsion unstable
Sodium sulfonates and calcined soda (1 : 2)	0.5 0.25 0.125	10.2 10.3 10.3	34.85 39.77 47.13	250—200—150—50 135—85—75—50 110—60—50—50	0.6156 0.5464 0.5000	The Same
Detergent sample No. 1	0.5 0.25 0.125	9.9 10.1 10.3	32.05 39.88 45.47	175—125—50—50 175—85—50—50 125—75—60—50	0.7880 0.7199 0.6691	»
Detergent sample No. 2	0.5 0.25 0.125	10.7 10.6 10.5	34.80 38.6 47.10	150—100—75—50 85—35—25—25 75—25—25—25	0.6691 0.6157	»

TABLE 3. Detergency of Aqueous Solutions of Sodium Sulfonates and Formulations Prepared on the Basis of the Sulfonates

Name of sample	Concentration of solution, %	% whiteness of soiled fabric	% whiteness of laundered fabric	% soil removed	Detergency, % in relation to water
Sodium sulfonates	0.5	22.6	27.2	4.6	98
	0.25	22.8	27.3	4.5	97
	0.125	22.8	27.3	4.5	97
	Water	22.4	27.1	4.7	100
Sodium sulfonates and calcined soda (1 : 2)	0.5	22.9	32.8	9.9	210
	0.25	22.8	31.9	0.1	194
	0.125	22.6	29.8	7.2	153
	Water	22.4	27.1	4.7	100
Detergent sample No. 1	0.5	25.1	39.3	14.2	302
	0.25	25.4	39.5	14.1	291
	0.125	25.2	36.9	11.7	250
	Water	25.1	29.8	4.7	100
Detergent sample No. 2	0.5	25.3	38.1	12.8	272
	0.25	25.4	36.3	10.9	232
	0.125	24.9	34.1	9.2	196
	Water	25.1	29.8	4.7	100

The sulfochlorides present in the products from the oxidative chlorination of the concentrate were converted to the sodium salts of sulfonic acid by saponification with 20% sodium hydroxide solution at 80°C. For complete saponification, in the majority of the experiments, it was sufficient to take 3 moles of sodium-hydrolyzable chlorine. The saponification of the products from the oxidative chlorination of the concentrate gave two layers – an oil layer and a water layer. The sodium salts of the sulfonic acids were extracted from the oil layer at 80°C with water in 1 : 1 or 0.5 : 1 ratio to the oil layer.

Most of the sodium sulfonates were extracted from the oil layer in the first extraction. The quantity of sulfonates in the second extraction in most of the experiments amounted to 2% of the content in the first extract. If the volume ratio of water to oil used in the extraction was decreased twofold, there was a twofold increase in the concentration of sodium sulfonates in the first extracts. In addition, the less concentrated solutions of these salts as a rule were heavily contaminated with oil. The yield of sodium sulfonates based on the concentrate amounted to 22-28% for the unpurified concentrate and 6-8% for the purified concentrate.

The sodium sulfonate solutions were purified by diluting them with an equal volume of methanol, separating off the oil that precipitated on standing, and distilling off the solvent from the purified solution. With such purification, the sodium sulfonates obtained from the first extracts of all the experiments contained 61.05% active substance, 8.32% sodium chloride, 29.87% moisture, and 1.69% oil.* Aqueous solutions of this sample had a low pH and were characterized by reduced foaming and emulsifying power. However, adding calcined soda and other useful additives† to these solutions led to an improvement in the surface tension and wetting and foaming power.

*The figures in the Russian text add up to a total of 100.93%. – Tr.

†Such as builders' suspending agents. – Tr.

On the basis of the sodium sulfates in combination with useful additives, two samples were prepared and tested, with the following composition: Sample No. 1 – sodium sulfonates 1 part, calcined soda 1 part, sodium tripolyphosphate 1 part, sodium perborate 5.5%,* and carboxymethylcellulose 2%*; Sample No. 2 – sodium sulfonates 1 part, calcined soda 1 part, sodium metasilicate 8%,* and carboxymethylcellulose 2%.*

The surface-active properties of aqueous solutions of the sodium sulfonates and of the detergent samples are shown in Table 2, and their detergency in Table 3.†

From an examination of Table 3, it follows that the aqueous solutions of the sodium sulfonates in combination with the useful additives possess satisfactory detergency.

Experimental

Oxidative Chlorination of Concentrate of Organic Sulfur Compounds

A 0.5-liter four-necked round-bottom flask fitted with stirrer, thermometer, barometer, and reflux condenser was charged with a mixture of 100 g of the concentrate and 100 ml of water, and, with effective stirring, chlorine was passed in. The rate of passage and consumption of chlorine were measured with a flowmeter. The oxidative chlorination reaction was exothermic, so the reaction flask was cooled to maintain the required temperature. After completion of the reaction, the water and oil layers were separated, and their densities at 20-25°C were determined by a hydrometer. After repeated water washing of the oil layer, its content of hydrolyzable chlorine was determined.

The materials taken for the oxidative chlorination were a concentrate containing tars (experiments 1-31) and a concentrate freed of tars by vacuum distillation (experiments 33-42). The reaction conditions and the results of the experiments on oxidative chlorination of the concentrate are shown in Table 4.

Oxidative Chlorination of Hydrocarbons

The oxidative chlorination of decalin, o-xylene, 1-methylnaphthalene, and biphenyl was carried out in a water–methanol medium (volume ratio 1 : 1) at 0-20°C up to complete saturation of the reaction mixture with chlorine. After completing the reaction, the excess chlorine was displaced by passing nitrogen through the reaction mixture; then it was diluted with water, and the reaction product was extracted with ether. The ether extract was washed with water, neutralized with sodium bicarbonate, again washed with water, and dried over anhydrous sodium sulfate. The residue after driving off the ether was characterized by density, refractive index, and chlorine content.

The characteristics of the products of oxidative chlorination of the hydrocarbons studied are shown in Table 5.

Saponification of Products of Oxidative Chlorination of Concentrates of Organic Sulfur Compounds

A 0.5-liter four-necked round-bottom flask fitted with a stirrer, thermometer, dropping funnel, and reflux condenser was charged with a 20% sodium hydroxide solution and heated to 80°C; then, with efficient stirring, the product of oxidative chlorination of the concentrate was added dropwise, after which heating was continued for some time longer. In experiments 2-11, the sodium hydroxide for saponification was taken in twice the theoretical quantity, and the saponification was continued for 10 hours.

In experiments 26-31 and 37-42, the sodium hydroxide for saponification was taken in stoichiometric quantity; after 3 hours of saponification, the reaction of the aqueous layer was determined; if it was acid, sodium hydroxide solution was added to the reaction mixture in quantity stoichiometric to the content of hydrolyzable chlorine remaining in the oil layer. The addition of sodium hydroxide solution in the saponification process was continued until the reaction of the aqueous layer became neutral or weakly alkaline. After completing the saponification reaction, the water and oil layers were separated, and their densities at 20-25°C were determined by a hydrometer. After it was established by analysis that the dry residue obtained after evaporating the water layer was 95% sodium chloride, the water layer was discarded without analysis.

*Calculated on the total weight of the sample.

†The tests of the surfactants and detergents were conducted at the Moscow branch of VNIIZh.

TABLE 4. Reaction Conditions and Results of Experiments on Oxidative Chlorination of Concentrates of Organic Sulfur Compounds

Experiment No.	Content of sulfide sulfur* in concentrate, %	Reaction conditions				Content of hydrolyzable chlorine in reaction products, %	Density of layers obtained as a result of reaction	
		Temp., °C	Time, h	Chlorine feed rate, liters/h	Chlorine consumption, % of theoretical		Oil	Water
2	3.7	0—20	4.5	—	800	2.8	1.2	1.185
3	3.7	0—10	1.33	17	300	2.3	1.03	—
4	3.7	0—20	2.0	17	400	2.88	1.07	—
5	3.7	0—10	4.16	17	1000	3.53	1.225	1.19
8	3.7	20—30	5.0	17	1100	2.7	1.11	—
9	3.7	30—40	5.0	17	1100	3.3	1.09	1.18
10	3.7	10—20	5.0	17	1100	2.5	1.07	1,16
11	3.7	0—10	5.0	17	1100	3.7	1.08	1,15
13	3.7	0—20	5.0	17	1100	2.45	1.10	1,16
14	3.7	0—10	3.33	30	1280	3.52	1.13	1.16
15	3.7	20—30	3.33	30	1280	3.32	1.13	1.17
16	3.7	30—40	3.33	30	1280	3.02	—	1.17
17	3.7	0—10	5.0	17	1100	3.30	1.15	1.17
18	5.8	10—20	2.5	30	600	5,70	1.24	1,19
19	5.8	0—10	3.5	30	1000	5.73	1.26	1,19
20	5.8	0—10	6.2	17	1000	5.16	1.26	1.19
21	5.8	10—20	4.5	17	600	5.40	—	—
22	3.01	5—15	0.5	17	200	2.38	—	—
23	4.07	8—12	0.9	28	300	2.40	—	—
24	4.07	6—12	0.3	28	100	2.42	—	—
25	4.07	6—14	3.57	28	1250	3.81	—	—
26	4.5	0—20	0.64	28	200	2.4	1.06	1.12
27	4.5	0—20	1.3	28	400	3.7	1.17	1.168
28	4.5	0—20	6.5	28	1000	4.5	1.27	1.19
29	4.5	20—25	1.6	28	800	4.5	1.265	1,19
30	4.5	20—25	1.9	28	600	4.5	1.215	1.18
31	4.5	20—25	0.3	28	200	2.75	1.06	1,13
33	3.02	0—20	2.5	28	1230	2.8	—	—
34	3.02	0—20	2.5	28	1230	2.84	—	—
35	3.9	0—20	2.5	28	950	4.7	—	—
37	3.9	10—20	0.75	28	300	4.45	—	—
38	3.9	10—20	0.3	28	150	3.0	—	—
39	3.9	30—40	0.3	28	150	1.7	—	—
40	3.9	30—40	0.75	28	300	4.7	—	—
41	3.9	30—40	1.0	28	420	4.6	—	—
42	3.9	10—20	1.0	28	420	4.65	—	—

*Sulfide sulfur was determined by potentiometric titration

TABLE 5. Characteristics of Products of Oxidative Chlorination of Certain Hydrocarbons

Initial hydrocarbons			Products of oxidative chlorination		
Name	η_4^{20}	d_4^{20}	η_4^{20}	d_4^{20}	Chlorine content, %
Decalin	1,4770	0,8872	1,4850	0,9414	8,4
o-Xylene	1,5060	0,8798	1,5339	1,1102	29,5
1-Methylnaphthalene . . .	—	—	1,6051	1,4550	11,2
Biphenyl	—	—	—	—	15,3

TABLE 6. Conditions and Results of Experiments on Saponification of Products of Oxidative Chlorination of Concentrate

Expt. No.	Saponification conditions		Density of layers after saponification		First extract			Second extract				Yield of NSSK based on concentrate, %
	Time, h	NaOH consumed, moles/mole of hydrolyzable Cl_2	Oil	Water	NaOH consumed (mg) in neutralizing 10 g of extract	Concentration of NSSK,* %	Concentration of oil in NSSK, %	NaOh consumed (mg) in neutralizing 10 g of extract	Concentration of NSSK, %	Concentration of oil in NSSK, %	Relative content of NSSK in 2nd extract compared to 1st, %	
2	10	4	—	—	18,8	31.4	18,1	9,6	0,87	270	2,3	18,6
3	10	4	1,0	1.185	15,2	30,4	12,1	5,6	0,35	200	1.67	21,3
4	10	4	1,1	1,18	16,4	29,9	13,7	8,4	1,1	200	1.76	25,4
5	10	4	1,185	1,20	18.0	30.0	18,3	7,6	1.07	177	1,78	28,8
8	10	4	1,05 † 1,20	1,19	22,4	31.0	24.8	9,2	0,96	240	2.13	21,6
9	10	4	1.03	1.19	21.6	30.2	23,1	0,2	1,01	227	1,96	23,6
10	10	4	1.07	1,19	17.6	27.5	18,9	5,6	3,1	45.1	6,83	22,9
11	10	4	1,12	1.18	22.0	29.5	25,1	7.2	1,1	163	1,76	28,8
26	9	2,5	—	—	—	15.4	78,0	—	—	—		17.0
27	10	3,47	—	—	—	16.0	54,2	—	—	—		21,7
28	9.5	2,66	—	—	—	15.5	72,9	—	—	—		22,6
30	9,5	4,4	—	—	—	16.2	57,4	—	—	—		21.7
31	6,5	3,1	—	—	—	14.6	19,1	—	—	—		19,4
37	10	3.0	—	—	16,0	16.5	40,0	—	—	—		7.2
38	9.0	3.0	—	—	20,0	15.3	53.0	—	—	—		7,5
39	6,0	3,0	—	—	19,0	15,4	45.0	—	—	—		6.3
40	6,0	2,93	—	—	22,0	6.7	103,0	—	—	—		6,2
41	6,0	2,82	—	—	18,2	14.8	60.1	—	—	—		6,9
42	1,2	2,0	—	—	19,6	16,1	58,0	—	—	—		8,2

*NSSK – sodium sulfonates.

† After saponification, two oil layers were obtained.

TABLE 7. Results of Purification of Sodium Sulfonate Solutions to Remove Oil

Purification method	Analysis of purified solutions	
	NSSK content, %	Oil content, % of NSSK
I	46.2	3.3
II	50.89	2.96
III	48.3	2.27
IV	49.72	16.0

The sodium sulfonates (NSSK) were extracted from the oil layer by two extractions with water at 80°C. The extraction consisted of shaking for 10 minutes and settling for 2 hours. In experiments 26-31, the volume ratio of water to oil in the extraction was 1 : 1; in the other experiments, it was 0.5 : 1.

The content of sodium sulfonates and oil was determined on the extracts by methods described for the analysis of NChK [7]. The reaction conditions and results of the experiments on saponification of the products of oxidative chlorination of the concentrate are shown in Table 6.

Purification of Sodium Sulfonate Solution to Remove Emulsified Oil

The initial solution, containing 15.5% sodium sulfonates and 60% oil, was purified by four methods.

Method I. A 50-ml portion of the sodium sulfonate solution was diluted with an equal volume of methanol. The lower oil layer formed in 12-hour settling was separated off, the alcohol and water were driven off from the upper layer, and the content of sodium sulfonates and oil was determined on the residue.

Method II. The purification was similar to Method I except that the alcoholic layer before distillation was subjected to five extractions with 50-ml portions of petroleum ether.

Method III. The purification was similar to Method I except that the initial solution of sodium sulfonates was subjected to three extractions with 100-ml portions of petroleum ether.

Method IV. The water and oil were distilled off from the sodium sulfonate solution, 50 ml of distilled water was added to the residue in the distillation flask, and the distillation was continued. This addition of water and subsequent distillation was repeated three times.

The results of the experiments on purifying the sodium sulfonate solutions are shown in Table 7.

Summary

1. A study has been made of the relation between yield of sulfochlorides and temperature and the rate of passage and consumption of chlorine in the oxidative chlorination of a concentrate of organic sulfur compounds recovered from a diesel distillate.

2. It has been established that saponification of the product of oxidative chlorination of the concentrate with sodium hydroxide solution gives sodium sulfonates in yield 23-28% based on the weight of the concentrate.

3. The sodium sulfonates obtained by saponification of the product of oxidative chlorination of the concentrate are surfactants and, in combination with useful additives, possess satisfactory detergency.

Literature Cited

1. F. Azinger, Chemistry and Technology of Paraffinic Hydrocarbons (Gostoptekhizdat, 1959).
2. D. Haresnape, F. A. Filder, and R. A. Lowry, Ind. Eng. Chem., 41, 2691 (1949).
3. R. H. Brown and S. Mayerson, Ind. Eng. Chem., 44, 2620 (1952).
4. R. D. Obolentsev and B. V. Aivazov, Collection, Chemistry of Organic Sulfur Compounds Contained in Petroleums and Petroleum Products, No. 1, Izd. Bashkirskogo filiala AN SSSR, Ufa (1958).
5. R. D. Obolentsev, B. A. Aivazov, G. V. Galeeva, and E. N. Chelov, Collection, Chemistry of Organic Sulfur Compounds Contained in Petroleums and Petroleum Products, No. 3, Izd. Bashkirskogo filiala AN SSSR, Ufa (1960).
6. C. A. Gustavson, 45, 6991 (1951).*
7. B. M. Rybak, Analysis of Petroleums and Petroleum Products, Chapter II (Aznefteizdat, Baku, 1948).

*Citation incomplete in the Russian text.

RESULTS OF LABORATORY STUDIES AND OIL-FIELD TESTS ON SURFACTANTS IN TURKMENIYA

M. A. Akhmetshin

At the present time, the oil fields in commercial development in Turkmeniya are the Kum-Dag, Negit-Dag, Dagadzhik, Aligul, Zapadnyi Cheleken, and Kotur-Tepe.

The oil content is confined to the Pliocene deposits, mainly the red-colored stratum. The Apsheron and Akchagyl'sk deposits are oil-bearing in Kotur-Tepe, Nebit-Dag, and Kum-Dag.

All of the oil accumulations are of the multiformation type, with different conditions prevailing in each individual group of formations.

The main workable horizons of the oil deposits occur at a depth of 700-2500 m. Oil accumulations at a depth less than 700 m are encountered in the central part of the Nebit-Dag field and partly in Kum-Dag and Dagadzhik.

One of the features of the Turkmen deposits is the high initial formation pressures, which exceed considerably the hydrostatic head of the water column, and the relatively high formation temperature: in Kum-Dag T_f = 35-65°C, in Nebit-Dag T_f = 40-75°C, in Dagadzhik T_f = 50-65°C, and in Aligul and Zapadnyi Cheleken T_f = 80-90°C [1].

The productive part of the majority of the deposits is represented by an alternation of sands and clays, lithologically variable in thickness. The collectors of the productive part of the formations of the Nebit-Dag and Kum-Dag oil deposits are expressed as fine- and close-grained, weakly consolidated sands with a high content of clay fraction (up to 50%) and have a relatively low permeability (50-300 mC). The porosity M = 15-25%, and the quantity of connate water is 20-25% [2].

The productive section of the deposits of the Cheleken structure is represented by an alternation of clays with fine- and close-grained sands and sandstones, porosity from 18 to 25%, permeability varying from 20 to 200 mD [3].

The crudes are represented mainly by paraffinic and naphthenic hydrocarbons with low content of aromatic hydrocarbons. Part of the crudes are paraffinic-naphthenic, and part naphthenic-paraffinic. The paraffinic-naphthenic crudes are low-asphalt and waxy. The naphthenic-paraffinic crudes contain less wax and more asphalt and also a small quantity of naphthenic acids.

The deposits with naphthenic-paraffinic crudes are located to the north of the regional breaks of the Pre-Balkan region: in the middle and upper parts of the red-colored stratum of Dagadzhik and Nebit-Dag, Eastern Kum-Dag, and also the Akchagyl'sk and Apsheron stages of the Nebit-Dag and Eastern Kum-Dag.

The most asphaltic and practically wax-free crudes lie in the red-colored stratum of Dagadzhik and Central Nebit-Dag [4].

The formation waters of the oil deposits are primarily highly mineralized (1138-9470), of the calcium chloride type [5], excepting the waters of certain horizons of the deposits of Dagadzhik, Aligul, and Nebit-Dag, where the waters have lower mineralization (less than 1000 meq/liter) and are mainly bicarbonate type waters.

TABLE 1

Water	meq/liter								Density, g/cm³	Salinity, °Be	Alkalinity, meq/liter
	Cl′	SO_4''	HCO_3'	n.k.*	Ca··	Mg··	Na·+K·	Σ			
Industrial, Bala-Ishemsk water intake	122	4	2	—	20	12	96	256	1,004	0.7	2.00
Sea water	166	58	2	—	16	56	154	452	1,009	1.3	2.0
Well 371, Horizon B (Kum-Dag)	2900	3	1	—	231	181	2492	5808	1,105	14,3	1.0
Well 586, Horizon Zh (Kum-Dag)	2525	2	4	—	212	147	2172	5062	1,093	12,3	4.0
Well 87, Horizon II (Dagadzhik)	520	2,7	16,5	1,8	6	6	529	1082	1,021	2,98	18.3
Well 194 (Dagadzhik)	623	4	18,2	1,0	9,8	14,7	621,7	1292	1,027	3.79	19.2
Well 169 VIII member (Aligul)	243,3	25,1	12,1	2.7	5,7	13.5	264.3	576	1,0092	1,29	14,8

*Translation unknown; apparently refers to "other anion."

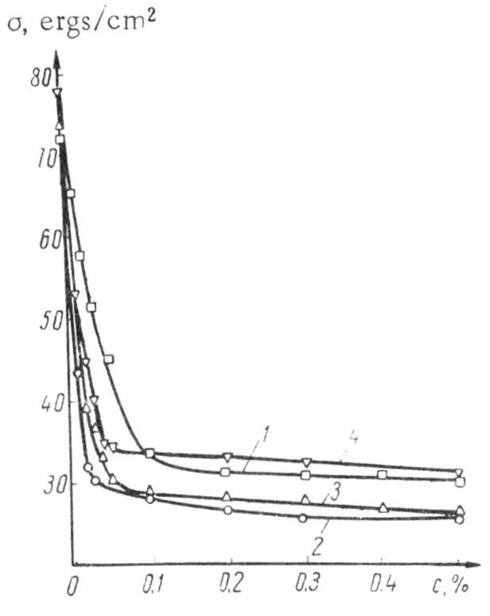

Fig. 1. Surface tension isotherms of aqueous solutions of Sulfonol NP-1 vs air. 1) Distilled water; 2) in industrial water; 3) in sea water; 4) in formation water.

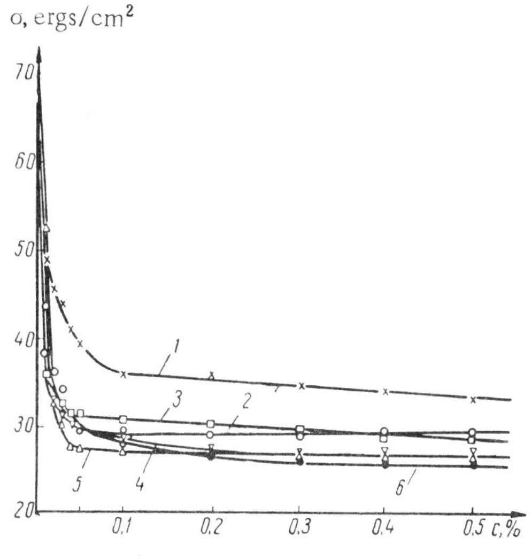

Fig. 2. Surface tension isotherms for solutions of various surfactants in industrial water. 1) Leveling Agent A-20; 2) Leveling Agent A; 3) OP-10; 4) OP-7; 5) DB; 6) NP-1.

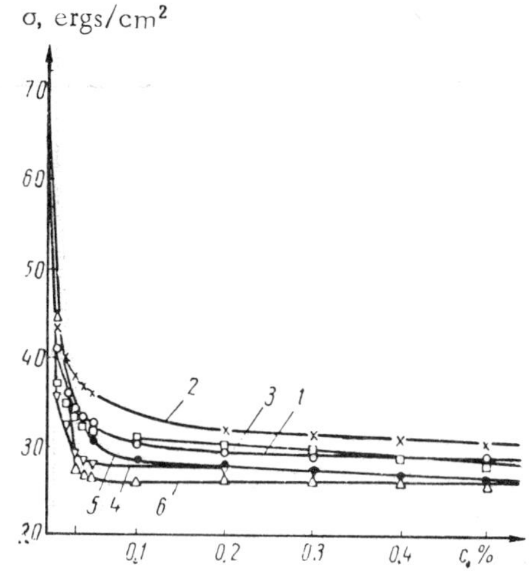

Fig. 3. Surface tension isotherms for solutions of various surfactants in sea water. 1) Leveling Agent A; 2) Leveling Agent A-20; 3) OP-10; 4) NP-1 (Sulfonol); 5) OP-7; 6) DB.

In the oil fields of Western Turkmeniya, the water used for industrial purposes includes that of the Bala-Ishemsk water intake (0.5° Be) and sea water (1.3° Be) of the Caspian Sea.

The salt composition of the waters are shown in Table 1.

It is well known that certain surfactants, used as part of treatments of the bottom-hole zone of the formation, interact with the formation water and form insoluble compounds, resulting in a decrease in the activity of the surfactant solutions.

A check on the solubility of certain surfactants in the above described waters showed that the nonionic surfactants OP-7 and OP-10 and a cation-active substance – Leveling Agent A – dissolve completely in the water regardless of the degree of mineralization.

The solubility of Wetting Agent DB and Leveling Agent A-20 in water depend on the degree of mineralization; the lower the salt content, the greater their solubility.

Visual observation of solutions of Sulfonol NP-1 in industrial, sea, and formation (14.3° Be) waters shows that the solution contains a water-suspended floc that is the product of partial salting out of the Sulfonol NP-1 by electrolytes, resulting in a 25-30% decrease in activity in sea and industrial waters and a 40-45% decrease in formation waters.

As highly mineralized formation water is added to a solution of Sulfonol NP-1 in distilled water, the solution becomes cloudy owing to the formation of a fine floc, which in time precipitates as a sediment. This phenomenon furnishes a basis for the hypothesis that when a solution of Sulfonol NP-1 prepared with industrial water is pumped into a formation there will be additional formation of floc as the solution mixes with the highly mineralized connate water.

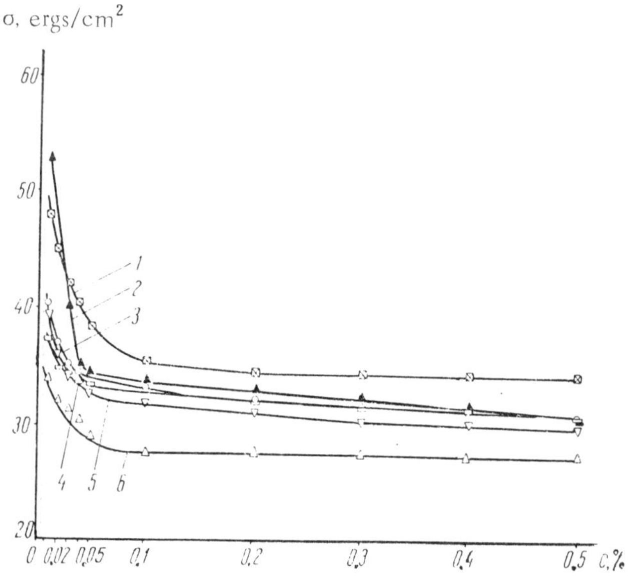

Fig. 4. Surface tension isotherms for solutions of various surfactants in formation water. 1) Leveling Agent A-20; 2) Sulfonol NP-1; 3) Leveling Agent A; 4) OP-10; 5) OP-7; 6) DB.

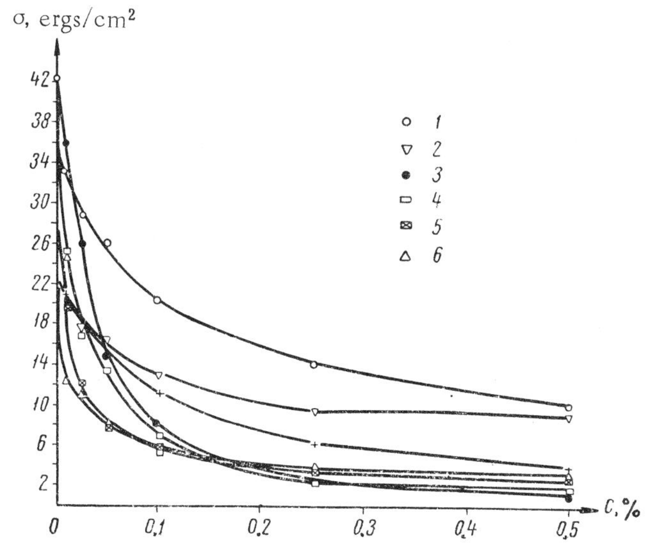

Fig. 5. Interfacial tension isotherms for solutions of surfactants in distilled water vs kerosene. 1) OZhK; 2) Wetting Agent DB + UFE_8; 3) Sulfonol NP-5; 4) Sulfonol NP-1; 5) OP-7; 6) OP-10.

It is subsequently determined that the nonionic surfactants OP-7 and OP-10 are stabilizers with respect to Sulfonol NP-1 in water, i.e., they prevent coagulation of the floc. For example, an 0.3% solution of Sulfonol NP-1 in distilled water remains transparent on pouring in formation water (12.30° Be) if the initial solution contains 0.05% OP-10.

Hence, it must be assumed that mixtures of different (ionic and nonionic) surfactants may find wide application in various engineering processes in the extraction of oil.

Figure 1 shows surface tension isotherms $\sigma = f(c)$ for solutions of Sulfonol NP-1 in distilled, industrial (Bala-Ishemsk water intake), sea, and formation (14.3° Be) waters.

TABLE 2

Concentration, %	Type of surfactant							
	Sulfonol NP-1		Sulfonol NP-5		OP-7		OZhK	
	θ_a°	θ_r°	θ_a°	θ_r°	θ_a°	θ_r°	θ_a°	θ_r°
0,00	—	43	—	—	—	—	—	—
0,025	95	18	68	26	50	28	90	26
0,05	110	18	103	28	48	30	108	23
0,1	95	20	78	29	44	28	108	21
0,2	43	29	94	43	44	31	80	23
0,3	42	29	34	45	44	33	70	25
0,5	25	30	26	40	37	—	67	20
1,0	20	30	16	40	30	—	—	—

Note. θ_a° – advancing angle – drop of water or aqueous solution of surfactant on quartz in kerosene medium. Determination made 10 min after depositing drop. θ_r° – receding angle – drop of petroleum on quartz in medium of solution of surfactant in distilled water.

TABLE 3

Time of interaction of sand with OP-10 solution, h	Temperature at which adsorption took place	
	20°C	60°C
0	30.85	–
24	31.60	34.90
48	32.20	36.20

Note. σ was determined at 20°C.

As is evident, Sulfonol NP-1 in mineralized water actually loses activity, but only slightly. Figure 2 shows surface tension isotherms $\sigma = f(c)$ for solutions of OP-7, OP-10, Wetting Agent DB, Leveling Agent A, Leveling Agent A-20, Sulfonol NP-1, and Katapin A in industrial water; Fig. 3 shows similar data for these same surfactants in sea water, and Fig. 4 in formation water.

From an examination of these isotherms $\sigma = f(c)$, the surfactants can be arranged in a series according to the degree of reduction of their surface activity in mineralized formation waters: Wetting Agent DB, OP-7, OP-10, Leveling Agent A, Sulfonol NP-1, Leveling Agent A-20, and Katapin A.

It should be noted that Sulfonol NP-1 stands in second place in degree of activity (after Wetting Agent DB) in the case of industrial and sea waters, even though it loses some of its activity in these waters.

Hence, with low mineralization of formation waters, Sulfonol NP-1 is a sufficiently surface-active substance. Figure 5 shows interfacial tension isotherms of aqueous solutions of certain surfactants vs kerosene at 20°C (density of refined kerosene 0.811 g/cm^3 at 20°C), from which it is evident that at the boundary of two liquid phases Sulfonols NP-1 and NP-5, beginning with a concentration of 0.13%, also reduce σ the same as the nonionic materials OP-7 and OP-10.

Changes in the preferential contact angle of wetting a quartz plate with water, as a function of the content in the water of Sulfonol NP-1 and other surfactants, are shown in Table 2.

Since the distinctive capability of surfactants is their capability to be adsorbed at the oil –water–rock phase boundaries, determinations were made of the magnitudes of adsorption by rock of the surfactants OP-7, OP-10, DB, and Sulfonol NP-1, in order to calculate the required concentration and consumption of surfactant solutions to obtain the most effective results.

TABLE 4

Time of interaction of sand with OP-10 solution, h–min	Surface tension (vs air) of OP-10 solution in industrial water before and after adsorption at the surface quartz (solution/sand ratio 2 : 1)					
	Concentration of OP-10 in water					
	0,01	0,04	0,05	0,1	0.2	0,5
0—0	37,40	30,60	31,05	30,98	30,50	30,25
0—3	41,05	31,40	31.53	—	30,65	—
0—6	42,20	31,40	—	31.75	30.90	—
0—10	44,00	33,33	31,30	32,00	—	—
0—15	56,22	—	31.30	32,00	—	—
0—30	64,50	—	31,30	32,00	—	32,25
1—00	66,80	—	31,81	—	—	32,25
2—00	68,16	33,5	—	32,10	—	—
3—00	—	—	—	32,00	—	—
4—00	69,34	33,95	—	—	—	32,50
5—00	—	—	—	—	—	32,60
24—00	69,39	36,60	33.36	32,00	31,20	33,00
48—00	69,40	36,60	—	32,00	31,17	—

TABLE 5

Type of surfactant	Adsorption after 24 h, mg/g		
	Solvent		
	Industrial	Sea	Formation
OP-7	1.70	3.44	7.6
OP-10	2.84	1.76	9.65
Wetting Agent DB	3.66	$\gg$3.86	$\gg$3.86
Sulfonol NP-1	1.50	1.92	9.53

The experiments were conducted on pulverized unfractionated rock from the red-colored stratum of the Monzhukly outcrop.

The specific surface of the rock was equal to 475 cm^2/g, the moisture content 0.15-0.25%. The solid and liquid phases were taken in weight ratio of 1 : 2. The adsorption of surfactants on the rock took place at a temperature of 60°C. This temperature corresponds to that of the formation. The surface tension (vs air) of the OP-10 solution in industrial water before and after adsorption on the sand is shown in Table 3.

From Table 3 it is evident that increasing the temperature can give some increase in the amount of adsorption of some surfactants on sand.

The amount of adsorption was determined on the basis of the change in surface tension (vs air) of the surfactant solutions before and after adsorption on the sand.

It was established by preliminary experiments (Table 4) that a 24-h adsorption time is entirely adequate to establish adsorption equilibrium.

Results are shown in Table 5 from determinations of maximum value of adsorption, calculated according to the adsorption isotherms of the surfactant solutions in industrial, sea, and formation waters.

It was established that the presence of electrolytes in the solution (solution in formation water) gives rise to increased adsorption of all the test surfactants at the rock surface.

The maximum value of adsorption of Sulfonol NP-1 from its solutions in industrial and sea waters was lower than the maximum adsorption from solutions of nonionic surfactants in those same waters.

However, it must be noted that the values of adsorption on sand for surfactants representing a mixture with different molecular weights are overstated [8]. Taking this circumstance into account, the value of adsorption on sand for the Sulfonol NP-1, which is the sulfonation product of 250-320°C alkylbenzenes, is taken by us to be approximate.

Considering the high surface activity of an aqueous solution of Sulfonol NP-1 at the interface with a hydrocarbon liquid and its low adsorption on rock, and also the possibility of obtaining it, this material was recommended by TFVNII for widespread introduction into many engineering processes in petroleum production.

A description is given below of experience in the use of Sulfonol NP-1 in the oil fields of Turkmeniya in 1961.

Sulfonol NP-1 represents the product from the alkylation of benzene with dodecane (propylene polymer) with subsequent fractionation, sulfonation, and neutralization. The resulting dodecylbenzenesulfonate ($C_nH_{2n+1}C_6H_4SO_3Na$, where n = 12-18) is a surfactant with 50-60% content of active material, produced in the form of a light-yellow powder, transportable (placed and stored in paper bags), and entirely nonhazardous in use.

In the oil fields of the NPU [Oil Field Administration] Kundagneft' and Chelekenneft' (Dagadzhik field), wells are as a rule flooded ("choked") with water or clay mud prior to carrying out repair operations (replacement of deep-well pump, repair-isolation operations) and also in preparing wells for carrying out various methods of action on the bottom-hole zone of the formation. This is explained by the fact that the oil deposits of

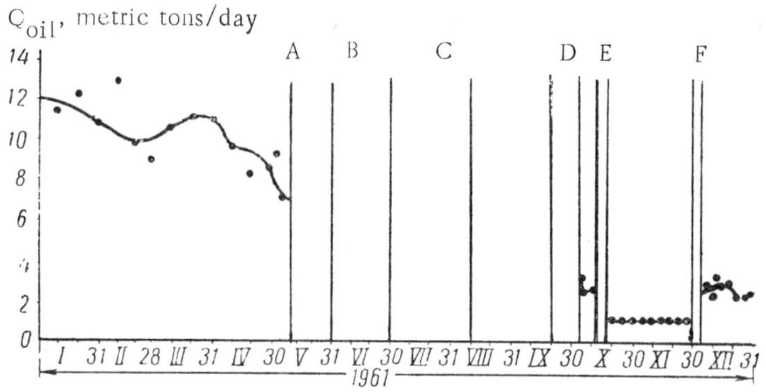

Fig. 6. Nature of operation of well 435 before and after operations on major overhaul and treatment of bottom-hole zone of formation with aqueous solution of Sulfonol NP-1. A) Choking with water, hoisting pump, major overhaul, washing with water, and inspection of casing: perforation in interval 1379-1383 m in clay mud with $\gamma = 1.25$ g/cm^3; washing out well bore to a depth of 1386 m; lowering of NGV-1-43 pump to a depth of 1126 m; absence of delivery; replacement of mud with water. B) Replacement of pump; absence of delivery; choking with water; hoisting of pump; lowering of new depth pump; lowering of pump with washing; seizure of pipes with case of pump. C) Clearing of well bore of foreign objects; unscrewing of pipes and subsequently hoisting them; fishing operations to extract pump body; lowering of pump; absence of delivery; replacement of pump; replacement of pump two times; absence of delivery; hoisting of pump. D) Lowering of pump; absence of delivery; treatment of well with Sulfonol NP-1 solution; replacement of pump; delivery obtained. E) Choking with water and replacement of pump because of absence of delivery. F) Treatment with Sulfonol NP-1 solution; replacement of pump.

Kum-Dag and Dagadzhik are developed with bottom-hole pressures considerably lower than the saturation pressures; therefore, when the rocker unit is stopped, gas comes out of the fluid column with subsequent gas blowout or short-term flowing.

The oil collectors of the Turkmen deposits are represented mainly by loose and weakly cemented sands or sandstones, often resulting in the formation of a sand plug during well operation, which in turn results in frequent plunger jamming in the deep-well pump. All of this necessitates frequent replacement of the deep-well pump (in the oil fields of NPU Kumdagneft') with flooding of the well with industrial water prior to each replacement.

The sand plugs (NPU Kumdagneft' and Nebitdagneft' are also washed out using low-mineralization industrial water of the Bala-Ishemsk water intake as the working fluid.

When choking wells or washing out sand plugs with water, water may possibly penetrate into the bottom-hole zone and drive back the oil into the depth of the formation.

The effect of water penetrating into the bottom-hole zone of the formation on bringing in a well has been examined in detail in the work of G. A. Babalyan [6]: "If water has penetrated into the bottom-hole zone, then when reverse displacement into the well occurs a certain quantity of the water remains in the pore space. We will term this water residual. The more residual water in the bottom-hole zone, the lower its oil permeability and hence also the lower the productivity of the well after bringing in."

This position is supported by data on the operation of wells 435 and 544 (Western Kum-Dag), 670 and 671 (Klin section of Kum-Dag deposit), 76 (Dagadzhik), and others.

Well 435 (horizon III, filter at 1383-1396) was operated by the deep-well pump method. On May 10, 1961, the well was flooded (choked) with water for the purpose of perforating the casing. On May 15, the per-

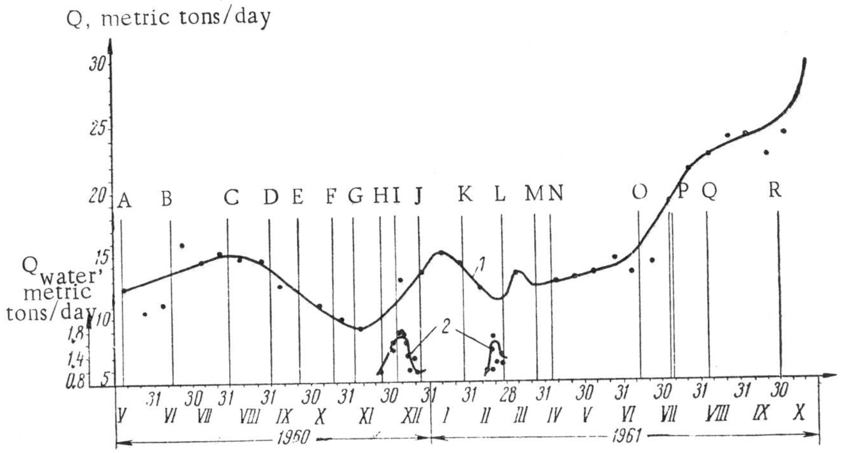

Fig. 7. A, B, C, D, E, H, Q) Choking with water and replacement of pump; F, G, I, K, M, N, O, R) pumping hot oil through; J) treatment of bottom-hole zone with Sulfonol NP-1 solution and replacement of pump; L) pumping hot oil through, choking with water, replacement of pump.

foration was carried out in the interval 1379-1383 m in clay mud of density $\gamma = 1.25$ g/cm^3. On May 16, after the well bore was washed, an NGV-1-43 deep-well pump was lowered to a depth of 1126 m. However, there was no delivery. The pump was hoisted, and the well bore was again washed out. A new pump was lowered; there was no delivery. The well was again washed with water and the pump replaced. During lowering of the pump, seizure occurred.

In July 1961, fishing operations were carried out to extract the pump-compressor pipes and pump case remaining in the well. After the well bore was cleaned of the foreign objects, deep-well pumps were lowered three times, owing to the absence of delivery. Five months after perforating the filter [casing], on September 20, the well bore was washed out with water containing Sulfonol NP-1, with subsequent forcing into the formation of 3-4 m^3 of aqueous solution of Sulfonol NP-1 (consumption of Sulfonol NP-1 was 14 kg). After treatment, the well entered into periodic operation with a production rate of 2.5-3 metric tons of crude per day.

The nature of the well operation before and after carrying out the performation of the casing and the treatment of the bottom-hole zone with Sulfonol NP-1 solution is shown in Fig. 6.

It is very indicative that when the well was flooded with water unenriched with Sulfonol NP-1 the oil production dropped to 1 metric ton/day. At this production rate the well was operated 1.5 months, and only after it was again treated with Sulfonol NP-1 solution (November 30, 1961) did it begin to operate with an average production rate of 2.5 metric tons/day. The concentration of Sulfonol NP-1 in the latter treatment was low (0.05%).

The nature of operation of well 435 illustrates graphically the picture of the adverse effect of water and low-quality clay mud on the bottom-hole zone of the formation, and its is also evidence of the effectiveness of treating the well with aqueous Sulfonol NP-1 solution after carrying out repair operations.

Well 544, drilled in 1955 in operating objective B (filter in the interval 679-724 m), up to November 1959 was operated as a flowing well with an oil production rate of 20-30 metric tons/day, but the rate decreased sharply and amounted to 4-8 metric tons/day at the end of the flowing period.

In November 1959, the well was converted to deep-well pump operation, after which it began to be flooded with water before each replacement of the deep-well pump. As a result of these floodings, in November 1960 (after exactly one year), water appeared in the production in amount of 1.2-1.4 metric tons/day (10% of the total fluid production rate).

On December 23, 1960, the well was treated with an aqueous solution of Sulfonol NP-1 (12 kg NP-1 in 18 m^3 of water); the well accepted (adsorbed) without back pressure on the well head. After treatment with

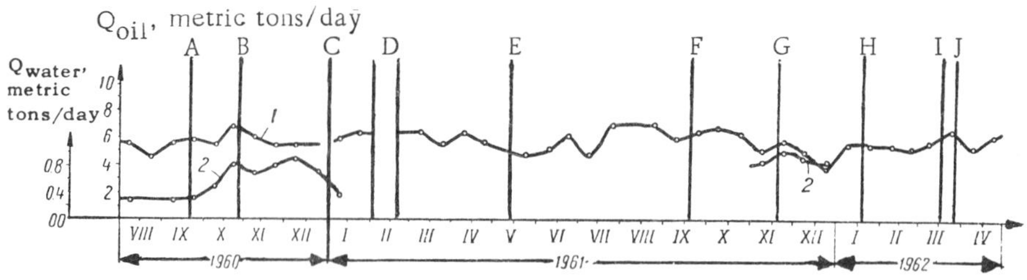

Fig. 8. A, B, E, F, H, I, J) Choking with water and replacement of pump; C) choking with aqueous solution of Sulfonol NP-1 and replacement of pump; D) choking with water, hoisting of pump, inspection of casing, perforation of column in clay mud ($\gamma = 1.25$ g/cm^3), and lowering of pump; G) treatment with aqueous Sulfonol solution and replacement of pump.

water (over 1-2 days), the well went over to producing straight oil, resulting in an additional yield of 1.2-1.4 metric tons/day of oil.

After 1.5 months, in February 1961, water again appeared in the well production at an average of 1.2 metric tons/day. On February 26, before replacing the pump, the well was flooded with water enriched with Sulfonol. The consumption of Sulfonol NP-1 here amounted to 14 kg. The well at each flooding up to the well head adsorbed a certain quantity of water into the formation.

After the water treatment, the well again went over to producing straight oil. From March 1961 to January 1962, there was no water in the well production.

At the end of January 1962, water again appeared in the well production to the extent of 10-20% of the daily fluid production. Up to January 1962, after the last treatment of the well with Sulfonol, there had been five pump replacements with "choking" with water before each replacement.

In 1962, the well continued to operate with a production rate of 2.5-5 metric tons per day (10-15% water).

Figure 7 shows the nature of operation of well 544 (graph of the well productivity as a function of operating time is plotted according to the data of daily production measurements).

Since the reverse displacement of water by oil from the bottom-hole zone of the formation takes place over a rather prolonged time, and since in some cases the residual (pumped) water is not displaced at all because of the manifestation of capillary forces and the formation of water–oil emulsion, then in the bottom-hole zone of the formation there is created a zone of increased water saturation, which naturally leads to reduction in the oil production rate.

From Fig. 7 it is evident that the appearance of water in the production of the well (curve 2) twice preceded a considerable drop in the oil production (curve 1). Only after two treatments with Sulfonol NP-1 solution was the water displaced from the bottom-hole zone by oil, which lead to a gradual increase in the oil production rate.

In July 1961, the well went over to a semiflowing regime of operation (pump delivery coefficient $\eta > 1$), and in November the oil production rate reached 34 metric tons/day.

It should be noted that the operation of the formation did not depend on changing the parameters of the operating regime of the deep-well pump.

The average daily oil production rate for 5 months before treatment, on the average, according to measurements, amounted to 12.5 metric tons/day.

For well 544, during the 10-month period from March to December 1961, the additional production was about 1950 metric tons of oil.

Well 670 (producing from the Apsheron deposits in the interval 855-921 m), after it had gone over to pumping operation, began to be flooded (choked) with water during repair operations, resulting in the appear-

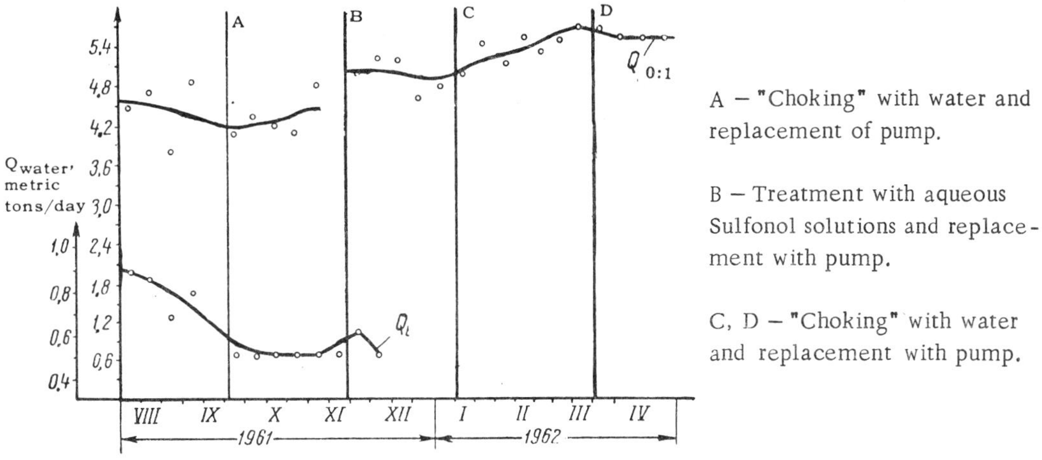

Fig. 9. Nature of operation of well 671 before and after treatments of bottom-hole zone of formation with aqueous solution of Sulfonol NP-1.

ance of water in the well production. From February to December 1960, the quantity of water in the production gradually increased from 5 to 15%.

In January 1961, the well was choked with a 0.05% aqueous solution of Sulfonol, after which the well went over to producing straight oil (Fig. 8, curve 1). In February, the casing was shot with a perforator. However, after this operation, there was still no water in the well production.

The well produced straight oil for 9 months (including a total of three pump replacements), and only in November did water again appear in the well production.

On November 18, the well was again treated with Sulfonol solution (consumption of Sulfonol NP-1 was 12 kg), after which, over the course of one month, water appeared periodically in amount of 20-30% of the fluid production rate (Fig. 8, curve 2). At the end of December 1961, the well again went over to producing straight oil.

Considering that the average daily production rate of oil and water during the last months before treatment were 5.5 and 0.8 metric tons/day, respectively, and that the rate of drop in formation pressure at well 670 in 1961 averaged 10 kg/cm^2, the amount of additional oil production during the water-free period of well operation up to January 1, 1961 amounted to about 200 metric tons.

An analogous picture was also observed for well 671 (operating objective "Apsheron" of the Klin section). The well was converted to pumping operation on February 9, 1960, and water in amount of 10% of the daily production rate (0.8-1.0 metric tons/day of water) appeared in the well production at the end of December 1960.

On November 18, 1961, the well was choked with water with the addition of Sulfonol NP-1 (12 kg in 18 m^3 of water); at each pumping of water, the well absorbed a certain quantity of water into the formation. After flooding with water with Sulfonol NP-1, water appeared in the well production for 0.5 month.

Subsequently, the well went over to producing straight oil (Fig. 9).

In the majority of the wells in the Kum-Dag deposit, the formation pressure is considerably lower than the hydrostatic pressure; therefore, on flooding with water there is observed a lowering of the water level in the well bore and, naturally, the driving back of oil by water into the depth of the formation.

The disappearance of water in the well production after its treatment with Sulfonol NP-1 was also observed at other wells (631, 390).

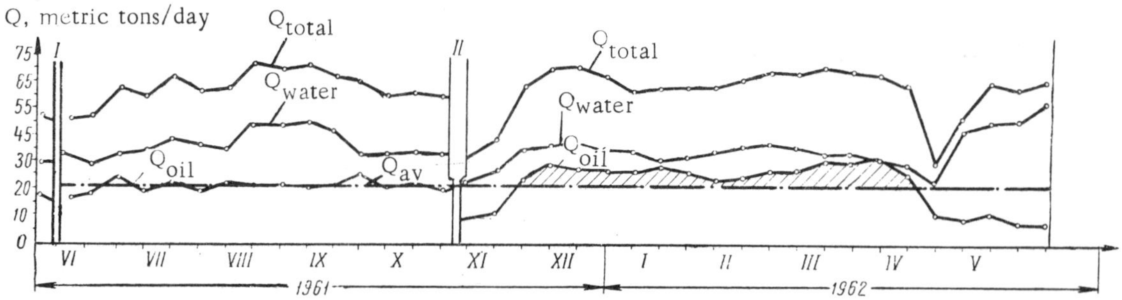

Fig. 10. Nature of operation of well 76 before and after treatment of bottom-hole zone of formation with solution of Sulfonol NP-1.

We have examined in the foregoing material the effect of residual water and aqueous solution of Sulfonol NP-1 on the nature of operation of the bottom-hole zone of the formation.

Let us examine now the nature of operation of a well after treatment with Sulfonol NP-1 in the case in which the well is watered with its own formation water.

Well 76 (filter in the interval 501-1562 m) was operated as a flowing well up to May 1961 with an oil production rate of 17-20 metric tons/day and a water cut of 50-60% at the end of the flowing period.

On May 20, 1961, an NGN-2-56 deep-well pump was lowered into the well. After 15 days, there was no fluid delivery from the pump. It was necessary to replace the pump: On June 6, the well was choked with sea water and the old pump was hoisted; on June 8, a new NGN-2-56 pump was lowered to a depth of 493 m, and the rocker unit was put into operation with parameters $l_x = 1.8$ m and n = 7 strokes per minute.

From June 8 to June 12 inclusive, the well supplied water with a film of oil, and only on June 13 was the well production directed to the collector. Thus, for a 5-day period, there was no oil in the well production.

An analogous picture was noted in 1962 in wells 194, 237, and 86 (Dagadzhik).

After 4.5 months, i.e., on November 3, 1961, it was again necessary to replace the pump because of no delivery. Before hoisting the pump, the well was choked with sea water enriched with Sulfonol NP-1 (solution concentration 0.1% by weight).

On November 5, an NGN-2-56 pump was lowered to the previous depth, and the rocker unit was put into operation with the previous parameters.

After 18 hours, the well began to yield oil at about 9 metric tons/day.

It must be noted that the oil of well 76 is high-asphalt and wax-free. The nature of operation of well 76 before and after treatment of the bottom-hole zone with Sulfonol NP-1 solution is shown in Fig. 10, from which it is evident that in the course of 4.5 months before treatment the average daily oil production was 22 metric tons/day, and after treatment 26-27 metric tons/day over the course of almost 5 months.

Hence, the additional oil production at this well during operation with increased production rate (cross-hatched area in Fig. 10) was about 600 metric tons.

Well 76 absorbed during the flooding with water, which naturally led to a prolonged water treatment (5 days).

Having taken account of the length of treatment of well 76 with water after pump replacement, the geological service of Oil Field No. 2 of NPU Chelekenneft' prohibited the choking of the well with water unenriched with surfactant.

In 1962, in the Dagadzhik field, the addition of Sulfonol NP-1 to the water during choking of a well was introduced as a required engineering operation.

The technology of well treatments with aqueous solution of Sulfonol NP-1 consists of the following.

The solution was prepared in a metal tank mounted and transportable on a caterpillar truck. The tank volume was 16-18 m^3.

Powdered Sulfonol NP-1 was charged to the tank, filled with pure water, and was mixed until completely dissolved.

In the majority of cases, the treatments were combined with replacement of the depth pump and therefore were carried out without hoisting the pump-compressor pipes and the deep-well pump from the well.

The solution was pumped into the well by a PA9-80 unit; when the solution reached the interval of the filter, the gate valve at the well head was closed, and a certain amount of the Sulfonol solution was forced into the formation in the absence of absorption by the formation.

Before calling for admission of fluid to the well, a time of 18 to 24 hours elapsed.

Summary

1. Flooding a well with water with the addition of a surfactant during pump replacement is a preventive measure to avoid the reduction in oil permeability of the bottom-hole zone of the formation.

2. Treatment of the bottom-hole zone of the formation with an aqueous solution of Sulfonol NP-1 leads to a lowering of the water saturation of the porous medium and hence to an increase in the oil productivity of the well.

3. Treatment of the bottom-hole zone of the formation with aqueous solution of Sulfonol NP-1 makes it possible to bring in wells that are temporarily existing in conservation and also to shorten the time of well treatment (especially "washing out") with water after replacement of the deep-well pump.

4. The operation of well treatment with Sulfonol solution is simple, relatively inexpensive, and effective.

Literature Cited

1. I. I. Lyal'ovich, "Properties of the oil field formations of Turkmeniya," Trudy TfVNII, No. 3 (1961).
2. I. F. Zelepukin, "Generalization of experience in the development of the Kum-Dag oil deposit and proposals for increasing the efficiency of its operation," Report for 1961-1962, Fondy TfVNII.
3. A. V. Slavyanova, "State of hydrogeological studies of the western part of Central Asia," Report of VSEGINGEO on Topic No. 87 (1956), Fondy UGiOH Turkmenskoi SSR, Ashkhabad.
4. I. S. Starobinets, "Technical and commercial characteristics of the principal types of crudes of the Pre-Balkan region," Report No. 9 for 1955, Fondy TfVNII.
5. Yu. V. Dobrov, Hydrochemistry of formation waters of Southwestern Turkmeniya in the light of prognosis of its oil and gas content. Dissertation in competition for university degree of Candidate of Geological-Mineralogical Sciences, TfVNII (1960).
6. G. A. Babalyan, "Certain problems in the physics of the process of bringing in oil wells," Bashkirskaya neft', No. 3, Ufa (1958).
7. L. V. Lyutit, "Effect of surface-active components of crudes on the process of oil production," Report No. 161 for 1961, Fondy VNII.

THE USE OF SURFACTANTS IN THE DEVELOPMENT OF LOWER CRETACEOUS WELLS IN THE KARABULAK-ACHALUKI FORMATION

E. M. Tosunov, R. B. Shaevich, and G. Z. Zhovner

Experimental treatment with surfactant solution in the petroleum fields of Chechen-Ingush ASSR gave satisfactory results on the whole. However, some of the treatments gave no results and even negative results. The reason for the latter is that a given surfactant may give opposite effects, depending on its concentration and other conditions of its use [1, 2]. For this reason, the possibility and desirability of using a reagent under particular conditions can only be decided after its basic technological characteristics have been investigated.

With this in mind, we started some experimental work on the conditions in the Lower Cretaceous wells of the Karabulak-Achaluki formation. A preliminary appraisal of surfactants was made, based on laboratory data on demulsification and surface tension. Tests were made on Sulfonol NP-1, OP-10, $KAUFE_{14}$, and also on reagents produced by the GrozNII from local crude, based on pyrolysis tar (alkylarylsulfonate), sulfonated scrubber condensate (an oxidation product of petrolatum), and secondary unsaponified products from the SZhK processes (sulfonated alcohols). Solutions of these reagents were prepared with industrial water.

Tables 1 and 2 give the compositions of the industrial water and the Lower Cretaceous petroleum with which the tests were carried out.

Laboratory Investigation of Surfactants. Experiments on Demulsification

By the entry of water or the filtrate from clay suspension into the prefilitration zone of the stratum, it is possible, under certain conditions, to form stable emulsions of the "water-in-petroleum" type [5]. It must be assumed that some of this emulsion will be extracted from the stratum in the process of development and exploitation of the wells, but some will remain in the stratum and considerably reduce its productive capacity [3, 4].

On this basis, in order to estimate the demulsifying properties of a reagent for the conditions of the Lower Cretaceous wells of the Karabulak-Achaluki field, which give a clean petroleum containing only 0.2 to 0.3% of water, experiments were carried out with artificial emulsions of the water-in-petroleum type. These emul-

TABLE 1. Composition of the Industrial Water

Sampling date	Dry residue, g/liter	Alkalinity, ml N HCl	Hardness, °H	Naphthenic acids, ml H NaOH	$Fe_2O_3 + Al_2O_3$, g/liter	SiO_2, g/liter	Anions, g/liter			Cation, g/liter		
							Cl'	SO_4'	CO_3''	$Na^{\cdot} + K^{\cdot}$	$Ca^{\cdot\cdot}$	$Mg^{\cdot\cdot}$
April 28, 1960	0.322	5.0	14.7	No data	0.005	0.035	0.010	0.019	0.150	0.001	0.075	0.017

TABLE 2. Composition of the Lower Cretaceous Petroleum

ρ_4^{20}	Water content, %	Acid number, mg KOH/g	Content							
			Sulfuric acid resins, % v/v	Silica gel resins, % w/w	Asphaltenes, % w/w	Paraffin, % w/w	Sulfur, % w/w	Nitrogen, Kjeldahl	Chloride, mg NaCl per liter of petroleum
0.8298	0.2	0.09	12.5	3.42	0.49	7.4 (51°C)	–	–	4.0

sions were prepared by mechanical mixing of the components in a paddle mixer with an electric drive. The water to petroleum ratio was 1 : 9. In all these tests we determined the type of the emulsion and its stability at 20 and 90°C over a 24-h period.

The effect of Sulfonol in preventing emulsion formation was investigated first. In these experiments the petroleum was mixed with Sulfonol solutions, varying in concentration from 0 to 1.2%. Figure 1, curve a, shows the relation between the amount of water separated and the concentration of Sulfonol solution at 90°C.

It was established that the stability of the emulsion diminished with increasing Sulfonol concentration up to 0.15%. However, the effect of the reagent decreased, i.e., the stability of the emulsion increased, with a further increase in emulsifier concentration.

The results of these tests showed that the optimum Sulfonol concentration was 0.08 to 0.24% (settled at 90°C).

The residual water was less in experiments at 20°C than at 90°C.

Experiments on the breakage of emulsions by surfactants were carried out at the same temperatures, 20 and 90°C. Figure 1, curve b, shows the results of demulsification by Sulfonol at 90°C. In these tests the most effective breakage of the emulsion was at relatively low concentrations, from 0.08 to 0.18%, showing that Sulfonol was a good demulsifier. However, there was a tendency for the demulsifying action to diminish at higher concentrations, and with 3% of surfactant the emulsion was more stable than in the absence of surfactant. It is clear that at 3% concentration the Sulfonol was in this case acting as an emulsifier.

In the breakage of emulsions by other reagents, the optimum concentration ranges were wider (Table 3). The maximum amount of water separated was either the same, or rather more, than in the experiments with Sulfonol. The demulsifying effects of these other reagents also diminished when their concentrations were increased above the optimum values.

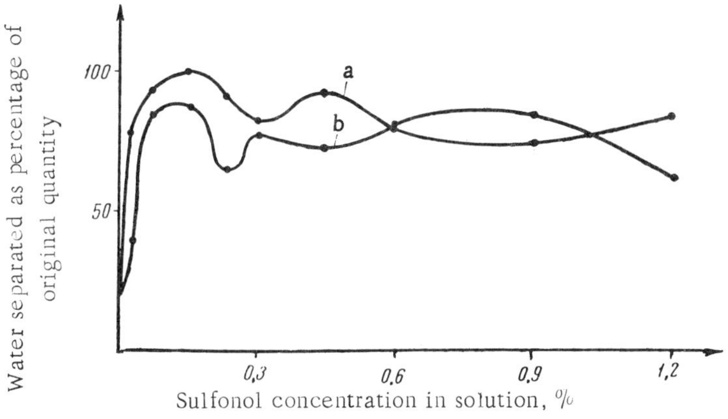

Fig. 1. Prevention of formation and breakage of emulsions by Sulfonol at 90°C. a) Prevention of formation; b) breakage.

TABLE 3

Surfactant	Surface tension σ, ergs/cm²		Demulsification		Recommended concentration, %
	Optimum concentration, %	Reduction in surface tension as compared with original value	Optimum concentration, %	Maximum settled water at 90°C, %	
Sulfonol NP-1	0.18	24.0	0.10-0.18	90	0.18
OP-10	0.30	4.2	0.15-0.30	90	0.30
$KAUFE_{14}$	0.24	3.2	0.15-0.45	90	0.30
Reagent based on pyrolysis resin . .	0.30	7.0	0.15-0.30	95	0.30
Reagent based on secondary unsaponified products	0.30	3.3	0.30-0.60	90	–
Reagent based on scrubber condensate	0.30	2.3	0.10-0.21	94	–

The surface tensions of the surfactant solutions were determined at 20°C, using the apparatus designed by P. A. Rebinder. Isotherms were recorded for the surface tensions of the solutions at the boundary with petroleum. The surfactant concentrations were varied from 0 to 0.3%. Figure 2 shows surface tension isotherms for the reagents used.

The results showed that the greatest lowering of surface tension was obtained with Sulfonol. The surface tension was reduced by a factor of 24 with a 0.18% solution of this reagent. A considerable reduction was also obtained with the local reagent based on pyrolysis resins. This reduced the surface tension by a factor of 7 at a concentration of 0.3%. The other reagents, at the same concentration, reduced the surface tension by a factor of 2.3 to 4.2.

On the basis of the results of laboratory investigations on demulsification and reduction in surface tension, shown in Table 3, Sulfonol NP-1 and the reagent based on pyrolysis resins, at the indicated concentrations of 0.18 and 0.3% respectively, were recommended for experimental treatment of the filtration zones of the stratum.

However, field trials were carried out with Sulfonol only, because the reagent based on pyrolysis resins was not available in sufficient quantities.

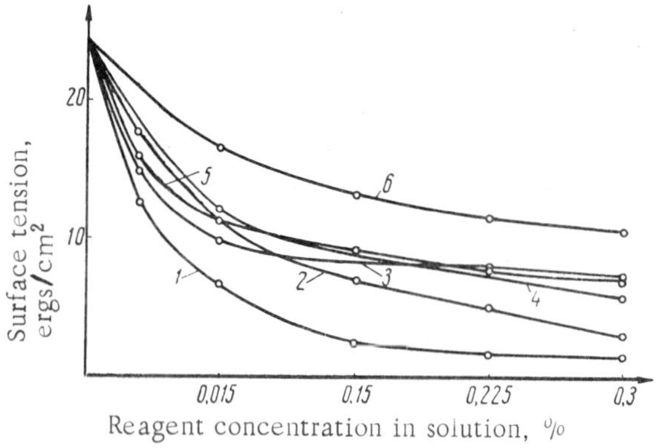

Fig. 2. Isotherms of surface tension at the boundary with petroleum for aqueous solutions of surfactants: 1) Sulfonol; 2) reagent based on pyrolysis resins; 3) $KAUFE_{14}$; 4) OP-10; 5) reagent based on secondary unsaponified products; 6) reagent based on scrubber condensate.

Experimental Treatment with Sulfonol Solutions

Experimental treatment with Sulfonol solutions was practiced during the development and exploitation of the Lower Cretaceous wells of the Karabulak-Achaluki field.

The Lower Cretaceous formations in this field consist mainly of sandstone, bonded with argillaceous carbonate material. The stratum pressure is about 390 atm, and the facial temperature is about 110°C, at a depth of 3000 m below the surface. The permeability of the trap is very low (0.001 to 1.5 millidarcys (m D)) according to tests on core material, but quite high (6.7 to 484 mD), according to tests on wells. In all, 22 treatments were carried out in seven wells, during the testing and exploitation of 14 different sections of the Lower Cretaceous.

The technology of the treatment was as follows. The Sulfonol solution in technical water was injected into the stratum through tubes, whose sockets were mounted in the upper opening or in the middle filter. The Sulfonol concentration was varied from 0.55 to 1.50 kg/m^3, with a mean value of 1.0 kg/m^3. The quantity was based on 1 m^3 for 1 m perforation interval. The solution was injected under a surface pressure of 140 to 300 atm. The wells were normally put into service 24 h after treatment.

This experimental injection of Sulfonol into the Lower Cretaceous wells gave positive results in 15 cases, corresponding to 60% of the total number of treatments. This is no worse than the success obtained by other methods for increasing the yield of petroleum, such as hydrochloric acid treatment, varying the pressure, etc.

The tests showed that treatment of the facial zone of a stratum with Sulfonol solution, using the technique described above, was one of the most effective and economical means for developing and increasing the productivity of the Lower Cretaceous wells of the Karabulak-Achaluki field.

In addition to this, we made an analysis of existing industrial data on the treatment of filtration zones of strata in the wells of other fields of the Republic. These treatments were mainly carried out with Sulfonol solution in wells draining old, exhausted, and flooded strata of the Karango-Chokraksk formation. Positive results were obtained in a number of cases. For example, treatment with Sulfonol was applied to the strongly flooded wells of the October field by the NPU Oktyabr'skneft'. The flooding of the wells subjected to treatment was normally more than 90%, and in some cases reached 98 to 99%.

Under these conditions other known methods of intensifying petroleum production had little effect, and only Sulfonol treatment gave positive results. The recovery of petroleum increased by a factor of 1.5 on the average, and this persisted for 2 months or more. Moreover, several wells were brought back into production from a state of inactivity and temporary conservation.

However, success in the treatment of such wells was not very high (below 48%).

Summary

1. The results of preliminary laboratory investigations indicated that Sulfonol, at a concentration of 0.10 to 0.18%, was the most effective of the reagents investigated for treating the facial zones of the strata in the Lower Cretaceous wells of the Karabulak-Achaluki field.

2. Experimental treatment with Sulfonol solution of the Lower Cretaceous wells of the Karabulak-Achaluki field gave positive results.

The tests showed that such treatment was one of the most effective and economical methods for developing and raising the production of Lower Cretaceous wells.

3. In planning a rational technology for the treatment of filtration zones of a stratum with surfactant solution, special attention should be directed to the choice of reagent and to determination of its optimum concentration under the particular conditions.

Literature Cited

1. P. A. Rebinder, "Surfactants and their uses," Khimicheskaya nauka i promyshlennost', IV, 5 (1959).
2. P. A. Rebinder, "Surface-active substances: their importance and applications in the petroleum industry," Proceedings of the First All-Union Conference on the Use of Surface-Active Substances in the Petroleum Industry and on Their Production (Gostoptekhizdat, 1961).
3. V. A. Amiyan, Increasing the Productivity of Wells (Gostoptekhizdat, 1961).
4. C. E. Clason, "Chemical surfactants, 1-3," Oil and Gas J., No. 7, 10, 12, p. 54 (1955).
5. Sh. N. Aliev and G. A. Babalyan, The Mobility of Petroleum Contours (Aznefteizdat, 1956).

RESULTS OF INVESTIGATION AND PROSPECTS FOR THE USE OF SURFACTANTS IN THE RECOVERY OF PETROLEUM AND GAS FROM THE FIELDS OF THE SARATOV REGION OF THE LAND ALONG THE VOLGA

V. P. Goncharov, A. I. Babayan, and I. D. Mineeva

It is well known that the fundamental problem in the development of petroleum and gas fields is to ensure the maximum flow rate and total yield from the stratum.

This can be achieved by introducing a rational system of development and intensification methods, one of which is to use surfactants in the various technological processes for recovering petroleum and gas.

The Soviet scientists Academician P. A. Rebinder and Professor G. A. Babalyan have provided the theoretical basis for the use of surfactants in the petroleum industry.

The use of surfactants in the Saratovsk wells is based on the need for reducing the water content and increasing the petroleum yield from flooded wells, restoring the productivity of wells which have become choked by water in the course of repair work (or preventing a reduction in productivity caused by such choking),

Fig. 1. Surface tension isotherms for aqueous solutions of surfactants at the interface with petroleum: I) DS–Na; II) Nekal; III) OP-10; IV) Bunigal; V) Sapal; VI) Mylonaft; VII) NChK; VIII) Sulfonol; IX) SNS.

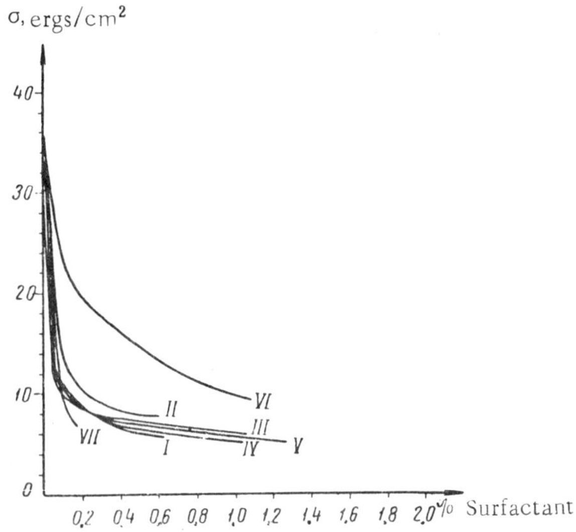

Fig. 2. Surface tension isotherms for solutions of surfactants in formation water at the interface with petroleum: I) DS–Na; II) Nekal; III) OP-10; IV) Bunigal; V) Sapal; VI) NChK; VII) Sulfonol.

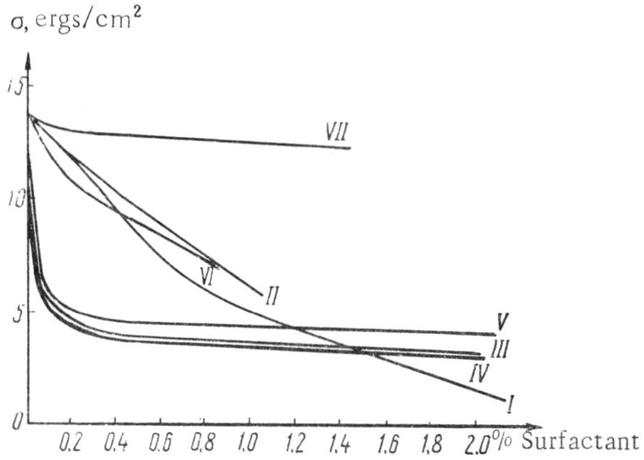

Fig. 3. Surface tension isotherms for solutions of surfactants in 0.5% hydrochloric acid at the interface with petroleum: I) DS–Na; II) Nekal; III) OP-10; IV) Bunigal; V) Sapal; VI) NChK; VII) Mylonaft.

increasing the effectiveness of acid treatment, developing formulations for the working liquids for hydraulic fraction of strata in gas wells, and searching for effective and cheap reagents to demulsify petroleum.

As a basis for the use of surfactants, the Laboratory for the Technology of Petroleum and Gas Recovery of the NV NIIGG carried out, during 1959 to 1961, an investigation of some of their properties: solubility in water and petroleum, adsorption, surface tension of solutions.

There was also a special investigation to formulate a working liquid, containing the surfactant OP-10, to be used for the hydraulic fracture of strata in gas wells.

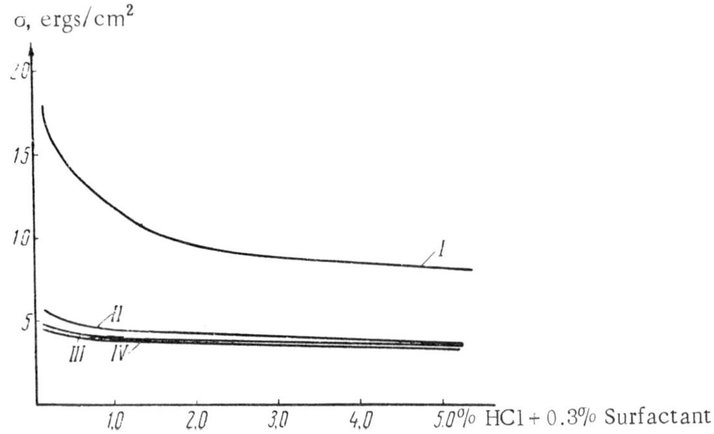

Fig. 4. Surface tension isotherms for 0.3% solutions of surfactant in hydrochloric acid of varying concentration at the interface with petroleum: I) without surfactant; II) Sapal; III) Bunigal; IV) OP-10.

The following types of surfactant were tested: DS–Na, Bunigal, Nekal, Sapal, OP-10, Mylonaft, NChK, sodium sulfonates (SNS), Stearox-6.

Results for the surface tension measurements on solutions and for the solubilities of the surfactants are given in Fig. 1, 2, 3, and 4, and in Table 1. These suggest the following conclusions.

1. All the surfactants greatly reduced surface tension.

2. OP-10, DS–Na, Bunigal, Sapal, and Nekal were the most active; Mylonaft, NChK, and SNS were less active.

3. At a fresh water–petroleum interface Nekal gave a marked reduction in surface tension at a concentration of 0.2% in the water; DS–Na did this at 0.4% concentration, but a concentration of 1.5% was required for the boundary between 0.5% hydrochloric acid and petroleum. For either interface, Bunigal, OP-10, and Sapal caused a marked reduction in σ at a concentration of 0.2%.

4. Some surfactants–DS–Na, Nekal, SNS, and Sulfonol– dissolved readily in fresh water but incompletely in formation water to give turbid solutions. Sulfonol gave a copious flocculent precipitate with formation water.

5. Surfactant addition to hydrochloric acid solution reduced the surface tension at the interface with petroleum. This should improve extraction from the stratum by partially unreacted acid and thus increase the effectiveness of acid treatment of the facial zone.

Table 2 shows the results of laboratory tests in the search for working liquids to be used for hydraulic fracture of gas-bearing strata. The suitability of each liquid was determined by checking the extent of restoration of the gas permeability of a core after the working liquid had been filtered through it.

On the basis of these tests we developed a recipe for a hydrophilic water-condensate emulsion with OP-10, for the hydraulic fracture of gas strata, corresponding to the requirements of the hydraulic fracture process (high sand-retaining capacity, low filtration, adequate viscosity) and the subsequent exploitation of wells after fracture (complete restoration of the gas permeability of cores with various lithological characteristics).

For comparison, Table 2 also shows some data on the investigation of other working liquids, particularly water thickened with VETs or Karbozolin S. The results of these tests indicated that, of viscous liquids, a hydrophilic emulsion with OP-10 gave the best results for the restoration of the gas permeability of cores with various lithological characteristics, after the liquid had been filtered through.

We give below the results of some field trials on the treatment of the facial zones of wells with surfactant solutions. The choice of surfactant was mainly determined by availability at the field.

Petroleum production well No. 74 of the Sokolovorog gas–oil field showed a reduction in output as the result of filtration of water into the stratum due to choking by water before replacement of the submerged pump. The output of the well was 28.2 tons/day before reconditioning but dropped to 18 tons/day afterwards. The well was operated at this yield for 5 months. The water content of the product was 33%.

In June 1961, in order to restore the productivity of the well the facial zone of the stratum was treated with the petroleum-soluble surfactant Stearox-6. The Stearox (100 kg) was dissolved in 1.5 m^3 of solar oil and heated to 70°C. Then the Diesel oil with Stearox was mixed with 20.5 m^3 of petroleum. The resulting mixture was injected into the well and forced by petroleum into the stratum at 130 atm pressure (with the lifting pump not in use), after which the well was allowed to stand and react for 24 h before it was brought back into production by turning on the submerged pump.

The well productivity was restored by this treatment. Measurements after June 1961 showed that the well was producing 28.8 tons of petroleum and 30.8% of water.

In October 1961 and in March 1962, treatment of the facial zone of the stratum with ammoniacal NChK (from Yaroslavl') was carried out in flooded wells No. 39, 139, and 146 of the Sokolovorog field. These wells exploited the productive stratum D_3–I, which consists of a dense sandstone with layers of clay. In all cases the amount of NChK used was 0.5 m^3 per m depth of stratum exposed. The NChK was forced into the stratum by petroleum and water at 150 atm pressure.

Surfactant	Surfactant concentration, %	Fresh water–petroleum		Salt water–petroleum		0.5% hydrochloric acid–petroleum	
		Surface tension σ, ergs/cm²	Solubility	Surface tension σ, ergs/cm²	Solubility	Surface tension σ, ergs/cm²	Solubility
—	—	36.39		34.87		13.8	
DS–Na	0.05	23.29	Dissolves readily in fresh water, clear solution	14.57	Turbid solution	13.08	Readily soluble
	0.1	19.30		10.74	The same	12.86	
	0.5	8.96		52	Very turbid solution	8.48	
	1.0	5.54		—		5.0	
	2.0	3.48		—		1.67	
Nekal	0.05	13.12	The same	22.10	Very sparingly soluble, forms deposit	13.02	Throws down a flocculent precipitate
	0.1	8.40		12.86		12.90	
	0.3	3.02		—		—	
	0.5	1.53		7.95		9.56	
	1.0	—		—		6.08	
OP-10	0.05	9.16	" "	12.78	Readily soluble, solution clear	6.42	Readily soluble
	0.1	7.74		9.91		5.50	
	0.5	5.58		7.47		3.88	
	1.0	5.03		6.42		3.57	
	2.0	4.26		—		3.30	
Bunigal	0.05	8.24	" "	12.10	The same	6.15	The same
	0.1	6.47		10.47		5.06	
	0.5	4.90		6.54		3.72	
	1.0	4.43		5.40		3.42	
	2.0	4.20		—		3.14	
Sapal	0.05	10.46	" "	12.10	" "	7.11	" "
	0.1	8.54		10.57		5.68	
	0.5	6.28		6.87		4.58	
	1.0	5.90		5.99		4.32	
	2.0	5.55		—		4.16	
Mylonaft	0.05	23.53	Solution turbid	—	Does not dissolve in stratum water regardless of temperature	13.66	Solution turbid, forms deposit
	0.1	20.71		—		13.39	
	0.5	13.88		—		12.72	
	1.0	10.26		—		—	
	2.0	6.97		—		—	
NChK	0.05	24.24	Dissolves readily, solution clear	27.94	Forms turbidity and precipitate in stratum water	12.98	Forms deposit
	0.1	20.48		22.43		11.90	
	0.5	14.80		16.47		8.80	
	1.0	9.86		9.88		—	
	2.0	8.02		—		—	

Surfactant	Surfactant concentration, %	Fresh water–petroleum		Salt water–petroleum		0.5% hydrochloric acid–petroleum	
		Surface tension σ, ergs/cm^2	Solubility	Surface tension σ, ergs/cm^2	Solubility	Surface tension σ, ergs/cm^2	Solubility
Sulfonol	0.05	8.92	Solution green, clear	19.80	Does not dissolve in cold stratum water but the solubility increases with rising temperature; forms flocculent deposit	–	Forms copious deposit
	0.1	5.60		8.94		–	
	0.5	3.99				–	
	1.0	4.20				–	
	2.0	4.18				–	
SNS	0.05	20.38	Solution clear	–	Completely insoluble in cold water, but dissolves incompletely on heating to 65°C	–	No data
	0.1	18.04		–		–	
	0.5	10.86		–		–	
	1.0	10.26		–		–	
	2.0	7.96		–		–	
Stearox-6 (solution of Stearox-6 in petroleum)	0.05	–	Forms a colloidal solution in water, but dissolves readily in Diesel oil to give a clear solution	–	No data	–	
	0.1	12.46		15.2		–	
	0.5	6.33		7.03		–	
	1.0	2.82		3.34		–	
	2.0					–	
Technical hydrochloric acid	0.5	13.8	Solution clear	11.37	Solution clear	No data	No data
	1.0	11.8		11.75			
	2.0	9.48		12.45			
	3.0	8.83		–			
	5.0	8.31		14.31			
	7.0	7.67		–			
	10.0	8.34		16.71			
	12.0	8.64		–			
	14.0	15.23		22.45			

TABLE 2

No.	Working liquid	Lithological characteristics of cores	Gas permeability, millidarcys		Gas permeability restoration, %	Purging time, h–min	Pressure drop (Δp), atm.	Notes
			Initial, before separation with working liquid	Restored, after saturation with working liquid				
1	2	3	4	5	6	7	8	9
1	Fresh water	Sandstone, fine-grain, light gray, quartzitic, not calcareous, weakly bonded	228	228	100	15	0.5-3	–
2	The same	Light gray sandstone, fine-grain, well sorted, micaceous, bonded	27.4	26.4	96.3	27	0.5-3	–
3	" "	Sandstone, light gray to white, well sorted, stratified, micaceous	1494	1471	98.4	3-15	0.5-1.0	–
4	" "	Gray-white sandstone, fine-grain, moderately bonded, with minute spangles of mica	311.25	269.75	86.6	10-15	3	–
5	Condensate	Sandstone, light gray almost white, fine-grain, well sorted, bonded	227	207.4	91.3	13-45	0.5-2	–
6	"	Quartzitic sandstone, fine-grain, bonded by calcerous cement	134.7	93.9	70	12	3	–
7	Diesel oil	Light gray sandstone, fine-grain, well sorted, moderately bonded, with spangles of mica	160	130	81.2	7	3	–
8	Sokolovorog petroleum, stratum D_2–V	Light gray sandstone, weakly bonded, aleuritic, with calcareous-argillaceous cement	142.7	18.7	13.7	9	3	–
9	The same	Fine-grain sandstone, light gray, not calcareous, uniform, weakly bonded	508.5	122	24	36	3	–
10	Sulfite-alcohol residue, μ = 200 sec SPV-5 at 20°C	Gray sandstone, varying grain size, micaceous, weakly bonded	2532	911	35.9	9-30	1	–
11	2% aqueous VETs, μ = 47 sec SPV-5 at 20°C	Light gray sandstone, quartzitic, fine-grain, weakly bituminous, bonded, not argillaceous	214	180	84.2	8	5	The core was washed with 1.5 liters of water after the VETs solution had been filtered through it.

No.	Working liquid	Lithological characteristics of cores	Gas permeability, millidarcys		Gas permeability restoration, %	Purging time, h–min	Pressure drop (Δp), atm.	Notes
			Initial, before separation with working liquid	Restored, after saturation with working liquid				
1	2	3	4	5	6	7	8	9
12	2% aqueous VETs, $\mu = 47$ sec SPV-5 at 20°C	Gray-white sandstone, fine-grain, strongly bonded, with spangles mica	41	29.1	71.0	21	5	Without subsequent washing of the core by water
13	6% aqueous Karbozolin S, $\mu = 59.5$ sec SPV-5 at 20°C	Light gray sandstone, fine-grain well sorted, moderately bonded, with spangles of mica	572.4	286.4	50	5	5	The core was washed with 2.0 liters of water after the Karbozolin solution had been filtered.
14	The same	Sandstone, light gray almost white in color, fine-grain, well sorted, bonded	137.0	80.6	59.0	21	5	The same
15	" "	Light gray sandstone, fine-grain, well sorted, moderately bonded, with spangles of mica	105.0	76.2	72.5	6	15	" "
16	" "	Light gray sandstone, fine-grain, well sorted, moderately bonded	611.6	450.4	73.7	7	5	Without subsequent washing of the core by water
17	" "	Light gray sandstone, fine-grain, well sorted, moderately bonded, with spangles of mica	106.3	21.6	20.3	22-30	10-15	Without subsequent washing of the core by water
18	Hydrophilic emulsion: 80% condensate, 20% water, 0.5% OP-10, $\mu = 120$ sec at 20°C	Sandstone, light gray almost white, fine-grain, well sorted, bonded	197	183	92.6	8	3	–
19	The same	White-gray sandstone, fine-grain, well sorted, strongly bonded	87.7	82.4	94	9	5	–
20	" "	Light gray sandstone, fine-grain, bonded, with spangles of mica	51.6	49.0	95	5	3	–
21	" "	Light gray sandstone, fine-grain, bonded, with spangles of mica	232.8	186.2	80	6	3	–

TABLE 2 (Continued)

No.	Working liquid	Lithological characteristics of cores	Gas permeability, millidarcys		Gas permeability restoration, %	Purging time, h–min	Pressure drop (Δp), atm.	Notes
			Initial, before separation with working liquid	Restored, after restored tion with working liquid				
1	2	3	4	5	6	7	8	9
22	Hydrophilic emulsion: 80% condensate, 20% water, 0.5% OP-10, $\mu = 120$ sec at 20°C	Light gray sandstone, fine-grain, with occasional inclusions of gypsum, with spangles of mica, strongly bonded	174.7	174.7	100	15	3	The core was washed with 1.5 liters of 0.5% OP-10 solution after the emulsion had been filtered through it.
23	The same	Gray-white sandstone, fine-grain, weakly bonded, with minute spangles of mica	90.7	85.7	94.5	19	3	The same
24	" "	Light gray sandstone, fine-grain, moderately bonded, with plentiful inclusions of gypsum grains	149.4	135.0	90.4	7	5	" "
25	" "	Gray-white sandstone, fine-grain, weakly bonded, with minute spangles of mica	72.85	72.85	100	9	5	" "
26	" "	Light gray sandstone, fine-grain, moderately bonded, with plentiful inclusions of gypsum grains	58	58	100	11	5	" "
27	" "	Light gray sandstone, fine-grain, with occasional inclusions of gypsum, strongly bonded	92.2	92.2	100	9	5	" "
28	Hydrophilic emulsion: 75% condensate, 25% water, 0.5% OP-10, $\mu = 100$ sec at 20°C	Gray sandstone, quartzitic, fine-grain, bonded, not calcareous	53.8	37.7	70.1	15	3	–
29	The same	The same	58.6	57.4	98.0	9	3	–

The result of the treatment was that the total liquid yield from the wells increased by 30 to 40%, due to an influx of water, but in the course of a month the yield dropped again to the value before treatment.

At the Bagaevsk gas and petroleum field, treatment was carried out of the productive petroleum-bearing stratum of the Bobrikovsk level at wells 62, 67, and 80.

Well 62 had a casing 1123.5 m deep, the productive stratum was a dense sandstone with layers of clay, and the perforation interval was 1113 to 1115 m. Treatment of the facial zone with NChK was carried out on July 4, 1961, when the following were injected into the stratum in turn: 3 m^3 of petroleum, 3 m^3 of NChK, and 3 m^3 of petroleum. The yield before treatment was 0.1 to 0.2 tons/day with 99% of water. After treatment, the yield from July 5, 1961 to August 4, 1961 was 7 to 8 tons/day of petroleum with 50% of water. During the next month the petroleum yield dropped to 0.6 tons/day, and the water content increased to 95%. By the beginning of September 1961 the yield had dropped to the original value of 0.1 to 0.2 tons/day with a water content of 99%. The total increase in petroleum yield as the result of treatment was 220 to 230 tons.

Well 67 was treated on September 5, 1961, when the following were injected into the stratum in turn: 10 m^3 of petroleum, 3 m^3 of NChK, and 6 m^3 of petroleum. Before treatment the well yielded 2.8 to 3.5 tons/day of petroleum, with 89% of water. After treatment the well at first gave water only, but the petroleum yield gradually increased over a period of 10 days to its previous value of 3 tons/day. During the next 5 days the petroleum yield further increased to 10 to 12 tons/day, and the water dropped to 66%. However, by November 26, 1961 the petroleum yield had dropped again to its original value of 3 to 3.5 tons/day, and the water had increased to 85%. The total increase in petroleum yield as the result of treatment was about 240 ton:

Well 80 was treated on November 21, 1961, when the following were injected into the stratum: 25 m^3 of petroleum, 3 m^3 of NChK, and 10 m^3 of petroleum. There was little change in yield after treatment, so that the effect, if any, was difficult to calculate.

In March 1962 a field trial on hydraulic fracture of the stratum, using the surfactant OP-10, was carried out at gas well 15 of Bagaevsk gas and petroleum field. The stratum was opened up with a hydrophilic acid condensate emulsion of the following composition: 75% of condensate and 25% of 10% hydrochloric acid, containing 0.5% of OP-10, of viscosity 120 sec SPV-5 at 20°C. A 0.8 to 1 mm sand fraction was injected into the stratum to the extent of 10 tons. After treatment the gas yield increased by a factor of 3 to 4 (working under different conditions).

It should be noted that before this fracture of the stratum with OP-10 emulsion the well had been treated twice (in 1960 and 1961) using clear condensate from the same locality as both fracturing liquid and sand-carrying liquid. In both cases, owing to the high filtration of the condensate and the inadequate injection rate of the sand into the well face, it was not possible to inject more than 1 ton of sand, and the treatment was ineffective.

However, when the fracture was carried out with an emulsion to which the surfactant OP-10 had been added, and with the same amount of aggregate, the process proceeded smoothly and proved to be highly effective.

There are also prospective advantages in the use of surfactants for hydrochloric acid treatment and development of petroleum, gas, and injection wells after drilling. This has been confirmed by the results of laboratory tests. Insufficient field trials have been carried out because of the nonavailability of surfactants at the oil fields.

In order to solve the problems associated with the effective use of surfactants in the processes of petroleum and gas production it is necessary to extend both laboratory work and field trials, but extensive field trials will only be possible if adequate supplies of surfactants are available.

THE USE OF SURFACTANTS IN THE PETROLEUM INDUSTRY OF THE PERM' REGION

R. G. Blyum and V. G. Kunov

In the Perm' region, the establishment of wells after drilling and the restoration of the productivity of wells which have been suppressed is extremely difficult. A number of wells which have given good results during drilling in recent years have not been established, although other wells in the neighborhood, with the same stratal pressure, have produced successful gushers (well No. 92 in Yarino and well No. 113 in Pavlovka).

In the Yarino undertaking, after 18 underground repairs in 15 wells (involving the cleaning of compressor pump scrapers) under unchanged exploitation conditions, output decreased on the average from 73.9 to 45.8 tons/day, i.e., by 38%. In three of these wells no gusher could be obtained. The wells were blocked with 30-60 m^3 of fresh or stratal water.

Analogous phenomena have been observed in other deposits. Thus well No. 43 in the Tanypskii region gave 24 tons/day (4 mm connecting pipe) before repair work. After suppression of the well with fresh water and removal of the scraper wire, the well could not be re-established for two months. It then became operational with an output of 4 tons/day, which increased periodically to 22-24 tons/day and again decreased to 6-7 tons/day for a period of four months. A steady output of 50-56 tons/day (7 mm connecting pipe) was established only six months after the repair operations.

The above data show the importance of the use of surfactants capable of preventing or decreasing the undesirable influence of water in the end-face zone of operational wells.

The effect of surfactants is due to their ability to decrease the surface tension at the oil– water interface, to prevent the formation of emulsions and to break down emulsions previously formed, to improve the selective wetting of the rocks by the oil, to reduce the swelling of clay material in water, etc.

Laboratory studies have made it possible to select the most effective materials for the petroleums of our region. The materials chosen for study as the most typical were oils from the Early Carboniferous terrigenous deposits of the Kamennolozhskii and Pavlovka oil fields. The first of these contains a small quantity of asphaltenes and tars, while the second belongs to the class of petroleums with high tar and asphaltene contents.

The physicochemical characteristics of the oils are given briefly in Table 1.

TABLE 1

Petroleum	Density, ρ_4^{20}	Viscosity in centistokes		Concentration, %			
		At 20°C	At 50°C	Neutral tars	Asphaltenes	Paraffin wax	Sulfur
Kamennolozhskii	0.814	4.46	2.65	5.13	0.19	4.66	0.47
Pavlovka	0.884	24.04	9.53	20.04	4.10	2.83	2.20

TABLE 2. The Influence of the Addition of Surfactants to Water on Its Ability to Form Emulsions with Petroleum

Surfactants	Pavlovka petroleum		Kamennolozhskii petroleum
	Smallest quantity of water with which an unstable emulsion is formed when surfactants are added, %	Concentration of surfactants in water, %	Concentration of surfactant at which an unstable emulsion is formed with 20% water, %
OP-20	20	0.25	0.0025
Wetting Agent NB	40	0.5	0.25
Alkamon OS-2	40	0.05	0.005
Leveling Agent A	40	0.05	0.005
Katapin A	40	0.05-0.1	0.01
Sulfonat	40	0.5	0.005
OZhK	60	0.005	0.005 (90%)
OP-10	60	0.05	0.01
SSB	60	0.1-0.5	Ineffective
Azolyat	60	0.25	0.005 (75%)
"Progress"	60	0.5	Ineffective
Sulfonol	60	1.0	0.05
SNS	60	1.0	0.05
DS–RAS	80	0.1	0.5
NChK	80	3.0	3.0

Note. Where the emulsion did not separate completely into layers the quantity of water separating is given in parentheses.

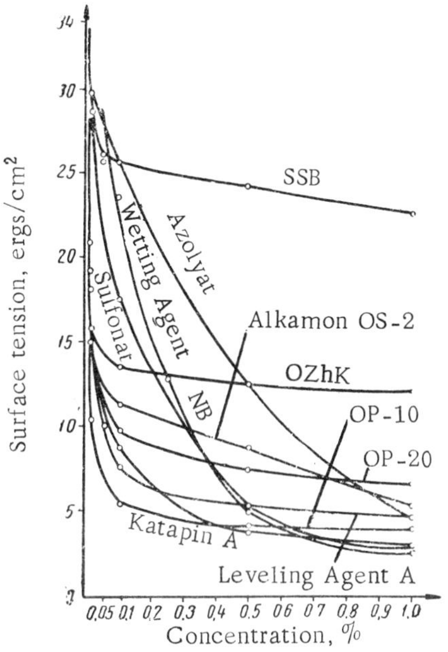

Fig. 1. Surface tension isotherms for aqueous solutions of different surfactants at their boundary with Pavlovka petroleum.

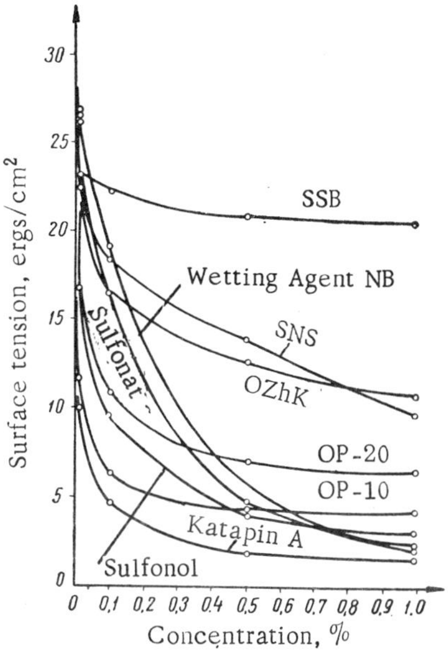

Fig. 2. Surface tension isotherms for aqueous solutions of different surfactants at their boundary with Kamennolozhskii petroleum.

TABLE 3. Optimum Concentrations of Surfactants (in g/100 ml of emulsion) at which Previously Prepared Emulsions Separate into Layers

Surfactants	Pavlovka petroleum				Kamennolozhskii petroleum
	% water in emulsion				% water in emulsion
	80	60	40	20	20
OZhK	0.0025	0.005	0.01	0.0025	0.0025
Alkamon OS-2	0.005	0.005	0.01	0.05	0.005
Katapin A	0.01	0.01	0.01	0.05	0.005
OP-10	0.01	0.05	0.05	0.05	0.01-0.005
Sulfonat	0.05	0.05	0.05	0.10	Ineffective
Wetting Agent NB	0.05	0.05	0.1-0.25	0.25	0.01-0.005
Sulfonol	0.01	0.05	0.05	0.25 (75%)	0.005
Leveling Agent A	0.1	0.1	0.1 (83%)	Ineffective	0.1
OP-20	0.1	0.01	0.5	0.5 (75%)	0.005
DS–RAS	0.05	0.10	0.5 (32%)	Ineffective	0.01-0.005
Azolyat	0.05	0.05	1.0	Ineffective	0.005
SNS	0.5	0.5	0.5	0.5 (85%)	0.5
NChK	0.5-1	1.0	1.0	1.0 (75%)	0.5
"Progress"	0.5	0.25	Ineffective		0.05
SSB	3		Ineffective		

Note. Where the emulsion did not separate completely into layers the quantity of water separating is given in parentheses.

Because of the acute shortage of surfactants, special attention was paid to a study of materials produced by the chemical industry of the Perm' region. Of these materials the following deserve attention: Wetting Agent NB – an anionogenic product (the sodium salt) from monobutylnaphthalenesulfonic acid $C_{14}H_{15}SO_3Na$; Alkamon OS-2 and Leveling Agent A – cationogenic products; and sulfite-alcohol waste (SSB).

We also tested the following surfactants: the nonionogenic materials OP-10, OP-20, and OZhK (a laboratory specimen from the VNII NP); the cationogenic material Katapin A; the anionogenic materials Sulfonat, Sulfonol, DS–RAS, and Azolyat, and "Progress," SNS, and NChK.

Laboratory experiments showed that the greatest decrease in surface tension at the boundary with petroleum for small quantities of surfactant (up to 0.01%) are given by nonionogenic and cationogenic materials; the anionogenic materials showed no significant influence. At concentrations of 0.5-1.0%, however, the anionogenic materials occupy the leading position (Figs. 1 and 2).

In the estimation of the demulsifying power of surfactants it was found necessary to use different methods of preparing the emulsions. In the case of the Kamennolozhskii petroleum it was possible to obtain a stable emulsion by mixing for 30 minutes at a concentration of water in the mixture not exceeding 20%, whereas the Pavlovka petroleum with high asphaltene content emulsified readily when mixed for a short period with up to 80% water. The comparative characteristics of the demulsifying properties of specimens with Pavlovka petroleum are therefore given for emulsions prepared by mixing for periods of 10 minutes with 20, 40, 60, and 80% of water, while the data for Kamennolozhskii petroleum are given for emulsions prepared by mixing for 30 minutes with 20% water. In the studies we used a model simulating the filtrates from industrial liquids, prepared by adding stratal water from the Carboniferous horizon to fresh water until the density of the mixture reached 1.008 g/cm^3.

The results of the studies are given in Tables 2 and 3. Table 2 characterizes the ability of the surfactants to prevent the formation of emulsions, and Table 3 characterizes the ability to separate previously formed emulsions into layers.

TABLE 4

	<0.1 mm	0.1-0.25 mm	0.25-0.5 mm	>0.5 mm	Specific surface, cm^2/g
Clay	61.2	28.3	10.2	0.3	1390
Sludge (carbonate)	—	53.4	24.1	22.4	97

The materials are listed in order of their decreasing activity with Pavlovka petroleum. The greatest ability to prevent the formation of emulsion is shown by OP-20, Wetting Agent NB, cationogenic materials, and Sulfonat. A noteworthy feature is the position occupied by SSB, whose efficiency with Pavlovka petroleum is almost as great as that of OP-10. In the experiments on the separation of previously prepared emulsions into layers, best results were obtained with OZhK, Alkamon OS-2, Katapin A, OP-10, and Wetting Agent NB. Sulfonat, which has a high efficiency in the separation of Pavlovka petroleum emulsion into layers, is ineffective in the case of Kamennolozhskii petroleum emulsions.

It is obvious that in the choice of additives for flushing liquids used to open up oil-bearing strata and liquids used to suppress wells, preference must be given to those surfactants which have the greatest ability to prevent the formation of emulsions. In order to establish the quantity of material to be added to the flushing liquid, it is necessary to know not only its optimum concentration in the liquid phase (Table 2) but also its adsorbability on the solid phase — clay and carbonate sludge. We characterized the adsorbability of surfactants from their surface tension isotherms before and after adsorption. It is known that the method for determining the magnitude of adsorption from the change in surface tension is not very accurate. Thus, according to L. V. Lyutin's data (VNII), the accuracy of the determination of the adsorbability of OP-10 sand amounts to 15%. The determination of the quantity of surfactant which must be used during drilling and control of its concentration in the liquid do not demand a high accuracy, however.

In the experiments, we used clay from a clay quarry in the Kamennolozhskii area and sludge from one of the wells, obtained during the drilling of a carbonate stratum with water. The granulometric composition (in percentages) and specific surface of the adsorbents are given in Table 4.

We also determined the adsorbability of OP-20 and OZhK by a clay suspension from one of the wells in the Kamennolozhskii area. The suspension was treated with calcined soda, so that the degree of dispersion of the clay particles was higher. The ratios of the solid to liquid phases in the clay suspensions and the sludge were the same.

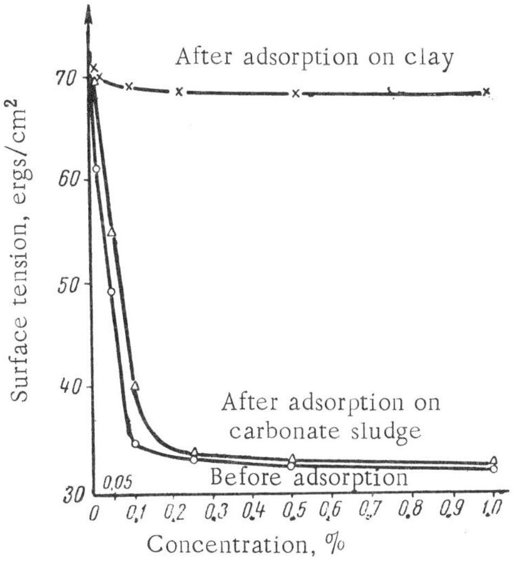

Fig. 3. Surface tension isotherms for Wetting Agent NB before and after adsorption on clay and carbonate sludge.

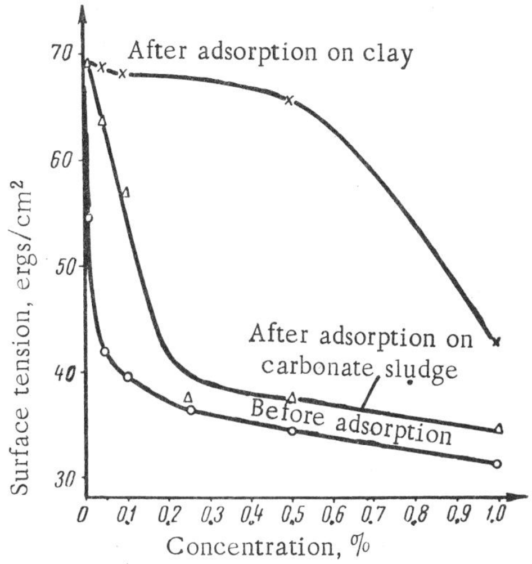

Fig. 4. Surface tension isotherms for OP-20 before and after adsorption on clay and carbonate sludge.

TABLE 5

Surfactant	% surfactant in clay suspension	Residual concentration of surfactant after adsorption, %		Residual concentration of surfactant in sludge
		In clay suspension (without chemical treatment)	In clay suspension treated with calcined soda	
Alkamon OS-2	1	0.015	–	0.17
Leveling Agent A	1	0.009	–	0.15
Katapin A	1	0.003	–	0.40
OP-20	1	0.04	0.004	0.45
OP-20	2	0.31	0.245	–
OP-10	1	0.035	–	0.5
OZhK	1	0.015	0.004	1.00
OZhK	2	–	0.006	–
Wetting Agent NB	1	0.50	–	0.25
Sulfonat	1	0.49	–	0.11
Sulfonol	1	0.11	–	0.06
DS–RAS	1	0.69	–	0.35
"Progress"	1	0.35	–	0.23
SNS	1	0.35	–	0.16
SSB	1	0.17	–	0.18
Azolyat	1	0.91	–	1.00

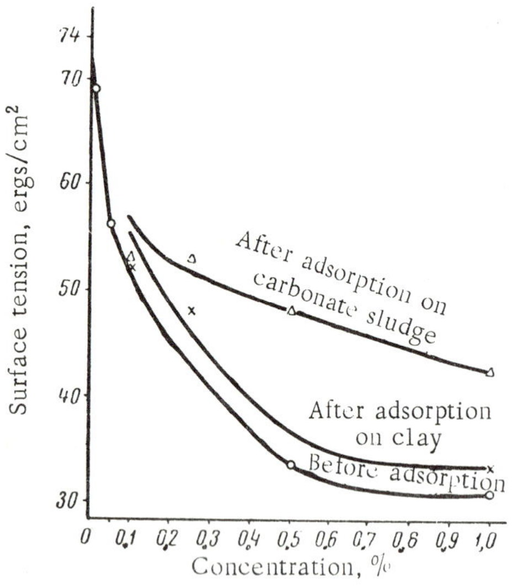

Fig. 5. Surface tension isotherms for Katapin A before and after adsorption on clay and carbonate sludge.

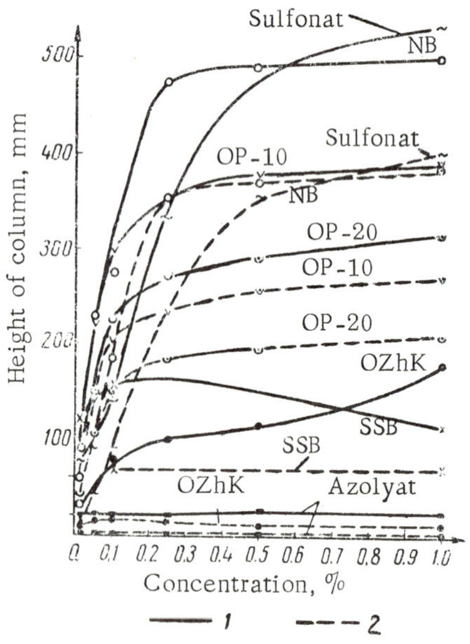

Fig. 6. Foam numbers of aqueous solutions of different surfactants. 1) Initially; 2) after 10 min.

TABLE 6. Foaming Power of Clay Suspensions

Surfactant	Density of clay suspension, g/cm^3	Concentration of surfactant in clay suspenion, %	Volume of clay suspensions with foam, as a percentage of the initial volume	
			Initially	After 10 min
OP-20	1.21	1.0	100	100
OP-20	1.16	1.0	115	109
OP-20	1.16	2.0	146	146
Wetting Agent NB	1.21	1.0	500	500
OZhK	1.21	1.0	100	100
OZhK	1.16	1.0	100	100
OZhK	1.16	2.0	109	109
SSB	1.21	0.5	105	100
SSB	1.21	1.5	112	103
SSB	1.21	3.0	120	112
SSB	1.16	1.5	112	100
Azolyat	1.21	1.0	133	133
Azolyat	1.21	2.0	136	125

The nature of the adsorability for one of the nonionogenic, cationogenic, and anionogenic materials is shown in Figs. 3, 4, and 5, respectively.

The studies showed that the adsorbability of the nonionogenic and cationogenic substances on clay is very high. For anionogenic materials, as expected, a higher adsorbability on carbonate sludge is observed.

An exception is Azolyat, whose adsorption on both clay and carbonates is slight (Table 5).

It is obvious that the use of cationogenic surfactants for opening up an oil-bearing stratum by drilling is possible only with a flushing liquid which does not contain clay (drilling using water). For example, the adsorption of Katapin A (molecular weight approximately 430) on the exchange complex of the clays which we used is equal, on the average, to 32 mg-eq per 100 g and the density of a 1.2 g/cm^3 clay suspension amounts to 50 kg/m^3 (approximately 5%).

The adsorption of nonionogenic surfactants on clay is lower than that of cationogenic surfactants but is nevertheless considerable. The equilibrium concentration in the liquid phase of a light clay suspension (density 1.16 g/cm^3), treated with calcined soda at a dose of 2%, amounted to 0.245% for OP-20 and 0.06% for OZhK (Table 5).

In order to decrease the consumption of nonionogenic surfactants, it is advantageous to open up the stratum with a natural carbonate suspension, whose solid phase contains only a limited clay fraction (not greater than 30%). The use of anionogenic materials is best combined with the use of a clay suspension containing the minimum quantity of carbonate sludge.

An important factor in the estimation of the suitability of a given material as an additive to clay suspensions is its foam-forming power, particularly when no effective antifoaming agents are available. Comparative characteristics of the foam-forming power of aqueous solutions of certain surfactants are given in Fig. 6, and data for the foam-forming power of clay suspensions are given in Table 6. The lowest foam-forming power is shown by OZhK and SSB. Azolyat, whose foam numbers are close to zero, gives a slight but stable foam with clay suspension. The high value of the foam numbers and the comparatively low foaming power of clay solutions containing added OP-20 is due to the adsorbability of the material on the solid phase of the suspension (Table 5).

The tests showed that it is possible, without particular danger, to add OZhK and SSB to the clay suspension during turbine drilling. The use of other more active specimens requires the use of antifoaming agents.

In 1962, the productive zone of one of the wells in the Pavlovka area was opened up with a clay suspension containing sulfite-alcohol waste. The formation of a small quantity of foam did not interfere in the normal drilling process. The pressure at the pump did not drop below 90 atm. The well was established without complications. It has not yet been examined.

Other noteworthy industrial tests include the treatment of well No. 83 in the Pavlovka area with a 0.1% solution of Alkamon OS-2 in petroleum, in order to reopen the well after repairs. The output from the well before repair work was 15-20 tons/day. After suppression of the well with 30 m^3 of fresh water and removal of the scraper wire, it was found impossible to re-establish the operational gusher. Repeated flushing of the well with petroleum, injection of 20 m^3 of petroleum into the stratum, and continued treatment with a compressor, did not give the desired results. The well began to operate after the injection of Alkamon solution (2 m^3 per meter of thickness of productive stratum). The maximum pressure during injection reached 100 atm (hydraulic rupture in wells in this area takes place at pressures of approximately 300 atm).

Satisfactory results were obtained in the treatment of well No. 81 in the Yarino area by a solution of Leveling Agent A in petroleum (concentration 0.35%). A well which had ceased to function was made operational again with an output of 30 tons/day. The pressure during injection did not exceed 80 atm.

It should be noted that a 3% aqueous solution of NChK (Orsk Petrochemicals Plant) has been used for several years in the Kozubaevskii deposit for the suppression of oil wells. The previously observed marked decrease in productivity after suppression was not observed when NChK was used.

The testing of surfactants in the petroleum industry in the Perm' region is only beginning. When the necessary materials are obtained in 1963, the experiments will be continued on a larger scale.

INVESTIGATION OF SURFACTANTS AS INHIBITORS OF HYDROGEN SULFIDE CORROSION OF STEEL IN THE ELECTROLYTE-HYDROCARBON SYSTEM

A. A. Gonik

One of the effective methods of combating corrosion of the inner surfaces of gas pipelines along which aggressive petroleum gases are transported is the use of corrosion inhibitors.

The introduction of inhibitors into industrial petroleum gas pipelines is a simple and effective measure for combating hydrogen sulfide corrosion in small gas pipelines, where the organization of purification of the gas from hydrogen sulfide is uneconomical.

The use of corrosion inhibitors in the petroleum industry, and recently in the gas industry, has been continously expanding. There is information on the use of inhibitors of various types to combat hydrogen sulfide corrosion of the equipment and supply lines in the production and transportation of oil and gas in the reprocessing of oil and gas at petroleum reprocessing and gasoline plants, and in the storage and transportation of petroleum and petroleum products.

Considering the available data on the use of corrosion inhibitors in the petroleum industry, as well as the protective effect thereby obtained, we must keep in mind that with all other conditions equal, the electroyte-hydrocarbon system, which is formed in particular, within industrial petroleum gas pipelines [1],is not at all equivalent in its corrosion activity to the oil-stratum water system, which is extracted from oil wells.

This is due to the fact that the high-molecular resinous substances that are always present in one amount or another in the oil to be extracted, especially in the so-called "active" oils [2], sometimes sharply lower its aggressiveness, even in spite of the presence of free hydrogen sulfide and water in the oil.

An explanation of this phenomenon was given by Hackerman and Shock [3] on the basis of the assumption that the decelerating effect in this case is exerted by natural inhibitors dissolved in these oils. The indicated hypothesis was then confirmed by other researchers [4], who used it to explain the different corrosion activity of extracted and reprocessed petroleum. Subsequent investigations [5] revealed that in weakly aggressive petroleums, the basic function of natural inhibitors is performed by nitrogen bases. There are reports [6] that nitrogen bases isolated from petroleum products have proved to be extremely effective decelerators of corrosion in the aggressive media encountered in petroleum reprocessing.

In addition, the stratum water, which represents a highly mineralized electrolyte, is usually distributed in the moving stream of oil in the form of a dispersed phase, forming a stable emulsion. This leads to different and inconstant conditions of delivery of the aggressive medium to the surface of the corroding metal, which is usually manifested in a local character of the corrosion damages to petroleum equipment and supply lines.

G. A. Babalyan [2], considering the basic factors determining the probability of the contact of water with the surface of transport hoses during the process of evacuation of the mixture of oil, water, and gas from an oil well,from the qualitative standpoint, showed that under these conditions not only the system of motion of the oil and water in the column of transport pipes, but also the degree of wetting of the metal surface (hose) by these liquids exerts an influence.

The role of wetting of the surface of the steel by the oil and water in the case of hydrogen sulfide corrosion, and the influence of inhibitors on the corrosion fatigue of steel in stratum waters of petroleum deposits, are indicated in the investigations of E. A. Andreeva, I. L. Faerman, and V. F. Negreev [7].

Hence, corrosion inhibitors effective in the medium of the extracted petroleum may not manifest an analogous inhibiting action in all cases in the aggressive medium found within industrial petroleum gas pipelines.

There is still no single point of view on the question of the inhibition of corrosion of metals in a system of two immiscible liquids with the aid of corrosion inhibitors, as well as the mechanism of their protective action.

Some researchers believe [8] that since corrosion of the metal in the electrolyte-hydrocarbon system is ultimately produced by the aqueous solution, the inhibitors should be introduced into the aqueous phase of the medium. They consider the mechanism of their protective action in the light of the principle that has been established for the action of organic inhibitors in acid and neutral media.

Other researchers [9, 10, 11, 12] believe that in view of the presence of a second, nonelectroconducting phase in the corrosion medium, the inhibitors should be introduced into the hydrocarbon liquid, and their protective action is due to the formation of a special shielding film on the surface of the metal, upon which a thin layer of hydrocarbon (petroleum), which intensifies the action of the inhibitor, is formed.

Thus, the existing opinions on the question of the deceleration of metal corrosion by means of inhibiting additives in the two-phase, electrolyte-hydrocarbon system give evidence of a uniqueness and specificity of their protective action under the conditions considered. It is obvious that the study of this uniqueness cannot be conducted in isolation from the generally accepted hypotheses that have by now accumulated on the mechanism of the protective action of corrosion inhibitors in general, and of organic inhibitors in particular.

More and more reports have recently appeared on the use of complex high-molecular organic substances as corrosion inhibitors – compounds of the type of amines and mercaptans, heterocyclic nitrogen-containing compounds, derivatives of urea, thiourea, sulfides, and aldehydes.

As is well known [13], many of these substances are among the typical surfactants having the distinguishing characteristic of an asymmetrical (diphilic) molecular structure of polar and nonpolar groups, capable of being intensively adsorbed on the surfaces of solids. It is clear that when a long hydrocarbon radical is present in such molecules, qualitatively new possibilities are created for the manifestation of a more complex protective action by these inhibitors.

Hackerman and Makrides [12], considering the problem of the inhibition of acid corrosion by high-molecular organic compounds, expressed the hypothesis that the functional groups in these substances play a significantly smaller role in the protective action than the functional groups in low-molecular organic compounds. Hence, when high-molecular organic compounds are used as inhibitors, certain properties become especially pronounced, such as their capacity for oriented adsorption on the surface of the metal to be protected and their ability to form stable films of hydrophobic character; both of these prevent the penetration of the aggressive medium to the corroding surface when their layer is sufficiently saturated.

At present, most workers in the field adhere to the conception that when a metal and inhibitor interact, it is primarily chemosorption phenomena that occur [14]. These views are particularly prevalent in areas where consideration is being given to the character of the protective action of high-molecular compounds used as corrosion inhibitors in various media.

As to the nature of the bond of the polar groups of organic inhibitors to the metal to be protected itself, there is still no single viewpoint on this question. However, many researchers believe that this bond, as a rule, is of a coordination character. Notwithstanding, the very fact of the existence of such a strong bond is of extremely great significance for an understanding of the unique property of this group of organic decelerators – the ability to form oriented layers on the surface of the metal to be protected.

It is well known that the problem of the oriented adsorption of surfactants with a diphilic asymmetric structure on the surface of solids, as applied to the conditions of the flotation of minerals, has found reflection in a number of fundamental investigations of Academician P. A. Rebinder and his school [15].

George and Hackerman [16] believe that the protective action of the shielding film depends on the properties of the inhibitors that promote an increase on the thickness of this film, on the density of the packing of the molecules that compose it, and the most favorable manifestation of the binding forces that act between the adsorbed inhibitor and the metal. These concepts have subsequently been reflected in almost all studies related to the use of high-molecular organic corrosion inhibitors in the petroleum and gas industries.

The possibility of the formation on the metal surface of a special barrier of oriented molecules of the organic substance, the hydrophobic hydrocarbon ends of which are oriented outward, and of an influence of these molecules introduced into the double layer on the electrochemical and corrosion processes of metal is indicated by Z. A. Iofa [17].

Subsequently, in the use of predominantly petroleum-soluble inhibitors, the hypotheses was advanced [10] that these substances exert a strong inhibiting effect only in the case when a film of hydrocarbon (petroleum) is formed on the inhibited metal surface. This hypothesis was soon confirmed experimentally. In an investigation of the protective action of weakly soluble organic sulfonates in mineralized media containing H_2S and CO_2, it was established [11] that the introduction of hydrocarbon liquid, together with the inhibitor, is extremely vital for the manifestation of high decelerating properties by these substances. In the absence of the hydrocarbon, the effectiveness of the inhibitor was sharply reduced.

Hackerman and Schmidt [18], considering the influence of the hydrocarbon medium on the protective action of organic inhibitors, believe that this medium provides only great impermeability of the layer of inhibitor adsorbed on the metal, i.e., plays a passive role. Subsequently Hackerman and Makrides [12] approached this problem from another direction. Assuming that the degree of coating of the metal by the organic molecules increases with increasing length of the radical chain, they believe that the weak solubility of the inhibitor in electrolytes (within definite limits) is a favorable prerequisite for effective deceleration.

Thus, the question of the mechanism of the protective action of high-molecular organic inhibitors with a long radical chain (surfactants) in the electrolyte-hydrocarbon two-phase system still remains largely unelucidated, which in most cases hinders a substantiated selection of inhibitors for combating the corrosion of metals under the conditions considered. Hence, the expansion of investigations along this line and the further accumulation of experimental data are of considerable practical and scientific interest.

We conducted an experimental study of inhibitors of the corrosion of carbon steel in the electrolyte-hydrocarbon two-phase system in the presence of H_2S along the following lines.

1. Determination of the protective action of a number of water-soluble and hydrocarbon-soluble organic compounds in a medium of two immiscible liquids.

2. Investigation of the influence of hydrocarbon-soluble surfactants on the kinetics of the selective wetting of carbon steel under conditions of metal-electrolyte-hydrocarbon contact.

3. The study of the corrosion behavior of metal in an aqueous solution of a two-phase medium in the presence of decelerating additives of hydrocarbon liquid.

The indicated investigations were aimed at studying the influence of additions of water- and hydrocarbon-soluble organic compounds on the corrosion and electrochemical behavior of carbon steel in an electrolyte-hydrocarbon system containing hydrogen sulfide.

Samples of brand St. 3 carbon steel, prepared in the form of polished plates, with dimensions 60 x 25 x 1 mm, as well as cylindrical samples, 10 mm in diameter and 100 mm long, were the basic objects of investigation.

The experiments studying corrosion inhibitors were conducted in individual beakers, placed in the reaction chamber, where an aggressive atmosphere of hydrogen sulfide, water vapors, and hydrocarbon was maintained during the experiment. The organic additives were introduced into the aqueous or hydrocarbon phase of the medium in suitable concentrations. Gasoline, withdrawn from an operating gas pipeline, and chemically pure isooctane were used as the hydrocarbon liquid. A 3% solution of NaCl, acidified with hydrochloric acid to pH = 2.0, was used as the electrolyte. The experiments lasted for 30 to 60 days. Parallel tests were conducted on three to five samples. The experiments were conducted at room temperature (20-25°C) without any special thermostatic control.

Primarily high-molecular organic compounds, soluble in water or hydrocarbons, possessing clearly pronounced surface-active properties, were investigated as inhibitors of the corrosion steel in a two-phase medium containing hydrogen sulfide. Of these compounds, we tested water-soluble nonionogenic surfactants – various alkylphenols, quaternary ammonium bases – Karbozolin S, Katapin A, Arquad T-50; hydrocarbon-soluble dialkyldimethylammonium chloride (Arquad 2C), a high-molecular amine (Amine C), a diamine (Doumine T), as well as a number of other preparations.

Investigations of the influence of hydrocarbon-soluble organic compounds with a long radical chain upon the kinetics of the selective wetting of steel under conditions of metal-electrolyte-hydrocarbon contact were conducted by measuring the contact wetting angles θ_{12}.

For this purpose, the surface-active additive was dissolved in the hydrocarbon liquid (gasoline) in the required concentration, and the solution was saturated with hydrogen sulfide. Concentrated solutions of certain preparations, poorly soluble in gasoline, were prepared on the basis of kerosene for subsequent dilution. For the measurements, the steel plate was placed in a horizontal position in a cuvette with plane-parallel walls, it was filled with hydrocarbon containing the surfactant being tested, and the metal was exposed for 20-30 min, necessary for the establishment of adsorption equilibrium. After this, a drop of the electrolyte was applied on the surface of the plate through the layer of hydrocarbon liquid, and the reading of the contact angles was immediately begun. The duration of the experiments, depending on the rapidity of the change in shape of the drop, varied from several minutes to 1 h. The values of $B_{12} = \cos\theta_{12}$ were calculated according to the data obtained, and the so-called selective wetting diagrams – the time variation of B_{12} – were constructed according to them.

The corrosion behavior of carbon steel in the aqueous phase of the electrolyte-hydrocarbon system in the presence and in the absence of decelerating additives in the hydrocarbon liquid was studied in the following way.

A polished cylindrical sample in a conical flask was suspended on a cork on nylon thread. The cork contained openings for placing a glass tube, designed for withdrawing samples of the aqueous solution, a short tube for filling the vessel with hydrocarbon liquid, and mixers. The electrolyte in these experiments was a 3% NaCl solution, acidified to pH = 2.0. The volume of the aqueous phase in all the experiments was constant, equal to 250 ml, which guaranteed constancy of the area of contact of the metal with the electrolyte at the same sample dimensions.

The corrosion of the part of the sample in contact with the electrolyte was calculated according to the content of iron that passed into solution with the metal. Several parallel determinations were performed during the sampling. The experiment was continued until an iron precipitate was deposited from solution. Two to four samples were tested simultaneously in individual vessels. The corrosion of shortened samples, placed entirely in the aqueous phase of the medium, was studied under the same conditions. The area of contact of the metal with the electrolyte was the same in the long and short samples.

As has been established [1], the sharp intensification of the corrosion of carbon steel with time in the electrolyte-hydrocarbon two-phase medium in the presence of hydrogen sulfide is due to a change in the selective wetting of the corroding metal with time, in the direction of increasing hydrophilic character, and to the formation on its surface in contact with the hydrocarbon phase of a thin layer of electrolyte, in which a voluminous precipitate of iron sulfide gradually accumulates. The latter also stimulates the corrosion process, since it is the cathode in the iron-sulfide couple.

The use of corrosion inhibitors in a system of two immiscible liquids in the presence of hydrogen sulfide encounters definite difficulties, since up to 95% of the total loss in weight of the metal occurs in this case on the surface of the steel in contact with the hydrocarbon liquid in a thin layer of electrolyte. Hence, although the corrosion of steel can be slightly reduced by introducing effective inhibitors into the aqueous phase of the aggressive medium, the possibility of decelerating the corrosion of the metal in the thin layer of the electrolyte seems extremely debatable.

As a verification of this, we first tested water-soluble surfactants of the nonionogenic type (condensation products of various fractions of cooled phenols with ethylene oxide) which had shown a high inhibiting effect in a number of cases, as well as a number of cationic surfactants (salts of quaternary ammonium bases): Arquad T-50, Karbozolin S, etc., in the optimum concentrations. For comparison we tested formaldehyde, long known as an effective inhibitor of hydrogen sulfide corrosion.

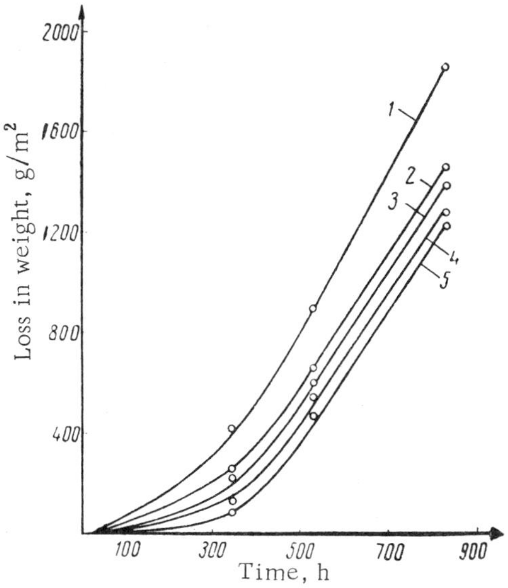

Fig. 1. Influence of certain water-soluble additives on the corrosion of carbon steel in a two-phase medium (in the aqueous and hydrocarbon portions of gas condensate) containing hydrogen sulfide. 1) Without additive; with additive: 2) formaldehyde 0.005%; 3) formaldehyde 0.1%; 4) $UF_{169}E_{8.5}$ – 0.05%; 5) $UF_{169}E_{8.5}$ – 0.1%.

The results obtained are presented in Fig. 1. We can see from them that the total effect of the deceleration of the corrosion of carbon steel in a two-phase medium by water-soluble additives is small. This is probably explained by the insufficient concentration of the inhibitor in the thin layer as a result of the hindrance to its approach to the corroding surface of the metal. It is possible that other factors also exerted an influence in this case. In any case, the introduction of inhibitors into the aqueous phase of the electrolyte–hydrocarbon system proved relatively ineffective for decelerating corrosion in the thin layer.

Obviously, the prevention of the formation of a thin layer of electrolyte on the metal to be protected, under which the most intensive decomposition process occurs in the presence of hydrogen sulfide, would be the most effective method of inhibiting corrosion under these conditions.

As has already been indicated earlier, the increase in hydrophilic character of the surface of the corroding metal and the appearance of a thin layer of electrolyte on it are closely related to the phenomenon of selective wetting on the boundary of the solid and the two immiscible liquids. A consideration of this circumstance is of vital significance in the case under consideration.

It is well known that the most fundamental investigations related to the problem of changing the conditions of selective wetting on the boundary of two liquids of opposite polarity under the influence of oriented layers of surfactants have been conducted by Academician P. A. Rebinder and his school [13, 15].

These investigations have established that under the conditions of selective wetting, solids with a hydrophilic surface—those predominantly wet by water in the presence of hydrocarbon liquid—best adsorb surfactants from the nonpolar phase, while solids with a hydrophobic surface adsorb surfactants from the aqeuous phase better under these conditions. In the first case the orientation of the surfactants in the adsorbed layer always occurs in such a way that their polar groups are turned toward the surface, while the hydrocarbon radicals are directed into the surrounding medium, evening out the polarity difference [13] and making the solid hydrophobic. In the second case, the orientation of the surface-active molecules is opposite in character: the polar hydrophilic groups from the outer layer, covering the entire surface with a hydrated film, as it were, create a gradual transition to the aqueous phase.

Thus, as a result of the oriented adsorption of surfactants with a long radical chain, originally hydrophilic solid surfaces become hydrophobic, and, on the contrary, originally hydrophobic surfaces become hydrophilic, the effect of increasing hydrophobic character being especially sharply pronounced, since it is intensified by a chemical bond – fixation of the polar groups of the surfactants on the corresponding portions of the solids. As P. A. Rebinder indicates, such fixation is achieved on account of a surface chemical reaction between the polar groups of the reagent and the surface atoms (ions) of the solid in the crystal lattice. This reaction is the secondary stage of adsorption and leads to the formation of insoluble films on the solid surfaces, hydrophobic as a result of orientation of the surface-active molecules with their hydrocarbon radicals into the surrounding medium.

A great role in the processes of oriented adsorption is played by the structure of the surface-active molecules. It is known [15] that surfactants with the most asymmetrically constructed molecules possess the greatest orientation ability. This property is especially sharply intensified in homologous series with lengthening of the hydrocarbon radical of the polar molecule, which increases its surface activity (in accord with the Duclot-

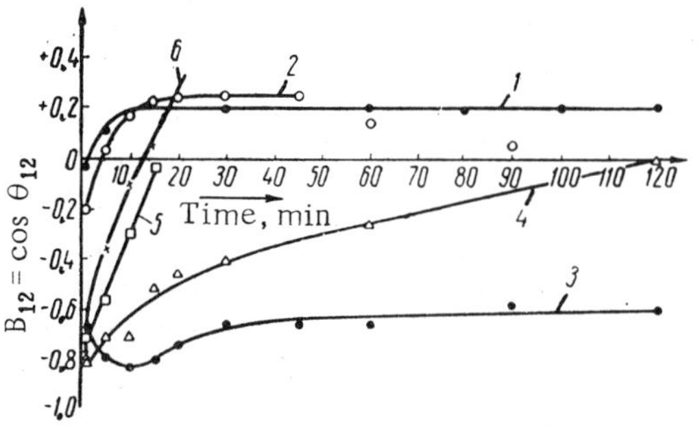

Figs. 2. Variation of the selective wetting B_{12} with time as a function of the concentration of the surfactant Arquad 2C in the hydrocarbon liquid. 1) Without additives, in the absence of hydrogen sulfide; 2) without additives, in the presence of hydrogen sulfide; in the presence of H_2S, additive of surfactant: 3) 0.0035%; 4) 0.015%; 5) 0.0350%; 6) 0.070%.

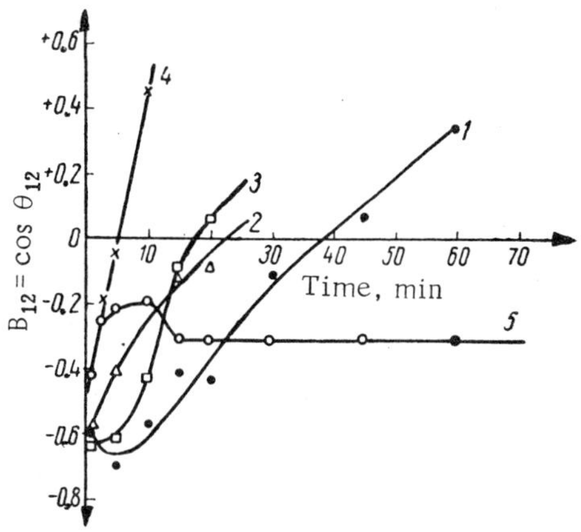

Fig. 3. Variation of the selective wetting B_{12} with time as a function of the concentration of the surfactant Amine C in the hydrocarbon liquid. In the presence of H_2S, additive: 1) 0.005%; surfactant: 2) 0.0010%; 3) 0.0025%; 4) 0.005%; 5) in the presence of hydrogen sulfide without additions of surfactant in the case of preliminary adsorption.

Traube rule). Hence, surface-active molecules with long hydrocarbon radicals, rather sparingly soluble in water, possess the greatest waterproofing properties, especially pronounced on originally hydrophilic surfaces. This principle is reflected in the well-known Rebinder-Semenchenko rule [13] on the reversible relationship between the surface activity of molecules and the solubility of one substance or another in water; in general outlines, it proves correct even for those surfactants whose polar groups are chemically bonded to the surface of a solid.

A change in the selective wetting in the presence of surface-active contaminants in cracking gasoline and press distillate have been observed in the corrosion of iron in polyphase systems [19]. In this case it was established that the adsorption of such contaminants and the stable water-proofing of the surface of iron produced by this results in the creation of a layer impermeable to the solution of hydrogen sulfide on the iron surface.

D. I. Mirlis and P. A. Rebinder [20] have shown that the introduction of fatty acids (for example, oleic) into the benzene phase of the medium produces a strong waterproofing of the surface of iron under conditions of selective wetting, as a result of the influence of a chemically fixed adsorption layer of oleic acid, the hydrophobic nonpolar radicals of which are oriented outward.

Evidently, in our case also, the most effective inhibitors were additives that can not only form a protective film on the metal, but can also produce stable waterproofing of the surface to be protected.

Recently, cation-active surfactants – mainly organic amines and their derivatives – have been used widely to protect ferrous metals from corrosion. Nitrogen-containing groups are usually used as the hydrocarbon portion of such molecules.

The hypothesis [14] that the most effective corrosion inhibitors in acid media are organic compounds that form an insoluble film of a surface-chemical compound with metal ions and the corrosion medium, which insulates the metal from the action of the aggressive electrolyte, is now extremely well substantiated. Moreover, it has been shown that in the case of iron, the reaction products of its atoms (ions) with amines are complex chloroferrates – compounds of the type of double salts which, being sparingly soluble in the surrounding medium, form a protective film in the surface layer of metal.

From this viewpoint, the possibility is not exluded that the essence of the so-called chemical fixation of the polar group of the organic molecule on the surface of the solid, to which such great significance in the creation of a stable oriented layer is attributed [13], as applied to iron or low-carbon steel, consists of the formation of complex surface compounds, including the polar groups of the high-molecular amine and the metal atoms. This may also promote such a distinct appearance of waterproofing properties of certain cation-type inhibitors.

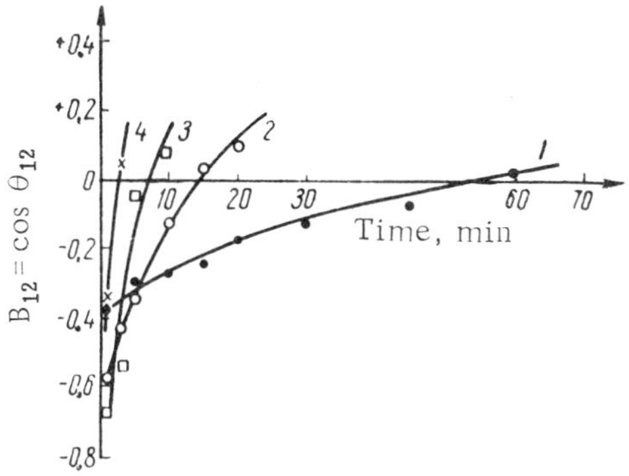

Fig. 4. Variation of the selective wetting B_{12} with time as a function of the concentration of the surfactant Doumine T in the hydrocarbon liquid. In the presence of H_2S, with an addition of surfactant: 1) 0.00025%; 2) 0.0005%; 3) 0.001%; 4) 0.0025%.

Figures 2, 3, and 4 present curves of the time variation of the selective wetting of steel ($B_{12} = \cos \theta_{12}$) in the presence of hydrogen sulfide as a function of additives of cationic surfactants to the hydrocarbon liquid: Arquad 2C, Amine C, Doumine T. We can see from Fig. 2 that without additives (curve 2), the value of the selective wetting varies regularly with time in the direction of increasing the hydrophilic properties of the metal surface. On the contrary, even in the presence of small additives of surfactants, the value of the selective wetting is sharply displaced into the hydrophobic region, producing an improvement of the selective wetting of the metal by the hydrocarbon liquid. As for the next time variation of the selective wetting curves in the direction of increasing the hydrophilic propertoes of the metal surface, this apparently is already related to the reverse orientation on the hydrophobic film, i.e., to the phenomenon of the so-called re-oiling [13].

It is well known that the most favorable conditions for the formation of a hydrophobic layer on an originally hydrophilic surface of oxidized metal are created when a surface-active reagent is introduced into the hydrocarbon liquid, since its adsorption occurs on the surface of separation with the greatest difference in polarities [13].

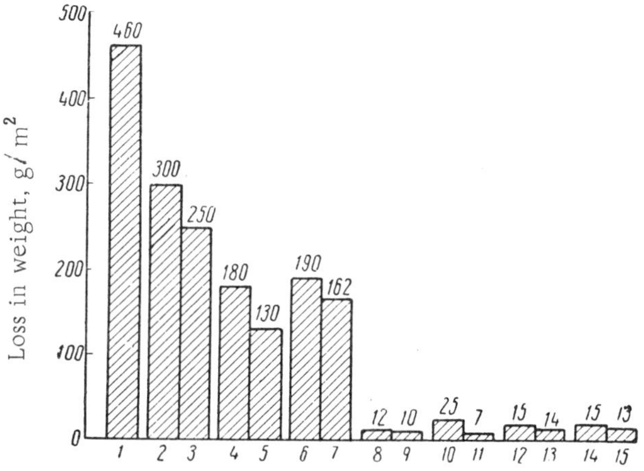

Fig. 5. Influence of certain additives on the stability of steel in an electrolyte-hydrocarbon system in the presence of hydrogen sulfide. Change in weight in 336 h. 1) Without additive; additives in aqueous system; 2) formaldehyde, 0.05%; 3) the same, 0.1%; 4) nonionogenic preparation $UFE_{169}E_{8.5}$ – 0.05%; 5)the same, 0.1%; 6) cation-active preparation Arquad T-50, 0.05%; 7) the same, 0.1%. Additives to the hydrocarbon phase of the system: 8) cation-active preparation Amine C, 0.5%; 9) the same, 0.1%; 10) the same, 0.1%; 11);* 12) cation-active preparation Armak C, 0.05%; 13) the same, 0.1%; 14) preparation Arquad 2C, 0.05%; 15) the same, 0.1%.

*Not given in Russian text.

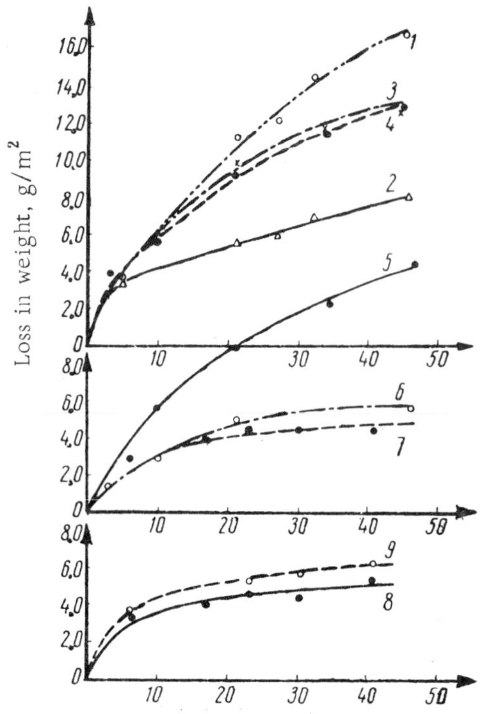

Fig. 6. Corrosion of carbon steel in a two-phase electrolyte-hydrocarbon system in the presence of surfactants in the nonpolar phase (surfactant concentration 1%). Samples cylindrical, in contact simultaneously with the two phases of the system: 1) without additives; 2) in the presence of Arquad 2C; 5) without additives; 6) in the presence of Doumine T; 8) in the presence of Amine C. Samples shortened, in contact only with the electrolyte: 3) without additives; 4) in the presence of Arquad 2C; 7) in the presence of Doumine T; 9) in the presence of Amine C.

This circumstance makes the use of hydrocarbon-soluble surfactants well founded in our case.

In accord with this, for experiments on the protection of carbon steel from corrosion in a system of two immiscible liquids we used cation-active preparations of the type of high-molecular aliphatic amines and diamines in concentrations from 0.001 to 0.1% (Arquad 2C, Amine C, Armak C, and Doumine T). The indicated reagents were introduced into the hydrocarbon phase of the aggressive medium in the presence of hydrogen sulfide.

The results of corrosion tests for 336 h are presented in Fig. 5. For comparison these figures present data on water-soluble substances. As can be seen from the data presented, the summary effect of deceleration of corrosion by hydrocarbon-soluble reagents even at small concentrations proved to be extremely high. The corrosion rate in this case decreased 30- to 40-fold in comparison with the original value. Many phenomena accompanying hydrogen sulfide corrosion of steel in a system of two immiscible liquids, such as increasing hydrophilic properties of the metal surface, with the formation of a thin layer of electrolyte, accumulation of a residue of iron sulfide, stimulating corrosion, and the appearance of hydrogenation bubbles – disappear entirely. The surface of steel in contact wit the medium containing inhibitors was coated with a thin film, translucent up to the metal, poorly wet by water.

We can conclude from the data presented that hydrocarbon-soluble additives not only are more effective than water-soluble additives, but also exert a protective action in extremely small concentrations (0.01-0.1%).

In addition, we also tested the cation-active preparation Arquad T-50, introduced into the hydrocarbon medium. However, this preparation exerted no positive effect. This may be due to its ability, when introduced into the hydrocarbon phase, to pass very rapidly into the aqueous phase in contact with it, i.e., to be distributed primarily in the latter. In view of this, its adsorption on the metal surface will probably occur mainly from the aqueous phase, which explains its low protective effect in the inhibition of corrosion in a two-phase medium.

Thus, in the case of contact of carbon steel with two immiscible liquids in the presence of hydrogen sulfide, the greatest protective effect was exerted by organic inhibitors of the cationic type, primarily soluble in the nonpolar phase and capable of forming a hydrophobic film, weakly permeable for the electrolyte, on the surface to be protected, as a result of adsorption of surfactants with long hydrocarbon chains on the metal surface.

In conducting experiments on the determination of the protective effect of hydrocarbon-soluble corrosion inhibitors in a two-phase medium, we found that in their presence the decomposition of the lower portion of the samples, in contact with the aqueous phase of the medium, is appreciably reduced. Moreover, the metal surface proved to be covered with a protective film, analogous in appearance to the film formed on the metal in the hydrocarbon phase of the medium. We should mention that the vessels that contained the samples with

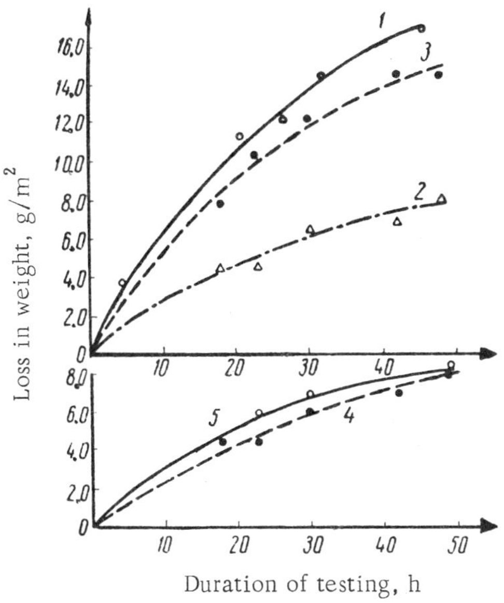

Fig. 7. Corrosion of carbon steel in a two-phase electrolyte-hydrocarbon system in the presence of surfactants in the nonpolar phase (surfactant concentration 0.01%). Samples of cylindrical, in contact simultaneously with the two phases of the system: 1) in the presence of additives; 2) in the presence of Doumine T; 4) in the presence of Amine C. Samples shortened, in contact only with the electrolyte; 3) in the presence of Doumine T; 5) in the presence of Amine C.

two immiscible liquids were filled in such a way so as not to permit flowing of the hydrocarbon liquid in which the inhibitor was dissolved below the boundary of phase separation.

In view of this, it was proposed that surfactants dissolved in the hydrocarbon exert an inhibiting action on the corrosion of the steel in the aqueous phase of the electrolyte-hydrocarbon system. To verify this hypothesis, as well as to determine how the penetration of surface-active molecules from the hydrocarbon phase into the aqueous phase is accomplished, we conducted experiments under comparable conditions with long and shortened samples, in contact in one case simultaneously with two immiscible liquids, and in the other with one of the two liquids (electrolyte). The shortened samples were placed in the electrolyte in the horizontal position, their area being the same as the area of the portion of the long sample that was found in the aqueous medium. Surfactants with long hydrocarbon radicals, practically insoluble in water – Doumine T, Amine C, in concentrations of 1.0 and 0.01%, and Arquad 2C in a concentration of 1.0%, were used as the inhibitors in these experiments.

The experimental results are presented in Fig. 6 and 7. The values of the weight loss of the samples plotted along the y-axis were calculated from the content of iron atoms that passed from the surface of the metal into solution.

As it follows from curves 1 and 5 in Fig. 6, in the absence of inhibiting additives in the hydrocarbon liquid, the corrosion of steel samples in a weakly acidic electrolyte increases regularly with time. Active evolution of hydrogen bubbles begins on the surface of the steel, gradually being weakened only toward the end of the experiment (on the fourth day). Moreover, shortened samples, submerged in the electrolyte in the horizontal position, corrode with a somewhat lower intensity (Fig. 6, curve 3) than long samples, which is probably related to the different conditions of delivery of the electrolyte to the surface of the metal, with equal area of contact with the aqueous medium.

When the investigated surface-active additives are present in the hydrocarbon, the corrosion of the portion of the long samples submerged in the electrolyte is appreciably reduced (Fig. 6, curves 2, 6, 8). The liberation of hydrogen, intense at the beginning of the experiment, slows down with time, which gives evidence of the formation of a protective film on the metal surface. This can also be judged by the slow darkening of the surface of the steel, extending from top to bottom, and the appearance of a larger and larger number of portions on the sample inert with respect to hydrogen liberation.

The corrosion on shortened samples, submerged in the electrolyte and not in contact with the hydrocarbon liquid, on the other hand, is decelerated only in the presence of Doumine T and Amine C at concentrations of 1.0% (Fig. 6, curves 7 and 9). Under these conditions Arquad 2C exerts no inhibiting influence (Fig. 6, curve 4), which apparently is evidence of its ability to decelerate metal corrosion in an electrolyte only through the surface of the sample connecting the hydrocarbon and aqueous phases.

In a comparison of curves 6, 7, and 8, 9 in Fig. 6, we can conclude that Doumine T and Amine C are capable of migrating from the hydrocarbon medium to the metal submerged in the aqueous phase, not only along the surface of the sample but also through the volume of the electrolyte.

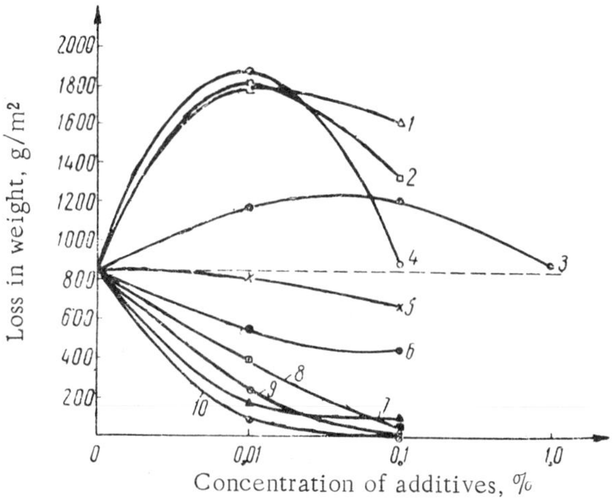

Fig. 8. Influence of additions of water- and hydrocarbon-soluble surfactants on the corrosion of steel in a two-phase electrolyte-hydrocarbon system in the presence of H_2S. Change in weight in 1000 h. Water-soluble additives: 1) Arquad T-50; 2) OP-10; 3) Karbozolin S; 4) UFE_8. Hydrocarbon-soluble additives: 5) Arquad T-50; 6) OP-4; 7) Arquad 2C; 8) Doumine T; 9) Armak C; 10) Amine C.

At a concentration of 0.01%, Doumine T ceased to exert a decelerating action on the corrosion of shortened samples submerged in the aqueous phase of the medium (Fig. 7, curves 1, 2, 3) but kept its ability to penetrate into the electrolyte along the surface of the metal (Fig. 7, curves 2 and 3). On the other hand, Amine C, practically insoluble in water, is capable of penetrating to the surface of the metal from the hydrocarbon medium through the volume of the electrolyte even in such a small concentration (Fig. 7, curves 4 and 5).

According to P. A. Rebinder [13, 15], the displacement of surface-active molecules along the metal from the hydrocarbon liquid into the electrolyte can be explained by displacement of water by the nonpolar liquid under the influence of the waterproofing orientation of the molecules on the solid surface.

As for the mechanism of the penetration of hydrocarbon-soluble molecules to the metal through the electrolyte, here much still remains unclear. It is relatively improbable that this phenomenon is related to the usual solution of surface-active molecules in the aqueous medium, since the indicated reagents are soluble in it to a negligible degree. It is possible that the phenomenon of solubilization – highly colloidal solubility of the nonpolar liquid in the aqueous phase – may exert an influence in this case.

On the basis of the hypotheses [13] of the mechanism of the action on metal of the so-called soluble oils, widely used in industry, it can be hypothesized in our case that the microdroplets of hydrocarbon liquid, arising around the boundary of separation of the two phases and stabilized by the surfactant, diffuse into the electrolye and reach the metal surface. Since the long-chain molecules contained in the microdroplets (micelles) are adsorbed by the oxidized metal surface with their polar groups and in many cases are fastened to it with a strong surface-chemical bond, under these conditions an oriented layer of molecules can form on the metal, upon which the nonpolar liquid flows as a result of coalescence of the droplets of the hydrocarbon and selective wetting. This leads to the formation on the metal of a shielding film, which prevents the access of the aggressive medium to its surface.

It is possible that the principle established by a number of researchers [11], consisting of the fact that the presence of a small amount of hydrocarbon liquid in an aggressive electrolyte containing hydrogen sulfide is extremely vital for the manifestation of a high protective effect by high-molecular inhibitors of the type of organic sulfonates, is related to the phenomenon of solubilization and selective wetting.

Water-soluble additives, as has been shown, do not slow down hydrogen sulfide corrosion in a two-phase medium. Such action of these additives can be explained on the basis of the concepts of selective wetting. According to [13], the adsorption of surfactants on a given solid surface always produces an improvement in the selective wetting of the liquid from which the adsorption occurs.

In the adsorption of surfactants from the aqueous medium of an electrolyte-hydrocarbon system, their orientation evidently will be such as to cause an improvement in the selective wetting of the surface of the solid by water. Moreover, molecules with a long surfactant chain will be adsorbed on the metal with their polar groups in such a way that their hydrocarbon radicals will be directed into the surrounding medium in a rather saturated adsorption layer. However, considering the fact that such a surface possesses substantial adsorption capacity with respect to the aqueous medium, then according to the rule of balancing of polarities [13], it should again adsorb surfactants from the aqueous medium, but this time with their opposite orientation. As a result of this, the metal proves to be coated with a hydrophilic film and will be selectively wet by water.

From this it follows that the introduction of water-soluble additives as inhibitors into the aqueous medium of an electrolyte-hydrocarbon system will promote an increase in the hydrophilic properties of the metal surface, forming a thin layer of electrolyte on it, in which an intensive corrosion process develops in the presence of hydrogen sulfide. When the inhibitor concentration is insufficient, stimulation of the corrosion process might even be expected in the thin layer.

An appreciable intensification of the corrosion of iron in kerosene containing water as a separate phase was established [21] in the presence of sodium benzoate, in spite of complete inhibition of corrosion in the aqueous phase. An analogous phenomenon was noted after the introduction of inhibitors into an electrolyte-liquid fuel system to combat corrosion [8].

The stimulation of hydrogen sulfide corrosion in a two-phase, electrolyte-hydrocarbon system in the presence of water-soluble additives was also noted under our conditions.

For the test we selected a number of surfactants investigated earlier as corrosion inhibitors. In this case we studied the inhibiting action of the two extreme protective concentrations – 0.01 and 0.1%. We used chemically pure iso-octane as the hydrocarbon liquid in these experiments, in hopes of excluding the influence of extraneous impurities that might be contained in the hydrocarbons we investigated earlier on the selective wetting of the metal. We used a 3% solution of NaCl, acidified to pH = 2.0, as the electrolyte.

The experimental results are presented in Fig. 8. From them we can see that water-soluble organic substances, being introduced into the electrolyte under protective conditions and practically suppressing the corrosion of steel in the aqueous phase, not only do not reduce it in the hydrocarbon liquid, but even intensify hydrogen sulfide corrosion if the inhibitor concentration in the electrolyte is relatively low. In an examination of corroded samples, we found that under these conditions the volume of corrosion products formed on the metal in the hydrocarbon phase is substantially greater than that in the control samples.

In view of this we might assume that at a small content of water-soluble additives in the volume of the electrolyte, their decelerating influence in the thin layer will be practically negligible; however, these same additives, facilitating the process of increasing the hydrophilic properties of the metal surface in the presence of H_2S, will promote the development of hydrogen sulfide corrosion of the steel.

Under these conditions, as can be seen from the curves presented, hydrocarbon-soluble additives are substantially more effective than water-soluble additives in decelerating hydrogen sulfide corrosion. Moreover, the protective action of OP-4 – a hydrocarbon-soluble preparation with a short radical chain – is appreciably less than the protective action of inhibitors with a long radical chain (Arquad 2C, Amine C, Armak C, and Doumine T).

It is interesting that the cation-active preparation Arquad T-50, introduced into the hydrocarbon liquid in a 0.01% concentration, exerts a small decelerating action on hydrogen sulfide corrosion of steel in a two-phase medium. On the contrary, when dissolved in the electrolyte, it appreciably stimulates corrosion of the metal.

Thus, water-soluble additives, while exerting the most complete protection of the metal in the aqeuous phase of the medium(where its corrosion losses are relatively great) are rather ineffective; they sometimes even stimulate hydrogen sulfide corrosion of carbon steel in an electrolyte-hydrocarbon system, since they promote an increase in the hydrophilic properties of the metal under conditions of selective wetting.

Hydrocarbon-soluble surfactants under these same conditions prevent an increase in the hydrophilic properties of the metal surface and the formation of a thin layer of electrolyte on it, which in itself sharply reduces the corrosion losses. In addition, the inhibitors considered exert an inhibiting influence on the development of metal corrosion in the aqueous phase of the medium, which appreciably increases the effectiveness of the hydrocarbon-soluble additives. This may explain the fact that petroleum-soluble corrosion inhibitors are receiving predominant distribution in the petroleum and gas industry in comparison with water-soluble inhibitors.

Literature Cited

1. A. A. Gonik, Dokl. AN SSSR, 135, 2, 381 (1960); 135, 3, 671 (1960); Gazovaya Promyshlennost', 3, 41 (1960); 4, 41 (1960).
2. G. A. Babalyan, Problems of the Mechanism of Oil Extraction (Azerbaidzhan Press for Petroleum Literature, 1956); Trudy UFNII, No. VI, Ufa (1960).
3. N. Hackerman and D. A. Shock, Ind. Eng. Chem. 39, 863, 1283 (1947).
4. P. L. Kalish, J. A. Rowe, and W. H. Rogers, Corrosion, 9, 1, 25 (1953).
5. L. P. Hatch, Petrol. Refiner., 35, 8, 147 (1956).
6. A. K. Efimova and A. M. Shatunova, and M. B. Vol'f, Trudy BashNII NP, No. 3 (1960).
7. E. A. Andreeva, I. L. Faerman, and V. F. Negreev, Azerb. Neft. Khoz., No. 12 (1950).
8. I. V. Krotov and T. M. Khachadurova, ZhPKh, 26, 5, 605 (1953).
9. L. Cavallaro, Corros. et anticorros., 7, 12, 417-433 (1959); Abstract Collection: Corrosion and Protection of Metals, No. 15, GOSINTI (1960).
10. P. L. Deverter and A. M. Jasek, Corrosion, 11, 6, 261 (1955).
11. A. H. Rocbuck, et al., Corrosion, 13, 11, 55 (1957).
12. N. Hackerman and A. C. Makrides, Ind. Eng. Chem., 3, 523 (1954).
13. P. A. Rebinder, Khimicheskaya Nauka i Promyshlennost', 4, 5, 554 (1959).
14. I. N. Putilova, S. A. Balezin, and V. P. Barannik, Inhibitors of the Corrosion of Metals (Goskhimizdat, 1958).
15. P. A. Rebinder, Physical Chemistry of Flotation Processes (Metallurgizdat, M. – L., 1933).
16. R. A. George and N. Hackerman, Corrosion, 11, 6, 249 (1955).
17. Z. A. Iofa, Collection: Corrosion Inhibitors, No. 2 (Profizdat, 1957).
18. N. Hackerman and H. K. Schmidt, Corrosion, 5, 7, 237 (1949).
19. D. I. Mirlis, A. I. Glukhova, and E. P. Litvinova, ZhPKh, 12, 4, 504 (1939).
20. D. I. Mirlis and P. A. Rebinder, Dokl. AN SSSR, 3, 123 (1936).
21. I. N. Putilova, et al., ZhPKh, 26, 2, 148 (1953).

THE USE OF SURFACTANTS IN THE DRILLING OF OIL AND GAS WELLS AND FOR THE OPENING UP OF PRODUCTIVE STRATA

S. Z. Zaripov

In recent years, laboratory studies and industrial tests of surfactants in the UFNII have been carried out for the most part in the following four areas:

a) the opening up of productive strata with water and clay suspensions;

b) the production of hydrophobic properties on the walls of wells and on drilled rock;

c) increasing the wetting power of flushing liquids;

d) improving the properties and stabilization of flushing liquids.

The results of theoretical and laboratory studies of surfactants and industrial tests on the nonionogenic surfactant UFE_8 for opening up the productive stratum have been dealt with earlier [3, 4].

Industrial tests on the opening up of a productive stratum by an aqueous solution of the nonionogenic surfactant OP-10, carried out in 1961 in four wells of the Arlanskii deposit, showed that the use of OP-10 is effective for these purposes: all the wells opened up with aqueous OP-10 solution had initial outputs equal to those of wells opened up with clay suspensions and drilled in comparable sections and conditions.

In the same deposit the VNII NP carried out work on the opening up of a stratum by an aqueous solution of the ionogenic surfactant Sulfonol. An analysis carried out by the UFNII showed that in wells where the productive stratum was opened up with an aqueous Sulfonol solution, the productivity of the wells was even lower than that in wells opened up with a clay flushing liquid with high water loss. This is apparently due to the fact that Sulfonol, which is an anionogenic surfactant, not only increases the surface tension but also reacts in solutions with highly mineralized Artinskii water of the calcium chloride type and in mixtures with Carboniferous stratal water to form voluminous flocculent precipitates which probably decrease the penetrability of the stratum.

However, since drilling station No. 3 of the Tuimazaburneft' Trust had at their disposal Sulfonol NP-1, the UFNII carried out studies to find possible methods of using this surfactant in the Arlanskii deposit, which might reduce even partly the undesirable consequences of the salting out of the ionogenic surfactant. As a result of this work the UFNII was able to give to drilling station No. 3 a method for preparing aqueous solutions of Sulfonol NP-1 for opening up productive strata in the presence of the highly mineralized waters of the Arlanskii deposit. The essential feature of the procedure is the preliminary salting out of aqueous solutions of Sulfonol NP-1 in measuring tanks by the stepwise addition of water from the Carboniferous horizon and Artinskii level, with subsequent coagulation of the resulting flocculent precipitates. The Sulfonol NP-1 solution, freed from the precipitates by filtration or decantation, is ready for use in the opening up of productive strata and is supplied to the pump intake.

These recommendations made by the UFNII are merely the outcome of the position created by the lack of nonionogenic surfactants and in no way represent a scientific solution of the problem of using surfactants in the Arlanskii deposit. The scientific solution of the problem, according to the UFNII studies, is to use, in the

Arlanskii deposit, nonionogenic surfactants including the oxyethylated phenols and alkylphenols UFE_8 and OP-10, which show no change in surface activity and do not form precipitates in mineralized media.

The work which has been carried out has shown that the choice of a given surfactant is determined by the specific geological conditions of the deposit being drilled. Under the conditions of the Arlanskii petroleum deposit, which is characterized by the presence of high-pressure, mineralized, calcium chloride stratal waters of the Artinskii horizon in the well sections, the most effective materials are the nonionogenic surfactants UFE_8 and OP-10.

In view of this, subsequent work in the UFNII on the opening up of the productive stratum by water will be continued only with surfactants of the nonionogenic type (OP-10, UFE_8, etc.).

Industrial tests have also shown the lubricating and hydrophobizing power of a flushing liquid formed by water containing surfactants of the nonionogenic type OP-10 and UFE_8. Their high lubricating power is indicated by the operational characteristics of the bits during drilling with these flushing liquids: the characteristics are better than those observed during drilling when the end face is flushed with water and clay suspension. Thus, during drilling in the Arlanskii deposit, when the end face was flushed with water containing OP-10, the mechanical velocity was greater by a factor of 2.5 and the distance cut per bit was 1.5 times greater than the corresponding average operational characteristics of bits in 32 wells drilled in the same deposit, where the end face was flushed with clay suspension.

Another characteristic feature is that compared with drilling in which the end face is flushed with water, the mechanical velocity, distance cut per bit, and duration of a single cut with flushing by water containing OP-10 are increased by 16, 30, and 13% respectively (average bit operational characteristics from seven wells).*

The slight hydrophobizing influence of aqueous solutions of the surfactants OP-10 and UFE_8 on clay rocks is indicated by the fact that the drilled mud obtained when these aqueous solutions are used contains larger particles than that obtained during drilling with other flushing liquids – water or clay suspensions treated with alkaline reagents.

In addition to the studies of aqueous surfactant solutions, laboratory and industrial tests are being carried out in the UFNII on the opening up of productive strata by clay solutions containing added surfactants.

Work carried out in the UFNII in 1959 showed that the addition of the nonionogenic surfactant UFE_8 to a clay suspension from Talalaevskii clay treated with calcined soda makes it possible to obtain a suspension with a filtrate with low surface tension and a controllable flocculating medium, which is an advantage in suspensions used to open up the productive stratum. A disadvantage of these solutions, however, is the formation of foam when surfactants are added. Work on the selection of antifoaming agents for clay suspensions containing UFE_8, carried out in the Surfactant and Flushing Liquids Laboratories, showed that for industrial tests it is possible to recommend the Devonian petroleum from the Tuimazinskii deposit and the tarry oils from the alcohol recovery plant in the Asha Wood-Chemicals Combine. At a concentration of 0.75% these materials gave complete suppression of foam in a clay suspension containing 2% UFE_8, and when petroleum was added (from 3-5%) the foam practically disappeared. It should be noted that the best foam suppression was achieved when the petroleum was added to the clay suspension before the UFE_8.

Laboratory studies are now being continued to find antifoaming agents for a clay suspension containing the nonionogenic surfactants UFE_8 and OP-10.

Industrial tests with a clay suspension containing the added surfactants UFE_8 were carried out in well No. 1332 of the Tuimazinskii area. Before the opening up of the roof in this well, unstable Mikhailovskii clays were used for the clay suspension. Down to a depth of 1701 m (the roof of the Pashiiskii strata) the well was drilled with a clay suspension treated with peat-alkali solution (without added surfactant) with the following parameters: density 1.30 g/cm^3, viscosity 30-47 sec, water loss 8-11 cm^3, crust thickness 1.0-2.5 mm. Before the surfactant was added to the clay suspension, a circulation system was introduced in well No. 1332, so that

*Further details are given in the paper by E. G. Ivanov, N. I. Skomorovskaya, and V. I. Tokunov in the collection: "The use of solutions of nonionogenic surface-active substances for opening up oil-bearing strata."

the flow of flushing liquid was steady, since the clay suspensions have a tendency to produce foam when surfactants are added.

3% of Devonian petroleum as antifoaming agent and 1.5% of the surfactant UFE_8 (percentages expressed relative to the volume of flushing liquid) were added directly to the intake during circulation of the flushing liquid. The petroleum was added in a fine stream over a period of 30 min, in such a way that the stream was broken up by the flow of clay suspension from the drilling pump outlets, and the UFE_8 was added similarly over a period of 2 h.

Throughout the whole of this period, in which the drilling was carried out with a clay suspension containing added UFE_8, a systematic check was made of the drilling process and the parameters of the clay suspension, including the concentration of UFE_8 in the suspension.

During drilling, the specific gravity of the suspension decreased as a result of foam formation, and its viscosity increased, in spite of the addition of the carbonate-alkali reagent and calcined soda solution. At a depth of 1728.5 m, before the opening of roof DII (stratal pressure 180 atm), when the specific gravity of the solution had decreased to 1.1 g/cm^3, further drilling with flushing by a clay suspension containing UFE_8 was discontinued.

The drilling characteristics when a clay suspension containing the surfactant UFE_8 was used were unsatisfactory, since two U8-3 pumps could develop a pressure of only 40-50 atm (because of the formation of a stable foam in the suspension), which was clearly insufficient for effective operation of the turbodrills at the end face of the well [3].

Mention was made above of the slight hydrophobizing influence of surfactants on clay rocks: the particle size of the drilled rock during drilling when the end face was flushed with water containing surfactants was greater than that when the drilling was carried out with water or clay suspension alone.

In order to investigate the hydrophobizing influence of surfactants, studies were carried out in the UFNII on the stability of argillites from the Ashinskii formation in aqueous solutions of various surfactants — Alkamon OS-2, Karbozolin O, UFE_8, and OP-10—by two methods [1]. This work showed that argillites are more stable in solutions of these surfactants than in mains water. The argillites showed the greatest stability in aqueous solutions of UFE_8 and OP-10. It is difficult to say anything about the behavior of argillites from the Ashinskii formation in aqueous surfactant solutions under practical conditions, since no industrial tests have as yet been carried out. When aqueous solutions of the surfactant UFE_8 were used to open up the productive stratum in the Tuimazinskii deposit, the stability of clays of the Mikhailovskii horizon (which are similar in character to the argillites of the Ashinskii formation) showed no appreciable increase compared with their stability during drilling when the end face was flushed with water.

The presence of surfactants as one of the components improves the lubricating power of the suspensions, increases their stability at increased temperatures and their hydrophobizing properties, and also decreases the surface tension of the filtrate and thus reduces its undesirable influence on the productivity of the stratum. Because of these advantages, work has begun in the UFNII on the development of formulae for clay suspensions containing surfactants for drilling deep wells. In the suspension formula which has been developed, the main surfactant is UFE_8 (up to 2.3 vol. %), the stabilizer is KMTs (1 wt. %), and the antifoaming agent is Sivushnii oil (not less than 2 vol. %).

The quality of suspensions of untreated clay powders from the Kuganakskii plant is particularly high when these components are added: the stability is good, the water loss equal to 6.0-6.5 cm^3, the viscosity 33.5-41.0 sec, and the SNS after 10 min 41.7-67.4 mg/cm^2. At the same time, the suspensions showed practically no foam formation; the density of the foaming suspensions was 1.18 g/cm^3, compared with a density of 1.20 g/cm^3 for the original suspensions. The studies showed that clay suspensions containing the surfactant UFE_8 are more stable to the action of high temperatures and aggressive stratal waters of the calcium chloride type; e.g., the water loss from suspensions containing UFE_8 after the addition of 10% stratal water was equal to 21 cm^3, whereas the water loss for suspensions containing no UFE_8 increased to 36 cm^3 for the same quantity of stratal water (or increased by a factor of almost 2 compared with the initial value).

The hydrophobizing properties of suspensions treated with surfactants are indicated by the fact that the stability of argillite specimens from the Ashinskii formation in the filtrates from these solutions is higher than that in filtrates from clay suspensions treated with KSSB or KMTs, or in filtrates from calcium suspensions [2]. On the basis of these studies, suspensions containing the surfactant UFE_8 were accepted as promising flushing liquids for drilling deep wells in crumbling argillites in the Bashkir Republic.

Work on the search for effective flushing liquids containing surfactants for the opening up of productive strata and for drilling deep wells will be continued in the future in the UFNII. It was pointed out above that the addition of the surfactant OP-10 to water improves its lubricating properties. The improvement in the lubricating power of clay suspensions treated with KSSB – a colloidal surfactant – is indicated by the results of an analysis of the operation of drill bits in wells drilled with suspensions treated with KSSB. The surfactants increase the distance cut per bit, compared with the corresponding value for suspensions treated with electrolytes (calcined soda, caustic soda), by 25-30%.

Work aimed at increasing the lubricating properties of flushing liquids will be continued in the UFNII not only on the estimation of the lubricating power of given surfactants in a flushing liquid but also on the development of the principles of application of given surfactants in order to control the lubricating power of flushing liquids containing surfactants.

Data obtained in the UFNII show that the action of surfactants on the rheological properties of suspensions is different for different surfactants even when these are of the same type. In particular, it has been found that the structural and mechanical properties of clay suspensions are higher when OP-10 is added than when UFE_8 is added.

Treatment of clay suspensions by other surfactants of the colloidal type, e.g., KMTs, leads to a sharp decrease in their structural and mechanical properties. In this connection, studies of the rheological properties of flushing liquids containing surfactants, which are to be carried out in the UFNII in 1962-1963, are of particular interest.

Surfactants of the colloidal type are widely used for stabilizing inhibited flushing liquids (calcium suspensions) and flushing liquids designed for drilling in sections with highly mineralized stratal waters (stabilized suspensions). Thus in the calcium suspensions used in the Bashkir Republic for drilling deep wells in crumbling Ashinskii and Bavlinskii argillites, surfactants of the colloidal type – KMTs and KSSB – are used as stabilizers. In the stabilized suspensions used for drilling wells in the Arlanskii petroleum deposit in the Bashkir Republic, one of the components used as stabilizer is again KSSB.

Work is to be carried out in the UFNII on the use of a new sulfonated cellulose ether – SETs, synthesized by the MINKhiGP – as a stabilizer in calcium suspensions.

It can be seen from the above that the UFNII is at present carrying out a considerable volume of work on the use of surfactants for drilling oil and gas wells and for opening up productive strata. This work will be extended even further in the future.

Literature Cited

1. N. Kh. Bogdanov and S. Z. Zaripov, "The drilling of deep wells," UFNII Report on Theme 2059 A, UFNII Fund (1960).
2. S. Z. Zaripov, "The study of flushing liquids for deep well drilling conditions and the search for thickening agents to replace barytes," UFNII Report on Theme 2360, UFNII Fund (1961).
3. E. G. Ivanova, N. I. Skomorovskaya, and E. K. Latypov, "The search for flushing liquids for opening up productive strata," UFNII Report on Theme 2659, UFNII Fund (1960).
4. N. I. Skomorovskaya and E. G. Ivanova, "The use of surface-active substances, during the opening up of the productive stratum of oil wells, to preserve the natural oil permeability of the end-face zone of the stratum," Collection: The Use of Surface-Active Substances in the Petroleum Industry (Gostoptekhizdat, 1961).

THE ECONOMIC EFFICIENCY OF THE USE OF SURFACTANTS IN THE PETROLEUM INDUSTRY

I. B. Genkin

The use of surfactants is a promising method for intensifying the development of petroleum deposits and technological processes in the petroleum-producing industry. Many laboratory studies and industrial tests have shown [1, 2] that the use of surfactants may provide an extremely powerful method of increasing the efficiency of capital investments, decreasing costs, and increasing productivity.

A resolution adopted at the All-Union Conference on the Use of Surface-Active Substances in the Petroleum Industry and Their Production (February 10-13, 1960) defined the wide range of application of surfactants in the petroleum industry and recommended the following surfactants for initial use: oxyethylated materials of the type OP, UFE_8, $KAUFE_{14}$, DS, Sulfonol, and many others [2].

The range of applications of surfactants in the petroleum-producing industry is extremely varied.

Surfactants can be used for the following purposes:

a) increasing the recovery of petroleum from reservoirs;

b) treating the end-face zone of operational wells in order to increase their productivity;

c) increasing the pickup of forcing wells;

d) drilling, opening up, and establishing new operational and exploratory wells;

e) repair and isolation work in old wells;

f) demulsification inside the well and petroleum pretreatment;

g) combating the deposition of paraffin wax in underground tubes and manifolds;

h) combating the corrosion of petroleum-producing equipment and contamination of the stratum by hydrogen sulfide;

i) cleaning effluents from the petroleum industry.

TABLE 1

Parameters	Year			
	1958	1959	1960	1961
Number of wells treated with surfactants	3	17	7	11
Including those giving an effect	3	16	6	9
Additional recovery from wells (without allowance for the transferred effect), thousand tons	–	18.8	10.0	8.8
Expenditure per treated well, thousand rubles	–	4.1	5.3	4.7

In spite of the fact that until recently surfactants were applied in the petroleum-producing industry only on a limited scale, due chiefly to the lack of the necessary quantities of surfactants of the required quality, the economic efficiency of the application of these materials cannot be doubted.

Let us examine some examples of the economically effective use of surfactants in the petroleum industry in the Bashkir Republic and other regions of the Soviet Union.

In order to increase the productivity of wells in the Tuimazaneft' NPU, in 1958-1961 the end-face zone of the wells was treated with surfactant solutions.

The results of this treatment are shown in Table 1.

The data in Table 1 show that the treatment of the wells with surfactants gives a high percentage of successful treatments and a considerable increase in petroleum recovery. In 1960 and 1961, however, the transferred effect was not taken into account in the calculations, so that the cost of the additional petroleum recovered for these years was found to be higher than the average figure for the NPU. In 1959 the transferred effect (to 1960) amounted to 34 thousand tons and the total effect in 1959 and 1960 to 52.7 thousand tons, i.e., an additional 3.1 thousand tons were obtained per well. As a result, the cost of one ton of additional petroleum was found to be lower than the average figure for the NPU. Thus the use of surfactants for the treatment of the end-face zone of the wells is economically efficient.

The use of surfactants of the Sulfonol and OP-7 type in the Tuimazaneft' NPU for isolating bottom waters during hydraulic rupture of the stratum gave satisfactory results. In 1958-1960 a number of flooded wells drilled in the water-flooded section of the stratum were returned to operational condition.

First experience in the use of added surfactants for the treatment of wells with hydrochloric acid made it possible to halve the number of ineffective acid treatments and to increase considerably the recovery of petroleum.

Studies carried out in the VNII and UFNII [2] have shown that the opening up and exploitation of the stratum in new wells with the addition of various surfactants increase the output from the wells by a factor of 1.5-2, leading to an increase in the efficiency of capital investment and a decrease in the cost of petroleum recovery.

The annual saving resulting from the use of surfactants (OP-10) to remove water and salts from Romashkinskii petroleum amounted to 170 thousand rubles, and the concentration of salts in the treated petroleum was decreased by a factor of 2.5 compared with petroleum treated with NChK.

Data obtained in studies of many forcing wells in various deposits in the USA have shown that the injection of water containing added surfactants into the forcing wells gives an increase in pickup of up to 10% [1].

One firm in the USA (Keiber-Coulter) created a margin by injecting a surfactant (Igepal-C) into four forcing wells. It was found that the cost of the additional petroleum was greater by a factor of 2.8 than the cost of the surfactant used, i.e., the use of surfactants was found to be economically inefficient [1].

The firm of Tibbet (USA) found that the cost of the additional petroleum amounted to only 16% of the cost of the injected surfactant [1]. Thus, in this case the use of surfactants was found to be economically efficient.

Studies have shown [1] that for increasing the oil-flushing properties of water injected into a stratum, nonionogenic surfactants have the highest economic efficiency. The experimental use of these materials in the Soviet Union has led to an increase in petroleum output, an increase in the productivity of wells, and a decrease in the cost of petroleum recovery.

It should be noted that the economic efficiency and success of the application of surfactants in the petroleum-producing industry will depend on the availability of cheap and highly effective surfactants and a decrease in their consumption for individual technological processes, calculated for each ton of petroleum recovered.

In order to maintain the stratal pressure by flooding (inside and outside the contour), 1.5-3 or more m^3 of water are used per ton of petroleum recovered. This ratio between the quantity of water injected and petroleum recovered is economically efficient.

TABLE 2

Parameters	Version I (without the use of surfactants)	Version II (using surfactants)
Geological petroleum reserves, thousand tons	35.0	35.0
Petroleum recovery coefficient for the strata, %	0.50	0.60
Petroleum recovery (extractable industrial reserves), thousand tons	17.5	21.0
Recovery of liquid, thousand m^3	82.0	55.0
Water injected, thousand m^3	82.0	55.0
Number of wells:		
Oil wells	1	1
Forcing wells	4	4
Period of extraction of industrial reserves, years	1.5	1.0
Consumption of surfactant (OP-10), tons	–	27.5
Method of exploitation	Gusher	Gusher

Note. All the expenditure is calculated on the new price scale.

In particular, the displacement of petroleum from strata by water treated with surfactants can be economically efficient only when the quantity used to bring about the process is small.

Studies have shown [1, 2] that in order to increase the petroleum recovery from strata, it is necessary to force into the stratum a small quantity of surfactant solution to create a margin which will move pure water through the stratum and thus increase the oil-flushing properties of the injected water.

Workers in the Scientific Research Institutes have proposed numerous different variations of possible technical solutions to the problem of using surfactants in the many varied technological processes of the petroleum-producing industry [2].

All these solutions require detailed and intensive development in design and technology, based on data obtained in laboratory studies and experimental industrial tests of surfactants in the various fields in which they are used. They cannot be introduced successfully into production, however, without highly qualified technical and economic estimation of the efficiency of the designs and proposed measures.

The methodologically correct determination of the economic efficiency of the use of surfactants, plus the economic justification of plans for their introduction, is of considerable importance.

It is necessary to develop and adopt an industrial procedure for determining the economic efficiency of the use of surfactants in the petroleum industry. The existence of such a procedure will facilitate the successful introduction of surfactants into production.

In order to increase the economic efficiency of processes involving the use of surfactants, it is necessary to take precautions to decrease expenditure on one particular operation. This reduction in cost depends to a considerable extent on the choice and industrial-scale preparation of a wide variety of economically effective surfactants, the development of a rational technology for their application, and an improvement of the organization of work on their use in oil wells and in petroleum plants.

The UFNII, in response to the requirements of the appropriate organizations, proposed and justified a technological scheme for the experimental industrial testing of surfactants for the treatment of water injected into a stratum in order to increase the petroleum recovery from strata to the maximum possible extent. It has been decided to carry out the experiment in one section of the Arlanskii petroleum deposit in the Bashkir Republic. This experiment should confirm calculations already carried out on the economic efficiency of this method.

Unfortunately,this experiment has not yet been started because of the lack of a sufficient quantity of inexpensive surfactants.

We can give some data from calculations providing an economic justification for the basic technological scheme for increasing the petroleum recovery from a stratum by the injection of water treated with surfactants.

The calculation was carried out on the basis of the following initial data (Table 2).

Calculation of the Once-Only Capital Investments

Two sets of operational and forcing wells were drilled in order to carry out the experimental industrial tests.

The expenditure on the drilling and equipping of these wells was as follows:

1) drilling of operational wells

$$2 \text{ wells} \times 50.0 \text{ thousand rubles} = 100 \text{ thousand rubles};$$

2) drilling of forcing wells

$$8 \text{ wells} \times 60.0 \text{ thousand rubles} = 480 \text{ thousand rubles};$$

3) equipping and framing of wells

$$10 \text{ wells} \times 16.0 \text{ thousand rubles} = 160 \text{ thousand rubles}.$$

Total expenditure: $100 + 480 + 160 = 740$ thousand rubles.

According to the project data, 50% of the wells (oil wells and forcing wells) will be used during subsequent exploitation of the deposit (return to upper horizons, conversion to forcing wells, etc.).

The expenditure on the drilling of wells which will subsequently be exploited is put down as amortization for the whole amortization period. The expenditure on the drilling of wells which cannot be subsequently exploited (written off) should be paid from special funds for experimental industrial work.

TABLE 3. Cost of Petroleum Recovery from Experimental Industrial Sections of the Arlanskii Deposit (in thousand rubles)

Items of expenditure	Version	
	I – Without the use of surfactants	II – Using surfactants
Expenditure on energy	5.7	3.9
Expenditure on maintenance of the wells	7.6	5.1
Amortization:		
Of oil wells	6.5	4.3
Of forcing wells	31.3	20.9
Of industrial equipment	2.4	1.6
Expenditure on the recovery, transport, and storage of the oil	4.0	4.8
Expenditure on pretreatment of the oil	13.1	8.8
Expenditure on the maintenance of the stratal pressure (without amortization of the forcing wells)	24.6	16.6
Expenditure on surfactant	–	9.6
General industrial expenses	4.0	2.6
Total exploitation expenditure	99.2	78.2
Petroleum recovery, thousand tons	17.5	21.0
Cost per ton of petroleum, rubles	5-68	3-72

TABLE 4

Parameters	Version I (without surfactants)	Version II (with surfactants)
Once-only capital investment (K_1 and K_2), thousand rubles	370.0	370.0
Current expenditure on experimental industrial work (C_1 and C_2), thousand rubles	99.3	78.2
Normative efficiency coefficient E_n	0.2	0.2
Total (reduced) expenditure ($Z = C + E_nK$), thousand rubles	173.2	152.2

Calculation of the Exploitation Expenditure

In the calculation of the cost of petroleum recovery according to regions, and the economics of the exploitational expenditure, we adopted specific expenditures obtained by calculating the cost of petroleum recovery according to the plan for the development of the Ashitovskii, Nagaevskii, and Aktanyshbashevskii sections of the Arlanskii petroleum deposit. These specific expenditures were calculated from the parameters of the second version of the plan for the whole period of development of these regions of the deposit, since the experimental sections are being developed entirely by the gusher method.

The following specific expenditures were taken for the calculation:

1. Expenditure on energy – 70 rubles per thousand m^3 of liquid recovered.

2. Expenditure on maintainenance of the wells (wages and salaries, current servicing of the wells, plant expenses, etc.) – 5.1 thousand rubles per well-year of exploitation.

3. Amortization of the oil wells and forcing wells – on the basis of a norm of 8.7% per year of the cost of the wells (including 6.7% for liquidation of the cost and 2% for capital repairs). The cost of the wells was taken as 50 thousand rubles for the oil wells and 60 thousand rubles for the forcing wells.

4. Amortization of the industrial equipment – 1.6 thousand rubles per well-year of exploitation.

5. Expenditure on the recovery, transport, and storage of the oil – 230 rubles per thousand tons of petroleum recovered.

6. Expenditure on the pretreatment of the oil – 160 rubles per thousand m^3 of liquid recovered.

7. Expenditure on the maintenance of the stratal pressure – 300 rubles per thousand m^3 of injected water (without surfactants).

8. Expenditure per ton of surfactant (OP-10) – 350 rubles, including transportation costs.

9. General industrial expenses – 2.6 thousand per well-year of exploitation.

On the basis of these specific expenditures and the above original data (see Table 2), the cost of petroleum recovery by the two methods (with and without the use of surfactants) is calculated.

The results of the calculations are given in Table 3. The version involving the use of surfactants gives a saving in exploitation expenditure as a result of the following factors:

a) the higher coefficient of petroleum recovery, leading to a higher average annual output from the operational well and to an increase in the total volume of petroleum recovery by 3.5 thousand tons;

b) a decrease in the period required to develop the section, as a result of the higher rate of filtration of the liquid during development as a whole;

c) the smaller quantity of liquid extracted from the well and the smaller quantity of water injected into the stratum.

The factor leading to an increase in the cost is the use of the surfactant (OP-10) on a large scale, and the high cost of the surfactant. The specific consumption of surfactant per 1 m^3 of injected water amounts to 0.5 kg.

Table 3 shows the total saving in exploitation expenditure amounts to 21.0 thousand rubles.

As a result of the increase in the coefficient of petroleum recovery from the stratum by 10% and the production of 3.5 thousand tons of additional petroleum in the second version, the saving in exploitation expenditure and the lower cost of the recovery of 1 ton of petroleum is accompanied by a saving in capital expenditure, determined from the following formula:

$$E_{cp} = \Delta Q (U_{pr} + U_{ct}),$$

where E_{cp} is the saving in capital expenditure in rubles; ΔQ is the additional petroleum recovered as a result of the increase in petroleum recovery from the stratum, in tons; U_{pr} is the specific capital expenditure on the exploration and prospecting per ton of extractable (industrial) reserves ot petroleum in the region, in rubles; and U_{ct} is the specific capital expenditure on the construction of new petroleum plant per ton of extractable reserves.

Since, according to studies carried out in the UFNII, the specific expenditure reaches the following levels: 0.40 rubles for the discovery and prospecting of new deposits and 1.60 rubles for the construction of petroleum plant per ton of extractable industrial reserves, the saving in capital expenditure amounts to:

$$E = 3.5\,(0.40 + 1.60) = 7.0 \text{ thousand rubles.}$$

In order to determine the economic efficiency of experimental industrial work, the total (reduced) expenditure on the process (Table 4) and the saving from its introduction* are calculated.

Thus, the saving from the work amounts to:

$$\begin{aligned} E &= [C_1 + (E_n K_1)] - [C_2 + (E_n K_2)] \\ &= [99.3 + (0.2 \cdot 370)] - [78.2 + (0.2 \cdot 370)] \\ &= 173.3 - 152.2 = 21.0 \text{ thousand rubles.} \end{aligned}$$

The experimental industrial injection of water treated with surfactants in an experimental section of the Arlanskii deposit, according to the proposed technological scheme, is thus economically efficient, and if satisfactory results are obtained it can be introduced throughout the whole region, provided that a sufficient quantity of suitable surfactants is available.

Literature Cited

1. G. A. Babalyan, I. I. Kravchenko, I. L. Markhasin, and G. V. Rudakov, The Physicochemical Principles of the Use of Surface-Active Substances in the Exploitation of Oil-Bearing Strata (Gostoptekhizdat, 1962).
2. Collection: The Use of Surface-Active Substances in the Petroleum Industry, Proceedings of the First All-Union Conference on the Use of Surface-Active Substances (Gostoptekhizdat, 1961).

*Procedure for Determining the Economic Efficiency of the Introduction of a New Technique, Mechanization, and Automization of Production Processes in Industry, Izd. AN SSSR, 1962.